Safer Homes, Stronger Communities

Safer Homes, Stronger Communities
A Handbook for Reconstructing after Natural Disasters

Abhas K. Jha
with
Jennifer Duyne Barenstein
Priscilla M. Phelps
Daniel Pittet
Stephen Sena

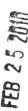

THE WORLD BANK

CONTENTS

The Process of Response and Reconstruction

DISASTER EVENT

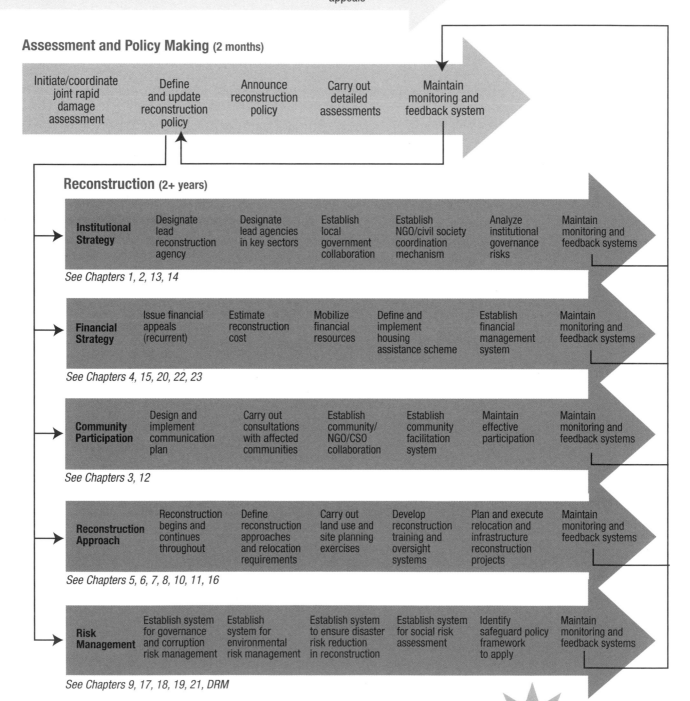

Initial Response (2 weeks)

| Establish coordination mechanism | Engage agencies | Conduct initial public communication | Carry out initial assessment | Define outline strategy | Issue rapid appeal and subsequent appeals | Launch/ orchestrate emergency response |

Assessment and Policy Making (2 months)

| Initiate/coordinate joint rapid damage assessment | Define and update reconstruction policy | Announce reconstruction policy | Carry out detailed assessments | Maintain monitoring and feedback system |

Reconstruction (2+ years)

| **Institutional Strategy** | Designate lead reconstruction agency | Designate lead agencies in key sectors | Establish local government collaboration | Establish NGO/civil society coordination mechanism | Analyze institutional governance risks | Maintain monitoring and feedback systems |

See Chapters 1, 2, 13, 14

| **Financial Strategy** | Issue financial appeals (recurrent) | Estimate reconstruction cost | Mobilize financial resources | Define and implement housing assistance scheme | Establish financial management system | Maintain monitoring and feedback systems |

See Chapters 4, 15, 20, 22, 23

| **Community Participation** | Design and implement communication plan | Carry out consultations with affected communities | Establish community/ NGO/CSO collaboration | Establish community facilitation system | Maintain effective participation | Maintain monitoring and feedback systems |

See Chapters 3, 12

| **Reconstruction Approach** | Reconstruction begins and continues throughout | Define reconstruction approaches and relocation requirements | Carry out land use and site planning exercises | Develop reconstruction training and oversight systems | Plan and execute relocation and infrastructure reconstruction projects | Maintain monitoring and feedback systems |

See Chapters 5, 6, 7, 8, 10, 11, 16

| **Risk Management** | Establish system for governance and corruption risk management | Establish system for environmental risk management | Establish system to ensure disaster risk reduction in reconstruction | Establish system for social risk assessment | Identify safeguard policy framework to apply | Maintain monitoring and feedback systems |

See Chapters 9, 17, 18, 19, 21, DRM

RECONSTRUCTION COMPLETE

Background

Safer Homes, Stronger Communities: A Handbook for Reconstructing after Disasters was developed to assist policy makers and project managers engaged in large-scale post-disaster reconstruction programs make decisions about how to reconstruct housing and communities after natural disasters.

As the handbook demonstrates, post-disaster reconstruction begins with a series of decisions that must be made almost immediately. Despite the urgency with which these decisions are made, they have long-term impacts, changing the lives of those affected by the disaster for years to come.

As a policy maker, you may be responsible for establishing the policy framework for the entire reconstruction process or for setting reconstruction policy in only one sector. The handbook is emphatic about the importance of establishing a policy to guide reconstruction. Effective reconstruction is set in motion only after the policy maker has evaluated his or her alternatives, conferred with stakeholders, and established the framework and the rules for reconstruction.

As international experience—and the examples in the handbook—clearly demonstrate, reconstruction policy improves both the efficiency and the effectiveness of the reconstruction process. In addition to providing advice on the content of such a policy, the handbook describes mechanisms for managing communications with stakeholders about the policy, for improving the consistency of the policy, and for monitoring the policy's implementation and outcomes. The handbook does not tell you exactly what to do, but it should greatly improve the likelihood that the reconstruction policy that is established leads to good outcomes.

Handbook Guiding Principles

1. A good reconstruction policy helps reactivate communities and empowers people to rebuild their housing, their lives, and their livelihoods.
2. Reconstruction begins the day of the disaster.
3. Community members should be partners in policy making and leaders of local implementation.
4. Reconstruction policy and plans should be financially realistic but ambitious with respect to disaster risk reduction.
5. Institutions matter and coordination among them improves outcomes.
6. Reconstruction is an opportunity to plan for the future and to conserve the past.
7. Relocation disrupts lives and should be minimized.
8. Civil society and the private sector are important parts of the solution.
9. Assessment and monitoring can improve reconstruction outcomes.
10. To contribute to long-term development, reconstruction must be sustainable.

The last word: every reconstruction project is unique.

Defining the Reconstruction Policy

The handbook begins with a statement of guiding principles (shown in the adjacent box). These guiding principles encapsulate the handbook's advice and reflect some of the key concepts behind it, including participation, collaboration, sustainability, and risk reduction.

Reconstruction begins the day of the disaster. Therefore, one of the principal challenges of the policy maker is to work quickly to establish the policy basis for reconstruction, while taking time to confer with stakeholders and plan the reconstruction properly. The purpose of this Note is to summarize some of the important parameters of the policy-making exercise and to provide a conceptual framework for the reconstruction policy.

Reconstruction policy needs to be defined in five principal areas: (1) the Institutional Strategy, (2) the Financial Strategy, (3) the Community Participation Approach, (4) the Reconstruction Approach, and (5) Risk Management. The handbook's flow chart (shown after the table of contents and in miniature below) graphically represents this process. It also includes one other critical activity, common to all of these policy areas: implementation of a monitoring and feedback system. Also shown in the flow chart are the other critical activities in each of the five policy areas. The following sections summarize the key policy issues within each of the components of the reconstruction policy.

Institutional Strategy. People make reconstruction happen, but they will act mostly through different types of organizations. Beginning with government itself, one of the most critical early steps for the policy maker is to identify who will do what and how the numerous organizations that may be involved in reconstruction will work together. A second critical set of decisions has to do with the rules under which reconstruction will take place, that is, what are the laws, regulations, and institutional arrangements, both formal and informal, that will apply and will regulate what reconstruction agencies do. For instance, will projects be subject to existing building codes or environmental law, or will exceptions be made? How will nongovernmental organizations (NGOs) be involved, and how formal should their agreements to provide assistance to affected communities be? The institutional strategy must also evaluate the capabilities of the organizations involved and decide how their activities will be coordinated. What reconstruction responsibilities will local governments handle, for instance, and how will they report back? Whether an effective, coordinated institutional strategy is established, and is then monitored and adjusted as reconstruction proceeds, can determine the success or failure of the entire reconstruction program.

The handbook provides guidance on institutional strategy in several chapters, covering everything from how humanitarian agencies and reconstruction agencies work together (Chapter 1) to providing guidance on institutional options for organizing reconstruction (Chapter 13). In each chapter, the Key Decisions section identifies the roles and tasks that need to be assigned and proposes the appropriate agency to assign them to. Ultimately, many of these decisions need to be made by policy makers and reflected in the institutional element of reconstruction policy. (The chapters that correspond to each policy area are shown in the flow chart.)

Financial Strategy. Without financial resources, there will of course be no reconstruction. But a shortage of resources is not the greatest risk in managing the financial aspects of reconstruction. Greater risks are found in the lack of control of financial resources and in the lack of effectiveness of the resources that are spent. Most of the resources spent on reconstruction are not government's. Yet once they are pledged to the reconstruction effort, good post-disaster financial management requires that these commitments be taken into consideration in planning and that their expenditure be tracked. This points out once again the importance of coordination among agencies involved in reconstruction, as well as the need for systems that will permit accurate programming and tracking of expenditures, no matter the funding source.

Resources must not only be mobilized, programmed, and tracked, but some must be allocated and delivered directly to those affected by the disaster. For this population, the design and execution of the assistance strategy for housing is their principal concern and may be the sole metric by which they evaluate the policy framework, since it is the decision that will most directly affect them. Yet for the policy maker, this is, in fact, a complex set of decisions that have social, economic, and logistical implications. The assistance strategy must be tailor-made to the country and the disaster, and take into account existing social policy, as well the social equity and development objectives established for reconstruction. Lastly, preventing the misuse of resources must be a priority of policy makers, and anticorruption measures must be planned for and implemented throughout the reconstruction process.

The handbook provides guidance on mobilizing finance, tracking expenditures, and allocating and delivering resources to households, and includes measures for qualifying recipients of assistance and redressing their grievances (Chapter 15). It also suggests criteria for designing the assistance scheme (Chapter 4) and explains how financial management and procurement are handled in World Bank projects (Chapters 22 and 23). An entire chapter is dedicated to measures to mitigate corruption (Chapter 19).

Community Participation Approach. What is the role of affected communities in reconstruction and who decides on that role? Government cannot control what people do after a disaster, but the reconstruction policy can establish an approach to communication and interaction with affected communities that puts them in the center, capitalizing on the community members' wisdom, experience, and energy, or, alternatively, an approach that frustrates and disempowers all involved. Engagement with affected communities begins with communication, and a two-way consultative form of communication is strongly encouraged. Affected communities should have the opportunity to participate in policy making, including in the institutional and financial elements described above. Working "as a community" in reconstruction will be a foreign concept in many places, and there can be capture of these processes and the resources that are provided. Therefore, facilitation and oversight are critical to ensuring that community-based efforts are effective, properly governed, and truly participatory.

Guidance is found throughout the handbook on putting and keeping affected communities in the driver's seat during reconstruction, recognizing that there will be pressures to establish a more "efficient," top-down approach and that the commitment to participatory reconstruction needs to come from the top. Chapters and case studies demonstrate different models of participation and advice on recruiting community facilitators (see in particular Chapter 12). Also provided are practical tools, such as guidelines for training (Chapter 16), participatory assessment (Chapter 2), and social auditing (Chapter 18).

Reconstruction Approach. If reconstruction begins the day of the disaster, what does it mean for government to define the reconstruction approach? This element of reconstruction policy addresses how physical reconstruction will be carried out at the community level, starting with the important issue of the role of affected households in the actual reconstruction. Depending on the respective roles of households and reconstruction agencies, different forms of support will be needed, whether it is finance, training, or community facilitation. One critical issue that the reconstruction policy must address is whether transitional shelter will be provided to affected households. Transitional shelter can smooth the transition from disaster to permanent housing, but it has difficult cost and logistical implications that must be analyzed.

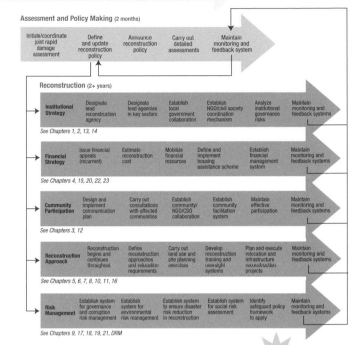

See full size version on page viii.

The reconstruction policy may need to create incentives to ensure the coordination of housing and infrastructure reconstruction. It will absolutely need to establish the goal of improving the safety of rebuilt housing and infrastructure, starting with defining standards to reduce the vulnerability to future hazards for all reconstructed and repaired housing and establishing the means for implementing these standards as broadly as possible. This must apply to housing built with public reconstruction funds, and should ideally extend to all rebuilt and repaired or retrofitted housing, no matter the funding source. In some cases, risk reduction will imply relocation of households or entire communities, and the policy must define the conditions for this. Again, coordination among agencies on these issues is key, so that households cannot circumvent the safety standards by seeking an alternative funding source.

Land use is almost always a difficult issue in reconstruction, and the reconstruction policy should anticipate this. The issues that may arise include, among others, (1) the need to replan land uses for housing and infrastructure, (2) the demand for tenure security, (3) the need for land for reconstruction, and (4) price escalation of land. In countries with extensive informal settlements and poor land administration and land use planning, solving these problems can hold up reconstruction. The policy should identify solutions or at least the means of reaching them.

Handbook chapters provide guidance on all policy issues related to the reconstruction approach, including the use of transitional shelter (Chapter 1), relocation (Chapter 5), land use and planning (Chapter 7), infrastructure reconstruction (Chapter 8), and housing design and technology (Chapter 10).

Risk Management. Finally, the reconstruction policy must ensure proper risk management in a number of areas, including (1) governance and corruption risk, (2) environmental risk, (3) disaster risk reduction, and (4) social risk. This is a disparate set of risks, but they share the common characteristic that poorly managing them during reconstruction can result in unforeseen or undesirable outcomes, delays, and loss of credibility for the individuals and institutions involved. These risks can be anticipated and measures can be taken to reduce vulnerability to them. This begins with establishing a culture of risk management in reconstruction and ensuring that tools to analyze and mitigate risks are widely understood and diligently applied.

Risk reduction begins with analysis and extends to project design and implementation. Each component of the policy should incorporate monitoring, which serves as a risk reduction tool by providing early signals that project design or implementation is poor or that communities are dissatisfied with outcomes. Some of the activities mentioned as good practice in the other policy areas also serve to reduce risk, for instance, the use of two-way communication, training, or good financial tracking. The handbook provides additional tools specifically identified as tools for risk management. These include methodologies for corruption risk assessment (Chapter 19), disaster risk management (Technical References), environmental management (Chapter 9), and social risk assessment (Chapter 4).

While it might seem logical that experts in disaster management would take a risk management approach to reconstruction, this is not always the case. However, the authors strongly encourage policy makers to do so, and to incorporate this approach through the inclusion of a risk management component in the reconstruction policy.

How Policy Makers Can Use the Handbook

Policy makers can use the handbook in a number of ways to help improve reconstruction outcomes.

The handbook can assist in the design of the reconstruction policy by offering a systematic approach and a comprehensive set of options. The handbook can serve as a shared frame of reference for specialists with diverse backgrounds who may be called on to assist government in proposing specific policies or in implementing reconstruction.

Policy makers can also encourage the use of the handbook by central and local government officials and officials of NGOs and civil society organizations to help them develop a common understanding of goals and the means to reach them and to improve the consistency of their interventions and, therefore, the efficiency of reconstruction. The assessment methodologies and other guidelines included in the annexes can serve as useful tools for joint action by a range of actors.

Please note that the house icon 🏠 is used throughout the handbook to alert the reader to related information elsewhere in the chapter or in another chapter.

The handbook is supported by a Web site, http://www.housingreconstruction.org, and a community of practice. The Web site contains additional materials related to each chapter and other relevant topics. Copies of the handbook can also be downloaded from the Web site.

The handbook will be updated periodically as comments are received from users and as the disaster reconstruction field and its best practices evolve. As you read and use the handbook, please feel free to comment on its contents at the Web site. User comments are most appreciated and will be taken into consideration to improve the next version of the handbook.

We sincerely hope that this handbook gives you the guidance you need to accomplish your goals and to provide the policy leadership in challenging post-disaster situations.

The Authors

Washington, DC
December 2009

A NOTE TO THE PROJECT MANAGER

Background

Safer Homes, Stronger Communities: A Handbook for Reconstructing after Disasters was developed to assist project managers and policy makers engaged in large-scale post-disaster reconstruction programs make decisions about how to reconstruct housing and communities after natural disasters.

As the handbook demonstrates, post-disaster reconstruction begins with a series of decisions that must be made almost immediately. Despite their urgency, these decisions—and the manner in which they are implemented—will have long-term impacts that will change the lives of those affected by the disaster for years to come.

As a project manager or task manager, you will be responsible for implementing government policy decisions and for making many operational decisions on the ground. The handbook provides information on the options that should be considered in various aspects of reconstruction and insight into what has worked elsewhere. It does not tell you exactly what to do, but it should improve the likelihood of good outcomes from the work that is done.

The handbook's flow chart (shown after the table of contents and in miniature below) graphically represents the entire reconstruction process.

Content of the Handbook by Chapter

The handbook begins with a statement of guiding principles (shown in the adjacent box). These guiding principles encapsulate the handbook's advice and reflect some of the key concepts behind it, including participation, collaboration, sustainability, and risk reduction.

The handbook is divided into four parts. Below is an overview of some of the key concepts and guidance presented in each part.

Part 1, Reconstruction Tasks and How to Undertake Them, provides both policy and practical advice on critical reconstruction issues. Part 1 contains three sections that correspond to the principal stages of reconstruction: (1) assessment and policy making, (2) planning, and (3) implementation. Below are summaries of the chapters contained in each of these three sections.

Section 1. Assessing Impact and Defining Reconstruction Policy

In *Chapter 1, Early Recovery: The Context for Housing and Community Reconstruction*, the handbook offers an overview of the institutional landscape project managers are likely to encounter in a post-disaster setting beginning with the disaster event, when humanitarian agencies are likely to be most prevalent, and of the sequence of events that are likely to unfold. It also describes the roles that affected populations and various agencies take on in the post-disaster environment. This chapter also presents the arguments in favor of and against providing transitional shelter. A number of common gaps or bottlenecks in the reconstruction process, including the funding gap, the planning gap, the implementation gap, and the participation gap, are described here. This chapter sets the tone for the rest of the handbook by arguing for a reconstruction approach that puts affected communities in the center, helping to set policy and organizing the entire reconstruction process.

Handbook Guiding Principles

1. A good reconstruction policy helps reactivate communities and empowers people to rebuild their housing, their lives, and their livelihoods.

2. Reconstruction begins the day of the disaster.

3. Community members should be partners in policy making and leaders of local implementation.

4. Reconstruction policy and plans should be financially realistic but ambitious with respect to disaster risk reduction.

5. Institutions matter and coordination among them improves outcomes.

6. Reconstruction is an opportunity to plan for the future and to conserve the past.

7. Relocation disrupts lives and should be minimized.

8. Civil society and the private sector are important parts of the solution.

9. Assessment and monitoring can improve reconstruction outcomes.

10. To contribute to long-term development, reconstruction must be sustainable.

The last word: every reconstruction project is unique.

Chapter 2, Assessing Damage and Setting Reconstruction Policy, discusses the assessment process and explains some of the common types of assessments. It focuses on three types of assessments: (1) multisectoral assessments (such as the damage and loss assessment); (2) housing sector assessments, which can be used to diagnose the land administration and affordable housing policy and institutional framework in the country and to identify capacity issues that may arise in reconstruction; and (3) local housing assessments, including housing damage assessments. Housing damage assessments are the door-to-door assessments that are often used to allocate housing assistance. This chapter shows how assessment results are used to define reconstruction policy and discusses the political economy of the reconstruction process. An outline for a reconstruction policy is provided, as are two good examples of reconstruction policies: those used in the aftermath of (1) the 2001 earthquake in Gujarat, India; and (2) the 2004 Indian Ocean tsunami in Tamil Nadu, India. The annexes in this chapter provide detailed methodologies on how to conduct (1) a housing sector assessment and (2) a housing condition assessment. The proposed housing condition assessment methodology recommends a number of activities to also assess the overall condition of the neighborhood, including the "village transect." This assessment should be conducted with the participation of affected communities.

Chapter 3, Communication in Post-Disaster Reconstruction, provides guidelines on the development of a comprehensive post-disaster communications strategy. This chapter encourages the use of continuous, two-way communications following a disaster to constantly monitor the relevance and quality of the outcomes of the reconstruction program. The communications strategy that was implemented following the 2005 North Pakistan earthquake is used as an example throughout this chapter, and case studies are presented on various aspects of this experience, including the importance of assessing the cultural context when designing communications activities and the use of beneficiary feedback as a monitoring and evaluation tool. Two annexes are included in this chapter: (1) a methodology for conducting a communications-based assessment and (2) a table that summarizes the cultural factors that affect communication.

Project managers will find information in **Chapter 4, Who Gets a House? The Social Dimension of Housing Reconstruction**, that will help ensure that housing assistance reaches its intended beneficiaries and has the desired social impact on the ground. This chapter presents a table of all the tenancy categories that might be assisted by a housing assistance program and includes matrices of criteria that can be used to design a housing assistance scheme. These matrices address the following questions: Who is entitled? What form of assistance are they entitled to? How much assistance should they receive? Each case study explains the logic of a different approach to providing reconstruction assistance. Annexes to this chapter cover (1) considerations for designing a social protection system for natural disasters and (2) a detailed methodology for conducting a social assessment of a disaster-affected community.

Chapter 5, To Relocate or Not to Relocate, is intended to guide project managers to minimize instances of relocation and to minimize the scope and impact when relocation is absolutely necessary. Relocation is frequently used as a risk reduction strategy even when the risks are not site-specific, because a rigorous disaster risk assessment is not conducted. This chapter should be read together with the Disaster Risk Management in Reconstruction chapter in Part 4, Technical References, which explains how a disaster risk assessment should be conducted and how to compare risk reduction options. Relocation is not the same as "resettlement" as defined in the policies of many international organizations, including the World Bank, and this chapter explains the difference in these concepts, the different types of displacement, and the implications for project design, such as the ways in which assistance may be provided. At the same time, an argument is made in favor of a relocation approach that carefully identifies both social and economic impacts for households and attempts to mitigate them, as provided in resettlement policies. Numerous case studies demonstrate the impacts of well- and poorly planned relocation, and the annex provides a systematic planning procedure for a resettlement project based on the International Finance Corporation resettlement policy.

Chapter 6, Reconstruction Approaches, presents a typology and a comparison of six of the most common approaches to housing reconstruction, ranging from full owner-driven (or owner-managed) to full agency-driven approaches. It explains the advantages and disadvantages of each one in particular situations and provides case studies for each of the approaches. The same tenancy categories are considered in this chapter as in the discussion of housing assistance in Chapter 4. It is explained that different circumstances make one or another reconstruction approach preferable. For instance, an owner-driven approach is probably infeasible for a high-rise urban apartment building, even if the residents are owners of the units. One of the approaches discussed entails reconstruction in a relocated site (agency-driven reconstruction in relocated site), and therefore has most of the same disadvantages as relocation in general, as discussed in Chapter 5.

Section 2. Planning Reconstruction

The section on planning reconstruction begins with ***Chapter 7, Land Use and Physical Planning***, which describes why planning of both sites and local land use is important, even in the post-disaster context. The content of a traditional land use plan is described, as are the challenges that arise in post-disaster planning, which include the lack of time, information, and capacity. This chapter explores the complex issues associated with the need for access to land and secure tenure in reconstruction and presents recommended solutions. Case studies include one that discusses the planning process Bhuj City, India, during which a number of residents were moved from the urban core to the periphery, and another that describes the innovative community-driven land adjudication process that took place in Aceh, Indonesia following the 2004 Indian Ocean tsunami. Annexes provide methodologies for planning (1) in situations where prior planning has taken place and institutions are experienced with planning methods and (2) in situations where this is not the case.

Chapter 8, Infrastructure and Services Delivery, explains both the short-term (lifeline) and longer-term (restoration/reconstruction) measures to restore infrastructure. It includes a chart that shows the most common types of damage from different types of infrastructure and the degree of severity of the damage. The coordination of housing and infrastructure reconstruction is difficult in a post-disaster environment, and some guidance is given on minimizing the risks of this situation. The chapter also presents an infrastructure planning methodology designed to improve the disaster resiliency of infrastructure during reconstruction or rehabilitation, and recommends that local service providers be strengthened both financially and in terms of their institutional capacity during the reconstruction process so that they are capable of maintaining the viability and the disaster resilience of rebuilt infrastructure over time. The case studies in this chapter explain various instances where infrastructure and housing reconstruction were not well coordinated and the practical solutions that were arrived at.

Chapter 9, Environmental Planning, alerts project managers to a range of environmental risks that may have been created by the disaster itself, or are likely to be encountered or created in the reconstruction process. It describes the types of environmental damage that are likely to result from different types of disasters and highlights common management problems, such as the handling of disaster debris and the planning of new settlements in a way to incorporate ecological considerations. A wide range of planning and analytical tools are described, including environmental risk assessment, eco- and hazard mapping, and environmental management plans, and links to resource information for these tools are provided. A summary on the risk of encountering asbestos in reconstruction and on the laws and regulations that govern its handling and transport is also included. Annexes provide instructions on (1) the development of a disaster debris management plan and (2) conducting an environmental impact assessment and preparing an environmental monitoring plan.

Chapter 10, Housing Design and Construction Technology, covers the range of critical issues associated with the design and construction of housing. Project managers will frequently need to decide, or help government decide, whether or to what extent construction methods will be upgraded in reconstruction, and this chapter is intended to provide support to those decisions. Issues covered include the choice of materials and building methods, the decision whether to repair/retrofit or rebuild, and the potential for incorporating universal design standards in reconstruction. Material is included on the use of vernacular construction methods, the controversies that surround this option, and the approaches that can be taken to improve their disaster resilience. To assist the project managers who may be deciding whether to support the use of local building methods in reconstruction, the chapter lists contact information for experts and institutions working to improve housing that uses local materials and vernacular building methods.

The Process of Response and Reconstruction

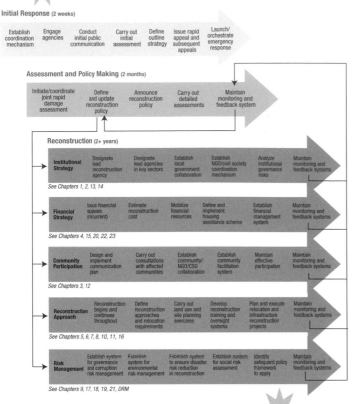

See full size version on page viii.

Chapter 11, Cultural Heritage Conservation, discusses the social and economic benefits for communities associated with including cultural heritage conservation in post-disaster reconstruction. The chapter explains that cultural heritage can include not only traditional historic sites, but historic housing, cultural landscapes, and aesthetic assets, such as the architectural style of housing. The chapter explains how cultural assets conservation fits into larger community reconstruction projects and discusses their social and economic value. If no planning for the treatment of cultural assets in a disaster has taken place beforehand, there are still interventions communities can carry out, and the chapter explains what some of these are. However, the text explains that the effort to salvage cultural assets can cause as much damage as the disaster itself, so expert support is likely to be needed, and extensive resource material is provided to assist communities in finding help. In reconstruction, there are also efforts that should be made, such as adopting building codes that are compatible with cultural assets and vernacular building practices and providing financial incentives to encourage the conservation of built vernacular heritage that may be in private hands but that may have public value.

Section 3. Project Implementation

The project implementation section is of particular value to project managers, due to its practical, operational focus.

The handbook authors strongly favor a community-based approach to reconstruction. **Chapter 12, Community Organizing and Participation**, provides guidance on how to operationalize this concept and empower communities to lead their own reconstruction effort. It includes an overview of the ways in which communities can manage the reconstruction process or otherwise participate in reconstruction, beginning with conducting participatory assessments and participating in the definition of reconstruction policy. The chapter emphasizes that communities need support to lead reconstruction, and it provides in annexes (1) a methodology for analyzing the existing organization and leadership structure of the community and the assets it has to contribute to the reconstruction process (the Community Participation Profile) and (2) a detailed description of the community facilitation process that has been used very successfully in Indonesia, beginning with the post-Indian Ocean tsunami reconstruction.

Chapter 13, Institutional Options for Reconstruction Management, addresses options for organizing the overall reconstruction program and explains the situations in which they are most suitable and their advantages and drawbacks. While the typology of options presented is focused on the overall reconstruction effort, the concepts (creating a new entity versus using existing agencies) are relevant to the housing sector and may be useful in organizing the institutional response, even in a single community. The chapter explains that the entity managing reconstruction needs a mandate, a reconstruction policy, and a reconstruction plan in order to be effective. This chapter also recommends that, wherever possible, a central role be given to local government in reconstruction and emphasizes the need to ensure coordination between local officials and officials managing the overall reconstruction effort. The case studies are correlated with the institutional typology to show how the various structures have worked in actual disaster situations.

From **Chapter 14, International, National, and Local Partnerships in Reconstruction**, project managers can gain insight into the requirements for successfully working with the variety of nongovernmental entities that are often at work in the reconstruction environment. While Chapter 2 explains the roles of agencies on a chronological basis, beginning immediately after the disaster, this chapter describes in more operational terms how these agencies can organize and coordinate their interventions. The chapter also explains how nongovernmental and civil society organizations get involved in reconstruction and provides guidelines on formalizing the relationship between central or local government and these organizations to help ensure that their actions contribute to larger development goals. One technique suggested is a registration system for nongovernmental and civil society organizations to improve transparency and accountability. Another is a process for formalizing the commitments of these organizations to help affected communities. Case studies provide a sampling of the numerous approaches nongovernmental organizations (NGOs) use to support the reconstruction effort.

While the parameters on the use of financial resources will be defined by policy makers, project managers can have an enormous influence on the effective use of these resources. ***Chapter 15, Mobilizing Financial Resources and Other Reconstruction Assistance***, provides guidelines on qualifying recipients, delivering financial and other resources, and tracking their use at the project level. The chapter reviews the various forms of assistance that may be provided to affected households, including, cash, vouchers, in-kind materials, and even whole houses. It also explains normal mechanisms that households use to support reconstruction, including microfinance and migrant remittances, both of which can be interrupted after a disaster and may need support. The annexes to this chapter are intended to assist project managers with two common issues: (1) whether to import or procure and distribute construction materials and (2) how to establish a grievance redressal system. The chapter points out the importance of coordinating and monitoring reconstruction finance, whatever its source, even at the project level, where agencies can inadvertently compete or duplicate efforts, both of which create disincentives for households and reduce the effectiveness of the overall reconstruction effort.

Chapter 16, Training Requirements in Reconstruction, provides instructions on developing a large-scale training program aimed at improving the quality of housing condition assessments and of reconstruction, whether the builder is a contractor, a homeowner, or a combination of the two. The approach described in this chapter incorporates the initial and detailed assessment of housing condition, the design of training materials, and the use of model buildings as a training tool. One of the most important concerns in implementing training or facilitation at the project level in a large disaster is scaling up these interventions to ensure that the reconstruction effort is not delayed. The methodology described in this chapter depends on the training of trainers, which allows the scale of the training and assessment system to grow quickly. A second important concern is quality control. This system entails using the trainers as inspectors once the initial training period is over, as a means to ensure quality control.

Part 2, Monitoring and Information Management, helps project managers with advice about technology use in reconstruction, project monitoring, and involving affected communities in project oversight.

Chapter 17, Information and Communications Technology in Reconstruction, describes the wide variety of technologies being used in post-disaster assessment and monitoring. The chapter explains that these technologies, and their use in the post-disaster environment, are constantly changing, but that currently they are being used to improve coordination, communications, assessment, planning, and monitoring. However, successful information and communications technology use should conform to protocols that improve the interoperability of equipment and the standardization of data. Tools that are described as useful for post-disaster communications include Web 2.0, mobile telephones, and ham radios. Detailed annexes to the chapter explain (1) considerations for procuring satellite data and (2) the organization of geographic information systems and their use in reconstruction.

Using ***Chapter 18, Monitoring and Evaluation***, project managers should be able to define the parameters of monitoring in reconstruction and decide whether and how to evaluate project impact. This chapter explains how monitoring and evaluation can be useful in reconstruction projects and the levels at which they may be conducted (household, project, program, and sector). The chapter advocates for the use of a mix of qualitative and quantitative data, including data collected in a participatory manner, and household survey data. Different agencies use different ways to organize and manage their monitoring data. The chapter contains explanations of two common systems. While impact evaluation of disaster projects is not common, a framework developed by the World Bank for the impact evaluation of slum upgrading projects is proposed as an important resource for those interested in evaluating post-disaster housing and community projects, as explained in the chapter's first annex. This annex also includes a table of potential monitoring indicators for housing and community projects. Annex 2 explains participatory performance monitoring, and includes a methodology for a social audit and a summary of other participatory performance monitoring methods. A case study summarizes of the results of the evaluation of a reconstruction project financed by the World Bank following the 1999 earthquake in Armenia, Colombia.

There is a significant risk of corruption in reconstruction, and *Chapter 19, Mitigating the Risk of Corruption*, contains guidance and a range of tools on mitigating this risk. It recommends emerging practices to reduce corruption, such as codes of conduct in the public sector and integrity pacts between the public sector and the private sector and NGOs. Mechanisms of social control are explained, such as systems to encourage whistleblowers and to protect their identity, explaining that these mechanisms can be used on a situational basis or as part of a more comprehensive integrity system. Extensive information is provided on the use of audits, along with an explanation of the different types of audits and the auditing standards they apply. The three annexes in this chapter provide the means to improve transparency in various aspects of post-disaster reconstruction. They are (1) instructions for developing a project governance and accountability plan, as defined by the World Bank; (2) guidelines for conducting a corruption risk assessment, which focuses on organizational financial controls; and (3) instructions for ordering a construction audit at the project level.

Part 3, Information on World Bank Projects and Policies, provides an overview of how the World Bank assists governments after disasters and of the policies and procedures that apply in World Bank projects.

Chapter 20, World Bank Response to Crises and Emergencies, explains Operational Policy/Bank Procedure (OP/BP) 8.00, including the forms of Bank response, the features of Bank response, and the processing steps for emergency operations.

Chapter 21, Safeguard Policies for World Bank Reconstruction Projects, includes a summary of the Bank's safeguards policies and an explanation of their application in normal and emergency operations.

Chapter 22, Financial Management in World Bank Reconstruction Projects, explains the Bank project cycle and the policies and procedures for financial management in Bank operations. The chapter also includes a summary of the financial management issues that can arise in emergency operations and means to address them, and includes a discussion of the financial management aspects of OP/BP 8.0.

Chapter 23, Procurement in World Bank Reconstruction Projects, provides an overview of Bank procurement rules and a summary of how the Bank assesses country procurement capacity at the country and agency levels. The chapter also describes the procurement issues that can arise in emergency operations and proposes ways to address them.

Part 4, Technical References, includes technical information that may be useful in various aspects of reconstruction. This part of the handbook includes a glossary and the following sections.

Disaster Types and Impacts describes global disaster impacts and the impact of disasters on poverty and includes historical disaster data. It is included to provide a longer-term economic context for decisions and discussions within government about disaster-related risk reduction, policy, planning, and public investment.

Disaster Risk Management in Reconstruction includes a framework for evaluating both short- and long-term mitigation options for housing and infrastructure and a comparative risk assessment methodology. This information is especially relevant to all chapters dealing with reconstruction planning. This chapter also provides guidance on how to organize a community-based hazard mitigation planning process and includes case studies on how disaster risk management has been used in specific disaster-related situations.

Matrix of Disaster Project Features is based on a matrix originally developed by the government of Pakistan to compare the policy decisions in a variety of disaster reconstruction projects between 2001 and 2005. It demonstrates the range of options governments select in these situations.

How Project Managers Can Use the Handbook

Project managers can use the handbook in a number of ways to help improve reconstruction outcomes. The handbook can assist project managers who are participating in policy decisions by offering a systematic approach and a comprehensive set of options to inform policy decisions.

Project managers can also share the handbook with affected communities at the project level and use the information it contains to make more sound decisions in consultation with them.

In addition, the handbook can be provided to local government officials and officials of nongovernmental and civil society organizations. Using the options and concepts presented in the handbook as a frame of reference, it should be easier to define common goals and the means to reach them, as well as to establish better systems for coordination. In particular, the assessment methodologies and other guidelines included in the annexes can be starting points for joint action by a range of actors.

Please note that the house icon is used throughout the handbook to alert the reader to related information elsewhere in the chapter or in another chapter.

The handbook is supported by a Web site, http://www.housingreconstruction.org, and a community of practice. The Web site contains additional materials related to each chapter and other relevant topics. Copies of the handbook can also be downloaded from the Web site.

The handbook will be updated periodically as comments are received from users and as the disaster reconstruction field and its best practices evolve. As you read and use the handbook, please feel free to comment on its contents at the Web site. User comments are most appreciated and will be taken into consideration to improve the next version of the handbook.

We sincerely hope that this handbook gives you the support you need to accomplish your goals and that it empowers your work as a project manager in any future post-disaster situation.

The Authors

Washington, DC
December 2009

ACKNOWLEDGMENTS

Gathering the current thinking on the best way to carry out post-disaster housing and community reconstruction from the experts on the subject—an itinerant group, often found in the remote corners of the world—was a challenging task. Nevertheless, many of them went out of their way to point out the pitfalls and to provide guidance and advice on the realities of post-disaster housing and community reconstruction. For their invaluable suggestions, interviews, participation in review meetings, comments, coordination, and case studies, we would like to thank:

Irvin Adonis, Jean-Christophe Adrian, Cut Dian Agustina, Kamran Akbar, Mr Mohammad Alizamani, Dr. Shoichi Ando, Raphael Anindito, Shahnaz Arshad, Raja Rehan Arshad, Anand S. Arya, Tency Baetens, Kraig H. Baier, Judy L. Baker, Lee Baker, Rick Bauer, Bakri Beck, Janis D. Bernstein, Chuck Billand, Judy Blanchette, Giovanni Boccardi, Lars Büchler, Sandra Buitrago, Ana Campos Garcia, Omar D. Cardona, Iride Ceccacci, P. G. Dhar Chakrabarti, Andrew Charlesson, Elena Correa, Saurabh Dani, Samantha de Silva, Sergio Dell'Anna, Katalin Demeter, Leslie Dep, Bruno Dercon, Rajendra Desai, Amod Mani Dixit, Alireza Fallahi, Wolfgang Fengler, María Fernández Moreno, Anita Firmanti, Thomas Fisler, Mario Flores, Francis Ghesquiere, Heiner Gloor, Joseph Goldberg, Sumila Guiliani, Iwan Gunawan, Manu Gupta, Maryoko Hadi, Muhammad Waqas Hanif, Paul Harvey, R. Ivan Hauri, Rasmus Heltberg, Carlos Herz, Andre Herzog, Montira Horayangura, Sushma Iyengar, Christopher Jennings, Saroj Jha, Rohit Jigyasu, Christianna Johniddes, Roberto Jovel, Dodo Juliman, Dr. Kalayeh, Hiroyuki Kameda, Sima Kanaan, Hemang Karelia, Shailesh Kataria, Rajesh Kumar Kaushik, Amir Ali Khan, Pankar Khana, Victoria Kianpour, Axel Kiene, Joseph King, Eberhard Knapp, Michael Koeniger, Tejas Kotak, Imam Krismanto, Jolanta Krispin-Watson, Santosh Kumar, Benny Kuriakose, Reidar Kvam, Robert Laprade, Joe Leitman, Esteban Leon, Stephen Ling, Barbara Lipman, Satprem Maïni, James Malakar, Ruby Mangunsong, Graham Matthews, Steve Matzie, Robert Maurer, Jock Mckeon, Vandana Mehra, Barjor Mehta, Vinod Menon, Provash Monda, Teungku Muhammad, Lulu Muhammad, Isabel Mutambe, Dr. Tatsuo Narafu, Ahmadou Moustapha Ndiaye, Rosanna Nitti, Kenji Okazaki, Klaus Palkovits, Al Panico, Parveen Pardeshi, Jeong Park, Ron Parker, Ayaz Parvez, Thakoor Persuad, Anna Pont, Esther Pormes, Robin Rajack, Fernando Ramirez Cortes, Priya Ranganath, Mohsen Rashtian, Vivek Rawal, Geoffrey Read, Kathryn Reid, Ritesh Sanan, C. V. Sankar, Shyamal Sarkar, Maoni Satprem, Graham Saunders, Charles Scawthorn, Tom Schacher, Theo Schilderman, Charles Setchell, Fatima Shah, Anshu Sharma, George Soraya, Maggie Stephenson, Wm. B. M. Stolte, Jishnu Subedi, Sri Probo Sudarmo, Parwoto Sugianto, Zeeshan Suhail, Aswin Sukahar, Bambang Sulistiyanto, Akhilesh Surjan, Fabio Taucer, John Tracey-White, Zoe Elena Trohanis, Etsuko Tsunozaki, Azmat Ulla, Simone van Dijk, Alexander van Leersum, Krishna S. Vatsa, Victor Vergara, Sandeep Virmani, Antonella Vitale, Gaetano Vivo, Hyoung Gun Wang, Doekle Geert Wielinga, Norbert Wilhelm, Berna Yekeler, Jaime Yepez, and Xiulan Zhang.

Special acknowledgement is due to those organizations who have agreed to be partners with the World Bank in the handbook project: Asian Disaster Preparedness Center, Earthquakes and Megacities Initiative, Habitat for Humanity International, Royal Institution of Chartered Surveyors, Shelter Centre, and World Housing Encyclopedia.

ABOUT THE AUTHORS

Mr. Abhas K. Jha (lead author and task manager) is a housing and urban finance specialist. He is currently Lead Urban Specialist and Regional Coordinator, Disaster Risk Management for the World Bank's East Asia and the Pacific Region. He is responsible for managing the Bank's disaster risk management practice in the region. Mr. Jha has been with the World Bank since 2001, leading the Bank's urban, housing, and disaster risk management work in Turkey, Mexico, Jamaica, and Peru, and serving as the Regional Coordinator, Disaster Risk Management, for the Europe and Central Asia region. He earlier served for 12 years in the Indian Administrative Service (the national senior civil service of India) in the government of India (in the Federal Ministry of Finance and earlier in the state of Bihar). Mr. Jha holds graduate degrees in finance from Johns Hopkins University and in economics from the University of Madras.

Principal Contributors

Dr. Jennifer E. Duyne Barenstein (anthropologist) is an expert in the socioeconomic and cultural dimensions of post-disaster reconstruction, rural infrastructure development, and water management, with extensive professional and research experience in Argentina, Bangladesh, India, Indonesia, Italy, Mexico, the Philippines, and Sri Lanka. She is the head of the World Habitat Research Centre of the University of Applied Science of Southern Switzerland in Lugano, Switzerland (http://www.worldhabitat.supsi.ch). She served as principal consultant for the handbook and contributed chapters on reconstruction approaches, cultural heritage, social issues, and resettlement.

Ms. Priscilla M. Phelps (project manager) is an expert in local development, affordable housing, and municipal finance who works for TCG International (TCGI) in Silver Spring, Maryland. She served as chief technical editor for the handbook and contributed chapters on finance, governance, and infrastructure. Her experience with post-disaster housing includes lending for post-disaster housing reconstruction after Hurricane Mitch in Honduras and the 2001 earthquakes in El Salvador while at the Inter-American Development Bank, and working for the Low-Income Housing Fund in San Francisco after the Loma Prieta earthquake. TCGI advises governments and donors on affordable housing, municipal and infrastructure finance, and community and economic development internationally and domestically in the United States (http://www.tcgillc.com).

Mr. Daniel Pittet (housing design and technology specialist) is a civil engineer with a master's degree in architecture and sustainable development from the Swiss Institute of Technology in Lausanne. Since 2005, he has worked as a researcher at the World Habitat Research Centre of the University of Applied Science of Southern Switzerland in Lugano, Switzerland (http://www.worldhabitat.supsi.ch). His areas of interest are sustainable architecture, housing environmental impact, and energy efficiency, issues that he has researched and applied in post-disaster reconstruction projects and in developing countries, including in Nepal and India, where he has worked extensively. He provides advice to municipalities on energy management and sustainability for the Swiss national program *SwissEnergy*. Mr. Pittet is a photographer with several photographic publications and exhibitions to his credit. In addition to serving as a member of the core team of consultants, he provided most of the photographs in the handbook.

Mr. Stephen Sena (information technology and Web site development) is an economic development consultant with TCG International (TCGI) in Silver Spring, Maryland, with a focus on information technology and communications. As a member of the core team of consultants, he contributed the chapter on information and communications technology in post-disaster reconstruction, coordinated the development of the handbook Web site, and managed design and production issues. TCGI advises governments and donors on affordable housing, municipal and infrastructure finance, and community and economic development internationally and domestically in the United States (http://www.tcgillc.com).

Other Contributors

Mr. B. R. Balachandran (urban planner) heads Alchemy Urban Systems in Bangalore, India (www.alchemyurban.com). He is an architect and urban planner actively engaged in professional practice since 1990. Beginning with his work on post-earthquake reconstruction of Bhuj (Gujarat, India) in 2001, post-disaster reconstruction planning has become one of his principal areas of specialization, including his current involvement in reconstruction from the 2008 flood in the Kosi region, Bihar, India. He contributed the chapter on land use and physical planning and provided advice on other planning issues.

Dr. Bernard Baratz (safeguards specialist) holds a Ph.D. in chemical engineering from Princeton University and has worked for 40 years addressing the environmental impacts of energy, industrial, and infrastructure projects. A World Bank staff member for almost 30 years, he was the first Europe and Central Asia Regional Safeguards Coordinator and one of the original authors of the World Bank policy for environmental assessment. Dr. Baratz has worked as an environmental specialist on disaster recovery and disaster management projects in China, Turkey, and Romania.

Dr. Camillo Boano (architect/urban planner) is an architect and urban planner who teaches, conducts research, and provides policy advice on topics of urban development, design and urban transformation, shelter and housing, reconstruction and recovery in conflicted areas and divided cities, architecture and planning in contested spaces, and linking emergency recovery and development. He is currently affiliated with the Development Planning Unit at the University College of London.

Mr. William Bohn (disaster risk reduction) is an engineer and GIS Specialist with Tetra Tech (http://www.tetratech.com/tetratech/). He has more than 10 years of experience in 19 countries providing disaster management services to the World Bank, the International Finance Corporation, the United States Agency for International Development, the U.S. Federal Emergency Management Agency, the U.S. National Oceanic and Atmospheric Administration, and the U.S. Environmental Protection Agency. He has authored numerous technical papers and publications on risk assessment, hazard analysis, vulnerability, mitigation, climate change, computer modeling, and geographic information systems.

Mr. Roberto Carrion (monitoring and evaluation) is an Ecuadorian architect and urban planner. He worked for CHF International in Honduras after Hurricane Mitch in 1998, and managed post-disaster housing reconstruction processes for and for the United States Agency for International Development in El Salvador after the 1986 earthquake, in Colombia after the 1999 earthquake in Armenia, and in the Dominican Republic after Hurricane Georges in 2000. Mr. Carrion contributed to the chapter on monitoring and evaluation and provided case studies from Latin America

Ms. Natalia Cieslik (communications) has worked in communications and journalism for the last 19 years, focusing on conflict and development issues in sub-Saharan Africa, Asia, and the Middle East. She has worked as a communications specialist for several nongovernmental organizations and is currently part of the World Bank team preparing the World Development Report 2011. Ms. Cieslik holds a BA in social and business communication from the University of Arts in Berlin and a master's in security studies from Georgetown University.

Dr. Tom Corsellis (partnerships and humanitarian assistance) is Co-Director of Shelter Centre, a nongovernmental organization (NGO) supporting the humanitarian community of practice by developing policy and technical guidance, training, and resources for and with the humanitarian shelter and reconstruction sector, informed by ongoing operational collaborations worldwide with donors, the United Nations (UN), international organizations, and NGO partners. Shelter Centre guidance developed with the UN Office for the Coordination of Humanitarian Affairs includes "Shelter after Disaster: Transitional Settlement and Reconstruction." Shelter Centre sector resources include a one-stop library of 2,000 publications from 200 agencies (http://www.shelterlibrary.org).

Mr. Ian Davis (shelter) has specialized in post-disaster shelter and reconstruction since 1972. He authored numerous publications on post-disaster housing; edited the first United Nations (UN) Guidelines on Shelter after Disaster; and provided consultancy advice to governments, UN agencies, and nongovernmental organizations in 38 countries. Mr. Davis has a special interest in disaster risk reduction and recovery at the community level.

Ms. Kirti Devi (infrastructure and services delivery) is an architect and urban planner who works as an infrastructure finance specialist with TCG International (TCGI) in Silver Spring, Maryland. She contributed to the chapter on infrastructure and services delivery and organized the field reviews of the draft handbook. In 2001, she provided technical assistance to the state government for post-earthquake infrastructure reconstruction in 14 towns, under the Asian Development Bank-funded Gujarat Earthquake Rehabilitation Project in Gujarat, India. TCGI advises governments and donors on affordable housing, municipal and infrastructure finance, and community and economic development internationally and domestically in the United States (http://www.tcgillc.com).

Mr. Jeff Feldmesser (editor) has been a writer/editor and publications specialist for more than 25 years. For the last 15 years, he has specialized in the production of publications in the field of international development.

Mr. Earl Kessler (community-based development) is an urban development expert who has developed shelter and urban programs since being a Peace Corps volunteer in Colombia in 1965. His post-disaster work with the United States Agency for International Development (USAID) includes the 1977 Guatemala earthquake reconstruction program; the Solanda, Ecuador housing project; and the assessment of USAID's role in Nevada de Ruiz after the explosion that buried Armero, Colombia. He directed USAID's Regional Urban Development Offices in Thailand from 1988 to 1993 and in India from 1993 to 1998. He also served as Deputy Executive Director of the Asian Disaster Preparedness Center. Currently, he is preparing a community-based disaster risk management guide for tsunami-affected communities in Thailand and is one of the authors of the World Bank "Climate Resilient Cities Primer for East Asia."

Mr. Richard Martin (monitoring and evaluation specialist) has specialized in housing in Africa for more than 30 years and has undertaken evaluations of a wide range of internationally funded projects. He has advised on housing in emergencies in Madagascar and South Africa, and has a special interest in the creative and effective involvement of communities. He contributed to the chapter on monitoring and evaluation.

Ms. Gretchen Maxwell (art director) is creative principal of GLM Design in Falls Church, Virginia. She designed and produced the handbook and was the contributing art director for the accompanying Web site. She has run her design studio for 15 years, focusing on publication design, including annual reports, research guides, and books.

Ms. Savitha Ram Mohan (research assistant) is an architect and urban planner with more than five years of experience in urban land development, environmental management, post-disaster reconstruction planning, and architecture. She contributed to India's first-ever climate change status report and also worked as a short-term consultant to the World Bank's South Asia Environment and Social Unit.

Ms. Brigitte Marti Rojas Rivas (research assistant) studied social anthropology in Switzerland and Ecuador, and has a degree in applied anthropology from the Universidad Politécnica Salesiana in Quito. She is working on various projects for the World Habitat Research Centre of the University of Applied Science of Southern Switzerland in Lugano, Switzerland (http://www.worldhabitat.supsi.ch), including researching the viability of resettlement as a hazard mitigation strategy in the City of Santa Fe, Argentina.

Dr.-Eng. Norbert E. Wilhelm (training) is a civil engineer with a Ph.D. in European standards for structural steelworks. Since 1982, he has worked in design, supervision, and training for development projects in the buildings and civil works sector in Africa, Asia, and Europe. He has worked on reconstruction projects for public buildings, education buildings, health facilities, roads, bridges, and housing. He has published papers on traditional building, appropriate technology, climatic appropriate design, and rehabilitation. As senior advisor for Grontmij-BGS Consultants (http://www.bgs-ing.de/), he has worked on reconstruction projects in India, Sri Lanka, and Pakistan with German financing though KfW that have provided a total of more than 20,000 units of housing.

Mr. Rajib Shaw (environmental management) is an Associate Professor in the Graduate School of Global Environmental Studies of Kyoto University, Japan. He works principally in Asia with the local communities, nongovernmental organizations, governments, and international organizations, including the United Nations, and conducts research on such topics as community-based disaster risk management, climate change adaptation, urban risk management, and disaster and environmental education. He has written books on disaster risk management, indigenous knowledge, and river basin management.

Mr. Fred Zobrist (disaster risk reduction) is an expert in international development and a professional engineer who has worked on projects in more than 100 countries on risk management related to housing and infrastructure and on hurricane and flood damage rehabilitation. He has served as a damage evaluation expert for hurricanes, floods, and tornados for the U.S. Federal Emergency Management Agency ; has developed floodplain maps; and served as risk management expert on disasters related to floods, tsunamis, earthquakes, high winds, high seas, and volcanoes for the U.S. Corps of Engineers.

ABBREVIATIONS

ADB	Asian Development Bank
ADPC	Asian Disaster Preparedness Centre
ADRIS	Agency-Driven Reconstruction in-Situ
ADRRS	Agency-Driven Reconstruction in Relocated Site
ALNAP	Active Learning Network for Accountability and Performance
ATC	Applied Technology Council
AusAID	Australian Agency for International Development
BP	Bank Procedure
BRCS	British Red Cross Society
BRR	Rehabilitation and Reconstruction Agency of the Government of Indonesia
CA	Cash Approach
CADRE	Centre for Action, Development, Research and Education (India)
CAP	Communications Action Plan
CBA	Communication-Based Assessment
CBO	Community-Based Organization
CDR	Community-Driven Reconstruction
CENOE	National Emergency Operation Center (Mozambique)
CEPREDENAC	Center for Coordination for the Prevention of Natural Disasters in Central America
CL	Cluster Lead
COSO	Committee of Sponsoring Organizations of the Treadway Commission
CPAR	Country Procurement Assessment Report
CPR	Conflict Prevention and Reconstruction Unit
CPZ	Coastal Protection Zone
CRC	Citizen Report Card
CRED	Centre for Research on the Epidemiology of Disasters
CSC	Community Score Card
CSO	Civil Society Organization
CWGER	Cluster Working Group on Early Recovery
DAC	OECD Development Assistance Committee
DAD	Development Assistance Database
DaLA	Damage and Loss Assessment
DCD	OECD Development Cooperation Directorate
DFID	Department for International Development (United Kingdom)
DRM	Disaster Risk Management
DRR	Disaster Risk Reduction
DTM	Digital Terrain Model
EA	Environmental Assessment
EC	European Commission
ECHA	Executive Committee for Humanitarian Affairs
ECLAC	United Nations Economic Commission for Latin America and the Caribbean
EERI	Earthquake Engineering Research Institute
EIA	Environmental Impact Assessment
EM-DAT	Emergency Events Database
EMI	Earthquakes and Megacities Initiative
EMMA	Emergency Market Mapping and Analysis
EMP	Environmental Management Plan
EO	Earth Observation
ERC	Emergency Relief Coordinator
ERL	Emergency Recovery Loan

ERRA	Earthquake Reconstruction and Rehabilitation Authority (Pakistan)
ESC	Emergency Shelter Cluster
ESSAF	Environmental and Social Screening and Assessment Framework
FAO	Food and Agricultural Organization
FEMA	Federal Emergency Management Agency (United States)
FHH	Female-Headed Household
FOREC	Reconstruction Fund for the Coffee Region (Colombia)
FORSUR	Fund for the Reconstruction of the South (Peru)
GAAP	Governance and Anticorruption Plan
GAC	Governance and Anticorruption
GAO	General Accountability Office (United States)
GCM	Global Circulation Model
GFDRR	Global Facility for Disaster Reduction and Recovery
GHP	Global Humanitarian Platform
GIS	Geographic Information System
GLTN	Global Land Tool Network
GPS	Global Positioning System
GSDMA	Gujarat State Disaster Management Authority (India)
GUDC	Gujarat Urban Development Company (India)
HC	Humanitarian Coordinator
HFA	Hyogo Framework for Action 2005–2015
HFHI	Habitat for Humanity International
HIC	Humanitarian Information Center
IAASB	International Auditing and Assurance Standards Board
IAEE	International Association for Earthquake Engineering
IASC	Inter-Agency Standing Committee
ICCROM	International Centre for the Study of the Preservation and Restoration of Cultural Property
ICOMOS	International Council of Monuments and Sites
ICRC	International Committee of the Red Cross
ICT	Information and Communications Technology
ICVA	International Council of Voluntary Agencies
IDA	International Development Association
IDB	Inter-American Development Bank
IDP	Internally Displaced Person
IEG	World Bank Independent Evaluation Group
IFAC	International Federation of Accountants
IFAD	International Fund for Agricultural Development
IFC	International Finance Corporation
IFI	International Financial Institution
IFRC	International Federation of Red Cross and Red Crescent Societies
IHSN	International Household Survey Network
INEE	Inter-Agency Network for Education in Emergencies
InterAction	American Council for Voluntary International Action
IPSASB	International Public Sector Accounting Standards Board
IRP	International Recovery Platform
ISDR	International Strategy for Disaster Reduction
ISDS	Integrated Safeguard Data Sheet
ISO	International Organization for Standardization
ITU	International Telecommunications Union
LENSS	Local Estimate of Needs for Shelter and Settlement
LICUS	Low-Income Country under Stress
M&E	Monitoring and Evaluation
MANGO	Management Accounting for NGOs
MFI	Microfinance Institution
MIS	Management Information System
MOU	Memorandum of Understanding
NCPDP	National Centre for People's Action in Disaster Preparedness (Kashmir)

NCRC	NGO Coordination and Resource Center (India)
NFI	Non-Food Item
NGO	Nongovernmental Organization
NZAID	New Zealand's International Aid & Development Agency
O&M	Operation and Maintenance
ODR	Owner-Driven Reconstruction
OECD	Organisation for Economic Co-operation and Development
OFDA	Office of U.S. Foreign Disaster Assistance
OHCHR	United Nations Office of the High Commissioner for Human Rights
OP	Operational Policy
PACS	Project Anti-Corruption System
PDNA	Post-Disaster Needs Assessment
PEFA	Public Expenditure and Financial Accountability
PFM	Public Financial Management
PIU	Project Implementation Unit
PPA	Project Preparation Advance
PPAF	Pakistan Poverty Alleviation Fund
RCM	Regional Circulation Model
REA	Rapid Environment Impact Assessment
RICS	Royal Institution of Chartered Surveyors
RRC	Rapid Response Committee
SAG	Shelter Advisory Group
SDR	Safeguard Diagnostic Review
SIFFS	South Indian Federation of Fishermen Societies (India)
SPP	Simplified Procurement Plan
SRTM	Shuttle Radar Topography Mission
SSG	Shelter Support Group (India)
TEC	Tsunami Evaluation Coalition
TNSCB	Tamil Nadu Slum Clearance Board
TOR	Terms of Reference
UN OCHA	United Nations Office for the Coordination of Humanitarian Affairs
UN	United Nations
UN-APCICT	United Nations Asian and Pacific Training Centre for Information and Communications Technology for Development
UNDAC	United Nations Disaster Assistance and Coordination System
UNDP	United Nations Development Programme
UNDPCPR	United Nations Development Programme Crisis Prevention and Recovery
UNEP	United Nations Environment Programme
UNESCO	United Nations Educational, Scientific and Cultural Organization
UN-HABITAT	United Nations Human Settlements Programme
UNHCR	United Nations High Commissioner for Refugees
UNICEF	United Nations Children's Fund
UNIDO	United Nations Industrial Development Organization
UNISDR	United Nations International Strategy for Disaster Reduction
UNNATI	Organization for Development Education (India)
UNOOSA	United Nations Office for Outer Space Affairs
UPLINK	Urban Poor Linkage Indonesia
USACE	United States Army Corps of Engineers
USAID	United States Agency for International Development
USGS	United States Geological Survey
WANGO	World Association of Non-Governmental Organizations
WEDC	Water, Engineering and Development Centre
WEF	World Economic Forum
WFP	World Food Programme
WHE	World Housing Encyclopedia
WHO	World Health Organization
WHRC	World Habitat Research Centre

This handbook is dedicated to all those women who
have had to put their households back together after
a disaster, in the hope that in the future fewer women
have to do the same, and that those who do feel more
empowered during the entire process.

PERSPIRE, HONDURAS, 1998
HURRICANE MITCH
PHOTO BY CHRISTOPHER JENNINGS

1   A good reconstruction policy helps reactivate communities and empowers people to rebuild their housing, their lives, and their livelihoods.

A reconstruction policy should be inclusive, equity-based, and focused on the vulnerable. Housing reconstruction is key to disaster recovery, but it depends on the recovery of markets, livelihoods, institutions, and the environment. Diverse groups need diverse solutions, but biases will creep in, so a system to redress grievances is a must.

2   Reconstruction begins the day of the disaster.

If traditional construction methods need to change to improve building safety, governments must be prepared to act quickly to establish norms and provide training. Otherwise, reconstructed housing will be no less vulnerable to future disasters than what was there before. Adequate transitional shelter solutions can reduce time pressure and should be considered in a reconstruction policy. Owners are almost always the best managers of their own housing reconstruction; they know how they live and what they need. But not all those affected are owners and not all are capable of managing reconstruction; so the reconstruction policy must be designed with all groups in mind: owners, tenants, and landlords, and those with both formal and informal tenancy.

3   Community members should be partners in policy making and leaders of local implementation.

People affected by a disaster are not victims; they are the first responders during an emergency and the most critical partners in reconstruction. Organizing communities is hard work, but empowering communities to carry out reconstruction allows their members to realize their aspirations and contribute their knowledge and skills. It also assists with psychosocial recovery, helps reestablish community cohesion, and increases the likelihood of satisfaction with the results. This requires maintaining two-way communication throughout the reconstruction process and may entail the facilitation of community efforts. A real commitment by policy makers and project managers is needed to sustain effective involvement of affected communities in reconstruction policy making and in all aspects of recovery, from assessment to monitoring.

4   Reconstruction policy and plans should be financially realistic but ambitious with respect to disaster risk reduction.

People's expectations may be unrealistic and funding will be limited. Policy makers should plan conservatively to ensure that funds are sufficient to complete reconstruction and that time frames are reasonable. Rebuilding that reduces the vulnerability of housing and communities must be the goal, but this requires both political will and technical support. Housing and community reconstruction should be integrated and closely coordinated with other reconstruction activities, especially the rehabilitation and reconstruction of infrastructure and the restoration of livelihoods.

5   Institutions matter and coordination among them improves outcomes.

Best practice is to have defined a reconstruction policy and designed an institutional response in advance of a disaster. In some cases, this will entail a new agency. Even so, line ministries should be involved in the reconstruction effort and existing sector policies should apply, whenever possible. The lead agency should coordinate housing policy decisions and ensure that those decisions are communicated to the public. It should also establish mechanisms for coordinating the actions and funding of local, national, and international organizations and for ensuring that information is shared and that projects conform to standards. Funding of all agencies must be allocated equitably and stay within agreed-upon limits. Using a range of anticorruption mechanisms and careful tracking of all funding sources minimizes fraud.

6   Reconstruction is an opportunity to plan for the future and to conserve the past.

What has been built over centuries cannot be replaced in a few months. Planning and stakeholder input help to establish local economic and social development goals and to identify cultural assets for conservation. Even a modest amount of time spent designing or updating physical plans can improve the overall result of reconstruction. Reconstruction guidelines help ensure that what is valued is preserved, while encouraging more sustainable post-disaster settlements. Improving land administration systems and updating development regulations reduces vulnerability and improves tenure security.

7   Relocation disrupts lives and should be kept to a minimum.

Relocation of affected communities should be avoided unless it is the only feasible approach to disaster risk management. If relocation is unavoidable, it should be kept to a minimum, affected communities should be involved in site selection, and sufficient budget support should be provided over a sufficient period of time to mitigate all social and economic impacts.

8   Civil society and the private sector are important parts of the solution.

The contributions of nongovernmental organizations (NGOs), civil society organizations (CSOs), and the private sector to reconstruction are critical. Besides managing core programs, these entities provide technical assistance, advocacy, and financial resources of enormous value. Government should encourage these initiatives; invite NGO, CSO, and private entity involvement in reconstruction planning; and partner in their efforts. Government should also require accountability and make sure that these interventions are consistent with reconstruction policy and goals.

9   Assessment and monitoring can improve reconstruction outcomes.

Assessment and monitoring improve current (and future) reconstruction efforts. Unnecessary assessments can be minimized if there are policies that require institutions to share assessment data and results. Local communities should participate in conducting assessments, setting objectives, and monitoring projects. Using reliable national data to establish monitoring baselines after the disaster increases the relevance of evaluations. Monitor both the use of funds and immediate physical results on the ground and evaluate the impact of reconstruction over time.

10   To contribute to long-term development, reconstruction must be sustainable.

Sustainability has many facets. Environmental sustainability requires addressing the impact of the disaster and the reconstruction process itself on the local environment. The desire for speed should not override environmental law or short-circuit coordination when addressing environmental issues. Economic sustainability requires that reconstruction is equitable and that livelihoods are restored. Livelihood opportunities in reconstruction should be maximized. Institutional sustainability means ensuring that local institutions emerge from reconstruction with the capability to maintain the reconstructed infrastructure and to pursue long-term disaster risk reduction. A reliable flow of resources is essential and institutional strengthening may be required.

The last word: Every reconstruction project is unique.

The nature and magnitude of the disaster, the country and institutional context, the level of urbanization, and the culture's values all influence decisions about how to manage reconstruction. Whether government uses special or normal procurement procedures, how it weighs the concerns of speed versus quality, and what it considers the proper institutional set-up and division of labor will also vary. History and best practices are simply evidence to be weighed in arriving at the best local approach.

PART 1

RECONSTRUCTION TASKS AND HOW TO UNDERTAKE THEM

RECONSTRUCTION TASKS AND HOW TO UNDERTAKE THEM

SECTION 1

ASSESSING DAMAGE AND DEFINING RECONSTRUCTION POLICY

Guiding Principles

1. A good reconstruction policy helps reactivate communities and empowers people to rebuild their housing, their lives, and their livelihoods.

2. Reconstruction begins the day of the disaster.

3. Community members should be partners in policy making and leaders of local implementation.

4. Reconstruction policy and plans should be financially realistic but ambitious with respect to disaster risk reduction.

5. Institutions matter and coordination among them improves outcomes.

6. Reconstruction is an opportunity to plan for the future and to conserve the past.

7. Relocation disrupts lives and should be kept to a minimum.

8. Civil society and the private sector are important parts of the solution.

9. Assessment and monitoring can improve reconstruction outcomes.

10. To contribute to long-term development, reconstruction must be sustainable.

The last word: Every reconstruction project is unique.

1 EARLY RECOVERY: THE CONTEXT FOR HOUSING AND COMMUNITY RECONSTRUCTION

Introduction

Reconstruction of housing and communities following a disaster is a continuous process that begins immediately after the disaster, and often lasts for years.[1] It is important to understand how affected populations and institutions will react after a disaster, and what roles and responsibilities stakeholders will take on throughout the post-disaster reconstruction process, so that institutions and affected populations can work in a coordinated and complementary way to accomplish their desired outcomes.

At the beginning of the response to a disaster, humanitarian agencies, including the United Nations (UN), are ordinarily the organizations that are most in contact with government, conducting initial assessments, mobilizing aid, and discussing options for how the recovery will be organized. The World Bank and other international financial institutions (IFIs), including regional development banks, may not be directly involved and may not commit resources this early in the process. However, it is essential that these organizations enter the process as soon as possible, especially so that they can be present during the early strategic planning with government that is normally led by the UN and the other humanitarian agencies, since—as discussed in this chapter—that planning will influence the entire reconstruction process.

The post-disaster reconstruction process almost always takes much longer than expected or planned. Except in life-threatening situations, compromises that ignore the need for integration, or for quality, safety, or good governance of the reconstruction, should not be made with the belief that they will save time. Time is rarely saved, and people will live for years with the consequences of those decisions.

This chapter introduces the context and process of reconstruction following natural disasters, referring to the guiding principles established at the beginning of this handbook, as well as to other handbook chapters. It offers guidance on assisting the entire affected population, both those who are displaced and those who are not displaced.

The chapter discusses such issues as the need to integrate housing and community reconstruction, the sequence of activities that reconstruction entails both for individuals and agencies, the roles and responsibilities of stakeholders and mechanisms of coordination, and the risk of losing continuity between the immediate response and long-term development and reconstruction.

Achieving People-Centered, Integrated Reconstruction

Post-disaster reconstruction is a complex process involving a number of interrelated activities. The level of complexity will vary, depending on the scale and nature of the disaster and the corresponding response of the population and the institutions involved. Like most humanitarian and development activities, the process tends to entail a cycle: assessment, planning, project development, implementation, and monitoring. Different project cycles are likely to be occurring simultaneously at different levels and for different purposes wherever people are organizing some element of the response. It can't be emphasized strongly enough that the affected population should be at the center of the reconstruction process and should have a preferential right to make the decisions that will affect their lives. In one increasingly accepted vision of how post-disaster reconstruction should work (which the authors generally subscribe to), government's first job after a disaster, with the help of humanitarian and development agencies, is to determine what the community wants to do and is capable of doing. The government should then do the rest.

This Chapter Is Especially Useful For:

- Policy makers
- Project managers
- Lead disaster agency
- Humanitarian agencies

1. What this handbook calls "housing and communities" is referred to by other names as well, including "shelter and settlements" and "housing and habitat." The meanings of these phrases are essentially identical, and they all seek to acknowledge that reconstruction entails not simply rebuilding a physical structure—the house—but restoring the entire social, economic, natural, and cultural environment in which the house and household was or will be located.

This may be an oversimplification of a very complex process; however, adopting this approach means that there are two overriding project cycles that are set in motion after a disaster: the one for the community's work and the one for government's work. Government has to conduct the macro-level assessments, set policy, coordinate nongovernmental organizations (NGOs) and humanitarian agencies who will support recovery, engage IFIs and other funders, organize the financing mechanisms, ensure all affected communities have adequate support, and so on. The community has its own work to do: assess its local needs, identify vulnerable members, salvage materials, develop a community-level plan, agree on housing designs and immediate infrastructure improvements, reconstruct its governance system, and plan how to manage reconstruction funds once they are available.

An integrated approach to reconstruction is one that harmonizes these efforts, simultaneously addressing both *what needs to be done* (with respect to land use, reconstruction approach, environmental management, infrastructure rehabilitation, choice of housing design and technology, and cultural and natural heritage conservation, for example) and *how it will be done* (including institutional roles, levels of citizen participation, and management of project financing). Each chapter in Part 2 of this handbook, Reconstruction Tasks and How to Undertake Them, covers one of these elements.

2. United Nations Office for the Coordination of Humanitarian Affairs (UN OCHA), Shelter Centre, 2010, *Shelter Aafter Disaster: Strategies for Transitional Settlement and Reconstruction* (Geneva: UN OCHA), http://www.sheltercentre.org/library/Shelter+After+Disaster.

The Steps in Response and Reconstruction

The experience from recent disasters shows that common steps are generally followed by government to organize a large disaster response, as shown below.[2] Steps where the affected population is likely to be involved are marked with *.

Activity in response timeline	Description of activity	Time frame
1. Coordination*	Development and maintenance of a coordination mechanism	From the disaster event through the end of reconstruction
2. Engagement*	Collaboration with stakeholders	From the disaster event through the end of reconstruction
3. Initial assessment*	Gathering of initial information and evaluation of local capacities	Week 1 following the disaster
4. Outline strategy*	Developing a framework for cooperation (see description below)	Week 1 following the disaster
5. Rapid appeal	First call for funding	Week 1 following the disaster
6. Emergency relief distribution	Coordinating emergency distribution based on the initial assessment activity	Throughout month 1
7. Program- and project-level work plan*	Specific shelter programs and projects	Periodic, starting in week 2
8. Program- and project-level implementation*	Implementation of the work plans based on work plan	Beginning week 2 through the end of reconstruction
9. Joint rapid needs assessment (such as Post-Disaster Needs Assessment [PDNA])*	Formally coordinated assessment based on initial assessment (see 📖 Chapter 2, Assessing Damage and Setting Reconstruction Policy, for a discussion of various assessment methodologies)	First 4-6 weeks
10. Full policy or strategy*	Detailed strategy built on outline strategy (see 📖 Chapter 2, Assessing Damage and Setting Reconstruction Policy, for a discussion of the parameters of a reconstruction policy)	First 4-6 weeks
11. Revised appeal	Further detailed calls for funding based on rapid needs assessment	First 4-6 weeks
12. Detailed assessments (generally sector-specific)*	Formally coordinated assessments building on rapid needs assessment (see 📖 Chapter 2, Assessing Damage and Setting Reconstruction Policy, for a discussion of various assessment methodologies)	Periodic, throughout reconstruction
13. Revised policy or strategy*	Revision of strategy based on detailed assessments	Periodic, throughout reconstruction
14. Public financing and additional appeals	Arrangement of multilateral and bilateral loans and grants, and ongoing humanitarian appeals	Periodic, throughout reconstruction
15. Achievement of agreed goals*	Completion of benchmarks set with government and communities in the strategies	End of reconstruction

By understanding and recognizing these common steps, different stakeholders can ensure better cooperation and coordination, which in turn will support a more consistent and efficient response that better meets the needs of the affected population.

Who Does What? Stakeholder Roles and Responsibilities

Following a major disaster, government frequently seeks external support, initially from the humanitarian community and later from IFIs, such as the World Bank. These same institutions are also involved in an increasing number of smaller-scale disasters, albeit in different ways. It is essential that these stakeholders work together in post-disaster reconstruction and that each understand the capacities, roles, responsibilities, and contributions of the others. Some specific efforts to improve coordination are mentioned in 📖 Chapter 14, International, National, and Local Partnerships in Reconstruction.

The Affected Population

People affected by a disaster are not victims; they are the first responders during an emergency and the most critical partners in reconstruction, undertaking the majority of work on their own recovery, without governmental, humanitarian, or IFI support. A good reconstruction strategy is one that focuses on empowering communities, families, and individuals to rebuild their housing, their lives, and their livelihoods. To make this work, community members should be partners in policy making and leaders of local implementation. They may need support to play these roles.

Real representation of the affected communities in the policy-making body and in all aspects of recovery is a must. At the same time, it is crucial that agencies do not succumb to the misconception that the affected population is a single entity, ignoring differences in needs and capacities. Communities are composed of numerous social and economic groups, each with its own characteristics, vulnerabilities, and ability to influence outcomes.

Key points about populations affected by disasters (with reference to the handbook's Guiding Principles [GPs]) include the following.
- People affected by a disaster need to secure shelter and rebuild their livelihoods. Infrastructure such as roads, schools, and power generation is as fundamental to recovery and livelihoods as housing is. Also important is the rebuilding of the sense of community and of social capital. Responses should reflect an understanding that reconstruction is not only about shelters and homes but also about reconstructing entire communities. (GP 1)
- People affected must have shelter during the time in which reconstruction takes place. While a tent, for example, only lasts a year, other transitional shelter options can be employed that last until permanent housing is available. (GP 2)
- For people who have not been displaced, reconstruction begins almost immediately, usually with the recovery of materials to recycle in building their shelter. (GP 2)
- People's expectations regarding the time frame for reconstruction are often overly optimistic; reconstruction and recovery will probably take a number of years. (GP 4)
- Some people will be displaced by the disaster and others won't be, and the ways to help these two groups may differ. At the same time, people may not wish to return to their pre-disaster circumstances, depending on changes in their lives and in their livelihoods.
- Some social groups are more vulnerable than others. The most vulnerable, poorest, and hardest to reach members of society are usually those most affected and most in need. Gender and age are also determining factors when assessing vulnerability.

Government

Central government is always responsible and accountable for managing a disaster response and for establishing policy to guide the reconstruction program. This does not mean government will do everything, but it does mean that defining a strategy that establishes "who will do what and how" is a governmental responsibility. Government, however, is not a monolith; it consists of different branches; public entities with different levels of autonomy; and usually different levels, e.g., central, state, provincial, local. Even if government's management capacity is adequate under normal circumstances, it can be overwhelmed immediately following a disaster, especially at the local level. These realities must be taken into account in developing the response and in defining the reconstruction policy. See 📖 Chapter 2, Assessing Damage and Setting Reconstruction Policy, for a discussion of the parameters of a reconstruction policy.

Dilemmas of Reconstruction

The most complex tasks for recovery managers are to determine and to implement the appropriate approach to reconstruction of buildings and infrastructure. Considerations include the wider political context, the operational requirements, and the expectations and preferences of the people most affected. Reconstruction poses many demands and dilemmas for officials. These include whether to emphasize short-term basic reconstruction needs or longer-term needs to reduce risk, whether to engage the affected population in rebuilding their own houses with technical guidance or engage professional building contractors to do the work, and whether reconstruction should be carried out in the original, disaster-prone location, or relocated to a new and possibly less vulnerable location.

Another important dilemma concerns the stages of shelter to employ before reaching permanent reconstruction. Experience demonstrates that it is generally better to avoid the process of building substantial temporary dwellings. Dialogue with the public may help identify more viable, and locally suited, immediate post-disaster shelter options. Without some intermediate step, extraordinary measures may be needed to accelerate the construction of permanent residential buildings.

None of these questions have easy answers, and much depends on the views of government officials responsible for the recovery process, relative to those of local people who will finally determine by their acceptance or rejection the success of any official decisions that are made.

Source: Ian Davis, 2007, adapted from *Learning from Disaster Recovery: Guidance for Decision Makers* (Geneva and Kobe: International Recovery Platform).

In certain situations, especially after a large-scale emergency, government may establish a dedicated organization or taskforce to coordinate, reinforce, or in some cases temporarily replace the responsibilities of line ministries. The taskforce can sometimes better coordinate tasks among ministries and departments. The taskforce is usually created for a specific period of time and will return responsibilities to the relevant line ministries, either gradually or when specific objectives are met. For a detailed discussion of these options, see 🔖 Chapter 13, Institutional Options for Reconstruction Management.

The National Military

The national military can sometimes be an effective partner in housing reconstruction. It may be able to quickly carry out initial rebuilding of bridges and essential infrastructure, and generally has better and faster logistics capability than any governmental entity, including rapid assessment capabilities and excellent communications. The military may maintain large stockpiles of goods and may be able to deliver materials even when roads are impassible by others. The military may also have high levels of local support, and can add a sense of security and order to early recovery.

There can also be challenges with military involvement. In some countries, the cost of the military's support is high and it may get charged against assistance budgets. The military is not always used to operating in the complex, multi-stakeholder environment of a disaster recovery situation, and may have little experience in listening to community concerns or accepting civil authority. Having the military run camps is usually not an appropriate long-term strategy, although the military's assistance in setting up these camps and their infrastructure can be crucial. Also, the military is generally not experienced in coordinating housing recovery and reconstruction, although there are notable exceptions, such as after the 2004 North Pakistan earthquake, where the military coordinated certain aspects of the inspection system for housing reconstruction. Lastly, where there is a prevalence of NGOs involved in reconstruction, conflicts may arise between the NGO culture (especially that of NGOs with pacifist origins) and the culture of the military.

The Humanitarian Community

Coordination of the response is the responsibility of government; however, support is often offered by the humanitarian community. Government usually establishes coordination mechanisms, and the humanitarian community, led by the Resident Representative or Humanitarian Coordinator assigned by the UN, often implements those mechanisms. A sector coordination team may involve information managers and technical specialists. There are two kinds of mechanisms used to establish coordination:

- Pre-agreed frameworks, such as those set up in contingency plans by government or the UN through the Inter-Agency Standing Committee (IASC) clusters system; and
- Ad hoc frameworks, such as those set up by government, the UN, other agencies, or communities at the national or local level when contingency plans do not exist and the cluster system is considered inappropriate.

Coordination within the humanitarian community has recently been reformed through the IASC and the creation of its 11 "clusters," such as for "Emergency Shelter" and "Early Recovery," which together constitute a framework of responsibilities at both global and response levels. UN agencies, international organizations, and the International Federation of the Red Cross and Red Crescent (IFRC) take the lead role in each cluster, with a series of partner agencies, representing other UN agencies, international organizations, and NGOs, supporting each cluster. Although this framework is intended to prevent overlaps and gaps in responsibility, operational coordination at the response level for reconstruction has not yet been clarified. A more detailed description of the cluster system, including a list of the cluster leads, is found in 🔖 Chapter 14, International, National, and Local Partnerships in Reconstruction.

United Nations agencies. In most countries, the UN maintains its own presence and that of the United Nations Development Programme (UNDP), under a Resident Representative. In large emergencies, the UN may expand its capacity by including a Humanitarian Coordinator, while agencies like the United Nations Children's Fund (UNICEF) and the United Nations Office for the Coordination of Humanitarian Affairs (UN OCHA) may also establish offices or increase capacity. Immediately following a disaster, a specifically mandated UN team often arrives to support coordination and assessment. The UN and the humanitarian community will agree together on a coordination structure and commitments as providers of last resort, in support of government.

Under the cluster system, UNDP works with government to coordinate "Early Recovery" activities following a disaster, including conducting a PDNA and developing an Early Recovery Framework. Assessments are discussed in 🖳 Chapter 2, Assessing Damage and Setting Reconstruction Policy.

EARL KESSLER

The Red Cross Movement and international organizations. Under the cluster system, the IFRC coordinates the activities of the Emergency Shelter Cluster following a disaster, although these responsibilities may be handed over to UN partners, such as the United Nations Human Settlements Programme (UN-HABITAT) and UNDP, for reconstruction and early recovery. For a detailed description of how the Emergency Shelter Cluster is mobilized following a disaster, see 🖳 Chapter 14, International, National, and Local Partnerships in Reconstruction.

National and international NGOs. A number of national and international NGOs increase capacity after a disaster, offering—along with the IFRC and other international organizations—support for implementation of response and reconstruction programs. Additional NGOs without an established presence in the country often arrive in the days immediately after a disaster and may, or may not, maintain a presence until reconstruction is completed. (See Mind the Gap, below.) NGOs often play a major role in facilitating the activities of communities or in serving as executing agencies for all funding sources.

Bilateral and Multilateral Donor Organizations

Technical and nontechnical representatives of bilateral donor organizations, such as the United States Agency for International Development (USAID) (directly or through the Office of U.S. Foreign Disaster Assistance [OFDA]) and the UK Department for International Development (DFID), and of multilateral donor organizations, such as the Organisation for Economic Co-operation and Development (OECD)-Development Cooperation Directorate (DCD), UNDP, and the World Food Programme (WFP), often arrive almost immediately following a disaster, participating in coordination structures from the outset of the response. These agencies can be important partners with IFIs in reconstruction. In some policy areas, bilateral agencies, working alone and in partnership, have also successfully taken the lead in developing, analyzing, and promoting post-disaster best practices.[3]

IFIs: The World Bank and Regional Development Banks

IFIs traditionally became involved in reconstruction after a number of months. However, this time frame is changing, and many IFIs now become involved in the early stages of a response. The resources and mechanisms offered by IFIs are also evolving, in order to support a diversity of responses. For a description of World Bank mechanisms, see 🖳 Chapter 20, World Bank Response to Crises and Emergencies.

Defining the Outline Strategy

Agreeing on a common strategy with government is key to ensuring that early decisions make a positive contribution to the longer-term reconstruction process, recognizing that reconstruction usually begins right away. In the absence of a common strategy, agencies that enter later in the process may be unaware or even discount the value of agreements made before they arrived; this is especially easy for larger agencies to do.

An outline strategy must be agreed to within the first weeks of the disaster for stakeholders to collaborate effectively and manage the needs of the affected population. It is generally developed by

3. Many examples of recent analytical work by bilateral and multilateral organizations are listed in the Resources sections throughout this handbook.

government and the lead disaster agency, in collaboration with affected communities and with the support of humanitarian agencies. This strategy is then reviewed and updated regularly as new and more detailed assessment information becomes available, until a full policy or strategy is defined, as described in #10 in the "The Steps in Response and Reconstruction" table, above. See Chapter 2, Assessing Damage and Setting Reconstruction Policy, for a description of the assessment process and the content of a housing and community reconstruction policy.

Reconstruction strategies are agreed to on a response-by-response basis. Within the humanitarian community, certain topics are commonly included in strategy documents. The same themes are common in situation reports ("sitreps") and funding proposals. Terminology will vary and additional topics are often included. The common elements of a reconstruction outline strategy include (1) introduction and context analysis, (2) goals (or strategic objectives), (3) needs assessment, (4) priorities for the sector, (5) activities, (6) projected outputs and outcomes, (7) projected impact (also called indicators), (8) intersector linkages, (9) timeline, and (10) resources.

While humanitarian agencies are likely to take the lead in helping government articulate the outline strategy, IFI participation (or, at a minimum, IFI review and revision of the strategy) is essential if IFI financing is expected to be utilized in reconstruction. It is only through early involvement that IFI knowledge and policy perspectives can be incorporated in early decisions.

Urban versus Rural Disasters

Disaster response and recovery in urban areas will be of larger scale, more concentrated, and more complex than in rural areas. Almost every aspect of reconstruction must be tailored to urban reality. Rural disaster programs pose their own unique problems. A disaster that has affected both urban and rural areas can be especially challenging to plan and execute.

Factors that influence the reconstruction approach in urban areas include:
- Higher population density and the resettlement options available to displaced persons
- More informal housing, much of it located in high-risk areas
- More multi-family housing and a larger proportion of renters
- Ownership and titling issues may require legal procedures to resolve
- More and generally more capable public sector organizations, including those responsible for disaster management, but often not used to working together
- Potential for disaster risk reduction (DRR) measures to be based on planning and regulation
- Higher income levels and living standards of the affected population, potentially requiring more generous assistance strategies
- Higher land values and less undeveloped land
- Unique and more challenging environmental risks
- Higher value and more infrastructure investments
- More complex social structures that are likely to give rise to conflicts and to complicate participation in reconstruction planning
- More clearly defined economic and social interests and more sophisticated political organizations
- Economic effects from the urban disaster that affect the rest of the country

Factors that influence the reconstruction approach in rural areas include:
- Lower land values
- Ownership and titling issues that can sometimes be resolved through negotiation
- The major role that the social structure plays in the dynamics of reconstruction
- The relative ease with which community participation can be achieved
- A higher sense of ownership
- The lack of institutional capacity for planning and regulation
- Housing that is usually designed and built by owners themselves or by masons, so DRR measures should be based on building awareness and on training construction workers

The differences between planning processes in urban and rural areas are discussed in more detail in Chapter 7, Land Use and Physical Planning. The case study on the 2003 Bam earthquake, below, describes the differential approach to reconstruction taken in urban and rural areas of Iran.

The Options Facing Displaced and Non-Displaced Populations

The process that people and households go through after a disaster to stabilize their housing situation can be quite lengthy, convoluted, and complex. People affected by the same disaster will be affected differently and will respond differently. Some will begin reconstruction of their partially damaged housing in the first days after the disaster, while others will be displaced for a period of time, even finding their situation changing from week to week for many months or even years. It is not uncommon that households affected by a disaster never again attain the level of prosperity and security they had before the event.

FEMA NEWS PHOTO

It is important to understand the range of options people face and not to impose artificial "phases" on diverse situations. These phases are sometimes more indicative of the bureaucratic practices and capacities of the agencies involved in response and reconstruction than they are of the priorities of the affected population. For example, in some past responses, support for reconstruction began only months after a disaster, after the affected populations themselves began rebuilding, because some agencies believed that reconstruction did not contribute to humanitarian objectives, or even distracted from them.

The following twelve options—six for displaced populations and six for non-displaced populations—are often used by the humanitarian community.

The Six Options for Displaced Populations

People displaced from their original location have different sheltering options that are important to consider in planning and implementing reconstruction programs. The six options for displaced populations are listed and described below.[4]

Settlement option	Description
Host families	The displaced are sheltered within the households of local families, or on land or in properties owned by them.
Urban self-settlement	The displaced settle in an urban area, occupying available public or private property or land.
Rural self-settlement	The displaced settle in a rural area, occupying available public or private property or land.
Collective centers	The displaced shelter in collective centers, or mass shelters, often transitory facilities housed in pre-existing structures.
Self-settled camps	The displaced settle independently in camps, often without services and infrastructure.
Planned camps	The displaced settle in purposely-built sites, where services and infrastructure are offered by government or the humanitarian community.

4. The six options for displaced populations were first described in Tom Corsellis and Antonella Vitale, 2005, *Transitional Settlement: Displaced Populations* (Cambridge: Oxfam), http://www.sheltercentre.org/library/transitional+settlement+displaced+populations. Displaced and non-displaced options are both described in United Nations Office for the Coordination of Humanitarian Affairs (UN OCHA) and Shelter Centre, 2010, *Shelter After Disaster: Strategies for Transitional Settlement and Reconstruction* (Geneva: UN OCHA), http://www.sheltercentre.org/library/Shelter+After+Disaster.

Following disasters, it is imperative to minimize the distance and duration of displacement, while keeping safety in mind. This allows people to better maintain their livelihoods and allows households to protect their land, property, and possessions. The displacement typology should not be perceived as describing a phase of resettlement, but instead as describing subcategories of the affected population. Displacement can continue long after post-disaster risks have receded, due to (1) the inability of households to document their property rights, which may be a prerequisite to reconstruction; (2) inappropriate reconstruction strategies, such as one that ignores the variety of needs within the affected population; or (3) the lack of resources and capacities of government and agencies to assist the displaced population.

 For access to additional resources and information on this topic, please visit the handbook Web site at www.housingreconstruction.org.

The Six Options for Non-Displaced Populations

Households that were not displaced or that have returned also will be found in diverse situations, especially in urban areas, where the proportion of tenants to owner-occupiers often exceeds 50 percent. Although the situation and context vary greatly from disaster to disaster, six options to describe the status of non-displaced populations are generally accepted within the humanitarian community.[5]

Settlement option	Description
House owner-occupant	The occupant owns his/her house and land, or is part-owner, such as when repaying a mortgage or a loan. Ownership may be formal or informal.
House tenant	The occupant rents the house and land, formally or informally.
Apartment owner-occupant	The occupant owns his/her apartment. Ownership may be formal or informal.
Apartment tenant	The occupant rents the apartment, formally or informally.
Land tenant	The occupant owns the house, and rents the land, formally or informally
Occupancy with no legal status (squatter)	The occupant occupies land or property without the explicit permission of the owner. Also called informal settlers.

Before the disaster, an affected household belonged to one of the categories listed above. After the disaster, it may move through one or more of the options for displaced or non-displaced population, and will eventually cycle back to one of the categories listed above. For example, the owners of an urban house that is badly damaged (house owner-occupant) may temporarily camp in another part of town (urban self-settlement), but eventually decide to resettle by buying an urban apartment (apartment owner-occupant). The number of paths through these options is almost infinite.

The ultimate goal of housing reconstruction is to ensure that all those affected by a disaster, whether they have been displaced or not, are eventually situated in a "durable solution." If transitional resettlement has relocated a significant percent of the population away from the affected area, an effort may be required to ascertain how many plan to return, so that the housing need is not overestimated. If reconstruction begins spontaneously and is being carried out in a way that creates unacceptable risks, the reconstruction strategy may need to compensate families for work already done in order to enlist their collaboration in improving the safety of construction.

The implementation of the housing and community reconstruction policy must incorporate and equitably support the needs of those in all categories. See Chapter 4, Who Gets a House? The Social Dimension of Housing Reconstruction, and Chapter 6, Reconstruction Approaches, to see how these categories can be used when defining the assistance strategy.

Resolving property rights issues often needs to be a high priority at the beginning of reconstruction programs, requiring considerable capacity from national and international stakeholders. See Chapter 7, Land Use and Physical Planning, for a list of strategies to address property rights issues and case studies of good practice.

The Transitional Shelter Approach

The transitional shelter approach can be used with both displaced and non-displaced populations.[6] Transitional shelter is not a phase of reconstruction, but is a philosophy that recognizes that reconstruction usually takes years to complete and that shelter is required throughout this period. The transitional shelter approach responds to the fact that post-disaster reconstruction can take a significant amount of time and that it is the affected population that does most of it.

Transitional shelter is used to house affected households with habitable, covered living space and a secure, healthy living environment with privacy and dignity during the period between a natural disaster and the availability of a permanent shelter solution. Communities have differing capacities to cope until permanent reconstruction is completed. The decision to employ transitional shelter should be made in consultation with the affected populations, keeping in mind that the preference for transitional shelter may be community-specific.

5. Because the handling of displaced populations is not a focus of this handbook, most of the handbook uses categories similar to those for non-displaced populations.

6. This section is adapted from United Nations Office for the Coordination of Humanitarian Affairs (UN OCHA) and Shelter Centre, 2010, *Shelter After Disaster: Strategies for Transitional Settlement and Reconstruction* (Geneva: UN OCHA), http://www.sheltercentre.org/library/Shelter+After+Disaster; *and* Tom Corsellis and Antonella Vitale, 2005, Transitional Settlement Displaced Populations (Cambridge: Oxfam), p. 41.

Transitional shelter provides incremental support from the moment recovery begins, and gives households mobility and autonomy. Advocates seek also to coordinate all shelter efforts from immediately after the disaster. It is distinct from temporary shelter (which is occupied immediately after a disaster and understood to be time-limited), in that it offers shelter on-site until the completion of reconstruction for those not displaced or throughout the displacement period for those displaced. Transitional shelter shares some characteristics with what is called "semi-permanent shelter," but, because it is generally moveable, may provide more flexibility as conditions change after the disaster and during the reconstruction period. For an example of an ambitious temporary shelter program, see the 🏠 case study on the temporary housing built following the 2009 L'Aquila, Italy earthquake, below.

Strengths of Transitional Shelter Programs

There are several potential advantages with transitional shelter as an assistance method from the point of view of executing agencies, including the following.

1. Transitional shelter programs can be implemented by humanitarian organizations without experience in transitional settlement or reconstruction. While the initial cost is similar to some traditional temporary solutions, such as tents, the operating costs may be significantly lower (compared to running a camp, for example).
2. Because the designs almost always use local materials, the resources that are spent in construction circulate in the local economy and help jump-start supply chains needed for the reconstruction phase, possibly reducing the need for purchasing and warehousing of building materials by agencies.
3. Production of shelters can start the process of educating builders and the public on hazard-resistant construction principles and techniques, which will later be employed in reconstruction.
4. The designs are sufficiently flexible to accommodate differences in family size, location, culture, and availability of materials.

Transitional shelter can be advantageous for the affected population as well, for the following reasons.

1. If transitional shelters are sufficiently durable to last until the completion of reconstruction, which may take a number of months or even years, the need for multiple moves by a family between the disaster and the completion of reconstruction is reduced.
2. Transitional shelters—being mobile, flexible, and under the control of the family—permit households to quickly return to the site where they have land rights or tenure, once it is deemed safe, allowing them to protect whatever assets still remain, to begin on-site reconstruction, to reestablish their livelihoods, and to preserve social networks.
3. The best designs allow the household to upgrade or incorporate the shelter into permanent reconstruction (for instance, as an extra room or a storage area), or permit the reuse of the majority of materials in permanent reconstruction.

The 🏠 case study, below, on reconstruction following the 2005 North Pakistan earthquake demonstrates how transitional shelter materials can be reused in permanent housing.

With respect to design, transitional shelters can usually be constructed quickly, with simple tools and local, relatively low-skilled labor, including that of the family itself. While designed for local materials and construction techniques, transitional shelter designs should also reflect agreed-upon standards that provide adequate safety and protection to the users. Often, the materials for transitional shelters are assembled and distributed as kits, which is helpful for affected families who may need to transport them. In fact, transitional shelter is designed to be disassembled and relocated. This may be advantageous if there are delays in the resolution of land rights or tenure, households can't return immediately to their land (until floodwaters recede, for example), decisions are pending as to whether a household must be resettled, or on-site reconstruction progresses to the point where the transitional shelter is in the way.

Weaknesses of Transitional Shelter Programs

There are risks and challenges involved in using transitional shelter as an assistance method, including the following.

1. A transitional shelter strategy does not exempt government from addressing the need for water, sanitation, and electricity at the sites where families locate their shelters. This may require providing interim services, such as water delivery and storage and latrines, until infrastructure and services are permanently restored.

The normal pattern in major disasters is for relief and emergency assistance—not reconstruction—to receive the overwhelming proportion of human, material, and financial resources.

2. Some families' land rights may not be readily resolved, leaving families settled indefinitely occupying the land where their transitional shelter is located, but with no legal status.
3. Government may become complacent with the transitional shelter solution and offer no other support, especially if resources for reconstruction fall short during the "transitional" period.
4. Even if local materials are used, production of the shelters may overtax supply channels, driving up prices or extending production times.
5. The transitional shelters may themselves represent a risk, especially if there is insufficient experience among those implementing the program. Units may be poorly constructed or fabricated of unsafe materials, unsafely sited, or located in areas with insufficient basic services.

The ⌂ case study, below, on the 2004 Indian Ocean tsunami reconstruction in Indonesia summarizes the findings of a transitional shelter program evaluation that showed that the positive economic impact of transitional shelter declined if it was occupied for too long.

International Experience

Transitional shelter is a rapidly evolving area of humanitarian assistance. Extensive technical resources are available and best practices continue to accumulate. Some knowledgeable organizations are listed in the Resources section of this chapter. Additional resources are available from the handbook Web site, http://www.housingreconstruction.org.

Unequal Distribution of Resources within the Post-Disaster Cycle

Viewing disaster management from a distance, one might think that each phase of the disaster cycle receives equal resources; this is not the case. The normal pattern in major disasters is for emergency response and relief to receive the overwhelming share of human, material, and financial resources. This can lead to funding shortfalls for reconstruction in certain regions or sectors over time and to bad decisions as agencies (including government) attempt to accelerate reconstruction, without sufficient planning, while resources for relief are still plentiful.

In addition, when resources are insufficient for reconstruction, funds for preventive and disaster risk reduction (DRR) measures may be severely constrained, so vulnerability is just built back. Overloading response and relief with resources, while short-changing DRR, is a serious reconstruction pitfall. Government must work with its funders to resolve this dilemma. The ⌂ case study on the 1963 Skopje reconstruction, below, recounts how temporary housing was occupied for so long after the earthquake it eventually affected the physical development of the city.

Minding the "Gaps"

The gaps between the emergency lifesaving effort and post-disaster reconstruction are of particular importance in housing and community reconstruction and deserve careful management. The transitional shelter and reconstruction approach is an attempt to bridge these gaps by acknowledging that for affected families the division is artificial; therefore, interventions should be planned to better integrate response.

The Funding Gap

The first gap that may appear is in the continuity of funding between the contributions of the humanitarian community and funding from IFIs, such as the World Bank. The implementation of programs by government and the humanitarian community may face interruptions if this occurs. Some humanitarian organizations may even be forced to withdraw.

Throughout reconstruction, funding from different sources will go through peaks and troughs. Emergency contingency funds, such as the UN Central Emergency Response Fund (CERF), are made available almost immediately, while money from public appeals will flow early on, although not immediately. Government often injects substantial funding initially, although this funding is often reduced as funding from other sources, such as IFIs, including the World Bank, is mobilized.

To avoid this gap and to ensure consistent availability of funds, it is imperative that relationships with all funding sources are carefully managed and that funds are carefully programmed and tracked.[7] The programming of funds should include, to the greatest extent possible, the funding of bilateral agencies and NGOs whose initial preference may be to operate outside of the government

7. Wolfgang Fengler, Ahya Ihsan, and Kai Kaiser, 2008, *Managing Post-Disaster Reconstruction Finance: International Experience in Public Financial Management*, Policy Research Working Paper 4475 (Washington, DC: World Bank). http://go.worldbank.org/YJDLB1UVE0.

coordination system, since this funding is sometimes the most flexible and readily available. Other solutions are discussed in 📖 Chapter 15, Mobilizing Financial Resources and Other Reconstruction Assistance. Arrangements available through the World Bank are described in 📖 Chapter 20, World Bank Response to Crises and Emergencies.

The Planning Gap

Poor or uncoordinated strategic planning—the "planning gap"—may result in unnecessary costs. For instance, if more than one plan is developed, shifts in policy and funding approaches may take place. In the process, commitments made to the affected population may be forgotten. Involving IFIs as early as possible in the planning process both resolves this strategic planning gap and helps resolve the funding gap, by ensuring that one continuous plan is produced, rather than developing two or more plans for different "phases" that may not integrate adequately.

The Implementation Gap

Another gap concerns specialist implementation capacity. Some hAnother gap concerns specialized implementation capacity. Some humanitarian agencies that are specialized in, or funded for, emergency lifesaving activities may need to withdraw after the post-disaster situation has stabilized. These entities should then hand over their responsibilities and caseloads to government or agencies involved in reconstruction, or the coordination, efficiency, and consistency of the response may be jeopardized. It is common that a number of agencies leave under these circumstances, so an effective coordination mechanism is essential to ensuring a smooth transition.

The British Royal Institution of Chartered Surveyors proposes, in its report, *Mind the Gap! Post-Disaster Reconstruction and the Transition from Humanitarian Relief*, that chartered surveyors, geographic information systems (GIS) technicians, disaster risk professionals, and other "built environment" experts with expertise in planning and management of complex projects can play a crucial role in closing the implementation gap between humanitarian relief and reconstruction and development.[8]

The Participation Gap

Another gap, which is often overlooked and is often the most significant, is in the capacity of the affected population itself to participate in the response. After the initial response, shock and trauma may limit the contributions of some members of the affected population. However, in the subsequent weeks, months, and years, the affected population constitutes the largest contributor of labor to the reconstruction effort. Yet as the population begins to recover its livelihoods, families affected by the disaster should not be forced to choose between reestablishing their financial independence and participating in reconstruction. This gap may come and go suddenly due to internal or external

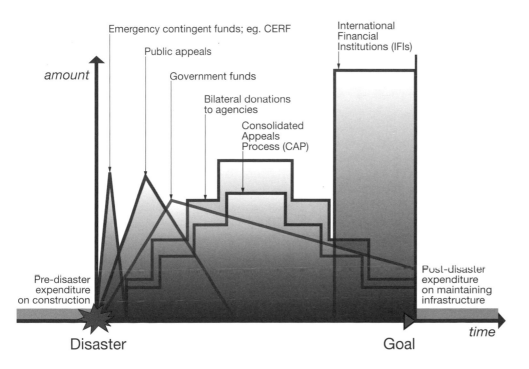

8. Tony Lloyd-Jones, 2006, *Mind the Gap! Post-Disaster Reconstruction and the Transition from Humanitarian Relief* (London: Max Lock Centre at the University of Westminster), http://www.rics.org/site/scripts/download_info.aspx?downloadID=1979.

Source: Shelter Centre Training, http://www.sheltercentre.org/node/3555.

DANIEL PITTET

factors, for example, the onset of a rainy season in a farming community or the decision to accept work opportunities elsewhere. Certain factors that may cause this gap, such as crop harvests, can be planned for. The Emergency Market Mapping and Analysis (EMMA) can be used to identify how the labor supply may fluctuate during the year.[9] An explanation of the EMMA methodology is included in  Chapter 15, Mobilizing Financial Resources and Other Reconstruction Assistance, Annex 1, Deciding Whether to Procure and Distribute Reconstruction Materials.

Good coordination by government, with support from the humanitarian community, can help mitigate these gaps. The indicators proposed by the OECD Development Assistance Committee (DAC) in 1991 for the evaluation of humanitarian programs (as updated) include an indicator for program "connectedness."[10] As explained by the Active Learning Network for Accountability and Performance in Humanitarian Assistance (ALNAP) in its evaluation guide based on these indicators, "connectedness" refers to the need to ensure that activities of a short-term emergency nature are carried out in a way that takes longer-term development requirements into account.[11] However, the guide also states that while the need for linkages among humanitarian action, recovery, and development are well understood, no consensus exists on how lifesaving humanitarian action should support longer-term needs, mentioning natural disasters as an example of where this objective is particularly difficult to achicvc.

Recommendations

1. Agencies should be cognizant and respectful of planning, strategies, and coordination mechanisms that have already been established when they begin operations in the disaster zone.
2. Agencies should realize that every disaster is unique in its complexity, impact, and cultural context, and should work toward an integrated approach that responds first and foremost to the needs of those affected.
3. Address the unique situations and requirements of those affected, whether displaced or not, without discriminating against any sub-group of the population.
4. Seriously consider implementing the transitional shelter approach, which addresses shelter needs over the entire period from the disaster to a permanent housing solution, rather than addressing needs in phases.
5. Make every effort to minimize the gaps between humanitarian programs and reconstruction and development, by getting all key funding institutions involved early in planning and by anticipating the transitions in delivery that will inevitably occur.
6. Expect reconstruction to take a long time, and encourage communities to think in those terms without undermining their determination to recover. Design interim shelter solutions based on realistic assumptions about time.
7. Don't shortchange reconstruction. Adequate financial and material resources must be made available throughout the response, recovery, and reconstruction process. Governments should work with funding sources to plan the distribution of available resources over a realistic reconstruction period.

Case Studies
1963 Skopje Earthquake, FYR Macedonia
The Influence of Early Decisions on Long-Term Recovery
More than 1,000 residents of Skopje, Macedonia, perished in the 1963 earthquake, and more than 3,300 persons were seriously injured. With the vast majority of its 36,000 housing units destroyed or damaged, 76 percent of Skopje's population was left without shelter. Decisions made immediately

9. The Emergency Market Mapping and Analysis tool, http://www.sheltercentre.org/meeting/material/The+Emergency+Market+Mapping+and+Analysis+tool+EMMA.
10. OECD-DAC, 1991, "DAC Principles for Evaluation of Development Assistance," http://www.oecd.org/document/22/0,2340,en_2649_34435_2086550_1_1_1_1,00.html.
11. ALNAP, 2006, "Evaluating Humanitarian Action Using the OECD-DAC Criteria. An ALNAP Guide for Humanitarian Agencies," pp. 20–21, http://www.odi.org.uk/alnap/pubs/pdfs/eha_2006.pdf.

 For access to additional resources and information on this topic, please visit the handbook Web site at www.housingreconstruction.org.

18   SAFER HOMES, STRONGER COMMUNITIES: A HANDBOOK FOR RECONSTRUCTING AFTER NATURAL DISASTERS

after the disaster had a major impact on Skopje's reconstruction. Vladimir Ladinski, an architect and urban planner who lived in the city throughout the recovery process, has written a detailed, 30-year longitudinal study of the transition from relief to full reconstruction. He notes that the relief operation was probably one of the most efficient on record, with minimal aid being wasted and with authorities having very clear priorities. But despite such achievements, Ladinski raises "serious doubts [about] whether the rapid initial planning decision on the location of new settlements was correct." His reference is to the decision, within a few days of the earthquake, to locate temporary housing in sites surrounding the city. It was naively expected that these temporary houses would be subsequently demolished to make way for permanent dwellings some nine months after the disaster. However, in modified form, many remained in place much longer and had a negative impact on the development of the overall city plan.

Sources: Ian Davis, 1975, "Skopje Rebuilt-Reconstruction following the 1963 Earthquake," *Architectural Design*, vol.11, pp. 660-663; V. Ladinski, 1997, in A. Awotona, *Reconstruction after Disaster, Issues and Practices* (Aldershot: Ashgate), pp. 73-107; *and* United Nations, 1970, *Skopje Resurgent* (New York: United Nations).

2004 Indian Ocean Tsunami, Indonesia
Evaluation of a Transitional Shelter Program
The 2004 Indian Ocean tsunami left more than 550,000 people displaced in the province of Aceh, Indonesia. The IFRC responded by implementing a transitional shelter program. From August 2005 to December 2007, some 19,920 transitional shelters were built for families that were still living in tents or shacks. The purpose of this transitional shelter program was to provide a solution that would "fill the gap" between an emergency shelter solution, such as tents, and permanent houses. The steel-framed houses with wooden walls and floor and a sheet-metal roof were intended to provide shelter for a period of 2–4 years. A study initiated by the IFRC and the Netherlands Red Cross showed that the program had a strong positive social impact and a slight positive economic impact on the beneficiaries. This positive impact continued even after the household moved to a permanent house, which shows that the benefits of transitional shelter can go beyond only being a temporary solution for housing after a disaster. However, the positive impact is seen only when a household lives in the transitional shelter for a relatively short period. Transitional shelter should be kept as a short-term solution because it does not meet the needs of a family over a longer period of time. There was a noticeable decrease in economic impact when a family lived in a transitional shelter for a longer period.

Source: S. G. van Dijk, 2009, *A Socio-Economic Impact Study on a Transitional Housing Program. Case Study of a Red Cross and Red Crescent Housing Program in Indonesia* (research report, Eindhoven University of Technology, in collaboration with the IFRC and Netherlands Red Cross).

2003 Bam Earthquake, Iran
Differences in Urban and Rural Shelter Approaches
The large-scale destruction caused by the 2003 earthquake in Bam, Iran, made it unlikely that affected communities would soon have permanent housing. Authorities estimated that at least two years of temporary shelter would be needed before permanent housing would be available, at least in urban areas. National relief agencies pushed to establish camps, despite the strong desire of the people to erect shelters on or close to their own land. The justification of the agencies was that camps would simplify the delivery of services and lower their costs. Yet worldwide, experience has shown that establishing camps for displaced people following

AP

disasters has negative socioeconomic impacts on reconstruction and long-term development, and only makes sense when concerns such as security make other alternatives impossible. Therefore, a national strategy for housing reconstruction was formulated and made public. It entailed (1) in urban areas, providing interim or transitional shelters—including prefabricated units on vacant urban lots or the family's land—that would address housing needs for a 2-year period for the entire affected population, and (2) in rural areas, building permanent housing on original housing plots as soon as practicable. Permanent shelter in urban areas would not be built until the city master plan was updated and reconstruction guidelines were approved.

Prefab houses measuring 18 square meters were provided to all urban households who could prove (using, among other things, the testimony of other residents) that they lived in the area prior to the disaster. For households with more than four people, two prefab houses were provided. An additional unit was available for purchase. The cost for a prefab house was around US$2,000, including transport. Each was installed on the land designated by the household.

Source: Victoria Kianpour, UNDP Iran, 2009, personal communication, http://www.undp.org.ir/.

2005 North Pakistan Earthquake, Pakistan
Reusing Transitional Shelter Materials in Permanent Homes
In October 2005, a strong earthquake and several aftershocks struck Pakistani-controlled Kashmir, northern Punjab, and the North West Frontier Province, devastating poor communities in towns and villages located in harsh, mountainous terrain. Even with winter approaching, many homeless families decided not to leave their villages for government camps because of concerns about their land and livestock, the sources of their livelihood. To help these families, Habitat for Humanity Pakistan (HFHP) introduced a dome-shaped transitional shelter made with materials that could be reused later in permanent houses. The shelters consisted of a tubular pipe structure, galvanized corrugated iron sheets attached using metal ties, and foam insulation. The transitional shelters were easy to assemble and cost about as much as a tent. Approximately 400 of them were erected in the mountainous areas around Balakot and Muzaffarabad. During the spring of 2006, HFHP began building new, earthquake-resistant homes, following guidance from the government's Earthquake Reconstruction and Rehabilitation Authority. Around Balakot and other villages in the Union Council area, HFHP implemented a new construction program that included recycling heavy timber from the destroyed houses into lighter wooden construction elements, using mobile sawmills that were transported from village to village. The galvanized corrugated iron sheets from the transitional shelters were used as roof elements in the new homes. More than 345 new homes were built and an additional 5,500 families were assisted with sawmill services.

Sources: Mario C. Flores, Habitat for Humanity International, 2009, personal communication; *and* Habitat for Humanity Pakistan Earthquake, http://www.habitat.org/disaster/active_programs/pakistan_earthquake_2005.aspx.

2009 Abruzzo Earthquake, L'Aquila, Italy
Temporary Housing Solution Implemented by the Italian Civil Protection
Soon after the April 6, 2009, Abruzzo earthquake, whose epicenter was near the city of L'Aquila, Italy, the Italian government began to analyze the feasibility of a temporary housing project. The first estimate after the earthquake was that 20,000–25,000 people would need temporary shelter. Later, it was agreed that 4,500 units would accommodate the temporary housing demand of all families with three or more members whose houses were destroyed or severely damaged by the earthquake.

Previous Italian experience with earthquakes had shown that reconstruction in a historical center can take 5–10 years, and sometimes longer, which creates a difficult situation for the affected population. While L'Aquila is in a cold mountainous area, the earthquake occurred in April, the beginning of six months of good weather for construction. This made the option of building comfortable temporary housing much more feasible. Another consideration was that, because there are approximately 15,000 students seeking housing in L'Aquila every year, the temporary apartments could eventually be reused as student dormitories.

Government approved the temporary apartments project, called Progetto C.A.S.E (Complesi Antisismici Sostenibili ed Ecocompatibili or Antiseismic, Sustainable, and Ecofriendly [housing] Complexes), at a cost of 700 million euro (US$929 million). The aim was to build energy-efficient, seismically sound temporary apartments in 3-story buildings. A total of 16 firms won design contracts. The new houses needed to be available within six months and to have expected useful lives similar to normal houses. While different designers used different materials (timber, steel, or concrete), all the units were prefabricated and had to meet a rigid time completion schedule. The design criteria and construction process were planned to allow this accelerated construction schedule.

The foundations of the houses are composed of a double platform: the lower one is a foundation plate that rests directly on the ground, and the upper platform lies on more than 7,000 seismic isolators mounted on steel columns fixed at the foundation plates. The design drastically reduces seismic forces and makes the buildings almost completely earthquake-resistant. The covered area between the two plates is designed for underground parking. Besides paying for Progetto C.A.S.E., government is covering the repair costs, including seismic retrofitting, of all permanent housing.

Source: Prof. Mauro Dolce, General Director of Seismic Emergency Unit, Italian Civil Protection, 2009, personal communication.

DANIEL PITTET

Resources

Bektaş, Esra. 2006. "A Post-Disaster Dilemma: Temporary Settlements in Düzce City, Turkey." Paper presented at I-REC 2006 international conference, Florence. http://www.grif.umontreal.ca/pages/BEKTAS_Esra.pdf.

IFRC, Sichuan Earthquake Support Operations. 2008. *Shelter Assessment Report Sichuan Province, China.* http://www.scribd.com/doc/9666212/sichuan-shelter-assessment-report-final-report?autodown=pdf.

International Federation of Red Cross Societies is the lead cluster agency with responsibility for shelter following natural disasters. http://www.ifrc.org.

International Recovery Platform (IRP) identifies gaps and constraints in disaster recovery and serves as a catalyst for the development of tools, resources, and capacity for resilient recovery. IRP has a range of resources that address reducing risks in recovery. http://www.recoveryplatform.org.

Lloyd-Jones, Tony. 2006. *Mind the Gap! Post-Disaster Reconstruction and the Transition from Humanitarian Relief.* London: Max Lock Centre at the University of Westminster. http://www.rics.org/site/scripts/download_info.aspx?downloadID=1979.

ProVention Consortium seeks to reduce the risk and social, economic, and environmental impacts of natural hazards on vulnerable populations in developing countries in order to alleviate poverty and to contribute to sustainable development. Excellent guidelines on earthquake recovery management. http://www.proventionconsortium.org.

UN-HABITAT promotes socially and environmentally sustainable towns and cities and adequate shelter for all. Its disaster management program helps governments and local authorities rebuild from war or natural disasters. http://www.unhabitat.org/categories.asp?catid=286.

United Nations Development Programme Crisis Prevention and Recovery (UNDPCPR) works around the world to restore the quality of life for people who have been devastated by natural disaster or violent conflict. http://www.undp.org/cpr/.

United Nations Development Programme. "Preliminary Post-Disaster Recovery Guidelines." http://www.proventionconsortium.org/themes/default/pdfs/social_analysis/UNDP_Recovery_Guidelines.pdf.

United Nations Development Programme. 2007. "Guidance Note on Early Recovery." http://www.humanitarianreform.org/humanitarianreform/Portals/1/cluster%20approach%20page/clusters%20pages/Early%20R/ER_Internet.pdf.

United Nations Office for the Coordination of Humanitarian Affairs (UN OCHA). 2010. *Shelter after Disaster: Strategies for Transitional Settlement and Reconstruction.* Geneva: UN OCHA. http://www.sheltercentre.org/library/Shelter+After+Disaster.

U.S. Federal Emergency Management Agency (FEMA) has extensive resources on disaster preparation and response and on training of local officials. http://www.fema.gov.

Information on Transitional Shelter Strategies

Organization	Area of expertise	Contact information
IFRC	IASC Emergency Shelter Cluster co-convener for natural disasters	http://www.ifrc.org
Shelter Centre	Sector support resources, including shelter library and global shelter community of practice	http://www.sheltercentre.org
UN OCHA	Humanitarian Reform Support Unit, Emergency Shelter	http://www.humanitarianreform.org/humanitarianreform/Default.aspx?tabid=77
UNHCR	IASC Emergency Shelter Cluster co-lead for conflicts	http://www.unhabitat.org

ASSESSING DAMAGE AND SETTING RECONSTRUCTION POLICY

2

Guiding Principles for Assessing Damage and Setting Reconstruction Policy

- For early, rapid assessments, timely presentation of assessment data takes precedence over exhaustive analytical precision. However, rapid assessments are generally followed by more detailed, sector-specific assessments.
- Joint (multi-donor) assessments and standardized assessment methodologies produce benefits in terms of efficiency, quality, and common understanding of the disaster situation.
- Data collected during assessments—whether multi-sectoral or sector-specific—should be shared, if possible, to reduce duplication of efforts.
- Consultation with affected communities is essential and is possible even in rapid-onset emergencies. Affected communities may want to conduct their own assessments.
- A detailed housing condition assessment is always necessary to estimate the total cost of reconstruction and to allocate the resources.
- Assessment should focus not just on bricks and mortar; the social condition of the people, their working ethos, their willingness to participate, and cultural values all affect reconstruction.
- The particular needs of different groups and individuals (e.g., men, women, the elderly, children) should be evaluated during assessments. Marginalized and vulnerable populations must be sought out and their needs and interests incorporated into reconstruction policy.
- The reconstruction policy is pivotal because it establishes the expectations of the affected community and provides the framework for intervention by local and international actors.
- Communicating the reconstruction policy effectively to those affected by it is almost as important as defining it well. The added value of communication is highest when included from the beginning.

This Chapter Is Especially Useful For:

- Policy makers
- Lead disaster agency
- Assessment teams
- Agencies involved in reconstruction

Introduction

Until the impact of a disaster is assessed, no significant or systematic response can be mobilized. For that reason, assessment is one of the most powerful tools in the disaster response tool kit. Assessments help to establish the extent of post-disaster damage, loss, and needs, and they come in many forms: rapid, detailed, multi-sectoral, and sector-specific. In housing and community reconstruction, a house-to-house assessment of housing damage should always be done. In addition, an assessment of the housing sector may be done. Many assessment methodologies exist; numerous efforts are under way to improve and standardize them.

The principal tradeoff in conducting a rapid assessment is timeliness versus accuracy and completeness. Early data will be more subject to revision over time, but having early information on damage and needs and estimates of reconstruction costs facilitates the initial appeals and response.

Once a disaster's impact is understood and can be quantified, reconstruction planning can begin. Ideally this is coordinated with government's definition and announcement of its reconstruction policy. Reconstruction policy lays out the "rules of the game" for reconstruction, especially the roles of various actors and how they will coordinate, the forms of support that will be provided, and the risk reduction measures that will be taken against future disasters.

This chapter presents the current state of the art of post-disaster assessments and provides some good examples of methodologies. It also explains what the scope and content of a post-disaster reconstruction policy should be and summarizes two examples.

Key Decisions

1. **Government** must designate the agency responsible for assessment; this is often the lead disaster agency, but it may be a statistical or technical agency in government or academia.
2. The **agency responsible for assessments** must decide how it wants the assessment process to be organized and coordinated, the assessment instruments it prefers, and whether and how assessment data will be shared. Humanitarian agencies usually provide assistance with rapid assessments in the early weeks.
3. The **agency responsible for assessments** should coordinate with **local government, agencies involved in reconstruction**, and the **affected communities** to define the rights of the communities with respect to assessment, including the management of their personal data, and their participation in the assessment process.
4. **Government** must designate the agency responsible for reconstruction policy, which will vary depending on the scale of the disaster and the institutional roles defined in national disaster policy. For a localized disaster where subnational government is strong, the responsible agency may be subnational government.
5. The **agency responsible for reconstruction policy** should decide how it will consult with stakeholders, including affected communities and agencies who wish to be involved in reconstruction, before announcing the reconstruction policy.
6. The **agency responsible for reconstruction policy** must establish the basic parameters of the reconstruction policy, including the household assistance strategy, before making its initial policy announcement, but may refine the policy over time.
7. **Agencies involved in reconstruction** should decide with **government** how to make project plans consistent with housing and infrastructure sector policies.

Public Policies Related to Assessment and Reconstruction Policy

If government has preplanned its disaster policy and institutional response for housing and community reconstruction, this plan—and the assessment process it contemplates—has only to be activated. If this has not been done, decisions on assessment procedures and policy will need to be made extemporaneously.

If a country has good social and economic data on the population and built environment affected by the disaster, the initial assessment can be greatly accelerated and its quality can be improved. Information and communications technology (ICT) is increasingly being employed in this way. The initial damage and loss assessment (DaLA) after the 2008 Wenchuan, China, earthquake was conducted exclusively

using government data sources and satellite imagery. This does not eliminate the need for on-the-ground assessments, but greatly accelerates initial assessments. Existing social and economic databases can also provide a baseline for post-disaster assessments, making the quantification of damage more reliable.

The reconstruction policy should take into consideration existing sector strategies and capital investment plans in the sectors affected by the disaster, such as housing, infrastructure, health, education, and transport. Government should coordinate with agencies involved in reconstruction to ensure that project plans based on assessments are also consistent with sector policies. The government of the Indian state of Tamil Nadu, for example, made a policy decision to include thousands of vulnerable households not affected by the 2004 Indian Ocean tsunami in the post-disaster housing reconstruction program and to require that sanitation was provided in all reconstruction sites, as described in the ⌂ case study below.

EARL KESSLER

Technical Issues

Assessment Types and Definitions

Type	Definition[1]
Damage assessment	An assessment of the total or partial destruction of physical assets, both physical units and replacement cost.
Loss assessment	An analysis of the changes in economic flows that occur after a disaster and over time, valued at current prices.
Needs assessment	An assessment of the financial, technical, and human resources needed to implement recovery, reconstruction, and risk management. Usually "nets out" resources available to respond to the disaster.
Rights-based assessment	An assessment that evaluates whether people's basic rights are being met. Has its origins in the United Nations Universal Declaration of Human Rights.[2]
Rapid assessment	An assessment conducted soon after a major event, usually within first two weeks. May be preceded by an initial assessment. May be multi-sectoral or sector-specific. Provides immediate information on needs, possible intervention types, and resource requirements.
Detailed assessment	An assessment undertaken after the first month to gather more reliable information for project planning. Often takes about a month to conduct, and is usually sector-specific.[3]
Housing damage assessment	A damage assessment that analyzes the impact of the disaster on residential communities, living quarters, and land used for housing (see details, below).
Housing sector assessment	An assessment of the policy framework for housing, the post-disaster housing assistance strategy, and the capability of housing sector institutions to carry it out (see details, below).
Communication-based assessment (CBA)	An assessment that analyzes how the context will affect reconstruction and the way in which communication with the affected community can support the reconstruction effort. It includes government and political risk analysis, stakeholder analysis; media, communication environment, and local capacity analysis; and social and participatory communication analysis.[4]

Convergence of Assessment and Analysis Methodologies

Experts working in the disaster field have been confounded in recent years by the array of post-disaster assessments, assessment terminology, and assessment methodologies they encounter. As a result, the United Nations (UN) clusters and other international agencies, including the World Bank, are engaged in various efforts to standardize and improve assessment and analytical tools at all phases of an emergency and to establish indicators, definitions, improved methodologies, standardized information requirements, and accepted thresholds for humanitarian action. A related effort is under way to build partnerships for joint assessments and information consolidation. All of these initiatives aim to address needs for better information for sectoral programming and more timely information at the onset of an emergency. Two of these efforts are especially significant, as discussed below.

Assessment and Classification of Emergencies. The UN Office for the Coordination of Humanitarian Affairs (UN OCHA) established the Assessment and Classification of Emergencies (ACE) project in 2008 in an attempt to map the various humanitarian assessment initiatives currently under way and to facilitate the development of an overarching approach to assessment. In February 2009, UN OCHA issued its "Mapping of Key Emergency Needs Assessment and Analysis Initiatives: Final Report," which analyzes the main assessment and analysis framework initiatives under way at the global level.[5] However, a wide variety of multi-sectoral and/or sector-specific tools used by particular organizations in the field were not analyzed, including those of donors.

The report organizes the various assessment initiatives in three categories:
1. standards-related initiatives, which serve as a foundation for assessment tools and data collection (for example, the Sphere Project[6]);
2. primary data collection, distinguishing between rapid and in-depth assessments (for example, the Local Estimate of Needs for Shelter and Settlement [LENSS], described below, being developed by the Inter-Agency Standing Committee [IASC] Emergency Shelter Cluster); and
3. analysis frameworks, where information and data generated by the two previous categories are integrated into a framework for analysis and/or planning (for example, the Post-Disaster Needs Assessment [PDNA] project, described below, being carried out by the UN, the World Bank, and the European Commission [EC]).

1. Charles Kelly, 2008, *Damage, Needs or Rights? Defining What Is Required After Disaster*, Benfield UCL Hazard Research Centre Disaster Studies and Management Working Paper No. 17 (London: Benfield UCL Hazard Research Centre), http://www.reliefweb.int/rw/lib.nsf/db900sid/FBUO-7HWHG9/$file/Benfield-Jul2008.pdf?openelement; *and* International Federation of Red Cross and Red Crescent Societies (IFRC), 2005, *Guidelines for Emergency Assessment* (Geneva: IFRC), http://www.proventionconsortium.org/themes/default/pdfs/71600-Guidelines-for-emergency-en.pdf.
2. United Nations, "The Universal Declaration of Human Rights," http://www.un.org/en/documents/udhr/.
3. Information on sector-specific assessments is found in several chapters of this handbook.
4. For a detailed explanation of a Communication-Based Assessment, refer to ⌨ Chapter 3, Communication in Post-Disaster Reconstruction.
5. UN OCHA, 2009, "Mapping of Key Emergency Needs Assessment and Analysis Initiatives: Final Report," http://www.humanitarianinfo.org/iasc/downloaddoc.aspx?docID=4927&type=pdf.
6. Sphere Project, 2004, *Humanitarian Charter and Minimum Standards in Disaster Response* (Geneva: Sphere Project), http://www.sphereproject.org/component/option,com_docman/task,doc_view/gid,12/Itemid,203/lang,english/.

As part of this effort, the ACE working group prepared a sequencing framework, which is useful for understanding when the various needs assessment initiatives (not all of which are yet in use) are being or would be applied within the emergency timeline. The timeline includes 24 separate assessments instruments or initiatives.

Needs Assessment Task Force. Since the issuance of the ACE report, a Needs Assessment Task Force (NATF) has been appointed, co-chaired by UN OCHA and the International Federation of Red Cross and Red Crescent Societies (IFRC).[7] NATF was created to strengthen decision making and to improve response by harmonizing and promoting cross-sector needs assessment initiatives that produce consistent, reliable, and timely data on humanitarian needs. Initially focusing its work on preparedness, Phase I (first 72 hours), and Phase II (first 2 weeks) in sudden onset emergencies, NATF will later work on Phase III (second 2 weeks) onward, including early recovery, and will address slow onset emergencies, as progress is made on the first phases.

If the effort to harmonize assessment methodologies is successful, future results will include (1) development of a consolidated needs assessment "tool box," including standardized tools, such as forms and questionnaires that can be adapted for specific contexts; (2) better data management and reduction in the unnecessary collection of similar information; (3) the development of a core set of indicators per sector, which would be consistently collected, thereby improving data aggregation, prioritization of needs across sectors, and equitable response; and (4) multi-sectoral needs assessment tools to collect core common data for decision making and immediate life-saving interventions. In the meantime, governments and agencies working in reconstruction will encounter a variety of assessment methods and tools, and should carefully evaluate the quality of the outputs from these methodologies before acting on them.

Post-Disaster Needs Assessment project. The PDNA project is a cooperative effort between United Nations agencies (led by the United Nations Development Programme as the Chair of the Cluster Working Group on Early Recovery[8] [CWGER]), the World Bank, and the EC to develop a practical guide to a multi-stakeholder PDNA and a recovery framework (RF).

The objective of this project is to develop a shared understanding of the impact of a natural disaster by integrating assessment methods used by international financial institutions (IFIs) (primarily the DaLA methodology developed by UN Economic Commission for Latin America and the Caribbean [ECLAC], which was published in 1991 and reissued in an updated format in 2003), which focus on macro-economic issues, and those used by the IASC humanitarian clusters, UN agencies, and nongovernmental organizations (NGOs), which tend to be sectoral and to have a humanitarian focus. It ultimately aims to improve coordination and capacity at national and international levels to conduct recovery-oriented needs assessments and to carry out recovery planning, in order to connect national plans with the delivery of recovery programs at the local level.

Expected outputs from the PDNA project include (1) protocols of cooperation between the United Nations, the World Bank, and the EC covering joint missions and capacity building; (2) a practical guide to multi-stakeholder PDNA and the RF; and (3) field-testing and training on the framework in high-risk countries with national and international recovery partners. In addition, sectoral assessment methods that are relevant to PDNA will be adapted to enable them to better determine early recovery needs in each sector.

Review of Selected Assessment Methodologies

Governments and agencies involved in housing and community reconstruction should be familiar with some of the common or especially useful assessment methodologies. The following section presents a brief description of some common assessment types, including multi-sectoral assessment (DaLA and community-led assessment), housing sector assessment, and community-specific assessment (LENSS and housing damage assessment).

Good practice in conducing assessments is universal, regardless of the type of assessment. This includes the need to compose assessment teams so that they incorporate the appropriate expertise and representation, including representation of the affected community, and the importance of properly training assessors in the use of the assessment instrument, the definitions of assessment terms, and the peculiarities of the assessment environment, so that the results are consistent.

Local development, housing, and land tenure issues that emerge after a disaster are often not new, but the disaster may exacerbate any weaknesses in the system. Reconstruction challenges such as widespread poverty, extensive informality in the housing system, or a large number of housing units that need to be reconstructed, will just make the problems more visible.

7. IASC, "Terms of Reference for the IASC Task Force on Needs Assessment," http://www.humanitarianinfo.org/iasc/downloaddoc.aspx?docID=4928&type=pdf.
8. Cluster Working Group on Early Recovery, "Early Recovery," http://www.humanitarianreform.org/humanitarianreform/Default.aspx?tabid=80.

Methodology Considerations

Multi-Sectoral Assessment

Damage and loss assessment

Rapid, joint, multi-sectoral – 1st month

The principal multi-sectoral preliminary assessment methodology used in recent years by IFIs, such as the World Bank, is the DaLA methodology developed by ECLAC. The assessment process is sometimes referred to as a "joint rapid assessment." This is generally conducted as soon as possible after the initial disaster response is over.[9] A DaLA is a detailed assessment methodology that estimates the direct economic impact (lost wealth), indirect economic impact (effect on gross domestic product), and secondary effects (fiscal impacts) of a major natural disaster. The methodology provides guidelines for social sectors, including housing, infrastructure, economic sectors, and damage assessment.[10] Numerous examples of completed DaLAs are available from the World Bank. A DaLA is a detailed yet rapid assessment that is conducted as early as possible after a disaster. It is not a substitute for either detailed, sector-specific assessments or a detailed, door-to-door housing condition assessment, both of which come later. DaLA results are often used by donors to establish initial financial commitments for housing and community reconstruction.

Community-led assessments

Detailed, multi-sectoral – 1st quarter

After any disaster, affected communities are the primary responders. Yet once organized relief operations get under way, communities may not be consulted on important aspects of the relief and recovery.[11]

Complementing traditional agency-led assessments with community-led assessments (CLAs) provides a more complete view of the needs and capacities of the affected population. CLAs will help capture the social and psychological impacts on a community, including livelihoods, and the resources available to survivors. Because these factors affect reconstruction, they should not be overlooked; reconstruction can only begin once the household is stabilized. The CLA team must include representation of all community groups in the assessment area and be coordinated by an entity trusted them all (e.g., local government, or local or international NGO).

The Community Damage Assessment and Demand Analysis (CDADA), developed by the All India Disaster Mitigation Institute, is a very good CLA methodology.[12] It is a detailed multidisciplinary, multi-sectoral, multicultural assessment that is adaptable to every disaster type, and can produce sector-specific outputs. The CDADA applies the Sphere Project principles and the IFRC Code of Conduct[13] and emphasizes the role of affected communities, local governments, and community organizations.

Housing Sector Assessment

Housing sector assessment

Detailed, sector-specific – 1st quarter

A housing sector assessment can be very useful after a disaster to analyze the capacity of an affected region's institutional framework for land tenure and housing and community development, its housing production and finance system, and the impact of the disaster on this system.

If it is conducted early (within the first few weeks of the disaster) in parallel with other assessments, the results of the housing sector assessment can be used in the formulation of the overall reconstruction policy and in defining the housing assistance strategy. If reconstruction has already begun and stakeholders are not satisfied with the results, a housing sector assessment will diagnose what is going wrong.

The importance of a housing sector and land tenure analysis may not be recognized early on. People may assume that recovery will not conform to "normal" procedures anyway, but will instead be done using "special" arrangements. However, this may not be the most sustainable reconstruction approach. Outside agency support to post-disaster reconstruction rarely runs long enough, or provides sufficient resources, for full recovery. Local development, housing, and land tenure issues that emerge after a disaster are often not new, but the disaster may exacerbate any weaknesses in the system. Reconstruction challenges—widespread poverty, extensive informality in the housing system, or a large number of housing units that need to be reconstructed—will just make the problems more visible.

A housing sector assessment can also help government and agencies involved in reconstruction identify longer-term housing sector reform initiatives. A detailed methodology for a post-disaster housing sector assessment is included in 🖱 Annex 1, How to Do It: Conducting a Post-Disaster Housing Sector Assessment.

Other Detailed Sector Assessments

Detailed, sector-specific – 1st quarter

Detailed sector assessments are likely to be carried out in other sectors as inputs to housing and community reconstruction planning, as discussed in other handbook chapters. These assessments can include, among others, environmental assessments, communications-based assessments, cultural heritage assessments, social assessments, and corruption risk assessments.

Local Housing Assessment

Local Estimate of Needs for Shelter and Settlement

Rapid, sector-specific – 1st month

The LENSS methodology is designed for rapid shelter and settlement needs assessment in the immediate aftermath of a disaster and before the recovery phase.[14] It provides a systematic assessment methodology and a series of extremely clear formats for collecting and organizing shelter data for a specific locality, which may be collected directly or extracted from other sources.

The tool kit is intended to be used to conduct a needs assessment of and by a locality, in whatever way the population is able to organize itself after the disaster, so that it is prepared to deal with agencies that offer to assist, but it could also be used by an agency itself. One innovation in the LENSS methodology is the use of a storytelling approach to explaining the shelter situation in the community.

9. For example, the 2009 Bhutan earthquake occurred on September 21, 2009. A joint rapid assessment was conducted by the government of Bhutan, the World Bank, and the UN, using a combination of the DaLA and PDNA methodologies, between September 30 and October 14, 2009, http://gfdrr.org/docs/Bhutan_Rapid_Needs_Assessment_Report_Oct_09.pdf.

10. UN ECLAC, *Handbook for Estimating the Socio-economic and Environmental Effects of Disasters* (Mexico: ECLAC), http://www.eclac.cl/cgi-bin/getProd.asp?xml=/publicaciones/xml/4/12774/P12774.xml&xsl=/mexico/tpl-i/p9f.xsl&base=/mexico/tpl/top-bottom.xsl.

11. An evaluation of the response to the 2004 Indian Ocean tsunami noted that the involvement of the communities in needs assessment, planning, and implementation was never made a priority. See Tsunami Evaluation Committee, 2006, *Joint Evaluation*

of the International Response to the Indian Ocean Tsunami: Synthesis Report (London: The Active Learning Network for Accountability and Performance in Humanitarian Action), http://www.alnap.org/initiatives/tec.aspx.

12. Mihir R. Bhatt and Mehul Pandya, 2005, *Community Damage Assessment and Demand Analysis* (Ahmedabad: All India Disaster Mitigation Institute), http://www.proventionconsortium.org/themes/default/pdfs/AIDMI_ELS-33.pdf.

13. International Red Cross and Red Crescent Societies, "Code of Conduct," http://www.ifrc.org/publicat/conduct/.

14. IASC Emergency Shelter Cluster, 2009, "LENSS Tool Kit, Field Version" (Nairobi: UN-HABITAT), http://www.unhabitat.org/pmss/getPage.asp?page=bookView&book=2738.

Methodology	Considerations
Housing damage (or condition) assessment *Detailed, sector-specific –1st to 2nd month*	A housing damage assessment is the necessary first step that will eventually permit the reoccupancy of residential buildings. It provides the evidence needed to support decisions about providing housing assistance, training, and technical assistance for reconstruction. The assessment process is made up of a predictable set of activities, and procedures for a number of them can be established ahead of the disaster to speed up the initiation of the post-disaster housing damage assessment process. A detailed methodology for a housing damage assessment is included in 📖 Annex 2, How to Do It: Assessing Post-Disaster Housing Damage.
	Beside demonstrating to citizens that recovery is beginning, housing assessments serve other purposes: (1) public safety: identify whether houses can be occupied during reconstruction (a housing safety inspection process may be required); (2) planning: to quantify the funds, time, and other resources required for recovery; (3) technical: provide information of the types of damage and the technical skills required in reconstruction; and (4) economic and social: to provide data on the impacts of the disaster at the household level.
	Other chapters of the handbook provide information related to damage assessments, including 📖 Chapter 5, To Relocate or Not to Relocate (decisions about relocation of housing and communities); 📖 Chapter 10, Housing Design and Construction Technology (how disasters damage housing and how design and technology affect housing disaster resistance); and 📖 Chapter 16, Training Requirements in Reconstruction (how reconstruction training is designed and executed using the housing damage assessment results.

EARL KESSLER

Data Management Issues in Assessment

Managing data. Different organizations and agencies collect post-disaster data independently at different periods and on different scales, often duplicating efforts and collecting data in a way that hampers data integration and comparison. Multiple assessments may fail to yield comprehensive, accurate, reliable, and timely assessments that are adequate to support a smooth transition between relief, recovery, and reconstruction. Geo-referencing is an example of a practice that improves the value of information and its ability to be shared, if it is collected using agreed-to standards.

Sharing disaster assessment data reduces duplication of effort and cost. UN OCHA is promoting the use of Humanitarian Information Centers (HICs), geographic information systems (GIS), data standardization, and other tools to make post-disaster data collection more efficient.[15] See 📖 Chapter 17, Information and Communications Technology in Reconstruction, for a discussion of HICs and other information and technology-related strategies relevant to reconstruction.

Managing assessment data is not without its risks. Some consider that assessment data should be treated effectively as a "public good," and the merits of this point of view are easily understood with respect to avoiding the duplication of data collection efforts. "Assessment fatigue" on the part of affected communities is frequently mentioned, and sharing data can help reduce this problem as well. However, data collected in assessments need to be handled and presented with care, since they represent personal information and in some cases may be of a nature that they are protected by confidentiality laws. The fact of the emergency should not override these rights. Information collected in assessments will also reflect the biases of both the informants and the assessors, and biases may affect the interpretation of data collected as well. If the assessors are not experienced, training will be needed before they conduct the assessment. Lastly, assessment data should be compared to baseline information, which may reside in government, but but be readily available.

Good assessment design, data collection protocols, and data management procedures can help control the risks mentioned above. Assessment design and data collection should anticipate how the information will be used. Rules for data confidentiality and disclosure should also be established. If a HIC or other common data management system is established, its functions can include review of assessment instruments, tabulation and interpretation of data, securing and management of baseline data, and definition of rules for data management and disclosure.

Ensuring data quality. For primary data, it may be advisable that data collection be organized at an interagency level and led by government, with one government department taking the lead in coordinating and managing data collection across departments and with agencies to ensure that:

15. A variety of tools related to information management are available at UN OCHA's Information Management Web site, http://www.humanitarianinfo.org/IMToolBox/index.html.

- data are collected on the basis of an agreed-to and mutually consistent analysis plan;
- damage classification criteria and categories are consistent across sectors;
- damage classification criteria are consistent within a sector, and across various administrative/geographical divisions;
- data are validated using empirical tools and plausibility checks; and
- baseline asset classification, such as definitions of various types of houses and categories of infrastructure, such as primary, secondary, and tertiary infrastructure, is consistent among assessments and with public accounts.

If independent assessment teams are concurrently determining damage levels or reconstruction needs, then guidelines and tools should be made available to ensure the consistency of the estimates of need, such as use of common rates and uniform reconstruction benchmarks for housing and infrastructure. Templates can be developed to ensure that damage data are being collected in a structured and uniform manner. Orientation sessions for assessors are essential to train them on the meaning of terms used in the templates, as well as on collection methods. Assessment teams should practice on damaged houses until their results are consistent. See 📖 Chapter 16, Training Requirements in Reconstruction, for more advice on training assessors.

The Needs of Vulnerable Groups in Assessments

Vulnerable groups include displaced people, women, the elderly, the disabled, orphans, and any group subject to discrimination. Vulnerable groups may be omitted from assessments unless an effort is made to ensure their involvement. This is not just a quantitative issue, but a qualitative one, since addressing the post-disaster needs of these groups may require that special measures be taken in reconstruction. Good practices include:

- involving vulnerable group members in assessment and in all stages of decision making;
- obtaining information about the needs of the affected group from both men and women;
- collecting data disaggregated by sex, age, health status, economic class, etc., and then using the disaggregated data in both program planning and monitoring;
- paying special attention in assessments to groups that experience social exclusion (such as the handicapped, widows, and female heads of household); and
- assessing disaster impact on the informal social protection systems that vulnerable groups depend on, not just the "bricks and mortar" impacts.

Defining Reconstruction Policy and Programs

Governments who have put emergency management plans, structures, and arrangements in place for preparedness and response are better prepared to define the institutional arrangements and reconstruction policy for any particular disaster. If the emergency management plan includes safeguard measures to help at-risk communities prepare for disasters, those communities not only are likely to be less affected by the disaster, they will be in a better position to manage reconstruction. For these reasons, it's critical that governments—especially those in vulnerable countries—make a serious commitment to implementing, or continuing to implement, the Priorities for Action of the Hyogo Framework for Action, shown at right.[16] Technical assistance is available from the Global Facility for Disaster Reduction and Recovery (GFDRR), the International Strategy for Disaster Reduction (ISDR), and other international agencies to design and implement disaster risk and emergency management plans.

> ### Hyogo Framework for Action Priorities for Action
>
> 1. Making disaster risk reduction a priority.
> 2. Improving risk information and early warning.
> 3. Building a culture of safety and resilience.
> 4. Reducing the risks in key sectors.
> 5. Strengthening preparedness for response.
>
> *Source:* International Strategy for Disaster Reduction (ISDR), 2006, "Words Into Action: Implementing the Hyogo Framework for Action Document for Consultation," http://www.preventionweb.net/english/professional/publications.

The flowchart entitled "The Process of Response and Reconstruction" included after the handbook's table of contents provides an overview of the types and sequence of decisions policy makers will be required to make in reconstruction. Individual chapters of this handbook discuss sector-specific policy options that should be considered, such as policies for environmental management, land use planning, and disaster risk management, to name a few. However, this section stresses the importance of elaborating an integrated reconstruction policy and strategy to guide the reconstruction program and communicating it broadly. While this is needed for all sectors, this section focuses specifically on the policy for housing and community reconstruction.

16. International Strategy for Disaster Reduction (ISDR), 2006, "Words Into Action: Implementing the Hyogo Framework for Action Document for Consultation," http://www.preventionweb.net/english/professional/publications/v.php?id=594.

The Six Dimensions of Governance

Governance is defined as the traditions and institutions by which authority in a country is exercised. This includes the process by which governments are selected, monitored, and replaced; the capacity of government to effectively formulate and implement sound policies; and the respect of citizens and the state for the institutions that govern economic and social interactions among them. WGI measures six dimensions of governance that correspond to this definition.

1. **Voice and Accountability:** the extent to which a country's citizens are able to participate in selecting their government, as well as freedom of expression, freedom of association, and a free media

2. **Political Stability and Absence of Violence:** the likelihood that government will be destabilized or overthrown by unconstitutional or violent means, including politically motivated violence and terrorism

3. **Government Effectiveness:** the quality of public services, the quality of the civil service and the degree of its independence from political pressures, the quality of policy formulation and implementation, and the credibility of government's commitment to such policies

4. **Regulatory Quality:** the ability of government to formulate and implement sound policies and regulations that permit and promote private sector development

5. **Rule of Law:** the extent to which agents have confidence in and abide by the rules of society, and in particular the quality of contract enforcement, property rights, the police, and the courts, as well as the likelihood of crime and violence

6. **Control of Corruption:** the extent to which public power is exercised for private gain, including both petty and grand forms of corruption

The political economy of reconstruction. In recent analytical work conducted on improving the results of policy reform related to poverty reduction, the World Bank has defined "political economy" as the study of the interactions between political processes and economic variables.[17] A political economy perspective provides insight into the dynamics of reform processes within a country or locality. Stakeholders' interests, and the power relations between social actors, influence their support or opposition to reforms. According to the Bank, the sequencing and timing of actions associated with policy reforms can also determine the level of tension and conflict, the duration, and ultimately the success or failure of reforms.

Reconstruction may not be viewed as policy reform per se, especially due to the accelerated nature of the reconstruction process. However, to the extent that the way in which reconstruction is carried out changes the power relationships or allocation of resources within society, it has many of the same effects as traditional policy reform. For instance, if tenure security is provided to affected communities in reconstruction (as this handbook recommends), there is an economic transfer to those communities, which, as a result, gain social standing and potential future influence.

Political economy factors will be brought to bear on the reconstruction process as economic and social interests vie for influence in many areas, including (1) setting the reconstruction agenda, (2) managing the message through communications with the public, (3) allocating the resources among social groups, and (4) gaining access to the resources being spent. Governments should analyze how political economy factors constitute risks or opportunities for the reconstruction program—including looking at how stakeholders are using their position to protect or strengthen their political or economic interests by building coalitions, negotiating, building consensus, and bargaining to generate outcomes that are favorable to them—and be prepared to manage this aspect of reconstruction. This may require the assistance of political scientists or political economy experts.

Because, inevitably, reconstruction benefits some more than others, and because government itself is part of a country's political economy, it is impossible to inoculate the reconstruction process from political economy influences. Good governance of the reconstruction process is the best antidote. Therefore, the goal in reconstruction should be to establish and orchestrate a reconstruction process whose outcomes promote social equity and reflect good governance practices. The dimensions of governance used in the Worldwide Governance Indicators (WGI) project,[18] shown in the text box, above, have become widely accepted. Good governance practice material is cited throughout 🏠 Chapter 19, Mitigating the Risk of Corruption, and its annexes.

Challenges in defining reconstruction policy. Data collected during assessments are critical evidence for establishing the reconstruction policy. However, it is highly likely that not all the necessary information will be available when the policy is first outlined and even announced publicly. For instance, a rapid assessment of housing damage may give government a estimate of the number of affected households, and perhaps of the extent of housing damage, but is unlikely to provide reliable estimates of the cost of repairing or the number of houses that will need to be demolished. That requires a housing damage assessment. This "information lag" creates a number of challenges that policy makers are forced to confront in defining and announcing reconstruction policy.

17. World Bank, 2008, The Political Economy of Policy Reform: Issues and Implications for Policy Dialogue and Development Operations (Washington, DC: World Bank), http://siteresources.worldbank.org/EXTSOCIALDEV/Resources/Political_Economy_of_Policy_Reform.pdf.

18. World Bank, 2009, "Governance Matters, 2009: Worldwide Governance Indicators, 1996–2008," http://info.worldbank.org/governance/wgi/index.asp; and Daniel Kaufmann, Aart Kraay, and Massimo Mastruzzi, 2009, Governance Matters VIII: Aggregate and Individual Governance Indicators,1996–2008 (Washington, DC: World Bank), http://papers.ssrn.com/sol3/papers.cfm?abstract_id=1424591.

🏠 **For access to additional resources and information on this topic, please visit the handbook Web site at www.housingreconstruction.org.**

Policy making challenge	Advice to policy makers
The reconstruction policy will be a work in progress that will need to be updated as more information becomes available.	Avoid announcing the details of assistance schemes before collecting relatively reliable data on the households affected, to avoid making commitments to the affected community that may become difficult to keep for logistical or financial reasons.
Affected communities and other stakeholders should be consulted with about the parameters of the reconstruction policy before those parameters are finalized. Avoiding this step can establish a dynamic of mistrust that will be difficult to overcome later.	Avoid presenting the reconstruction policy as final before a substantive dialogue concerning reconstruction has taken place with stakeholders. See the 🖭 case study on Tamil Nadu reconstruction, below, for examples of how stakeholder consultation was used before reconstruction policies were announced.
Decisions made early in the response may affect how reconstruction can be carried out. As explained in 🖭 Chapter 1, Early Recovery: The Context for Housing and Community Reconstruction, government—working with the humanitarian community in the first two weeks after the disaster—is likely to have conducted the initial assessment, announced a rapid appeal, and defined project level work plans.	Realize that early shelter decisions may affect the options available later in the reconstruction program and think carefully about the longer-term implications of short-term solutions. A decision to move the entire population to camps, as opposed to providing *in-situ* transitional shelter solutions, for example, could disperse an affected community to such a degree as to make a community-led reconstruction approach nearly impossible.
Announcing the assistance scheme before assessments are conducted may create an incentive for homeowners to damage their houses in order to receive the announced benefit, and result in multiple assessments and extensive processing of grievances.	Conduct at least an initial census and housing damage assessment before announcing housing assistance schemes.

The lack of involvement in these early assistance policy discussions of the IFIs who may provide the financing to government for reconstruction has been identified as an international coordination issue that needs to be addressed. This is one of the motivations for the efforts to harmonize assessment methods discussed earlier in this chapter.

The parameters of the reconstruction policy. Two 🖭 case studies of successful reconstruction policies are included below. While there is no template for a reconstruction policy, the chapters of this handbook represent the critical areas that need to be covered in such a policy, and each provides relevant advice for policy makers. Particularly important are the chapters in Section 1, Assessing Impact and Defining Reconstruction Policy, and Section 2, Planning Reconstruction. The scope of the policy, and the corresponding handbook chapters, include the following.

Policy section	Content	Reference in handbook
Background and Context		
Context for reconstruction	Overview of the response and early recovery. Roles and responsibilities of agencies involved to date. How early decisions and actions will be coordinated with reconstruction policy	🖭 Chapter 1, Early Recovery: The Context for Housing and Community Reconstruction
Assessment of damage	Concrete definition of the scope and nature of the disaster and disaster impacts that the reconstruction policy needs to address	🖭 Chapter 2, Assessing Housing Damage and Setting Reconstruction Policy
Goals and objectives	Physical, social, and economic goals and objectives of the reconstruction program	🖭 Chapter 2, Assessing Housing Damage and Setting Reconstruction Policy
	Based on the initial damage and loss assessment and sector-specific assessments as they become available	🖭 Chapter 4, Who Gets a House? The Social Dimension of Housing Reconstruction
Institutional Strategy		
Program implementation	Definition of the institutional arrangements for managing reconstruction, including the role of local governments	🖭 Chapter 13, Institutional Options for Reconstruction Management
NGO/CSO role and coordination	Definition of the role of international, national, and local NGOs and CSOs in reconstruction, and mechanisms for coordination	🖭 Chapter 14, International, National, and Local Partnerships In Reconstruction
Financial Strategy		
Housing assistance scheme	Special measures to address the housing and reconstruction requirements of vulnerable households and groups	🖭 Chapter 4, Who Gets a House? The Social Dimension of Housing Reconstruction

Policy section	Content	Reference in handbook
Financial mobilization, tracking, and management	Sources of financing and means for coordinating and monitoring expenditures and results	Chapter 15, Mobilizing Financial Resources and Other Reconstruction Assistance
Financial assistance delivery and materials facilitation	The financial assistance concept and the assistance scheme for different categories of the affected population	Chapter 22, Financial Management in World Bank Reconstruction Projects
	The policy for the facilitation and provision of construction materials at the locality or household level	Chapter 23, Procurement in World Bank Reconstruction Projects
Community Participation		
Communications strategy	The modes of communication to consult with affected communities and the general public, to receive feedback and suggestions, and to share reconstruction decisions and updates	Chapter 3, Communication in Post-Disaster Reconstruction
Community organization and participation	How households and communities will be mobilized to participate in reconstruction. The roles of households and communities in reconstruction	Chapter 12, Community Organizing and Participation
Reconstruction Approach		
Relocation policy	The basis for relocation decisions, including criteria, decision process	Chapter 5, To Relocate or Not to Relocate
Reconstruction approach (Defined in consultation with the affected households)	The role of households, contractors, and agencies in reconstruction	Chapter 6, Reconstruction Approaches
	Need for transitional or temporary shelter	
Land use and infrastructure reconstruction	Improving the disaster resilience of land use and construction	Chapter 7, Land Use and Physical Planning
	The regulatory framework to ensure the safety of reconstruction and the allocation of responsibilities between the public and private sector for compliance	Chapter 8, Infrastructure and Services Delivery
"Build back better" in housing reconstruction	The strategies to apply "build back better" principles in housing reconstruction	Chapter 10, Housing Design and Construction Technology
		Chapter 11, Cultural Heritage Conservation
Grievance redressal	Information on the means by which households and affected communities may have their grievances heard and addressed	Chapter 15, Mobilizing Financial Resources and Other Reconstruction Assistance
Technical assistance and training	The nature of the technical assistance and training to be provided to local governments, communities, and households	Chapter 16, Training Requirements in Reconstruction
Risk Management		
Environmental management	How the environmental impact of the disaster will be addressed and the environmental, social, and economic impact of reconstruction and relocation will be minimized	Chapter 9, Environmental Planning
Environmental and social safeguard policy framework		Chapter 21, Safeguard Policies for World Bank Reconstruction Projects
Disaster risk management	Means for establishing standards for disaster risk reduction in reconstruction for relocated and unrelocated communities and mitigation measures to be employed	Part 4, Technical References, Disaster Risk Management in Reconstruction
Anticorruption strategy	Define the measures to be taken by government other agencies to minimize corruption	Chapter 19, Mitigating the Risk of Corruption
Monitoring and evaluation	The information and communications technologies to be employed in reconstruction	Chapter 17, Information and Communications Technology in Reconstruction
	The means to ensure transparency, permit the reporting of any perceived corruption or other wrongdoing, and involve stakeholders in monitoring progress	Chapter 18, Monitoring and Evaluation

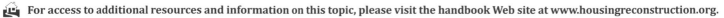

 For access to additional resources and information on this topic, please visit the handbook Web site at www.housingreconstruction.org.

32   SAFER HOMES, STRONGER COMMUNITIES: A HANDBOOK FOR RECONSTRUCTING AFTER NATURAL DISASTERS

Risks and Challenges

- No single competent government agency designated to manage assessments after the disaster; leaving it instead to individual agencies.
- Government overpromising in the early days after the disaster before sufficient information is available.
- Inaccuracies in the damage assessment caused by the lack of baseline data (original condition of infrastructure and housing) or inconsistencies of data from various sources.
- Proliferation of assessments and the resulting assessment fatigue among the affected population. False expectations being created in the assessed population.
- Failure of agencies to share assessment data.
- Inadequate assessment quality because assessors are not sufficiently trained.
- Local requirements not reflected because community or other local data are not incorporated in national assessments. Needs of vulnerable populations not highlighted in assessments.
- Owners who damage their own housing in order to qualify for housing assistance because the assessment of housing damages takes place after the assistance scheme is announced.
- Assessment data not objectively analyzed. Assessments carried out simply to justify agency decisions, not to inform them.
- Reconstruction policy that does not take sector investment plans and long-standing needs into account.
- Lack of stakeholder consultation in the process for establishing the reconstruction policy.

Recommendations

1. Conduct multi-donor assessments whenever possible, using standardized assessment methodologies.
2. In designing the assessment and data collection, take into consideration how the information will be used and shared, the biases of the assessors, and the need for training.
3. Treat national and sector-specific data collected during assessments as public information, while respecting principles of confidentiality, to reduce duplication of data collection efforts.
4. Evaluate the needs of different groups and individuals (such as men, women, the elderly, and children) during assessments. Seek out marginalized groups and evaluate their needs and interests as well.
5. Advocate for a Communication-Based Assessment at the beginning of the project cycle to ensure that the reconstruction program is designed based on its results.
6. Consult with the community regarding the need for information and consider using community-led assessments to complement the information gathered from traditional assessment methodologies.
7. Establish a clear system of damage categories for housing, and try to carry out housing damage assessments before announcing the housing assistance scheme.
8. Observe the warnings regarding the announcement of reconstruction policy, such as the need to consider the impact of short-term shelter decisions on longer-term reconstruction.
9. Understand that the reconstruction policy should be one of the primary messages to be passed along to the public through the communication plan. Remember that what's important is what people hear, not what is said to them.
10. If reconstruction is not going well, or there are concerns about institutional capacity for reconstruction, conduct a housing sector assessment to identify whether assistance may be needed.

Case Studies

2004 Indian Ocean Tsunami, Tamil Nadu, India

Tamil Nadu State Tsunami Reconstruction Policy

The 2004 Indian Ocean tsunami had a widespread impact on the fishing villages and towns along the coastline of Tamil Nadu, India. The state had never seen a calamity of this nature. More than 8,000 lives were lost and 1 million people were affected. Most of the 54,000 housing units affected were destroyed. More than 400 schools, health clinics, and other public buildings were destroyed and many more were damaged, as were roads and other infrastructure. The impact was spread over 13 districts and 350 towns or villages. The entire coastal economy of Tamil Nadu was affected.

The Tamil Nadu State Tsunami Reconstruction Policy[19] addressed a wide range of issues, focusing particularly on the environment, livelihoods, and shelter, and included all the measures the government of Tamil Nadu was taking to bring the lives of the affected people back to normal. The responsibility

19. Government of Tamil Nadu, 2008, "Tiding Over Tsunami, Part II," http://www.tn.gov.in/tsunami/TidingoverPART2.pdf.

for managing reconstruction was assigned to a Project Implementation Unit within the Revenue Administration, Disaster Management, and Mitigation Department of the state government. Numerous state agencies and the federal government collaborated in the effort. The cost of rebuilding in Tamil Nadu was estimated at US$880 million. Of this, US$566 million was borrowed from IFIs.

Assessment. A questionnaire was developed by the state and administered by district officials to ascertain the number of affected families in each zone, the type of construction, the ownership of the structures, the number of family members, etc. The survey covered 278,000 families who lived within 1,000 meters of the coast. The survey included families whose houses were not damaged, but whose livelihoods had been affected by the tsunami. The survey results formed the basis of the choices before government in terms of the area, the average size of houses to be built, the general nature of construction, and the approximate cost. The survey showed that tenancy was an issue mainly in the urban areas, so a policy decision on this aspect of reconstruction was postponed until further input could be gathered.

Housing sector policy. Within two weeks of the tsunami, the Revenue Administration, Disaster Management, and Mitigation Department had issued a government order[20] that announced a comprehensive village development model. The order promoted private participation in reconstruction, limiting the role of government to providing land, specifications for housing, and common amenities. The order included the parameters for the projects and solicited NGO proposals. Media advertisements were issued calling for support.

The specific policies established to guide reconstruction included extensive consultation with stakeholders, community choice on relocation decision, agency-driven reconstruction with NGOs providing resources and assisting communities, a strong role for district governments with support from the state, adherence to coastal zone regulations, safe rebuilding according to building codes and guidelines, and financial assistance for a core house with a choice of models.

Disaster risk reduction. Government acknowledged that most of the buildings damaged by the tsunami were built with construction practices that were not appropriate for the area, given the hazards it was exposed to. To mitigate future risks, the reconstruction policy for housing and infrastructure would strongly promote use of disaster-resistant technologies. The decision about whether the community would relocate was left to the community itself. Relocated communities were given free parcels of land in urban and rural areas.

Assistance packages. Assistance was provided by NGOs and was the same for all. The amount was sufficient to provide a core house and basic infrastructure. This approach was used both for equity purposes and because the property records would not have permitted a fair valuation of the property that was lost.

The state decided to adhere to the Coastal Regulatory Zone regulations under the Environment Protection Act, which regulate building activity up to 500 meters from the high tide line. The only exceptions were for fishers, who were allowed to stay if not willing to relocate beyond 200 meters, but who were not allowed to rebuild, only repair, their houses, and were not given housing assistance.

Building codes. The state relief commissioner's office set up a committee of experts to study the National Building Code and the guidelines developed in Gujarat after the earthquake. They suggested modifications based on the windy conditions prevailing along the Tamil Nadu coast. These were used to develop the core house designs.

Institutional arrangements. While owner-driven construction was permitted, in the end, NGOs, donors, and government built most of the housing with a high level of community involvement. District governments were given responsibility for coordination of the reconstruction, with significant financial and technical support from the state. NGOs of various kinds were invited to provide reconstruction resources and to assist communities. In those cases where NGOs or corporations did not come forward, reconstruction was coordinated by district collectors and financed by government after organizing the families into self-help groups.

20. Government of Tamil Nadu, 2005, "Revenue (NC.III) Department, Government Order Ms.No.25," http://www.tn.gov.in/gorders/rev/rev-e-25-2005.htm.

The government of Tamil Nadu committed itself to carrying out the following measures.

Commitment	Significance
1. Temporary shelters	Government provided a grant of Rs 8,000 (US$180) for temporary shelters to be built by government as well as by NGOs.
2. Core houses	Instead of adopting "compensation" as the basis for entitlement, a core house was provided irrespective of the size of the original house. NGOs were asked to spend an average of Rs 150,000 (US$3,500) for a house of 325 square feet, plus infrastructure and livelihood activities.
3. Building codes and guidelines	The state Public Works Department developed plans for the model core houses and made them widely available. Some were two floors.
4. NGO guidelines	The state provided district officials with guidelines to assess the genuineness of the NGOs operating in the post-disaster environment and a model format for a detailed memorandum of understanding (MOU) between the district and the NGOs.[21] Several respected NGOs reviewed the draft MOU and proposed clarifications. Local officials could make minor changes to suit local conditions.
5. Land acquisition	The state provided funds to districts for land acquisition up to 200 percent of value in order to relocate owners from the coastal zone and other high-risk areas. There were no lawsuits by property owners. Land for relocated houses was provided by the state.

EARL KESSLER

Insurance and title. The housing assistance included payment for 10 years of property insurance on the new houses. The ownership of the houses passed to both the husband and wife after construction was completed.

Communications with stakeholders. Several rounds of consultations were held by the districts with the community regarding the housing aspirations of the affected families, especially women. Five housing models were developed. When the original designs did not have a staircase, this was subsequently changed based on consultations.

The results of assessments, the names of assistance recipients, the reconstruction guidelines, and the housing reconstruction policy were widely publicized and made available on the Web sites of the districts and states.

2001 Gujarat Earthquake, India
The Gujarat Earthquake Reconstruction and Rehabilitation Policy

When an earthquake measuring 6.9 on the Richter Scale struck Gujarat, India, on January 26, 2001, and was followed by more than 500 aftershocks, the effect was devastating and somewhat unexpected, given the geological characteristics of the location where it struck. Approximately 13,800 people died and approximately 167,000 were injured. More than 1.2 million houses were damaged or destroyed and nearly all the civic facilities—schools, hospitals, health centers, and public buildings—were damaged, some extensively. The utility infrastructure, including water supply, electricity, and telecommunications, was completely disrupted.

The government of the state of Gujarat announced the Gujarat Earthquake Reconstruction and Rehabilitation Policy[22] four months after the earthquake. The policy document, only 30 pages long, included the creation of the Gujarat State Disaster Management Authority. (It had actually been created in the month following the earthquake; the policy document formalized the entity). It proposed different reconstruction approach for urban and rural reconstruction and in different regions of the state, depending on their seismic zone. The cost of rebuilding was estimated at US$1.77 billion, of which more than half was to be borrowed from IFIs.

Policy objectives. The stated objectives of the policy included building, retrofitting, repairing, and strengthening houses and public buildings, and improving the earthquake resistance of what was rebuilt. Other objectives related to revival of the local economy, reconstruction of community and social infrastructure, health support to those affected by the earthquake, restoration of lifeline and major infrastructure, gender empowerment, social attention to the poor, implementation of a comprehensive disaster preparedness and management program, and the need for long-term mitigation of a variety of risks to which the population was exposed.

21. Government of Tamil Nadu, 2005, "MOU for Public Private Partnership," http://www.tn.gov.in/tsunami/MOU_for_Public_Private_Parthership.pdf.
22. State of Gujarat, 2001, "Gujarat Earthquake Reconstruction and Rehabilitation Policy," http://www.gsdma.org/pdf/Earthquake%20Rehabilitation%20Policy.pdf.

Guiding principles. Among the guiding principles of the policy were the need to involve people and representative institutions in decision making; the strengthening of civil society institutions; the importance of ensuring that the needs of the vulnerable were addressed; the necessity to give people information to make informed choices in rebuilding, including about disaster risk reduction; and the importance of involving the private sector, NGOs, and expert institutions in the reconstruction program. Lastly, it called for the highest levels of transparency and accountability in the reconstruction program through the use of appropriate institutional mechanisms and practices.

Housing sector policy. The housing sector was defined as encompassing debris removal, salvage, and recycling; construction of temporary shelters; reconstruction of more than 230,000 houses; repairs and strengthening of more than 1 million houses; and reconstruction and repairs of government staff quarters. The policy established that there would be a community-driven housing recovery process, under which earthquake-affected communities would be given a range of choices from complete or partial relocation to *in-situ* reconstruction. While acknowledging that there existed a predominant sentiment for minimal relocation, this policy gave communities the responsibility for deciding on their preferred option, using a participatory process. Selection of new sites would be undertaken with the support of village officials and the NGO or other agency assisting the village. Other aspects of the policy included:

- delegation of technical and financial powers for the housing reconstruction process to the district administration or Area Development Authorities;
- use of a community-based, owner-driven approach, with technical assistance from engineers provided by government, building centers, NGOs, etc.; and
- basing reconstruction on a tripartite partnership, including the government of Gujarat, the private sector (including NGOs), and the beneficiaries.

The government of Gujarat committed itself to carrying out the following measures.

Commitment	Significance
1. Removal of rubble	Included the commitment to environmental management and recycling to reduce construction costs
2. Setting up of temporary/interim shelters	Shelters for urban and rural homeless, provided through government agencies or NGOs, or provision of shelter material
3. Full reconstruction of collapsed and demolished houses	Provision of financial entitlement package adequate for core house, to which owners could contribute additional resources from their own resources or by borrowing
4. Repair of damaged units	Provision of financial entitlement package for repair
5. Retrofitting of undamaged units	Technical assistance to owners or cooperatives wanting to retrofit their properties for earthquake or cyclone resistance
6. Rebuilding of social and community infrastructure	Reconstruction of minimum infrastructure for each village, including primary school, water storage, roads, electricity, and building of infrastructure at relocated site and repair/retrofit for *in-situ* reconstruction

Assistance packages. The government of Gujarat announced five packages of assistance for reconstruction, retrofitting, and repairs of approximately a million houses destroyed or damaged in the earthquake. The amounts varied depending on the type of house, the extent of damages, and the location.

- **Package 1:** For villages in seismic Zones IV and V, where more than 50 percent of the houses collapsed
- **Package 2:** For villages in Zones IV and V, which opted for *in-situ* reconstruction
- **Package 3:** Villages situated in areas other than Zones IV and V, where individual houses were destroyed or damaged
- **Package 4-A:** Reinforced cement concrete frame structures (low- and high-rise) in urban areas, which include municipal corporations, urban development authority areas, and other municipalities (excluding Bhuj, Bhachau, Rapar, and Anjar in the Kutch District)
- **Package 4-B:** Load-bearing structures in corporation areas, urban development authority areas, and municipalities (except Bhuj, Bhachau, Rapar, and Anjar in the Kutch District)
- **Package 5:** Rehabilitation in the four worst affected municipal towns of Bhuj, Anjar, Bhachau, and Rapar in the district of Kutch, with a stress on urban town planning

Urban rehabilitation. In Kutch, four towns—Bhuj, Bhachau, Anjar, and Rapar—suffered large-scale devastation. The collapse of a large number of multistory buildings and the limited availability of land in these towns called for a different strategy for rehabilitation. Congested inner towns were redeveloped, and the residents were given the option of relocation. A number of residents living in multistory buildings were asked to construct houses on new plots, in view of new town planning rules, development regulations, and a reduced floor space index.

A number of public buildings were also proposed for relocation. The urban infrastructure was to be expanded and upgraded. Construction in these towns was owner-driven. Government facilitated the process by providing technical guidance, material specifications, and technical supervision for building earthquake-resistant buildings. See the 🏚 case study on the planning process for the redevelopment of Bhuj in 🏚 Chapter 7, Land Use and Physical Planning.

DANIEL PITTET

Resources

American Society of Civil Engineers. 2009. "Post-Disaster Assessment Manual." http://www.asce.org/inside/TCERP_Manual_Final.pdf.

Bhatt, Mihir R. and Mehul Pandya. 2005. *Community Damage Assessment and Demand Analysis.* Ahmedabad: All India Disaster Mitigation Institute. http://www.proventionconsortium.org/themes/default/pdfs/AIDMI_ELS-33.pdf.

ECLAC. *Handbook for Estimating the Socio-economic and Environmental Effects of Disasters.* Mexico: ECLAC. http://www.eclac.cl/cgi-bin/getProd.asp?xml=/publicaciones/xml/4/12774/P12774.xml&xsl=/mexico/tpl-i/p9f.xsl&base=/mexico/tpl/top-bottom.xsl.

Global Risk Identification Program. http://www.gripweb.org/grip.php?ido=1000&lang=eng.

IASC Emergency Shelter Cluster. 2009. *LENSS Tool Kit, Field Version.* Nairobi: UN-HABITAT. http://http://www.unhabitat.org/pmss/getPage.asp?page=bookView&book=2738.

IFRC. 2005. *Guidelines for Emergency Assessment.* Geneva: IFRC. http://www.proventionconsortium.org/themes/default/pdfs/71600-Guidelines-for-emergency-en.pdf.

International Recovery Program Initiative on Early Recovery. "OCHA Disaster Response Preparedness Toolkit." http://ocha.unog.ch/drptoolkit/PPreparednessEarlyRecovery.html.

International Strategy for Disaster Reduction (ISDR). 2006. "Words Into Action: Implementing the Hyogo Framework for Action Document for Consultation." http://www.preventionweb.net/english/professional/publications/v.php?id=594.

Kelly, Charles. 2008. *Damage, Needs or Rights? Defining What Is Required After Disaster.* Benfield UCL Hazard Research Centre Disaster Studies and Management Working Paper No. 17. London: Benfield UCL Hazard Research Centre. http://www.reliefweb.int/rw/lib.nsf/db900sid/FBUO-7HWHG9/$file/Benfield-Jul2008.pdf?openelement.

National Development and Reform Committee. 2008. "The Overall Planning for Post-Wenchuan Earthquake Restoration and Reconstruction." http://en.ndrc.gov.cn/policyrelease/P020081010622006749250.pdf.

Tsunami Evaluation Committee. 2006. *Joint Evaluation of the International Response to the Indian Ocean Tsunami: Synthesis Report.* London: The Active Learning Network for Accountability and Performance in Humanitarian Action. http://www.tsunami-evaluation.org/.

Tsunami Global Lessons Learned Project. 2009. "The Tsunami Legacy: Innovation, Breakthroughs, and Change." Banda Aceh: Tsunami Global Lessons Learned Project Steering Committee. http://www.undp.org/asia/the-tsunami-legacy.pdf.

United Nations Human Settlements Programme (UN-HABITAT). "Disaster Assessment Portal: Assessments for Disaster Response and Early Recovery." http://www.disasterassessment.org/resources.asp?id=6&cid=1. This site includes a variety of assessment methodologies and other tools.

World Bank. 2007. *Tools for Institutional, Political, and Social Analysis of Policy Reform: A Sourcebook for Development Practitioners.* Washington, DC: World Bank. http://go.worldbank.org/GZ9TK1W7R0.

World Bank. 2008. *The Political Economy of Policy Reform: Issues and Implications for Policy Dialogue and Development Operations.* Washington, DC: World Bank. http://siteresources.worldbank.org/EXTSOCIALDEV/Resources/Political_Economy_of_Policy_Reform.pdf.

World Bank. n.d. "Guidelines and Sample Damage and Needs Assessments." http://go.worldbank.org/KWCRRCKA20.

Having a coherent understanding of an affected region's pre-disaster housing and community development system and the likely impact of a disaster on this system is often essential for developing an effective post-disaster housing and community reconstruction strategy or for diagnosing what is going wrong if reconstruction has begun and local actors are not satisfied with the results. Local development, housing, and land tenure issues that emerge in the aftermath of a disaster are often not new, but the disaster will exacerbate the weaknesses in the system, especially when there are challenges such as widespread poverty, extensive informality in the housing system, or a large number of housing units that need to be reconstructed.

When to Conduct a Housing Sector Assessment

A housing sector assessment should be carried out during the first few weeks after a disaster in parallel with other assessments and should be used in the formulation of the overall reconstruction policy and longer-term housing sector reform.[1] The importance of a housing sector and land tenure analysis may not be recognized early on, as people assume that the recovery process will not conform to "normal" processes, but instead will be carried out using a series of "special" arrangements. However, this is seldom the best or most sustainable reconstruction approach. The focused, humanitarian period of post-disaster reconstruction rarely runs long enough, or provides sufficient resources, for full recovery. As a result, the reconstruction process should be able to "run itself" after the formal reconstruction period is over. A more sustainable approach is one that improves on—but is still based on—normal reconstruction practices in the country. Such a strategy also mobilizes local actors, such as small-scale builders, and gives them training and livelihood opportunities. During this time, they can be enlisted in improving longer-term construction practices. Therefore, helping government gain the insight on how prior practices led to the disaster outcomes, and how they can be improved, can stimulate efforts to improve "normal" housing construction policies, procedures, and conditions.

Critical Elements of the Housing Sector

The critical elements of the process by which housing and communities are constructed and reconstructed are considered to be the following: (1) local governance, (2) land administration, (3) housing construction system and practices, (4) housing finance, and (5) local infrastructure construction and operation. While this guidance puts relatively equal emphasis on all of these elements, one or more of them may need to be emphasized in the assessment, depending on the prior conditions in the country.

Objectives of a Housing Sector Assessment

The general objective of the assessment is to assist government at all levels to improve the quality of outcomes from the response in the area of housing and community reconstruction. The assessment will increase the awareness of the agencies involved with reconstruction of the strengths and weaknesses in the local housing sector and land administration systems and show how they may affect recovery, while providing specific recommendations on short- and medium-term actions to be taken to improve the effectiveness of reconstruction program implementation that will contribute longer-term strengthening of the housing sector and improvements in the quality of the housing that is reconstructed.

The specific objectives of the assessment include the following:

A. Provide a comprehensive analysis of the country's policy and institutional frameworks for the housing sector and the land administration system, with particular emphasis on:
- the adequacy of these frameworks under normal conditions and their ability to be adapted to the demands of the post-disaster housing reconstruction process;
- the implications of any relevant policies announced since the disaster;
- the capacity of the organizations involved in the housing reconstruction; and
- the specific challenges that have already arisen, or may be expected to emerge, as the post-disaster housing reconstruction program is planned and executed.

B. Provide concrete and specific recommendations on how to improve the response to the disaster in such areas as: policy modifications, institutional roles and responsibilities, coordination mechanisms, and needs for institutional strengthening, including capacity-building activities, financial strategies, or other areas.

Methodology for a Housing Sector Assessment

The assessment should focus on the policy and institutional frameworks for housing and community reconstruction. It is not intended to be a housing damage assessment, although the extent and nature of the damage may affect the recommendations, so this data should be analyzed and taken into consideration in the assessment. Similarly, while the principal area of concern is the reconstruction of permanent housing solutions for the affected population, not temporary shelter solutions, the two cannot be analyzed in isolation. Therefore, the consistency between the temporary housing strategy (if any) and the permanent reconstruction strategy should be analyzed.

Housing reconstruction takes place on a very local and even personal basis. The concerns and perspectives of local actors should have a strong influence on the reconstruction approach.

The consultants should use a variety of data collection methods to capture different types of information and social perspectives, and it should have a bias toward capturing the perspective of households, local government officials, and other local actors. A reconstruction approach not based on local reality, and not seen as workable at the local level, is unlikely to succeed. For that reason, it is suggested that the consultants reside in the disaster area while conducting the assessment.

Expertise Required

Specialists should be hired to carry out this assessment, due to the complexity of the issues and the need to organize and interpret a wide range of information. The specific expertise may vary, depending on the disaster situation. In general, a team of 5–7 people will be required to carry out this assessment in a timely manner. The team should include members with expertise in housing policy, housing finance, post-disaster reconstruction, local government administration, and local service provision. The team leader should have post-disaster housing reconstruction experience. One member should be responsible for handling poverty and social safeguards issues, including the analysis of social policies related to housing provision for low-income and vulnerable populations and the differential effects of the disaster and the reconstruction policies being proposed.

A counterpart in government who understands the policy issues related to the work and who can facilitate contacts and access to information must be appointed. Ideally, this person is supported by a technical committee that includes representation from the affected population.

Sources of Information

1. **Documentation.** Previous studies of the housing sector should be reviewed and may serve as the starting point for this assessment. Key documentation includes, among others things, at the national level, disaster damage, loss, and needs assessments; pronouncements and policies related to the disaster; laws and other material related to the legal framework for housing and land; and national policy documents; and, at the local level, local damage, loss, and needs assessments; registers of affected persons; policy documents; land use plans and policies and related ordinances; capital investment plans; and procedures related to building permitting and inspection.

2. **Interviews.** National and regional government officials (including representatives from appropriate ministries); municipal authorities (mayor, technical experts, public service organizations, counsel members); social leaders and social movement representatives; the affected population and their representatives; locally active international organizations; civil society organizations and NGOs; academic institutions; and representatives of the private sector.

3. **Observations.** Time should be dedicated to observations in the field and to taking testimony from unofficial actors, both of which can reveal needs and problems that might otherwise be overlooked.

4. **Other.** Collect and provide photographs and other documentation that contribute to illustrating the principal findings and/or supporting the conclusions of the assessment.

Scope of a Housing Sector Assessment

Topic	Issues
A. Overview of the disaster and disaster zone	
	1. Present a list of the disaster-affected zones identified by local government jurisdiction.
	2. Analyze the socioeconomic characteristics of the affected area, including income levels, economic base, quality of major and basic infrastructure.
	3. Provide data for the disaster zones on distribution and type of housing and infrastructure damage, and numbers of housing units and population affected, by income level and other relevant social characteristics.
	4. Provide maps of the disaster zones showing distribution of affected infrastructure, housing units, and population.
	5. Analyze and describe the legal and institutional framework that defines the roles and responsibilities of the relevant agencies in reconstruction. Describe the lead agency responsible for reconstruction and any specific tasks identified for it related to local reconstruction. Include in an annex any relevant degrees, policy statements, announcements, etc.
B. Policy frameworks and organizational arrangements	
General	1. With particular emphasis in the zone affected by the disaster, provide an overview of the roles and responsibilities of central, local, and any intermediate levels of government; the state of both operational and fiscal decentralization in the country; and the normal mechanisms for fiscal mobilization and distribution.
	2. Describe the pre-disaster situation of the municipalities, including information on financial, human, and technical capacities.
	3. Provide information on losses and damages of the municipality caused by the disaster, the functioning of the municipality after the disaster, and coping strategies.
	4. Based on the roles and responsibilities defined in the legal framework, analyze the capacity and magnitude of local government to comply with its obligations.

Topic	Issues
Land use planning	5. Summarize the principal elements of land use policy, particularly with respect to planning; land use regulation; subdivision of land; risk management; and the roles of the central, regional, and local governments.
	6. Describe the framework for disaster risk management (DRM) as it applies to land use planning and regulation, and the effectiveness of its implementation.
	7. Identify specific land use issues caused by the disaster, including the need for relocation of housing or infrastructure.
Housing sector	8. Summarize the principal elements of housing sector policy, particularly with respect to housing construction and financing; and the roles of the central, regional, and local governments.
	9. Analyze government policy concerning the provision of housing to low-income and vulnerable populations, including any subsidy programs or direct provision efforts that might be relevant to the reconstruction process.
	10. Analyze and show graphically the normal process for land development and for both single- and multi-family housing construction, identifying common bottlenecks.
	11. Describe the procedures for approving and issuing building permits for housing construction and improvement, especially as it relates to housing quality and DRM in housing design, materials, siting, etc., and the effectiveness of its implementation.
	12. Analyze the engineering and non-engineered construction practices normally used in housing in the disaster zone, the practices for contracting construction, and the capacity of the construction industry.
Local infrastructure	13. Summarize the principal elements of policies that govern the provision of basic local infrastructure, particularly with respect to operations and financing (both capital and operational), and the roles of the central, regional, and local governments or other entities.
	14. Analyze the requirements and/or any programs already contemplated for post-disaster infrastructure reconstruction and the suitability of these programs for the reconstruction of local infrastructure in the communities affected by the disaster, for both *in-situ* reconstruction and relocated housing.
Land ownership and tenure	15. Summarize the principal elements of policies that govern the provision and ownership of land; especially with regard to the state of the private market; the formal and informal institutional arrangements for sale, titling, registry, and inheritance of land; and the roles of the central, regional, and local governments or other entities.
	16. Provide a typology of the official legal land tenure options in the country.
	17. Describe particular local socio-cultural customs regarding land ownership and titling and any problems with land ownership and tenure commonly experienced before the disaster.
	18. Analyze the impact of the disaster on these problems and any additional land tenure problems that have emerged since the disaster. This analysis in this and the prior item should cover problems related to the following issues: collective ownership; legal and illegal possession of private land; occupation of public land; tenancy, including problems related to inheritance and death from the disaster; land titling; land registration; loss of land from the disaster; rural versus urban land; and proof of ownership, including loss of records at the household or municipal level, among others.
Housing finance	19. Analyze and describe the systems used to finance housing construction by different social classes and for both single- and multi-family housing and identify the impact of the disaster on this system. Include the use of property insurance if a system exists in the country.
	20. Describe and analyze any financial assistance strategies announced or being contemplated by government to facilitate post-disaster housing reconstruction. Analyze the effectiveness of the strategies and their likely differential impact by type of housing, social class, or other relevant factor.

C. Post-disaster reconstruction process

1. Present a chronological summary of the concrete steps that have been taken to provide temporary or transitional shelter and/or to mitigate the housing-related impacts of the disaster on the affected population, identifying the agency responsible and the source of funding.

2. Describe the coordination mechanisms established between the central, regional, and local governments to organize the reconstruction program or to address land tenure issues.

3. Analyze intermunicipal linkages and describe collaboration among different municipalities, noting whether these bonds existed before the disaster or were created as a response to it.

4. Summarize the role of NGOs in the shelter sector and their anticipated role in reconstruction.

5. Describe any financial strategy for housing and community reconstruction announced at the national or local level and analyze its implementation to date.

Presentation of Findings and Recommendations

For each topic above, the consultants should provide a systematic summary of their findings and corresponding short- and medium-term recommendations that will improve the outcomes of the housing reconstruction program. The recommendations should be grouped in the way that the consultants believe will make them the most understandable during the review process and, in the final report, most useful for implementation. Once subject to an initial review, the recommendations should be presented in the final report as a work plan that identifies both the sequence of activities and the party or parties responsible for carrying them out. The work plan should include an initial budget for the implementation of the activities recommended.

The work plan and budget form the basis of an ongoing dialogue between government and the organizations that are providing financial support to the reconstruction program.

Expected Results and Outputs

The principal output is an in-depth housing sector and land tenure assessment for the disaster-affected area that contributes to a comprehensive understanding of strengths and limitations that are likely to influence the post-disaster housing reconstruction process, accompanied by related recommendations regarding policy and operational reforms that should be implemented in the short and medium term. In the initial report, the consultants will present their strategy, plan, and schedule for the consultancy. The assessment should be presented in draft and final forms.

Time will be of the essence in carrying out this assessment. The following schedule allows the consultancy to be completed in approximately two months. The following time intervals are ambitious, but can be adjusted, depending on the particular situation. Outputs will include:

- an initial report, within 7 days of the contract, in which the consultants present any recommendations for modification of the scope of work as well as a work plan and schedule for the presentation of outputs;
- a draft report, presented within approximately 21 days of the acceptance of the initial report; and
- a final report, presented within the earlier of 21 days of the receipt of comments on the draft report from the party or parties responsible for overseeing the assessment or 30 days of the presentation of the draft report.

The draft and final reports should be presented along with an executive summary or abbreviated version that can be widely circulated, in a language and format easily understandable by stakeholders.

An effective review process will help guarantee the success of the consultancy, and the consultants should take an active role in carrying it out, with assistance from government and the sponsor of the consultancy. This may entail various meetings with government, community, and other stakeholders; use of information technology; or other means to ensure wide distribution of the draft report and collection of feedback. Meetings may also be required once the report is finalized to more widely disseminate the findings and recommendations.

Annex 1 Endnote
1. The assessment methodology proposed here is based on *Land Ownership and Housing, Final Report* (Informe Final, Tenencia de la Tierra y la Vivienda), conducted in Peru to analyze the effect of the Ica/Pisco earthquake in 2008 by Centro de Estudios y Promoción del Desarrollo, under the supervision of UN-HABITAT and in collaboration with the Department for International Development and the Ministry of Housing, Construction and Sanitation.

For access to additional resources and information on this topic, please visit the handbook Web site at www.housingreconstruction.org.

CHAPTER 2: ASSESSING DAMAGE AND SETTING RECONSTRUCTION POLICY

41

A housing damage assessment is the necessary first step that leads to the eventual reoccupancy of buildings and that supports decisions about providing other housing solutions after a disaster. The assessment process is made up of a predictable set of activities, and procedures for a number of them can be established ahead of the disaster in order to speed up the initiation of the post-disaster housing damage assessment process.

Beside demonstrating to citizens that the recovery period is beginning, housing inspections serve other purposes, including (1) **public safety:** identify whether houses can be fully or partially occupied, or must be vacated until reconstruction takes place (generally the result of a separate housing safety inspection; see box, below); (2) **planning:** use the results to quantify the funds, time, and other resources required for recovery, particularly when damage to housing makes up a large component of reconstruction; (3) **technical:** provide information about the specific types of damage that have been

sustained and therefore the types of technical interventions, technical expertise, and training that will be required in reconstruction; and (4) **economic and social:** provide data on the impacts of the disaster at the household level.

Developing an appropriate methodology for housing damage assessment is one of the most critical aspects of the post-disaster response.[1] The process must be made transparent and participatory to establish trust with the affected community and to ensure that local knowledge is fully incorporated. It should contribute to disaster risk reduction (DRR), social inclusion, and gender neutrality. The tools should be tailored to local conditions and be designed to ensure reliability and accessibility by the affected population for both collection and review of data. The approach described here also has the benefit of providing a view of the situation from various perspectives. While the content and sophistication of the assessment tools will vary from one disaster to another, it is recommended that all of the following tools be employed in most cases.

Housing Damage Assessment Tools

Tool	How the tool is applied	Output
1. Initial reconnaissance walk	The initial reconnaissance entails a walk through the affected area to get a general sense of the type, extent, and range of damages. The intelligence gathered at this stage will help in the design of the household survey instrument and the damage classification system. ***Who does it? Assessors with engineers, local officials, and community members.***	Initial impression of types and extent of damage.
2. Habitat mapping[2]	Use habitat mapping to create a "bird's-eye" view of the disaster damage based on local information by identifying each house, locating it geographically, and providing an initial categorization of damage. The map shows how the damaged houses relate to each other and to public buildings and common areas. Mapping can be carried out using any technology, from hand drawing to high-resolution GIS data, so long as the needed information is gained, although local information will be lost by using only a high-tech approach. Information from the habitat map should be transformed into a list that is cross-checked against a cadastre or the civil registry database. One mapping technology can also be used to validate another (artisanal mapping against the cadastre or GIS data). In a community with caste or other social distinctions, this activity can be conducted by sector and aggregated later. ***Who does it? Trained assessors, some of whom may be local officials and/or community members.***	Visual representation of location of damaged and undamaged houses and initial damage categories. List of properties, addresses, and relation to built environment.
3. Village transect[3]	Use the village transect to identify patterns of housing damage and relate the damage to settlement patterns, the local geography, environmental features, and other land uses. Elevation drawings or other visual tools can be used to convey the degrees and types of damage as they relate to these features. This information is used to make decisions about environmental management, as well as relocation, resettlement, and the organization of the reconstruction process. ***Who does it? Trained assessors together with community members.***	Site-specific data and relation of damage to environmental features and land uses.

Tool	How the tool is applied	Output
4. Household-level survey	The household-level survey provides data for both administrative purposes (tenure of property, family characteristics, category of damage) and technical purposes (housing materials, location and specific nature of damage, potential for repair). These data are collected on a standardized form tailored to the disaster, and the data are later entered into a database for the project. Several examples of standardized short and long forms are available.[4] As part of this process, building damage levels are assigned. A wide range of persons can conduct the survey, if properly trained. However, even if engineers, architects, or building inspectors are brought in to conduct the surveys (their involvement is strongly recommended), they must be sufficiently trained and tested on the use of the survey instruments to ensure consistent results across surveyors. Involving in the surveying those who will later train builders is strongly recommended. The assessment must explain the physical mechanisms that caused the damage in order to provide data for reducing the vulnerability in designing reconstruction. ***Who does it? Trained assessors (chartered surveyors, engineers or architects) together with local officials and/or designated community members.***	Detailed property and household data.
5. Photographic documentation	Create a photographic database of each damaged house, ideally with the owner present in the photo. This helps to validate other data and can serve as the baseline for a visual monitoring system for the reconstruction process.[5] ***Who does it? Photographers trained in the documentation process (can be local).***	Visual documentation of damage at the household level.
6. System to number, classify, and label buildings	■ If no **numbering system** exists for lots in the affected communities, create a simple temporary numbering system for the purpose of managing the reconstruction process and assign numbers to houses during the household survey. ***Who designs it? Local officials with community input.*** ■ Develop the **classification system**[6] for levels of damage and train the surveyors in its use. Generally, there should be no more than three categories. The surveyors should be sufficiently trained and tested in the use of the classification system to ensure consistency in its application. ***Who designs it? Engineers/building surveyors with local input.***	Universe of numbered and classified houses.

Building Safety Inspections

Building safety inspections are a public safety measure that is taken very soon after a disaster, to reduce the risk of death and injury to users, residents, and passersby due to building collapse, falling interior or exterior materials and equipment, or other unsafe conditions. They are more necessary in urban contexts where population is dense and buildings have multiple stories.

A placard is commonly affixed to each house once it has been inspected. The placard shows that the house has been surveyed and warns residents and others if there are limitations on its use. A common labeling system uses green/yellow/red placards corresponding to the level of risk. The use of local language and/or visuals will be necessary in contexts where illiteracy or multiple languages are found. Public officials or private sector volunteers (engineers, inspectors) may carry out the inspections, but, as in the case of damage assessments, inspectors should be sufficiently trained to produce predictable results.

Safety inspections are provisional and are not meant to provide information about the value of the damage or the building's potential for being repaired. This information is gained during the housing damage assessment and/or later engineering studies.

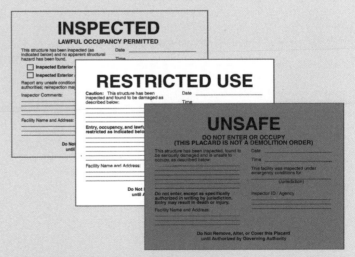

ATC-20/45 Post-Disaster Building Placards

Source: Japan Council for Quick Inspection of Earthquake Damaged Buildings, n.d., *Post-Earthquake Quick Inspection of Damaged Buildings.*

Next Steps

It is important that the data collected via the assessment process be properly validated from secondary sources, including through consultation with the residents and owners. Given its potential value, the data should be entrusted to professionals in data management to ensure that their reliability and safety are preserved. (Remember that because this data will form the basis of assistance schemes, there may be incentives to access and manipulate it, especially once the assistance scheme is announced.)

Depending on the construction technology in the area, engineering and architectural expertise will likely be needed to translate the assessment data into estimates of time and materials required to carry out at least minimum safety repairs. In addition to repairs, the program may cover retrofitting of buildings at risk of future damage. This work should be specified and the costs estimated as well.

Numerous critical activities can be initiated once the housing assessment has been conducted, the results analyzed, recommended DRM measures identified, and cost estimates made. These include, among others:

- decisions regarding the reconstruction approach that government will promote, and the need for relocation and for transitional shelter options to be provided;
- design of the financial assistance strategy;
- determination of technical assistance requirements for builders;
- design of the training program for builders and construction inspectors; and
- design of the communications plan related to the assistance program and DRR measures.

Chapter 16, Training Requirements in Reconstruction, explains how housing damage assessment data are used in developing training for builders.

The housing assessment process does not necessarily eliminate the need for individual homeowners to hire engineers, contractors, or both to provide specifications and cost estimates for their specific reconstruction projects, particularly for engineered buildings.

Preparing for the Next Disaster

In anticipation of a future disaster, central and local governments can establish many of the tools used in the housing damage assessment and safety inspection processes, including mapping and assessment methodologies, design of survey instruments, design and printing of placards, procedures and systems for the management of statistical and photographic databases, and a reconstruction monitoring system.

Annex 2 Endnotes

1. Vivek Rawal and Dinest Prajapati, 2007, "Assessing Damage after Disasters: A Participatory Framework and Toolkit" (Ahmedabad: Organization for Development Education [UNNATI]), http://www.unnati.org/pdfs/books/damage_assessment_toolkit.pdf. The UNNATI tool kit also provides methodologies for assessing damage to community infrastructure and the environment, and impacts on human life, livelihoods, health, and psycho-social status. This "How to Do It" section covers only housing.
2. Common participatory appraisal methods can be applied in carrying out this activity.
3. A transect is a line following a route along which a survey is conducted or observations are made. A transect is used to analyze changes in human and/or physical characteristics from one place to another. An urban transect usually follows one or more streets and will show changes in land use; the nature of buildings, such as houses and shops; or features such as schools, churches, community centers, and parks. A rural transect might follow a road, a section line, or a stream, and may show the kinds of crops in adjoining fields, farm buildings, vegetation, or changing features along a riverbank. For an explanation of the use of the transect in urban planning and zoning, see http://www.newurbannews.com/transect.html.
4. See Applied Technology Council, "ATC-20 Procedures for Post-Earthquake Safety Evaluation of Buildings" and companion "ATC-20-1 Field Manual: Post-Earthquake Safety Evaluation of Buildings, Second Edition," "ATC-20-2 Rapid Evaluation Safety Assessment Form," ATC-20-2 Detailed Evaluation Safety Assessment Form for Earthquake Assessment"; "ATC-45 Field Manual: Safety Evaluation of Buildings after Wind Storms and Floods"; "ATC-45 Rapid Evaluation Safety Assessment Form"; and ATC-45 Detailed Evaluation Safety Assessment Form, https://www.atcouncil.org/index.php?option=com_content&view=article&id=63&Itemid=80.
5. A good example of a household-level system for monitoring reconstruction is the Yogyakarta, Indonesia, reconstruction. Java Reconstruction Fund, "Community-Based Settlement Reconstruction and Rehabilitation Project for NAD and Nias," http://www.rekompakjrf.org (in Bahasa), and "Progress Report 2008, Two Years after the Java Earthquake and Tsunami: Implementing Community Based Reconstruction, Increasing Transparency," http://www.javareconstructionfund.org/ducuments/pdf/2008-07-07_JRF-2nd%20Progress%20Report_ENG.pdf.
6. See Chapter 9, Housing Design and Construction Technology, for a discussion of damage categories.

3 COMMUNICATION IN POST-DISASTER RECONSTRUCTION

Guiding Principles for Communication in Reconstruction

■ Effective communication in a reconstruction project is not about what governments and project managers "say," but what beneficiaries "hear."

■ Two-way information flow builds trust, consensus, and active participation, key factors for positive outcomes in development programs, and limits the potential for setbacks and misunderstandings.

■ An understanding of people's perceptions is crucial to designing a communications strategy since these perceptions can dominate behavior, whether or not it seems rational to an outsider.

■ The cultural and social context affects communications. Inadequate or improper understanding of this context can create risks to project implementation.

■ The largest benefits from communication are realized when it is made an integral part of a development or reconstruction project from the first day.

■ Communication experts should be at the table when decisions about reconstruction are made, giving them access to the information they will need to develop the external messages that support the desired outcomes.

■ The communication campaign is always a work in progress that will need to be adapted as additional input is received from stakeholders and results on the ground are monitored and evaluated.

This Chapter Is Especially Useful For:

■ Policy makers
■ Lead disaster agency
■ Communications specialists

Introduction

The task of rebuilding homes and communities is complex, challenging, and fraught with potential pitfalls. In post-disaster situations the status quo shifts constantly, a challenge that makes strategic communication a crucial element in the response and reconstruction environment. Two-way information flow facilitates recovery and limits the potential for setbacks and misunderstandings. Good communication also helps ensure understanding and buy-in from governments, agencies involved in reconstruction, and the affected population.

The messages that governments and project managers send out to the affected community about reconstruction have less influence over how the community behaves than the messages these communities receive, whether from government or other sources. In other words, if project leaders and communication specialists do not engage stakeholders in the process, they will not be able to formulate messages that will be understood by the people they want to help and the results may be unpredictable.

Strategic communication builds trust, consensus, and active participation, key factors for positive outcomes in development programs. It promotes credibility, transparency, legitimacy, and ownership for the project and ensures that the right messages are reaching all relevant stakeholders.[1] Particularly in a post-disaster situation, good communication is the foundation for acceptance, sustainability, and mutual understanding when rebuilding people's lives. This chapter shows why communication should be initiated as early as possible in reconstruction projects and provides project managers, partner organizations, and governments with tools and guidelines for development and carrying out a successful communications strategy. The communications strategy used after the 2005 North Pakistan earthquake is use to illustrate many of the points made in the chapter.

1. World Bank Independent Evaluation Group, 2006, *Hazards of Nature, Risks to Development*, (Washington, DC: World Bank), p. 116, http://www.worldbank.org/ieg/naturaldisasters/docs/natural_disasters_evaluation.pdf.

For access to additional resources and information on this topic, please visit the handbook Web site at www.housingreconstruction.org.

45

Key Terminology

Communication	**Communication** encompasses all forms of human interactions, from the interpersonal to the mediated, and from the one-way linear flow to the two-way dialogic process.[2] For development purposes, communication components include (1) external communication, (2) media relations, (3) grassroots communication, (4) institutional coordination, (5) capacity building, (6) community development, and (7) coordination with program implementation units.
Types of communication assessments	A **Communication-Based Assessment (CBA)** is an assessment in any sector that uses a variety of communication techniques to detect political risks, contextual issues, and perceptions in that sector that are not easily recognized by a normal assessment. The World Bank emphasizes the need for a CBA at the beginning of the project cycle. Although a CBA can be performed at any stage of the project, its value is highest when conducted early. A CBA uses two-way communication techniques (dialogue, focus groups, open questions, discussion groups) and generally provides qualitative findings. It takes about two weeks to conduct. A **Communication Needs Assessment (CNA)** is carried out to analyze the communication sector and understand its capacity and common practices. It focuses only on the media environment, infrastructure, communication policies, capacities, gaps, information flow, and networks. A CNA is part of a CBA.

Key Decisions

1. **Government** should decide on the lead agency to develop and coordinate the post-disaster communications strategy and assign staff to carry it out. Other **public agencies** and **levels of government** may also designate focal points for communications activities.
2. The **lead communications agency** should decide with the **lead disaster agency** whether there is a need for technical assistance or institutional strengthening in communications, how the communications strategy will be financed, and whether and how to mobilize additional resources.
3. The **lead communications agency** should decide with the **lead disaster agency** whether communications should be included as an element of the initial post-disaster assessment.
4. The **lead communications agency** should decide what assessments will be conducted before defining the communications strategy.
5. The **lead communications agency** should confer with **key stakeholders** and the **local private communications sector** to agree on the role of the community, local governments, nongovernmental organizations (NGOs), and the private sector in defining and carrying out the communications strategy.
6. **Agencies involved in reconstruction** and **other key stakeholders** should agree on the outcomes that are being sought from the community (disaster risk reduction [DRR], changes in construction practices, community participation, etc.) and on the messages and forms of communications that should be used.
7. The **lead communications agency** should collaborate with **agencies involved in reconstruction** and **other stakeholders** to design the monitoring and evaluation (M&E) plan for the project communications strategy and agree on feedback mechanisms to be used during reconstruction.

Technical Issues

One-Way versus Two-Way Communication

Although the value of communication in development projects is widely recognized, it remains an underutilized tool that often focuses too narrowly on informing people rather than communicating with them.[3]

Two divergent conceptions of communication predominate in the field.

- One-way communication: the practice of disseminating information. Messages are put out to inform recipients about the reconstruction process.
- Two-way communication: a model that allows project managers, governments, and all other stakeholders to communicate with each other about the reconstruction process.

2. Paolo Mefalopulos, 2008, *Development Communication Sourcebook, Broadening the Boundaries of Communication* (Washington, DC: World Bank), p. 8, http://siteresources.worldbank.org/EXTDEVCOMMENG/Resources/DevelopmentCommSourcebook.pdf.
3. Paolo Mefalopulos, 2008, *Development Communication Sourcebook, Broadening the Boundaries of Communication* (Washington, DC: World Bank), p. 8, http://siteresources.worldbank.org/EXTDEVCOMMENG/Resources/DevelopmentCommSourcebook.pdf.

These may seem to be opposing approaches, but they're not. In post-disaster situations, the need for information dissemination and dialogue are both pressing. Project managers and government officials should agree as early as possible on a communications strategy that includes–as appropriate to the communications culture of the location–both one-way and two-way elements.

Communication should also be viewed as contributing to other goals of the reconstruction program, including transparency, accountability, good governance, community participation, consensus, and trust-building and as mitigating risks, such as corruption, excess bureaucracy, and political and reputational risks for agencies such as the World Bank and government.[4] The 🖳 case study on the Nation of the Cree, below, demonstrates how inadequate communications can delay a development project.

EARL KESSLER

Communication Capacity within Government

Communications are used in various ways by government and the responsibility for communications may be found in various locations within the bureaucracy. The most visible locus for communications may be the public relations function of the Office of the President. Effective post-disaster communications is less about public relations and more about social communications; that is, meant not to simply publicize or create impressions, but to enlist certain groups to cooperate or change behavior. This distinction should be kept in mind when the lead communications agency is designated.

Ideally, the lead communications agency will have the internal expertise to coordinate all communication activities, including assessment, strategy definition, and implementation. However, assistance may be needed. Resource can include staff seconded from international organizations, or experts hired externally. The World Bank has development communication specialists who can help government conduct the CBA or CNA and define the communications strategy.

Communication experts should be part of the reconstruction decision-making process. They should develop protocols for communication with the affected population and should have access to the information that will allow them to develop the external messages that will support the desired outcomes.

The lead communications agency may also have a role in facilitating information flow within government and among government, donors, and other agencies involved in recovery and reconstruction. By developing protocols for government communication with partners and maintaining institutional relationships, the agency can help improve the consistency among institutions of both the messages to the affected community and the actual implementation on the ground.

It is important that government view the communication campaign as a work in progress that will need to be adapted as feedback is received from stakeholders and results are analyzed. The lead community agency should establish a knowledge-management system to process this feedback, which is then used to support timely corrective actions being taken on the ground.

Case Study: 2005 North Pakistan Earthquake, Pakistan

Assessing the Cultural Context before Defining Communications Strategy

Communities affected by the North Pakistan earthquake were spread out over 20,000 sq. km. of mountainous and rough terrain. Most communication infrastructure, including radio and television, was damaged or destroyed. Within weeks, the Earthquake Reconstruction and Rehabilitation Authority (ERRA) and the donor community realized that a large communication effort was necessary to start an efficient owner-driven rural housing reconstruction project. A CBA was conducted by ERRA with support from a communication specialist working with the World Bank-administered Water and Sanitation Program in India. The CBA concluded that people would need to be motivated to rebuild their lives and would have to be sensitized to new and safer methods of building homes in this disaster-prone area. Survivors were rooted in a very traditional and—from an outsider's point of view—conservative lifestyle. Traumatized by high levels of mortality and destruction, people feared that their value system was also threatened. The reconstruction strategy, therefore, had to address prejudices and fears over "new ways."

Source: Vandana Mehra, M. Waqas Hanif, and Moncef Bouhafa, 2008, "Strategic Communications and ERRA: Overall Approach," (Presentation, February 28, 2008) and conversation with Vandana Mehra, May 2009.

4. Nobuya Inagaki, 2007, "Communicating the Impact of Communication for Development. Recent Trends in Empirical Research," (working paper No. 120, Washington, DC: World Bank), http://www-wds. worldbank.org/external/default/ WDSContentServer/WDSP/IB/2007 /08/10/000310607_20070810123 306/Rendered/PDF/405430Comm unic18082137167101PUBLIC1.pdf; *and* Larry Hass, Leonardo Mazzei, and Donal O'Leary, 2007, "Setting Standards for Communication and Governance. The Example of Infrastructure," (working paper, No. 121, Washington, DC: World Bank), http://www-wds. worldbank.org/external/default/ WDSContentServer/WDSP/IB/2007 /08/10/000011823_20070810125 218/Rendered/PDF/405620Setting 018082137169501PUBLIC1.pdf.

Communication-Based Assessments

People will likely be affected by the disaster in many ways: lost homes; injured or dead family members, neighbors, and friends; destroyed livelihoods; food-insecurity and suffering from mental and physical health problems. These experiences will affect people's needs, opinions, and perceptions, which in turn will affect their ability to participate in the programs that will help them recover.

A CBA is used to identify knowledge, perceptions, fears, and expectations of main stakeholders, and contextual and situational knowledge, much of which cannot be easily detected with other forms of assessment. An understanding of stakeholders' perceptions is crucial to designing a communications strategy since these perceptions can dominate behavior, whether it seems rational to an outsider, or not. A CBA captures this qualitative information and can help government and other agencies tailor the communications strategy for the reconstruction program. But more than that, it also provides knowledge that can be used to improve the design of reconstruction projects. Using this type of information helps ensure broader impact and sustainability of interventions and helps mitigate political and reputational risks. See 📖 Annex 2, Culture and Other Contextual Factors in Communication.

The critical areas to analyze and understand in order to develop an effective communications strategy are listed briefly in the following table. For detailed instructions on conducting a CBA, see 📖 Annex 1, How to Do It: Conducting a Communication-Based Assessment.[5]

Area to analyze	Considerations
Political risk, challenges, and opportunities	Includes such issues as perceptions of government's disaster response and plans, and perceptions of how government and other service providers have performed since the disaster
Stakeholder analysis	Inventory and analysis of stakeholders who will be directly and indirectly affected by the reconstruction program
Media, communications channels, and local capacity[6]	Analyzes the range of ways groups communicate formally with one another in the society being assessed, as well as issues related to local capability and acceptability of media institutions
Social and participatory communication	Provides an understanding of informal systems and community communications practices

Who should conduct the assessment? A CBA can be conducted by communication specialists within the lead agency, outside consultants (hired locally or internationally), or qualified members of a donor project team. National experts with communications experience in prior emergencies may be available, even if they are no longer acting in an official capacity.

Timing the assessment. World Bank research shows clearly that the largest benefit of strategic communication is gained when communication is considered an integral part of the project or program from its inception.[7] Ideally, a CBA is conducted as an integral part of the initial multi-sector damage and loss assessment. When the CBA is conducted at a later stage or separately from the initial assessment, governments and project managers should be prepared to make revisions later so that the recommendations of communication specialists can be implemented. Various assessment methodologies are described in 📖 Chapter 2, Assessing Damage and Setting Reconstruction Policy.

Time pressure will be great and key actors may consider communications a peripheral concern. Be aware that a focus on speed and physical damage can be costly if perceptions, knowledge, attitudes, and expectations are not taken into account.

Designing the Communications Strategy

Once the CBA is completed, the objectives of a communications strategy should be formulated. The findings of the CBA may produce a numerous objectives that need to be pursued to achieve the project goals. However, they might not all need to be pursued at the same time.

The communications strategy for a program or project should answer the following questions: Which audiences need to be reached and which stakeholders need to be engaged? What is

5. Paul Mitchell and Karla Chaman-Ruiz, 2007, "Communication-Based Assessment for Bank Operations," Working Paper No. 119 (Washington, DC: World Bank), pp. 20-30, http://www-wds. worldbank.org/external/default/ WDSContentServer/WDSP/IB/20 07/08/10/000310607_2007081 0124552/Rendered/PDF/405610 Communic18082137165701PUB LIC1.pdf.
6. This element of the assessment is the CNA mentioned in the key definitions.
7. Paolo Mefalopulos, 2008, *Development Communication Sourcebook, Broadening the Boundaries of Communication,* (Washington, DC: World Bank), p. 9, http://siteresources.worldbank. org/EXTDEVCOMMENG/Resources/ DevelopmentCommSourcebook.pdf.

the required behavior change? What messages are appropriate? Which channels/tools of communication will be most effective? Over what time frame will implementation take place? How will implementation of the communication strategy be monitored and evaluated? Implementation includes all activities necessary to engage stakeholders (for example, design, production, and distribution of communication materials and training and hiring of staff).[8]

Who should design the strategy? The communication capacity of the lead communications agency needs to be evaluated and, if necessary, complemented by hiring staff or acquiring additional services, whether of outside consultants or staff on secondment from other agencies. An experienced strategic communication specialist should draft and design the strategy in cooperation with project managers and local counterparts with experience in the disaster field. It is advisable that the communication specialists who conducted the assessment be involved in designing the strategy.

Throughout the design phase, the lead communications agency should consult closely with communities to determine whether the strategy is addressing the right audiences and stakeholders with the right messages. The steps and associated activities and tools in the design of the communications strategy are shown in the following table.

Throughout the design phase, the lead agency should consult closely with communities to determine whether the strategy is addressing the right audiences and stakeholders with the right messages. The basic steps in designing the communications strategy are the following.

- Define and formulate the objectives (SMART: specific, measurable, achievable, realistic, and timely) and then transform those objectives into outcomes.
- Define primary and secondary stakeholders/audiences.
- Evaluate whether the changes sought are related to awareness, knowledge, attitudes, behaviors, mobilization, collaboration, or mediation. See 🖳 Chapter 16, Training Requirements in Reconstruction, to understand the importance of communications in builder training programs.
- Define whether the communication approaches/tactics are linear, interactive, or both.
- Select the appropriate media from among those available.
- Design key content/message and determine the most effective way to package it.
- Define realistic results for the strategy and develop a means to monitor and evaluate progress toward their achievement.

Implementing the communications strategy. A Communications Action Plan (CAP) guides the implementation of the communications strategy. The CAP covers institutional coordination, media relations, grassroots communication, capacity building, external relations, and coordination with program implementation units (environmental, resettlement unit, etc.). The CAP details the production, training, hiring, budgeting, and timing of all initiatives. The nature of a post-disaster intervention will lead to overlaps and doubling of messages. Lessons learned show that too much communication is better than too little. However, retracting or correcting information can be difficult. "Silence" promotes rumors that can be exploited for political or economic reasons, which can lower community participation levels. The 🖳 case study, below, on three earthquakes in Iran shows how a carefully planned community-based communications program overcame public unrest over the perceived lack of information from government.

A system to collect feedback should be implemented as part of the plan. Messaging, audiences, and tools will have to be adapted according to feedback from beneficiaries and implementing agencies. Agencies should also share the feedback they are getting. In the case of the Pakistan earthquake, the need for internal communication was quickly recognized, and a meeting schedule, knowledge management cells, and workshops were held to promote unity and synergies among all partners.

Who should implement the strategy? Depending on the institutional situation, a wide set of actors may be involved in implementing the communications strategy and tasks should be distributed to take advantage of the capabilities of various actors while being cost effective. The lead communications agency will be in charge of the overall approach and budget for the campaign. This agency is likely to define scopes of work for any outside services hired and to oversee procurement.

When task managers were asked what can be done to increase ownership of disaster prevention and mitigation components in natural disaster-related projects, they mentioned "Develop good communication strategies" more often than any other action.

World Bank Independent Evaluation Group, 2006, Hazards of Nature, Risks to Development.

8. Paolo Mefalopulos, 2008, *Development Communication Sourcebook, Broadening the Boundaries of Communication* (Washington, DC: World Bank), p. 129, http://siteresources.worldbank.org/EXTDEVCOMMENG/Resources/DevelopmentCommSourcebook.pdf.

The Communications Action Plan for the 2005 North Pakistan earthquake[9]

Activity	Details of approach
1. Review and confirm objectives	Motivate and make people aware of available assistance including eligibility Ensure that people build better/earthquake resistant houses Ensure that people know about training and information tools to qualify for assistance
2. Review and confirm primary and secondary audiences	Affected population, households
3. Activities and Approaches *What activities are needed (media production, message design, air time booking, translation, etc.)*	Advertisement in print media and electronic media Media coverage Media PR Road shows (live programs in affected areas) Billboards, posters, brochures (multilingual) Social mobilization with events at mosques, etc Helpline, website Information kiosks Grievance redress mechanism (helpline numbers -- an excellent two-way communication tool with beneficiaries providing valuable feedback, resulting in resolution turnaround within a week) Village reconstruction committees Workshops and seminars
4. Resources needed (human and material *Experts in audiovisual design and production (experts in training, related materials, etc.)*	Post-Earthquake Public Information Campaign: US$1.8 million (little more than 1% of the housing reconstruction total budget (US$1.4 billion) as of November 2007 WSP/World Bank communication specialist provided client support to kick start a public information campaign (including hiring firm for design and dissemination) The World Bank communication specialist worked with counterparts of the ERRA the central institution that was tasked with the response Outsourcing versus capacity building. Due to limited time (approaching winter and thousands still without shelter) the implementing agency hired skilled communication staff internally A knowledge management unit was established after about one year, headed by a senior manager and 5-6 researchers and writers Training sessions were held for government communications staff in media relations, interviews, case studies, presentation, and community participation
5. Party responsible (action promoter) *Who is the source and initiator for the action?*	ERRA established by the Pakistani government
6. Time frame *Sequence and time needed for each activity*	Phase 1: General messages on policy and rural housing program Phase 2: Motivate and mobilize people to access grants by rebuilding in better ways Phase 3: Advanced messages on training for safe reconstruction and culturally sensitive behavior change
7. Expected results from strategy	Affected Population: ■ New and safer houses ■ Adoption of new building behavior ■ Feeling informed of reconstruction project ■ Developing ownership Government, partners, donors, etc.: ■ Functioning communication protocol ■ Conducting and ongoing communication campaign to support the flagship rural housing program

9. Paolo Mefalopulos, 2008, *Development Communication Sourcebook, Broadening the Boundaries of Communication* (Washington, DC: World Bank), pp. 129-134, http://siteresources.worldbank.org/EXTDEVCOMMENG/Resources/DevelopmentCommSourcebook.pdf.

Depending on the scope of the disaster, the lead communications agency might delegate the distribution of specific messages to local organizations. For example, implementation at the grassroots level could be done by local NGOs partnering with local government or by an advertising agency hired to produce and distribute communications material in specific areas. Communities might be asked to select representatives or form committees that will function as intermediaries with their peers. Religious or tribal leaders may agree to distribute messages in meetings and through social networks.

Case Study: 2005 North Pakistan Earthquake, Pakistan

Ad Hoc Communications Precede a Communications Action Plan

In the aftermath of the devastating 2005 earthquake in Pakistan, ERRA and international donors quickly realized the immediate need for communicating with the surviving population, but also recognized that there was little time for developing a proper plan. The level of destruction and the difficult terrain made it hard to disseminate information and to engage with beneficiaries. The lack of information flow in both directions led people to feel frustrated over a perceived slowness in the response.

To bridge this gap, the Pakistani military used helicopters to reach remote areas to distribute information to beneficiaries and to assist with the initial needs assessment. Later, local NGOs partnered with the firm that was hired by ERRA to produce and disseminate information and assisted with implementation at the grassroots level. When radio and television were restored, the firm implemented all mass media aspects of the campaign. Culturally acceptable "heroes" were developed for educational radio shows. These characters were immensely popular. For example, a wise mason was created for posters, and he became a lead figure for "correct construction." In the early days of the campaign, the radio show addressed concerns of beneficiaries, expressed either to the local authorities or via a help line set up to answer questions.

This case demonstrates how, in some cases, the urgency of a situation does not permit development of a fully sequenced communication action plan. Sometimes initiatives have to be implemented ad hoc, while an actual plan (in this case, mainly the hiring of an advertising firm) is still being developed.

Source: Vandana Mehra, World Bank, 2009, personal conversation.

Human Resources and Professional Services

One of the most challenging tasks in a post-disaster response is finding qualified staff and support services fast enough. The procurement process recommended for long-term development initiatives might need to be adapted to the post-disaster conditions. Single-source selection and direct purchase of services and personnel might be the only viable option, particularly in the beginning of the project. However, other procurement methods can be introduced at later stages in the communications action plan, when there is more time. Procurement procedures for World Bank projects can be found online and are summarized in 🖳 Chapter 23, Procurement in World Bank Reconstruction Projects.[10]

Hiring consultants will be one of the first and most pressing tasks. The terms of references (TOR) must be specific to the disaster and the requirements of the project. Three main guiding principles should apply: (1) the TORs should contain sufficient background information on the project to enable consultants to present responsive proposals, (2) the scope of work should be consistent with the available budget, and (3) the TORs should take into account the organization of the client implementing the communication component and its level of technical expertise and institutional strength.[11] This chapter provides guidance for developing the scope of work.

The very nature of a post-disaster communication project will require personnel to show a high degree of flexibility and willingness to adapt to demanding circumstances. More than in non-disaster projects, the communication specialist to be hired should understand crisis communications, political risk management, and internal communications, and should have coordination skills. Other beneficial skills include stakeholder mapping and engagement, communications for operations, media management, spokesperson/presentation skills, and donor engagement.[12]

Monitoring and Evaluation

The M&E process should be ongoing and should mirror the project cycle. Communications outputs are best measured if there are constant feedback channels that include quantitative and qualitative indicators. For example, it is important not only to measure how many radio spots have been aired, but whether stakeholders have changed behavior and adopted new technologies. Project managers and government authorities should also pay attention to whether the attitudes, perceptions, and

10. World Bank, 2005, *A Toolkit for Procurement of Communication Activities in World Bank Financed Projects* (Washington, DC: World Bank), http://siteresources.worldbank.org/EXTDEVCOMMENG/Resources/toolkitENfinal.pdf.
11. World Bank, 2005, *A Toolkit for Procurement of Communication Activities in World Bank Financed Projects* (Washington, DC: World Bank), http://siteresources.worldbank.org/EXTDEVCOMMENG/Resources/toolkitENfinal.pdf.
12. Samples of TORs and RFPs can be found on World Bank, 2005, *A Toolkit for Procurement of Communication Activities in World Bank Financed Projects* (Washington, DC: World Bank), http://siteresources.worldbank.org/EXTDEVCOMMENG/Resources/toolkitENfinal.pdf.

FEMA NEWS PHOTO

fears that were examined during the assessment have been successfully addressed in the campaign.

Practitioners should establish ways to receive periodic feedback from stakeholders to be responsive to the highly contextual nature of the initiative. Often, the lead disaster response agency does not have the capacity to analyze the information it receives. It is important to set up an internal system of information sharing, ranging from an IT system to regular information exchange meetings that allow for vertical and horizontal flow of input. This will help making necessary and meaningful changes to the project and the communications strategy at an early stage and throughout the project cycle.

Case Study: 2005 North Pakistan Earthquake, Pakistan
Using Beneficiary Feedback for Monitoring and Evaluation

The challenging environment created by the 2005 North Pakistan earthquake and the need for fast dissemination of information made it difficult to develop complex standards and benchmarks for the evaluation process. However, ERRA realized that it was important to monitor feedback to detect information gaps and to adapt its information campaign. ERRA focused on analyzing calls it received at its Islamabad offices from the 24-hour help lines that had been established all over the affected area. The feedback, questions, and comments people provided helped ERRA determine the level of understanding on the side of beneficiaries and, indirectly, whether the information campaign and stakeholder engagement had been effective. The communication team at ERRA updated its Web site and other information material accordingly. The results were shared with program managers who were able to address the issues raised by callers. Very often a query or concern that was voiced by several people would be addressed through a very popular interactive radio show. ERRA also set up an internal knowledge management mechanism that allowed for easy access to information at all administrative levels.

Source: Raja Rehan Arshad, 2008, "Lessons and Experiences from Disaster Recovery in Pakistan," (presentation for "Workshop on Consultations and Strategic Communications in Water and Sanitation Sector in East and South Asia," Bangkok, March 31-April 7, 2008).

Risks and Challenges

- Missing the full picture in conducting the damage and loss assessment by asking only quantifiable questions and using only one-way communication.
- Government focuses on media relations and overlooks the social aspects of communication.
- Leaving post-disaster communications to a central disaster agency that lacks the capacity and skill to design and execute the communications strategy.
- Project teams believe that they know what the affected population wants without asking, and design reconstruction projects undesirable the affected population.
- Assuming that if the basic goal of a reconstruction program (rebuilding homes) is widely accepted, then all other aspects of the project (management, assistance policies, intended behavior change, and reconstruction approach) are widely accepted as well.
- Not understanding the context and contextual factors and how they affect reconstruction (for example, language barriers, perceptions of corruption, ability of religious groups to work together).
- Attempting to implement a communications strategy inconsistent with normal communications culture (for example, highly dialogue-driven in a country with a history of top-down communication).
- Underfunding communication activities.

Recommendations

1. See communications in housing reconstruction as a tool that can improve stakeholder participation and ultimately the suitability of the outcomes.
2. At the same time, realize that two-way communication (dialogue) is not only about achieving the project's objectives but also about giving voice and dignity to vulnerable and marginalized people.
3. Don't allow the urgency to implement to shortchange communications.
4. Adopt a multi-track, dialogue-driven communications strategy, which allows beneficiaries to provide input, ideas, and feedback, rather than employing a one-way (information dissemination) approach.
5. In developing the communications strategy, focus first on the messages that will be effective with people, before selecting media.
6. Adapt communication tools to the targeted audience and its preferred and trusted ways of communicating.
7. Tailor the communications strategy to reflect contextual variables.
8. Incorporate communications as early as possible in the process and sustain it throughout the project cycle. Be willing to redefine and adapt the strategy during the project as results are realized.
9. Ensure that communications within government and with other funders is open and results in a unified message to the affected population.
10. Incorporate feedback about the effectiveness of the communications strategy in a timely manner to improve reconstruction outcomes.

Case Studies
2003 Bam Earthquake, 2005 Zarand Earthquake, and 2006 Lorestan Earthquake, Iran
Community-Based Information Management and Communication

In the aftermath of the 2003 earthquake in Bam, Iran, there was a need for an active exchange of information and viewpoints between the affected communities and local authorities. To that end, the United Nations Development Programme (UNDP) supported a community-based information management and communications initiative. This initiative became particularly important after people's perception of a lack of information on the distribution of relief items provoked demonstrations in front of public offices early in 2004 in Bam. The aim of the communications initiative was to empower the affected communities through participation and enhanced access to information on recovery and reconstruction using information and communication technologies (ICT). Information on government policies and activities, updated damage reports, entitlements, land status, and rehabilitation schemes was made available, using

VICTORIA KIANPOUR

an information Web site in Persian, print and electronic information products, and ICT-based kiosks and information boards located throughout the affected areas. The project produced and published a biweekly newsletter with the help of local volunteers trained as journalists (all of whom have become professional journalists in the area). The Swiss Agency for Development and Co-operation supported the initiative, which was replicated later following the Zarand (2005) and Lorestan (2006) earthquakes, with initial support from the UNDP and subsequent support from the Housing Foundation of the Islamic Revolution. Activities in these cases also included information centers and notice boards, and distribution of such products as a pamphlet on "dos" and "don'ts" before, during, and after earthquakes.

Source: Victoria Kianpour, UNDP Iran, 2009, personal communication, http://www.undp.org.ir/.

2002 Hydro-Quebec vs. Nation of the Cree, Canada
The Cost of Not Communicating

In the early 1990s, after years of disagreement and diverging views over one of the world's largest energy infrastructure programs, the indigenous Cree population of Northern Quebec forced Hydro-Quebec, a leading company in the energy sector, to halt construction all together. According to John Paul Murdoch, Legal Counsel of the Cree Nation, Hydro-Quebec faced construction delays of almost 20 years and had to spend an additional US$268 million to adequately address communication gaps, concerns over mercury pollution, and potential loss of livelihoods to the Cree. Murdoch told an audience at a World Bank Energy Week in 2005 that the failure to communicate properly had become costly for the company. In 2002, Hydro-Quebec and the Cree Nation entered a "New Relationship Agreement" that addresses concerns over safety, economic and social benefits, and a mechanism for a permanent standing liaison committee, paving the way for the project to proceed.

Sources: Paolo Mefalopulos, 2008, *Development Communication Sourcebook, Broadening the Boundaries of Communication* (Washington, DC: World Bank). p. 136; *and* John Paul Murdoch, "The Value of Communications" (conference presentation, World Bank, March 14, 2005), http://irispublic.worldbank.org/85257559006C22E9/All+Documents/85257559006C22E985256FFF007255D2/$File/Mafia_EW05.pdf.

Resources

Hass, Larry, Leonardo Mazzei, and Donal O'Leary. 2007. "Setting Standards for Communication and Governance. The Example of Infrastructure." World Bank Working Paper No. 121. Washington, DC: World Bank. http://www-wds.worldbank.org/external/default/WDSContentServer/WDSP/IB/2007/08/10/000011823_20070810125218/Rendered/PDF/405620Setting018082137169501PUBLIC1.pdf.

Inagaki, Nobuya. 2007. "Communicating the Impact of Communication for Development. Recent Trends in Empirical Research." World Bank Working Paper No. 120. Washington, DC: World Bank. http://www-wds.worldbank.org/external/default/WDSContentServer/WDSP/IB/2007/08/10/000310607_20070810123306/Rendered/PDF/405430Communic18082137167101PUBLIC1.pdf.

Kalathil, Shanthi, John Langlois, and Adam Kaplan. 2008. "Towards a New Model: Media and Communications in Post-Conflict and Fragile States. Communication for Government and Accountability Program." Washington, DC: World Bank. http://web.worldbank.org/WBSITE/EXTERNAL/TOPICS/EXTDEVCOMMENG/EXTGOVACC/0,,contentMDK:21768613~pagePK:64168445~piPK:64168309~theSitePK:3252001,00.html.

Mafalopulos, Paolo. 2008. *Development Communication Sourcebook, Broadening the Boundaries of Communication.* Washington, DC: World Bank. http://siteresources.worldbank.org/EXTDEVCOMMENG/Resources/DevelopmentCommSourcebook.pdf.

Mazzei, Leonardo and Gianmarco Scuppa. 2006. "The Role of Communication in Large Infrastructure. The Bumbuna Hydroelectric Project in Post-Conflict Sierra Leone." World Bank Working Paper No. 84. Washington, DC: World Bank. http://siteresources.worldbank.org/EXTDEVCOMMENG/Resources/wpsierraleoneebook.pdf.

World Bank. 2005. "A Toolkit for Procurement of Communication Activities in World Bank Financed Projects." Washington, DC: World Bank. http://siteresources.worldbank.org/EXTDEVCOMMENG/Resources/toolkitENfinal.pdf.

Key Organizations with Best Practices and Research

Development Communications Evidence Research Network. "Impact of Communications in Development." http://www.dcern.org/.

World Bank. "Development Communication." http://web.worldbank.org/WBSITE/EXTERNAL/TOPICS/EXTDEVCOMMENG/0,,contentMDK:21460410~menuPK:490442~pagePK:34000187~piPK:34000160~theSitePK:423815,00.html.

A Communications-Based Assessment (CBA) for a post-disaster housing reconstruction project will take about two weeks and should be conducted either before or in parallel with other early assessments. Some information relevant to the reconstruction process may be readily available.

To avoid duplicating efforts, cooperate closely with the local and international relief community and government agencies (not only groups working in the shelter field). When the United Nations cluster system has been activated, the Humanitarian Information Centre (http://www.humanitarianinfo.org) will be both a source of information and a platform for sharing information that is collected.

The assessment should identify and analyze all relevant aspects of the social context. An open-minded approach at the beginning of the assessment is crucial for grasping the complexity of the entire situation. ⬛ Annex 2, Culture and Other Contextual Factors in Communication, suggests social factors that form part of the context in which communications takes place, and may be important to consider.[1]

The critical areas that must be analyzed and understood to develop an effective communications strategy and use communications to improve the project design include the following.[2]

Scope of a Communications-Based Assessment

Topic	Issues to analyze
Political risk, challenges, and opportunities	This includes such issues as perceptions of government's disaster response and plans, perceptions of how government and other service providers have performed since the disaster, mechanisms used by government to communicate with stakeholders in general and since the disaster, and key knowledge gaps.
	■ Challenges and risks can include the geography of the affected areas, high mortality, loss of livelihood, large displacement, poverty, resistance to behavior change (introducing new and unknown forms of building techniques), necessity to resettle, unclear land rights, and complex owner-tenant relations.
	■ Obstacles can include real and perceived corruption and mismanagement, lack of income opportunities, lack of credible communication channels, absence of community representatives, non-existing local fiscal capacity for distribution, lack of building material, and lack of know-how for new techniques.
	■ Opportunities can arise from the crisis. In disaster-prone environments, beneficiaries might be open to new technologies and improved building approaches.
Stakeholder analysis[3]	For the communications strategy, an analysis is needed of:
	■ primary stakeholders and audiences (the affected population, household and grassroots representatives, government officials, civil servants, national and international media, civil society, academic institutions, professional groups, religious groups, business community, NGOs, partner organizations, donors);
	■ "hidden" or secondary stakeholders (less-affected non-beneficiaries who might feel overlooked during the project and might act as spoilers);
	■ vulnerable groups (female-headed households, orphans, disabled, chronically ill, the extremely poor, and socially marginalized);
	■ public opinion leaders or allies (societies listen best to their own leaders);
	■ stakeholder perceptions, expectations, attitudes;
	■ socially relevant topics or controversies related to the reconstruction project (relocation, land rights issues, service delivery in new neighborhood, social/tribal/religious fabric of project area, environmental issues, cultural heritage, customs, and livelihoods); and
	■ past and ongoing stakeholder behavior in similar situations or projects.

⬛ **For access to additional resources and information on this topic, please visit the handbook Web site at www.housingreconstruction.org.**

Topic	Issues to analyze
Media, communications channels, and local capacity[4]	This element of the CBA should encompass the range of ways one group communicates with another in the society being assessed, as well as issues related to local capability and acceptability of each, including: ■ communication channels that stakeholders normally use to receive and disseminate information; ■ the degree of trust in each channel; ■ the availability of channels or limitations since the emergency; ■ looking beyond mass media at alternative communications channels (for example, SMS and social media); ■ options for face-to-face communication (particularly for affected population with trust issues and to communicate behavior change messages, for example DRR, environment issues, and new building techniques); ■ capability and experience of media organizations and consultants, including any involved in social communications, social marketing, market research, and public relations; ■ The nature of the relationship between government and the various media; and ■ communications channel, including electronic (TV, radio), road shows (live programs), advertising (billboards, posters, brochures, leaflets), *shuras*, ceremonial and cultural events, media coverage, and mobile phones (text messaging).
Social and participatory communication	Not all communications media are formally organized, and an understanding of informal systems and community communications practices is an important element of the assessment. This includes: ■ existing social communication mechanisms (such as schools, churches, markets, and social interactions); ■ networks (such as religious, tribal, neighborhood, professional, and school); ■ traditional forms of dialogue (such as meetings with elders, religious leaders); ■ formal and informal ways of designating community leaders and representatives; ■ decision-making mechanisms at the community level (are they producing communication products that can be used in a communications program?); ■ beneficiary consultation mechanisms or involvement in development initiatives (current, past, in other areas during the post-disaster phase); ■ prior initiatives to identify interests of or conflicts between community and/or subgroups; existing joint projects or plans of the community; and ■ familiarity with help lines, toll-free alert numbers (to report corruption, misuse, problems).

Expertise Required

A CBA can be conducted by communication specialists within the lead agency, outside consultants (hired locally or internationally), or qualified members of a donor project team. National experts with communications experience in prior emergencies may be available, even if they are no longer acting in an official capacity.

Annex 1 Endnotes
1. See ▣ Annex 2, Culture and Other Contextual Factors in Communication.
2. Paul Mitchell and Karla Chaman-Ruiz, 2007, "Communication-Based Assessment for Bank Operations," World Bank Working Paper No. 119 (Washington, DC: World Bank), pp. 20-30, http://www-wds.worldbank.org/external/default/WDSContentServer/WDSP/IB/2007/08/10/000310607_20070810124552/Rendered/PDF/405610Commun ic18082137165701PUBLIC1.pdf.
3. Stakeholder analysis is also discussed in ▣ Chapter 12, Community Organizing and Participation.
4. This element of the assessment is the CNA mentioned in the Key Definitions section earlier in this chapter.

Misunderstanding the social and cultural context can create risks in reconstruction. This misunderstanding may cause unintended consequences or make implementation more difficult. As a result, the social and cultural factors take on great importance when the communications strategy is being designed.

Both aid agencies and local people may have trouble identifying contextual factors. For outsiders they are difficult to detect; for insiders they are a "given." These factors are not problematic per se; problems only arise when assumptions made by those attempting to communicate with the population or to implement a reconstruction project understand the context to be different than what it really is.

The CBA is the opportunity to identify these factors, evaluate their importance, and understand how they affect both perception and behavior. They should be taken into consideration in communicating with the public and the affected population about recovery and reconstruction. Some of the contextual factors that might be evaluated include the following.

Contextual factors	Examples of how they may affect communications
Peculiarities of the disaster effect, for example: ■ Disproportionate loss of certain social groups ■ Affect of disaster on materials availability ■ Changes in labor market due to migration ■ Disaster history in the region	If many heads of household are lost in the disaster, non-traditional approaches to reconstruction may need to be promoted. If common local building materials are damaged, use of alternative materials will have to be explained and promoted. Repeated disasters may make the population reticent to rebuild, so motivational messages may be needed.
Institutional/governance context, for example: ■ Local/national government relations ■ Degree of sectoral and fiscal decentralization ■ Roles and responsibilities of governmental entities, levels of government ■ Maturity of community organizations ■ Trust in government and perceptions of corruption	A conflictive relationship between local and national governments could produce contradictory messages that confuse the public. Local governments may be suspected of corruption, so accountability measures may need to be improved to give assurance to the population. Newer community-based organizations may not have the credibility in the community to deliver certain information.
Political context, for example: ■ History of ongoing violence ■ Role of political parties ■ Level of social organization or activism	Concerns about violence may discourage community involvement. Opposition parties may politicize the disaster and affect the acceptance of messages. Well-organized communities may move faster than government and perceive later government involvement as "interference."
Sociological context, for example: ■ Demographic factors ■ Relationships of religious groups ■ Class, race, and status relationships of those affected ■ Gender relationships ■ Perception of rights of disabled	In societies where class, race, and/or status are polarized, communication may need to be tailored to specific groups. Members of religious groups may prefer that messages come from their religious body, rather than from government. Men may keep their wives from participating in projects that strongly promote gender equality.
Cultural factors, for example: ■ Cultural practices and values, such as perceptions of time ■ Aesthetic value systems, such as Feng Shui ■ Place of money in cultural life ■ Superstitions ■ Language barriers ■ Perceptions about social change	Cultural differences in the perception of time will affect planning efforts. Relationship to money and beliefs about accepting gifts differ enormously from one culture to another. Individuals may have beliefs about the orientation of houses, position of doorways, etc. that affect their interest in new houses. Resident satisfaction surveys may not reveal families' real opinions. Social judgments about who deserves assistance may be based on intangibles, such as a family's history in the community.

Contextual factors	Examples of how they may affect communications
Economic context, for example: ■ Wealth distribution in disaster area ■ Effect of disaster on economic base ■ Importance of homestead for livelihood ■ Migration and other work/living patterns ■ Role of remittances in local economy ■ Market culture	Cultural perceptions about gift-giving may affect rates of participation in assistance programs. Women may be financial decision makers of household, but not be exposed to communications media that are used. The inability for women to reach markets or to go to markets alone may affect use of assistance strategies, such as vouchers.
Territorial/land use issues, for example: ■ Specifics of disaster location (urban/rural) ■ Access into/out of disaster location ■ Ecological context ■ Legal status of land occupancy of affected population	Expectations about the standard of housing may be quite different in urban and rural communities, even in same country. Messages about land and tenancy need to be fine-tuned to local land-ownership practices. Perceptions about the natural environment vary between cultures, and affect environmental messages.
Housing/community culture, for example: ■ Household decision maker on housing issues ■ Adequacy of housing situation before the disaster ■ Role of communal spaces within and around the settlement ■ Relationship of housing styles and settlement layout to culture or climate	Where women don't attend community gatherings, opinions expressed in meetings may not represent the entire household. Perspectives about suitable housing assistance schemes will vary from one location to another. Localities where income segregation in housing is the norm may not be persuaded to relocate in "mixed income" communities. The disaster may change people's perceptions of the value of vernacular housing, in favor or against.

  SAFER HOMES, STRONGER COMMUNITIES: A HANDBOOK FOR RECONSTRUCTING AFTER NATURAL DISASTERS

4   WHO GETS A HOUSE? THE SOCIAL DIMENSION OF HOUSING RECONSTRUCTION

Guiding Principles for the Social Dimension of Housing Reconstruction

- The housing assistance scheme should support the objectives established for the reconstruction program in the reconstruction policy.
- Each disaster will require its own housing assistance scheme; there is no "one size fits all" approach.
- Decisions regarding eligibility criteria and housing assistance must be objectively applied and transparently disclosed.
- Post-disaster housing policy must consider the situation of people in all categories of housing tenancy, including squatters, although all members of all categories may not receive assistance.
- Assistance schemes should be tailored to different levels of damage. Avoid incentives to exaggerate damage that then result in overpayment.

Introduction

Pre-disaster housing conditions vary widely, from luxurious to ramshackle, but no type of housing is immune to the effects of disasters. In addition, in a post-disaster environment, households have different kinds and levels of resources to rely on for rebuilding; some can rebuild solely with their own resources, while others are totally dependent on government assistance. It may also be beneficial in a post-disaster environment for government to provide assistance to households that weren't even affected by the disaster.

When post-disaster housing assistance is being allocated, policy makers have to address the following critical questions[1]:
1. Who is entitled to housing?
2. What type of housing solution are they entitled to receive?
3. How much housing assistance will they receive?

These questions have no "right" answer. While all post-disaster housing assistance is intended to help recipients solve disaster-related housing problems, the approach must be fine-tuned to the circumstances, culture, and available resources. This chapter provides guidance on the factors to consider in making these decisions and discusses some of the consequences. The discussion is focused principally on assistance to help return housing to a safe and livable condition and is meant to address the needs of the affected population, in all tenancy categories, as shown below.

Tenancy categories	Party normally responsible for reconstruction
House owner-occupant or house landlord	Owner-occupant or landlord
House tenant	Landlord
Apartment owner-occupant or apartment landlord	Owners as a group or landlord
Apartment tenant	Landlord (public or private)
Land tenant	Tenant, unless tenure is not secure
Occupancy with no legal status (squatter)	Squatter, if status remains informal; otherwise moves to another category

1. A discussion of the delivery of housing assistance is found in ⬛ Chapter 15, Mobilizing Financial Resources and Other Reconstruction Assistance.

"What Is a House?" A Critical Question for Assessments and Program Design

How a house is defined in a given culture or location has important implications for post-disaster surveys, such as for the damage and loss assessments, and for program design. Fundamental to defining a house is gaining an understanding of the foundational social, cultural, and economic relationships among the disaster-affected people who live inside houses, i.e., households. This basic socioeconomic unit is the core metric used in designing shelter and settlements interventions in the wake of disasters.

Defining the number and composition of households, and the physical structures they occupy, is often quite difficult, particularly for foreigners, who may be unfamiliar with shelter and settlement patterns in disaster-affected areas. In some areas, for example, multiple generations live together, often necessitating separate living quarters within attached or detached structures. In other areas, one or more related households might live together as a family household, again in attached or detached structures. These and other "extended family" living arrangements might be extended further through such practices as polygamy, still common in many countries.

To the social complexity outlined above can be added economic complexities, for instance where structures such as granaries or workshops are located amidst or within living quarters, thereby combining to create a set of structures—often very similar in appearance—that together constitute a form of shelter called compound housing. This example, and many others, underscore the claim that how various structures defined and are (or are not) counted as houses when assessing damage in the wake of a disaster will largely determine the magnitude of the disaster, as well as the scale of any formal response efforts by civil society, local and national authorities, humanitarian actors, international agencies, and donors.

Source: Charles A. Setchell, Shelter, Settlements, and Hazard Mitigation Advisor, United States Agency for International Development Office of U.S. Foreign Disaster Assistance (USAID/OFDA), personal communication.

Public Policies Related to Housing Assistance and Beneficiary Eligibility

Few public agencies have policies on how to allocate and distribute post-disaster housing assistance. If they do, they will have to be adjusted to the particularities of the emergency at hand. However, there may be existing financial assistance programs related to housing and community development (such as down-payment assistance, low-interest loans, or ongoing community revitalization programs). These programs may have data on families that can be used to facilitate the qualification process. Or an administrative system that includes identification numbers may be in place that can be adapted to the reconstruction program. If other subsidy programs are already operating, public agencies should calibrate the level of assistance and qualification rules so that the housing assistance program is seen as fair and consistent with existing public policies (not providing disaster assistance in excess of other programs that seek to accomplish a similar goal). The agencies should also be prepared to explain publicly how the terms of the disaster program and other housing assistance programs relate.[2]

Specific laws may apply when the housing assistance is offered to an owner by government in exchange for the property, for instance, to acquire a house in a high-risk area so that the residents relocate. If the owner objects to government taking the property, and government can argue that the property is being taken for a public purpose (risk reduction, in this case), eminent domain law may be applied.[3] Governments will ordinarily avoid using eminent domain in a post-disaster situation because of the time and cost involved. Whether taken by eminent domain or another procedure, local law may stipulate the basis for the housing assistance, usually that the owner is "justly compensated" (often, paid fair market value) for his or her loss. Calculating the assistance on the basis of lost value, however, may not be equitable or politically palatable, since the wealthiest will receive the most assistance.

Technical Issues
Social Risk Management and Disasters

Post-disaster housing assistance by government is an example of a public arrangement for social protection or social risk management. Social risk management arrangements are generally categorized as follows: (1) informal arrangements, such as sale of personal assets or community self-help; (2) market-based arrangements, such as property insurance; and (3) public arrangements, such as assistance grants or other social safety nets. All families will use informal arrangements in their recovery and reconstruction, but they are unlikely to be sufficient. Only a select group will generally have access to market-based arrangements. The expectation after a disaster is that public arrangements, in this case housing assistance, will fill the gap that remains when informal arrangements and market-based arrangements are inadequate.[4] See ⬛ Annex 1, How to Do It: Considerations in Designing a Social Protection System for Natural Disasters.

Government as Insurer

In many countries, government acts as the principal insurer of housing after a disaster. This is common when there is an inadequate property insurance system, an insurance market that is unaffordable to some households, no sanctions against being uninsured or underinsured, or disaster damage exceeds whatever insurance coverage people may have had. But when government plays this role, the "insurance terms" are not defined until after the disaster, which creates uncertainty for those affected, and the expectation that government will provide assistance creates political and economic burdens for government.

2. For an example of an existing program whose criteria were adapted to provide assistance for post-tsunami reconstruction in Orissa, India, see the Indira Awas Yojana housing program of the Ministry of Rural Development, Government of India, http://rural.nic.in/iaygd2.htm.

3. Eminent domain (United States), compulsory purchase (United Kingdom, New Zealand, Ireland), resumption/compulsory acquisition (Australia) or expropriation (South Africa and Canada) is the inherent power of the state to seize or expropriate property or seize rights in property, with due monetary compensation, but without the owner's consent. The property is taken to devote it to public or civic use. Source: Wikipedia, "eminent domain," http://en.wikipedia.org/wiki/Eminent_domain.

4. Robert Holzmann, Lynne Sherburne-Benz, and Emil Tesliuc, 2003, *The World Bank's Approach to Social Protection in a Globalizing World* (Washington, D.C.: World Bank), http://siteresources.worldbank.org/SOCIALPROTECTION/Publications/20847129/SRMWBApproachtoSP.pdf.

The assistance policy after one disaster will be interpreted as a signal to property owners about what government will do in future disasters, but these interpretations may be incorrect, or government policy may change over time. Eventually, government may decide that the moral hazard created by repeatedly providing reconstruction funds is too great and that alternatives must be sought. Creating a private insurance market and requiring homeowners to participate is one step in the process of removing government from the role of insurer. Also important are land use restrictions that forbid the occupancy of high-risk areas. There may be situations when government decides not to provide housing assistance after a disaster; for instance, when homeowners have the opportunity to insure their property and do not do it, or have knowingly chosen to live in high-risk areas over other options available to them. These policies should be defined before a disaster so that people have the opportunity to adjust their decision making. But the policy option of not providing assistance at all—or only for some part of the affected population (having an income cutoff, for example, as discussed below)—is one that should be evaluated even after the disaster.

Reconstruction as Opportunity to Resolve Long-Standing Problems

As part of reconstruction policy, government must decide the degree to which reconstruction will be used to accomplish longer-term development objectives. A disaster is often viewed as an opportunity to resolve long-standing development shortcomings, and, with a significant inflow of external assistance, the potential for correcting inadequacies in pre-disaster housing and community services obviously increases. It is clearly sound policy to rebuild houses and infrastructure that is less vulnerable to future disasters ("built back better"). A more complex decision in development terms is whether to move disaster-affected communities "to the "head of the line" of all those waiting to have their basic needs met (e.g., providing sewerage systems or updated road configurations), thereby favoring affected communities with a standard of living higher than that in similar, but unaffected, communities. The savings of taking a comprehensive approach to reconstruction may justify it, even at the risk of political fallout. An example of when the reconstruction period was used as a time to address the vulnerabilities of undamaged housing is discussed in the case study on the Gujarat reconstruction policy in 📖 Chapter 2, Assessing Damage and Setting Reconstruction Policy. It explains how strengthening of housing not damaged by the disaster was defined in the reconstruction policy as an integral part of the reconstruction effort.

Reconstruction as Social Policy

A post-disaster housing assistance program will raise questions of equity, both among those affected and between the affected group and unaffected households with similar needs. Poor and vulnerable households are likely to need a disproportionate level of assistance after a disaster because they are otherwise less able to rebuild or reestablish their livelihoods, but they may not receive it. Assistance is likely to arrive from numerous sources—private, public, and official, national and international—and to be channeled through a range of entities. Each organization may define housing needs or rights differently, and an organization's imperative to establish a foothold in the disaster location can produce unexpected and inequitable outcomes. Government has the right and responsibility to ensure a consistent and equitable allocation of the available resources.' The challenge of doing so effectively increases with the number of agencies involved. A useful tool to reach a common understanding of the social impacts of the disaster is social assessment. 📖 Annex 2 of this chapter contains a step-by-step explanation of how to conduct a social assessment.

Choice of Criteria

The task of allocating housing assistance can have unintended consequences when applied in real-world situations. Applying criteria in a logical manner is not easy, and the reality of limited resources further complicates the task. Government should develop an assistance strategy that selects among and weights these (and perhaps other) criteria in a way that reflects both governmental objectives and social values. Arriving at the proper solution is likely to be an iterative process. Government should consult with the public, especially the affected communities, on the appropriate assistance policy. Once decided, it should be announced publicly and applied objectively. Government should also monitor and publicize the results of the policy and be willing to make any needed adjustments.

The tables below contain some housing-related questions that are commonly asked during the process of developing a disaster housing assistance strategy, some issues that should be taken into consideration in responding to those questions, and some recommendations on how to proceed.

Fundamental to defining a house is gaining an understanding of the foundational social, cultural, and economic relationships among the disaster-affected people who live inside houses....

Who Is Entitled?

Criteria	Questions	Issues	Recommendations
Threshold	Should all people who suffered housing losses be entitled to aid or should assistance be targeted only to specific categories of people? Is having legal status in the country a requirement? Should households not affected by the disaster be assisted if they have housing problems similar to those who were affected? How will those with a need for housing who have migrated into the disaster region *after* the disaster be treated?	Categories may be economic, geographic, or related to some aspect of pre-disaster housing condition, but any choice can create inequitable outcomes in certain situations. The 🏠 case study on reconstruction following the 2004 Indian Ocean tsunami in Tamil Nadu, below, demonstrates how persistence may be needed to establish eligibility for assistance.	The implementing agency must have sufficient resources and administrative capacity to carry out the qualification process and the program.
Unit of assistance	Is the unit of entitlement the house, the family, or the household? Is a single-person household treated differently? How is assistance calculated for a household with multiple families?	If pre-disaster housing supply was inadequate, multiple households or extended families may be sharing a single house unwillingly. Conversely, a single family may own or live in more than one house.	Make an early decision on the unit of assistance and the extent to which the goal is to address pre-disaster housing shortcomings.
Economic status	Is income below a certain level a qualification or do all income levels qualify?	Income records may be falsified, destroyed in the disaster, or nonexistent.	Ensure there is a feasible method for qualifying according to income.
Social characteristics	Do social characteristics, such as gender, caste, or incapacity, override income as a criterion in those cases where there is an income cutoff?	Women and members of other vulnerable groups may need housing assistance even when their income exceeds the cutoff. The 🏠 case study on reconstruction following Typhoon Durian in the Philippines in 2006, below, describes a multi-step targeting procedure that was used to identify the poorest and most vulnerable.	Consider using community members to help identify those who truly need assistance.
Renters versus owners	Who gets the assistance? Renters? Owners? Both?	It is equally important for rental housing to be rebuilt, yet during reconstruction renters may need assistance for temporary housing.	Consider requiring owners to let renters return at similar pre-disaster rents as a condition of owners receiving assistance.
Informal tenure-holders	Is a squatter or informal settler entitled to the same housing assistance as a property owner?	Squatters may need assistance in addition to housing. This assistance will require planning for a more comprehensive set of services. Squatters often move to a disaster area after a disaster just to obtain housing assistance.	Ensure sufficient resources are available to carry out a full-service relocation program. It may be necessary to exclude families that have migrated post-disaster.
Absentee owners versus owner-occupants	Should owners living elsewhere be entitled to housing assistance or only residents of the disaster area? Are owners of houses under construction entitled to assistance?	This issue is related to the question of the unit of assistance. If the primary motivation is to relocate residents, absentee owners may not qualify. If neighborhood stability is a concern, broader eligibility will help prevent the negative effect of abandoned properties. If the owners are migrants, the remittances they are earning elsewhere may be supporting other households in the affected area.	Try to use housing assistance as an incentive for owners to sell or rent.

What Type of Housing Solution Are People Entitled To?

The questions regarding "type of housing solution" and "amount of assistance" are closely related. The former address issues related to the physical result being sought; the latter address issues related to the resources needed to accomplish the physical result. Neither is related to the reconstruction approach; almost any type of solution can be provided using a range of reconstruction approaches.

Issue	Questions	Issues	Recommendations
Use of the assistance	For what purpose is assistance available? Options may include reconstruction, repair, retrofitting, purchase of housing or land, and even rental assistance or transitional shelter.	Important to avoid an incentive for homeowners to exaggerate the extent of damage or to deliberately damage their houses further. Assistance for both land acquisition and housing may be necessary, if current location is not safe. Transitional shelter solutions may allow families to remain on their land, thus saving other temporary housing costs.	Both the level and the purpose of the assistance should be related to the condition of the house. If repairs are feasible and location is suitable, assistance should be geared to that cost, even if the family prefers to relocate. Consider assistance for retrofitting a high priority if those not directly affected by the disaster are to be aided.
Standard solution	Is it best to give everyone a core house of standard size and features (or resources sufficient to build one) and let them modify it as they see fit?	This is the "core house" model, which has been used in both agency-driven and owner-driven projects. Experience shows owners usually spend their own resources to augment the minimum assistance.	The core house at a minimum should be built for disaster-resilience, although additional rooms may not. This can be a cost-effective reconstruction approach.
Minimum housing standard	Is it better for government to provide assistance at a level that will ensure a minimum standard of housing for everyone (e.g., persons/bedroom, square footage of common space per occupant) or a minimum level of safety?	Ensuring a minimum solution requires variations in total assistance levels according to household size. Defining an acceptable minimum level will be culturally and even neighborhood specific. Vernacular solutions and non-standard designs and materials may be rejected. Government may provide assistance only to rebuild a strong house structure, leaving it to owners to contribute the rest.	Consider applying the minimum standards approach for public infrastructure, even if some other approach is used for housing. Consider targeting housing assistance to building a better housing structure only.
Pre-disaster housing situation	Should those whose housing had a higher value qualify for more assistance than those whose housing had a lower value? Conversely, should those whose pre-disaster situation was substandard qualify for more?	This is related to the assessment of damage. Restoring pre-disaster housing status means that government is providing assistance for value—paying more to those who had more, rather than striving for equity. Giving more assistance to those whose pre-disaster housing had shortcomings than to those who had adequate housing means other social objectives are being pursued.	Realize that the assistance scheme may send an unintentional message about future assistance and the type of rebuilding that should be done. Consider conditioning the assistance (see note below on "Conditions on assistance") and make sure government's intentions regarding future assistance are clearly articulated and communicated.
Customized solution	Can the entitlement criteria be weighted to produce a socially and economically optimal allocation of resources based on the characteristics of the family?	Value judgments are required to select and weight the criteria.	Decide whether a single weighting system is acceptable or appropriate for all affected groups. Government should persuade outside agencies to align their assistance criteria with those of government.

What Amount of Housing Assistance Should Be Provided?

Quantifying the amount of assistance may be the policy issue that concerns decision makers even more than what result will come of it. The factors above, such as the types of solutions sought, influence the level of assistance. Below are other critical questions.

Options	Questions	Issues	Recommendations
Need	Should available family resources be considered in setting housing assistance? Should all households be expected to make a contribution (labor, cash, in-kind)?	Assistance may not be necessary if a qualified household is capable of acquiring the minimum housing solution with its own resources. If only the cost of the solution is considered, it implies no expectation of self-help. Experience with use of credit in reconstruction is limited. It is best to avoid lending by the public sector. Credit was used for reconstruction by all those above the poverty line after the 1999 Orissa Super Cyclone, as described in the 🗐 case study, below.	Establish a consistent policy about use of family' resources in rebuilding and decide whether all households will receive some housing assistance. Decide whether households with capacity to borrow should be encouraged to finance reconstruction with credit.
Housing assistance for different levels of capacity	How should households be assisted who have additional vulnerabilities or reduced capacity to manage rebuilding and therefore need extra help in acquiring a desirable housing solution?	A support system will assist households in using the housing assistance that they are provided. Providing extra housing assistance to these households to buy services, such as supervision of construction, is another option, but agencies may need to support them in any case to ensure that appropriate services are in fact received.	Ensure that the monitoring system keeps track of outcomes (appropriate housing solutions occupied by different types of households) as well as outputs (funds disbursed).
Replacement of other assets	Is the housing assistance only for housing? Or should it cover furniture and other household investments, such as equipment for home-based businesses that will permit the restoration of livelihoods?	The house may not be occupied or the household sustainable unless these other assets are replaced.	Funding agencies should understand that the household is not just a house. It's an economic system that needs to be rebuilt and the agencies should provide appropriate forms of funding.

Additional Considerations

Poverty and vulnerability. People's capacity to recover from a disaster depends on their socioeconomic status. The majority of the poor are women and children who may be isolated socially and who may have less access to physical, financial, and social capital. Members of vulnerable groups and the poor may not incur high losses in absolute terms simply because they own less, but they tend to be the most severely affected by disasters. These households often do not own the land or shelter they occupy. And their dwellings may be weaker and located in more vulnerable sites. In addition, if the house or land belonged to a husband or brother who has died in the disaster, women may be at risk of displacement and destitution. These issues need to be taken into consideration in designing assistance strategies. The vulnerability of households may be related to the loss of livelihood. The 🗐 case study on reconstruction following the 1993 Maharashtra earthquake, below, explains how the priority of preserving employment in the affected villages resulted in a decision to provide more assistance to larger land owners.

Family size and composition. Housing requirements are a function of, among other things, family size and composition. These characteristics change with time and vary among societies. For instance, an assistance strategy that ignores the requirements of extended families can weaken family ties that support livelihoods and that serve as informal social security systems. This is one problem with providing overly standardized housing solutions that are difficult to customize later.

Conditions on assistance. Government may decide to tie housing assistance to a requirement to comply with some condition that accomplishes a public purpose. The most common example is the requirement that the recipient improve the disaster resilience of the reconstructed house ("building back better"). Requirements could also address the reduction of environmental impact, improvement of fire safety, compliance with universal design standards for handicap accessibility in a multi-family building, co-ownership by a couple, or conformance with architectural guidelines in a historic district. Any of these conditions may be reasonable, depending on the circumstances. Governments that condition assistance in this way must have adequate controls to ensure compliance and even-handedness in the application of the requirement.

Land and housing tenure. The United Nations Office for the Coordination of Humanitarian Affairs (UN OCHA) recognizes six tenancy categories, shown in the Introduction section of this chapter.[5] In fact, the number of categories is much larger in some places. Rural households usually own the house they occupy and have tenure security through formal land titles or customary land rights. Urban and rural residents in the same country may have different tenure and occupancy options. Owners of housing may not own their land. In many reconstruction programs, only homeowners with clear title to their land have been entitled

DANIEL PITTET

to housing assistance. Those designing housing assistance strategies should make sure they understand all the categories of tenancy relevant to the affected population and craft an assistance program that considers them all. The United Nations Human Settlements Programme (UN-HABITAT) identified 31 different tenancy situations in the affected population in Peru following the 2007 Ica/Pisco earthquake.[6] 🏠 Chapter 7, Land Use and Physical Planning, contains a section on resolving land tenure issues in reconstruction.

Gender issues. Generally, women spend more time in their homes than men do, and they have clearer ideas about what they need. Yet women often do not participate in public consultations or express their views in the presence of men, which can lead to errors in developing the assistance strategy. (This may be particularly true of female-headed households.) Best practice would be to place special emphasis on the particular post-disaster situation of women and to organize separate women-only community consultations. Among the gender-related housing issues to consider in housing reconstruction are (1) legal (the differential legal status of women), (2) economic (women's low economic status and the prevalence of women's home-based enterprises), (3) security (safety issues related to housing and access to services and markets), and (4) social (children's access to schools).

Disaster-induced mortality and migration. If a disaster causes high rates of mortality or migration, it may not make sense to estimate housing requirements based on a pre-disaster census or to adopt a house-for-house assistance policy. More time and professional support may be required before an appropriate housing assistance policy can be defined.

The importance of social assessments. While the damage and loss assessment estimates physical damages and needs for reconstruction, a social analysis is required to understand the social dimension of housing and to design the assistance policy. The social analysis should include consultations with stakeholders and affected communities. The World Bank has experience and resources that provide conceptual and methodological guidance on conducting social analysis, as well as e-learning courses. Although none of these tools focuses specifically on social analysis in relation to disasters, they can be adapted for this purpose.[7] 🏠 Annex 2 to this chapter contains a step-by-step explanation of how to conduct a social assessment.

5. United Nations Office for the Coordination of Humanitarian Affairs (UN OCHA) and Shelter Centre, 2010, *Shelter After Disaster: Strategies for Transitional Settlement and Reconstruction* (Geneva: UN OCHA), http://www.sheltercentre.org/library/Shelter+After+Disaster.

6. Department for International Development and the Ministry of Housing, Construction and Sanitation, 2008, "Final Report, Land Ownership and Housing" ("Informe Final, Tenencia de la Tierra y la Vivienda").

7. World Bank, 2003, *Social Analysis Sourcebook: Incorporating Social Dimensions into Bank-Supported Projects* (Washington, DC: World Bank), http://go.worldbank.org/HRXPCILR30.

🏠 **For access to additional resources and information on this topic, please visit the handbook Web site at www.housingreconstruction.org.**

Examples of Recent Housing Assistance Schemes

Gujarat India Earthquake (2001)	Sri Lanka Earthquake/Tsunami (2004)	Indonesia Earthquake/Tsunami (2004)	United States Hurricane Katrina (2005)	Pakistan Earthquake (2005)
Not a uniform package, leading to equity issues. Assistance disbursed in three tranches. Compensation ranging from INR 5,000 to INR 90,000 (US$126 to US$2,277).	Uniform assistance package. Assistance of LKR 100,000 (US$880) disbursed in two tranches for partially damaged houses and LKR 250,000 (US$2,200) disbursed in four tranches for destroyed houses.	Uniform assistance package. Assistance of IDR 20 million (US$2,000) for repairable damaged house and IDR 42 million (US$4,200) for full reconstruction of destroyed house.	Not a uniform package. Assistance based on actual value of house and insurance coverage. Assistance of up to US$150,000 available for homeowner.	Uniform assistance package. Assistance of PKR 75,000 (US$1,250) for partially damaged house disbursed in two tranches and assistance of PKR 175,000 (US$2,917) for destroyed house disbursed in four tranches.

Source: Pakistan Earthquake Recovery and Reconstruction Agency, 2007, "Global Post-Disaster Housing Reconstruction Comparative Analysis," http://www.erra.gov.pk. (See full table in ⛭ Part 4, Technical References, Matrix of Disaster Project Features.)

Risks and Challenges

- Inappropriate or inequitable housing assistance program designs created by an inaccurate understanding of the social context or of local needs and capacities.
- Social conflicts created by a failure to establish sound and consistent program rules, apply them objectively and predictably, and communicate them clearly to the affected population.
- Creating an incentive for owners to overestimate damage or cause damage to their own house.
- Thinking that it is sufficient to create the assistance scheme and forgetting to monitor its effectiveness, including ease of access by target groups and impact on the ground.

Recommendations

1. Rather than borrowing from other disaster responses, develop a housing assistance policy consistent with the specifics of the situation and reflective of public policy and social values.
2. Base assistance policies on sound social analysis.
3. Involve local communities and stakeholders in defining entitlement policies and make a special effort to consult with women, privately if necessary.
4. Develop a policy that contributes to equity, risk reduction, and sustainability. At the operational level, fine-tune it to the needs and capacities of different categories of affected people and their household requirements.
5. A single post-disaster reconstruction program may include various approaches to housing assistance, depending on levels of damage from one location to another, household composition, the institutional context, and other factors. However, even if a range of approaches are employed, government should ensure the available resources are being well allocated overall, promote the use of consistent eligibility criteria among organizations, and establish minimum and maximum levels of assistance.
6. Make the assistance policy easy to understand. Publicize both the policy and any conditions on the access to funds.
7. Avoid paying more than is necessary for the level of damage. Also avoid indiscriminate distribution of free houses to avoid negative socioeconomic consequences.
8. Closely monitor outcomes from application of the assistance policy and communicate them publicly. Evaluate the program and be willing to adjust the policy over time.

Case Studies

2006 Typhoon Durian, Bicol, Philippines

Targeting during Post-Typhoon Reconstruction

Typhoon Durian hit the Philippines in November 2006, just when the country—especially the Bicol Region in the Luzon Island group—was recovering from a previous typhoon and from the eruption of the Mayon volcano. Durian, categorized as a super typhoon, caused mudslides, floods, and powerful winds that affected almost 650,000 households, displaced more than 19,000 households, and damaged approximately 540,000 houses, of which 214,000 were destroyed. Some 2,360 people were reported injured and 720 deaths were confirmed. Coordinating with government and municipalities, the nongovernmental organization Community Organization of the Philippine

Enterprise Foundation (COPE) decided to focus on relocation and construction of permanent shelter in Daraga and Legazpi City, two cities that were heavily affected by the typhoon. The multiple-step targeting process was designed to identify the poorest of the poor, using information from community associations and local government units. The criteria targeted people without access to any financial assistance for reconstruction, single parents or widows with at least four dependents, vulnerable individuals (orphaned, disabled, or ill), and poor families that had lost their major source of income. Home visits were carried out to validate beneficiary information provided by local governments. Focus groups were held to discuss relocation. Psychosocial therapy was provided to help the families overcome the disaster experience and prepare for reconstruction and relocation. To be selected, beneficiaries had to commit to provide counterpart labor during construction (the value of the labor ranged from US$60 to US$151). During the construction process, constraints included the unavailability of land for permanent shelter close to the original settlement and the constantly increasing prices due to high rates of inflation. Despite these problems, COPE provided 170 typhoon-resilient permanent housing units to the selected families.

Sources: Government of the Philippines, 2006, "Philippines: NDCC media update –Typhoon 'Reming' (Durian), December 13, 2006," http://www.reliefweb.int/rw/RWB.NSF/db900SID/EKOI-6WF8T8?OpenDocument&rc=3&emid=TC-2006-000175-PHL; *and* Myrna Abella-Llanes and Salve Alemania-Cadag, 2006, "Housing & Community Reconstruction, Bikol Super Typhoon" (COPE Foundation Case Study, unpublished).

2004 Indian Ocean Tsunami, Tamil Nadu, India
Identifying Eligible Families in an Urban Setting

Tamil Nadu was one of the Indian states most affected by the 2004 Indian Ocean tsunami. While a number of agencies were involved in the provision of temporary housing and in reconstruction in Chennai, it was principally the Tamil Nadu Slum Clearance Board (TNSCB) that worked with the fisher community. The city of Chennai had an ongoing initiative, funded by the World Bank, to replace slums with tenement housing and therefore had procedures in place that were helpful in planning the apartments required by the community. The key challenges for the TNSCB were (1) getting the fishers to concur with the design of houses, and (2) developing the list of eligible families. Because the houses were to be given away, many ineligible people tried to be declared eligible (including one person who claimed ownership of 32 structures!). Approximately 11,000 people claimed to be the owners of the 6,000 properties slated for replacement. However, the fishers resisted participating in the field survey that would validate their claims; the survey teams faced physical assault and required a police escort. When eventually the TNSCB completed the field enumeration, each family was surveyed and photographed in front of its property. The TNSCB used an eligibility matrix to award points for current residency on site, residency immediately after the tsunami, and documentary proof of residence (current and immediately after the tsunami). No family could receive more than one housing unit. Based on the scoring, an eligibility list was finalized and presented to the families for their review. Because of the transparent manner in which the survey was conducted, development of the eligibility list—a daunting task—was eventually accomplished and approved by all stakeholders. Although this process delayed the start of reconstruction by more than two years, it produced a detailed tool that could be used to streamline the eligibility process in future disasters.

Source: C. V. Sankar, India National Disaster Management Authority, 2009, personal communication.

Different states in India have adopted entirely different housing assistance policies, each of which reflects an interpretation of an affected community's socioeconomic conditions and housing needs. The case studies below show how housing assistance policies can exacerbate existing socioeconomic inequalities.

1993 Maharashtra Earthquake, India
Pre-Disaster Landholding as Basis for Assistance in Maharashtra

The 1993 Maharashtra earthquake caused damage in 728 villages, 37 of which were completely destroyed; the collapse of 25,000 houses; and damage to another 200,000 houses. A reconstruction program was executed, largely with resources from the US$221 million World Bank loan—the Maharashtra Emergency Earthquake Rehabilitation Project.

The affected villages were divided into three damage categories. Category B villages (22 villages, 10,000 houses) received financial assistance for reconstruction *in-situ.* Work was stalled in some cases while people lobbied for relocation. Construction and land purchases for these villages were done largely by nongovernmental organizations. Investment in amenities was modest, but satisfaction levels were high. Category C villages (180,000 houses) used owner-driven reconstruction for repair and retrofitting damaged houses, with materials distribution and extensive

DANIEL PITTET

supervision. The work in these villages started late, but went more quickly than the others. Satisfaction levels were high. Cash assistance to beneficiaries in Categories B and C were uniform: Rs 62,000 (US$2,000) for reconstruction, and Rs 17,000 (US$548) and Rs 34,500 (US$1,113) for repairs, depending on the level of damage.

The more complex situation had to do with certain villages that were classified as Category A (52 villages, 28,000 houses), including the Latur villages in Killari. In these villages, houses were relocated and full reconstruction took place. For the Category A beneficiaries, the size of the plots and new houses varied, depending on the original landholdings of the beneficiary. Landless and marginal landholders got a plot of 1,575 sq. ft. and 250 sq. ft. houses. Households owning between 1 and 7 hectares of land got 2,500 sq. ft. plots and 400 sq. ft. houses. Farmers owning more than 7 hectares of land got 5,000 sq. ft. plots and 750 sq. ft. houses. As a result, wealthier households benefitted more than poor households, regardless of their own endowments or requirements.

The justification for this approach had to do with the characteristics of the local economy. The Latur village economy in Killari consisted of a few large Patils who owned major land holdings and lived in the village center in large stone, mud, and wooden frame *gaddis*. Some had up to 1-acre plots with sprawling structures. Landless *dalits* who provided farm labor lived on marginal land in mud and thatch huts. However, the *gaddis* were not only residences, they were effectively agro-processing centers. On these properties, many productive activities took place: produce of the farms was stored, cattle was milked, sugar cane was converted into jaggery, fodder was dried, and grapes were converted into resins. As a result, dozens of landless workers were employed on the *gaddis*. Originally, the decision was made to give everyone equal housing assistance after the earthquake. But the Patil owners refused to accept this solution, saying that they would move their dwellings out of the village to large farm plots. If they had done so, it would have destroyed the village economy, because each large house employed dozens of landless workers. To find employment, the landless workers would then move to the city or have to move onto the Patil properties. The land owners argued that they lost the most and that to continue to live in the village they needed large houses to store and process the farm produce. Contractors were hired for all work, and amenities, including infrastructure, were extensive. There was limited community participation, which reduced the level of beneficiary satisfaction.

Landless dwellers in small huts, mostly squatters with uncertain titles, not damaged by the earthquake, received fixed houses of 250 sq. ft. on 1,500 sq. ft. plots with full ownership titles. The large *gaddi* owners received up to 5,000 sq. ft. plots, which were nearly half or one-third of their original household plots. Thus, the *gaddi* owners had less than what they had before, but, by remaining in the villages, they enabled agro-processing to subsist on, and the landless (who now owned small plots) retained their livelihood. If the large land owners moved to their individual farm lands, the landless small house owners would not have been able to stay in the village as there would have been no employment. The lesson from this experience, according to those involved, is that post disaster reconstruction can improve the lot of many, but cannot resolve all pre-disaster social inequities.

Sources: World Bank, "Maharashtra Emergency Earthquake Rehabilitation Project," http://web.worldbank.org/external/projects/main?pagePK=64283627&piPK=73230&theSitePK=40941&menuPK=228424&Projectid=P034162; "Latur District 1993 Earthquake," http://latur.nic.in/html/earthquake.htm; *and* Praveen Pardeshi, 2009, written communication.

1999 Orissa Super Cyclone, India
Beneficiary Assistance Varies by Poverty Level

The "super cyclone" that hit Orissa, India, in September 1999, affected 13 million people, killed nearly 10,000, and destroyed some 800,000 houses. Immediately after the disaster, all affected people received a minor grant. No comprehensive governmental reconstruction program was organized. Instead, government provided two types of housing assistance: free housing to 200,000 poor families through the Indira Awas Yojana, an ongoing social housing program targeting the scheduled castes and tribes and households below the poverty line, and loans to 175,000 families above the poverty line through the Housing and Urban Development Corporation. This policy reflected a recognition that the type of assistance provided to better-off households who could afford to repay the cost to rebuild should be different from the assistance provided to poor families.

Source: Commissionerate of Rural Development, Gujarat, n.d., "Indira Awas Yojana," http://www.ruraldev.gujarat.gov.in/iay.html.

2001 Gujarat Earthquake, India
Funds Allocated According to Damage Level

The earthquake in the state of Gujarat, India, destroyed 344,000 houses and damaged another 888,000. Using World Bank funds from the Gujarat Emergency Earthquake Reconstruction Programme, government offered financial, material, and technical support to all affected families based on the type of house they owned and the level of damage incurred. Families with completely destroyed *kuchcha* house (built with low-cost materials, such as mud and thatch) received a maximum grant of Rs 30,000 (US$630). Families with a completely destroyed *pukka* house (built with industrial materials, such as bricks and cement) received a maximum assistance of Rs 90,000 (US$1,900). While poor people received less assistance than rich people, the minimum assistance was sufficient to replace a *kuchcha* house with a higher-standard house; however, the maximum grant was not sufficient to replace houses of higher-income people. The housing rights of the homeless and tenants were also recognized.

Source: Jennifer Duyne Barenstein, 2009, "Who Governs Reconstruction? Changes and Continuity in Policies, Practices and Outcomes," in *Rebuilding after Disasters: From Emergency to Sustainability,* G. Lizarralde, C. Johnson, and C. Davidson, eds. (London: Taylor and Francis).

Resources

Centre on Housing Rights and Evictions. 1993. *Bibliography on Housing Rights and Evictions.* Utrecht, Netherlands: Centre on Housing Rights and Evictions. http://www.cohre.org/store/attachments/COHRE%20Sources%202.pdf.

Centre on Housing Rights and Evictions. n.d. *The Pinheiro Principles: United Nations Principles for Housing and Property Restitution for Refugees and Displaced Persons.* Geneva: COHRE. http://www.cohre.org/store/attachments/Pinheiro%20Principles.pdf.

Holzmann, Robert, Lynne Sherburne-Benz, and Emil Tesliuc. 2003. *The World Bank's Approach to Social Protection in a Globalizing World.* Washington, D.C.: World Bank. http://siteresources.worldbank.org/SOCIALPROTECTION/Publications/20847129/SRMWBApproachtoSP.pdf.

Skoufias, E. 2003. "Economic Crisis and Natural Disasters: Coping Strategies and Policy Implications." *World Development* 31/7: 1087–1102. http://info.worldbank.org/etools/docs/library/78330/3rd%20Workshop/Srmafrica/paristwo/pdf/readings/weather.pdf.

United Nations. 2008. *Transitional Settlement and Reconstruction after Natural Disasters.* Field Edition. UN OCHA: Geneva. http://www.sheltercentre.org/library/Transitional+settlement+and+reconstruction+after+natural+ disasters.

UN OHCHR. *The Human Right to Adequate Housing.* Fact Sheet No. 21. Geneva: United Nations. Office of the High Commissioner for Human Rights. http://www.ohchr.org/Documents/Publications/FactSheet21en.pdf.

World Bank. 2003. *Social Analysis Sourcebook: Incorporating Social Dimensions into Bank-Supported Projects.* Washington, DC: World Bank. http://go.worldbank.org/HRXPCILR30.

World Bank. 2003. *A User's Guide to Poverty and Social Impact Analysis.* Washington, DC: World Bank. http://go.worldbank.org/IR9SLBWTQ0.

World Bank. 2009. "Social Analysis at the World Bank." http://go.worldbank.org/UDVDOCK3X0.

Natural disasters are external shocks that can have a major impact on the social and economic welfare of populations and households. Social risk management (SRM) refers to the use of a range of social protection mechanisms to prevent and mitigate risk (ex ante strategy) or cope with its impacts after a shock such as a disaster has occurred (ex post coping strategy). In the context of poverty reduction, SRM is a set of tools that improve the management of vulnerability by households, and may even lead to poverty reduction. The focus of SRM in the post-disaster context is on restoring and rebuilding both assets and livelihoods of households and affected communities.

Social safety nets are a type of program within the broader range of social protection. Social safety nets generally refer to non-contributory transfers (in cash or in kind), targeted at both populations at risk of economic destitution and the permanently poor, designed to keep their income above a specified minimum. In a post-disaster situation, social safety nets are almost always publicly funded transfers that help households avoid irreversible losses and decline into poverty by providing basic income and employment support. Social safety net support is often accompanied by other public or private resources provided for reconstruction and recovery. (Other instruments of social protection and social policy include mechanisms as wide-ranging as labor market policies or pension schemes. None of these other mechanisms is addressed in this annex.)

This annex presents some of the issues to consider in designing a disaster-related social safety net program. While social protection and livelihood support have been considered an important part of post-disaster response for years, there has been little ex ante planning of these disaster interventions by government. Yet planning ahead to anticipate post-disaster demands has significant benefits, since trying to create an effective social safety net program from scratch immediately after a disaster is virtually impossible. At least four months is needed to design a quality social safety net program; the special challenges that arise in the aftermath of a disaster may require additional time. The World Bank can provide extensive technical and financial assistance to governments on designing social protection systems.[1]

The two best options for putting a post-disaster social safety net system in place are to adapt a system that is already operating or to create a system to provide a short-term response while simultaneously designing a better system to be implemented in the medium term.

Options for Implementing Safety Nets in the Context of a Disaster

Adapt existing systems

- Expand existing safety nets to provide a short-term option for offsetting the immediate effect of a natural disaster with minimum negative impacts on economic incentives.
- Provide immediate productive activities that lead to more sustainable activities in the medium term (phasing out).
- If necessary, temporarily relax standards, but maintain a minimum level of requirements.
- Expand existing monitoring systems to detect immediate impacts and problems in any program design adapted to the disaster.
- Set up response systems for future disaster risks during the reconstruction process.

Example: After Hurricane Mitch in 1998, the Honduras Social Investment Fund (Fondo de Inversión Social [FIS]) played a crucial role in rebuilding the country's infrastructure. Regional offices and technical experts quickly estimated the need to clean up the debris, repair water and sanitation systems, and provide access to roads, bridges, health centers, and schools. To respond to the urgency of the situation, the FIS simplified its subproject requirements while maintaining minimum standards. Within 100 days, more than 2,100 projects were approved for a total value of US$40 million. Labor accounted for 70 percent of the clean-up activities and 25–30 percent of the value of most subprojects. The FIS created 100,000 person-months of employment in the first three months after Hurricane Mitch.[2]

Provide a suboptimal immediate safety net while developing a more optimal longer-term system

- Be aware that time constraints and poor planning for disasters may result in suboptimal programs.
- Begin to build an effective safety net for the medium term.
- Put systems in place to monitor negative impacts of the disaster, such as indebtedness.
- Use rapid surveys and spot-checks to assess if assistance is reaching vulnerable groups.

Example: Increased indebtedness was identified in disaster-affected villages in Myanmar six months after the Cyclone Nargis. Villagers worried that they would not be able to meet loan obligations and satisfy consumption needs in the following year. Although relief assistance reached all villages, much more assistance was needed for communities to recover, particularly in the form of cash grants. Without a way for people to manage their indebtedness, there was a risk of a loss of family assets.[3]

Balancing Speed and Design Quality

It is crucial to evaluate the disaster impact on households while also considering pre-existing vulnerabilities. The impacts of a natural disaster are not uniformly distributed within a population, and the effects on different people—and on their ability to cope—are strongly correlated with their pre-disaster situation. The social protection response depends on the relative intensity of those impacts and needs. At the same time, disasters affect entire communities and tend to destroy the informal safety nets and personal arrangements that traditionally provide "insurance" for poorer households.[4] Since protection is a function of vulnerability, targeted programs are preferable to untargeted ones. The design process should include considerations of equity, cost-effectiveness, incentive compatibility, and sustainability.

Consideration	Recommendations[5]
Context and disaster impact	■ Analyze disaster impact and needs of the population. ■ Analyze impact of disaster on the economy and employment. ■ Evaluate markets and access to market. ■ Evaluate supply availability for key goods and inflation consideration. ■ Evaluate whether traders can respond to additional demand.
Country conditions	■ Analyze national priorities and needs. ■ Analyze available safety protection mechanisms, formal and informal, and program design, including targeting. ■ Identify safety nets structures that are flexible enough to cover the affected areas. ■ Identify programs that can be quickly scaled up and that can rapidly channel additional resources to vulnerable groups. ■ Use household-level data on program access, targeting, and benefit incidence.
Vulnerabilities of the population	■ Analyze vulnerabilities such as: 　■ hazardous locations, substandard housing; 　■ availability of ex ante risk management instruments; 　■ loss of jobs and income; 　■ lack of income-generating activities and resources for rebuilding income-generating activities (micro-finance, savings clubs, etc.); and 　■ lack of savings and other assets. ■ Focus on the chronically poor, the temporary poor, and people living in the affected areas (and, within these groups, children, orphans, the elderly, the disabled, and women).
Targeting beneficiaries according to vulnerabilities and defining eligibility criteria	■ Identify populations already covered by a safety net program and the eligibility criteria for those programs. ■ Identify targeting methods that can be used (geographic, demographic, community-based) to channel resources to the affected areas. ■ Identify eligibility criteria for affected populations that can be combined with existing targeting criteria. ■ Avoid criteria that could create friction between groups and grievances. ■ Develop criteria that are easy to explain and administer. ■ Consider criteria such as loss of assets for immediate support, shifting to poverty criteria for medium-term support.
Benefit level	■ Make sure level is adequate for subsistence. ■ Avoid benefit level that could jeopardize work incentives or distort markets or prices. ■ Provide larger amounts only as one-off compensation, for example, for loss of house.
Duration	■ May vary by target group and nature of emergency. ■ Provide cash or in-kind support for a limited period, longer only for the most vulnerable. ■ Consider large initial transfer to all those affected, followed by a second, smaller transfer for those who still need it (e.g., after three months). ■ Target later transfers to vulnerable/poor households. ■ In large emergencies, consider targeting all transfers. ■ Provide additional social services for the most vulnerable groups (such as orphans and disabled people).

Social Safety Net Program Options

Social safety net programs can be carried out (1) to support immediate household and livelihood needs following a disaster, (2) as part of a scheme to facilitate housing and community reconstruction, or (3) to provide a combination of the two types of support. The forms of assistance that can be provided are similar in the three cases. Because this handbook focuses on reconstruction, this annex is intended to complement the rest of the handbook by explaining the options for immediate support.

A detailed discussion of criteria to be used in allocating reconstruction assistance is provided above in this chapter. For a discussion of the options for mobilizing and delivering financial resources and other assistance to support reconstruction, see ⬛ Chapter 15, Mobilizing Financial Resources and Other Reconstruction Assistance.

The table below summarizes the three principle safety net options for providing immediate support to sustain household and livelihoods following a disaster and some considerations to take into account when choosing among them.

Safety Net Options for Immediate Support

Program Feature	Cash and Near-Cash Transfers	In-Kind Transfers	Public Works
Description	▪ Simplest way to channel resources to the most vulnerable households ▪ Increases households' real income ▪ Normally designed for a limited time until economic activities generate employment	▪ In-kind transfers (food, clothing, and temporary housing) preferable if markets are not functioning or supply of basic goods is limited	▪ Generates income in targeted areas while producing desired outcome: removing debris, opening roads, or restoring services ▪ Can be implemented at any time from response and reconstruction ▪ Should not be considered for long-term income support
Target	▪ Chronically poor working families ▪ People not expected to work: children, the elderly, the disabled ▪ Those needing temporary assistance ▪ All affected households or households selected by geographical targeting	▪ Chronically poor who cannot afford necessary commodities ▪ Highly affected people needing nutritional support, commodities (blankets, clothing) ▪ When beneficiary group is limited	▪ Unemployed at the margins of the labor market ▪ Temporarily poor, short-term unemployed ▪ Self-targeting is effective when wage is low[6]
Pros	▪ Low administrative cost ▪ Transfer can directly meet critical household needs ▪ Benefits can be tailored according to the level of need and household size ▪ Provides beneficiaries with a greater freedom of choice	▪ Effective in life-saving situation ▪ Compensates for food shortages, alleviates hunger, improves nutrition ▪ Mitigates temporary shortages of essential goods ▪ Can be used to provide tools to enable families to undertake reconstruction	▪ Needed infrastructure built or maintained ▪ Contributes to resumption of basics services (roads, hospitals) ▪ Politically popular programs
Cons	▪ Targeting methods can be information intensive, especially if the affected population is dispersed ▪ Risk of moving cash ▪ Transfers are fungible, subject to unintended usage	▪ High logistical cost in terms of storage, transport, and distribution ▪ Errors of inclusion, depending on the targeting methods ▪ Beneficiaries have no choice of commodities ▪ Procurement difficulties and long supply chains in remote areas	▪ Administratively demanding if linked to large-scale infrastructure programs ▪ Tradeoff between infrastructure development and poverty alleviation ▪ Serves vulnerable, able-bodied households, not those in which no one can work (children, elderly, disabled)
Context	▪ Only when markets are functioning and goods are available	▪ In emergency situations for life-saving interventions ▪ When prices are too high and markets are inefficient ▪ When markets are not accessible (transport, logistics) or affected areas are cut off	▪ When unemployment is high, after a disaster or the collapse of the labor market

✖ ▨

Program Feature	Cash and Near-Cash Transfers	In-Kind Transfers	Public Works
Challenges	■ Defining benefits levels for different types of beneficiaries ■ Reaching intended beneficiaries, including those in temporary shelters or camps	■ Reaching most needy (especially in very remote areas) ■ Procurement, storage, and avoiding waste, spoilage, and pilferage ■ Determining whether approach is needed	■ Setting correct wage rate (lower than alternative employment opportunities) ■ Setting the right labor intensity to make the program cost-effective ■ Identifying projects with high labor requirements ■ Maintaining projects if there is no community involvement in the planning and design or sense of local ownership ■ Cannot always be set up quickly due to adverse weather and other conditions
Recommendations	■ Program should be simple and easy to verify and should use available technology ■ Clear implementation arrangements should include eligibility criteria, payment amounts, and duration of payments ■ Transaction costs for beneficiaries should be kept to a minimum ■ Immediate cash delivery avoids the delays of opening bank accounts	■ Use for shortest term possible in order to avoid creating dependency and suppressing the resumption of economic activities	■ Target disaster-affected regions and produce infrastructure desired by local communities ■ Develop community-driven programs using participatory approach whenever possible ■ Ensure community ownership of assets and system for maintenance ■ Avoid displacing people from other economic activities (harvest or other employment) ■ Ensure participation of women, since their participation produces larger improvements in child welfare and family health[7]

Experience with cash transfer: After the South Asia earthquake in 2005, the government of Pakistan allocated a monthly cash grant of US$50 to each eligible household. The amount was established based on a calculation of needs for an average household of seven persons. A policy decision was made by government that the payment would be uniform for all beneficiary households and would continue for six months.[8]

Experiences with in-kind assistance: After Cyclone Nargis in Myanmar in 2008, people monetized some of the in-kind assistance given to them through exchange or sale.[9]

During the 1998 Bangladesh floods, in-kind food relief operations were aimed at increasing nutrition levels and avoiding starvation of targeted groups.

Experiences with public works: In Indonesia, some 18,000 participants were involved in public works programs in approximately 60 villages after the 2004 Indian Ocean tsunami.

Following the 2001 earthquakes in El Salvador, Catholic Relief Services and Caritas ran a 2-year program in which communities were organized to build 1,300 houses as well as schools, health centers, and roads in exchange for food.[10]

Literature review by Iride Ceccacci.

Annex 1 Endnotes
1. Margaret E. Grosh et al.,2008, *For Protection & Promotion: The Design and Implementation of Effective Safety Nets* (Washington, DC: World Bank), http://go.worldbank.org/I0JA2JIMV0.
2. Tara Vishwanath and Xiaoping Yu, 2008, "Providing Social Protection and Livelihood Support," World Bank, http://siteresources.worldbank.org/CHINAEXTN/Resources/318949-1217387111415/Social_Protection_en.pdf.
3. Tripartite Core Group (Government of the Union of Myanmar, Association of Southeast Asian Nations, United Nations), 2008, "Post-Nargis Social Impacts Monitoring," http://www.aseansec.org/CN-SocialImpactMonitoring-November08.pdf.
4. Ruchira Bhattamishra and Christopher C. Barrett, 2008, "Community-Based Risk Management Arrangements: An Overview and Implications for Social Fund Program," World Bank, SP Discussion Paper No. 0830, http://siteresources.worldbank.org/SOCIALPROTECTION/Resources/SP-Discussion-papers/Social-Funds-DP/0830.pdf.
5. Renos Vakis, 2006, "Complementing Natural Disasters Management: The Role of Social Protection," World Bank, SP Discussion Paper No. 0543, http://siteresources.worldbank.org/SOCIALPROTECTION/Resources/SP-Discussion-papers/Social-Risk-Management-DP/0543.pdf; Rasmus Heltberg, 2007, "Helping South Asia Cope Better with Natural Disasters: The Role of Social Protection," *Development Policy Review,* vol. 25, no. 6, pp. 681–98; *and* "The World Bank's Experience With Cash Support In Some Recent Natural Disasters," *Humanitarian Exchange Magazine,* Issue 40, http://www.odihpn.org/report.asp?id=2937.
6. Carlo Del Ninno, Kalanidhi Subbarao, and Annamaria Milazzo, 2009, "How to Make Public Works Work: A Review of the Experiences," World Bank, SP Discussion Paper No. 0905, http://siteresources.worldbank.org/SOCIALPROTECTION/Resources/SP-Discussion-papers/Safety-Nets-DP/0905.pdf. The use of multiple targeting methods makes the identification of the neediest more accurate and comprehensive. The pure self-selection might be insufficient in reaching vulnerable groups in poor areas or when the demand for participation is very large and some form of employment rationing is needed.
7. Carlo Del Ninno, Kalanidhi Subbarao, and Annamaria Milazzo, 2009, "How to Make Public Works Work: A Review of the Experiences," World Bank, SP Discussion Paper No. 0905.
8. Tara Vishwanath and Xiaoping Yu, 2008, "Providing Social Protection and Livelihood Support," World Bank, http://siteresources.worldbank.org/CHINAEXTN/Resources/318949-1217387111415/Social_Protection_en.pdf.
9. Tripartite Core Group (Government of the Union of Myanmar, Association of Southeast Asian Nations, United Nations), 2008, "Post-Nargis Social Impacts Monitoring," http://www.aseansec.org/CN-SocialImpactMonitoring-November08.pdf.
10. Sultan Barakat,2003, "Housing Reconstruction after Conflict and Disaster," *Humanitarian Practice Network Paper,* Overseas Development Institute, http://www.odihpn.org/report.asp?id=2577.

The implementation of any post-disaster reconstruction project can have technical, physical, environmental, economic, or social impacts. Some of the impacts are desired and planned, others are unforeseen. These impacts may become obvious immediately during the project implementation or show up months or even years later. While the technical and environmental impacts of projects have long been analyzed in detail during project preparation, only since the 1990s have international organizations such as the World Bank used social assessment (SA) to systematically analyze and adjust for the potential social impacts of projects.[1] Project outcomes improve when potential risks from social impacts are analyzed early while projects are still being designed and the findings are used to fine-tune project design. SA helps all involved understand the social and economic context, incorporate the perspectives and interests of those whom the project is intended to assist, anticipate the project's social impacts (both positive and negative), and prepare to mitigate them, when necessary.

Objectives of the Social Assessment

The general objective of SA is to improve the long-term social development outcomes of post-disaster reconstruction policies, programs, or projects by analyzing and managing their social impacts and by mitigating risks.

The specific objectives are to (1) analyze the contextual factors of a particular project or sector policy and information on how these socio-cultural, institutional, historical, economic, and political factors may influence development outcomes; (2) identify the project's social impacts on all relevant stakeholders, including beneficiaries and other populations affected, and their corresponding strengths, vulnerabilities, and risks; (3) analyze implementing institutions and the institutional framework; (4) identify opportunities and specific constraints the project may encounter; and (5) make concrete recommendations of actions that will mitigate any adverse social impacts or improve social outcomes during implementation and monitoring of the project or policy. The process of social assessment can itself enhance project equity and strengthen social inclusion and cohesion, by facilitating the participation of relevant stakeholders, including the poor and socially excluded, in project analysis, design, and/ or implementation.

Methodology for Preparing a Social Assessment

The success of SA depends on the ability and capacity of the expert team to capture the multiple dimensions of the community social reality and to use this information to estimate social impacts and possible mitigation measures. It is fundamental that the team have sufficient experience in both the qualitative and quantitative aspects of social analysis, ideally in a post-disaster or similarly volatile context, and is comfortable working under time pressure. Below is a list of recommendations for conducing SA.

- Specialists should be hired to carry out this assessment, due to the complexity of the issues and the need to organize and interpret a wide range of information. The team should consist largely of experts in the social sciences, such as sociologists, anthropologists, geographers, social psychologists, or other persons experienced in social data collection and analysis of complex socio-cultural structures, as well as experts in political science and law. The composition of the team will vary, depending on the nature of the disaster and the project being analyzed.
- A suitable counterpart in government should be appointed who understands the importance of the work and who can facilitate contacts and access to information.
- This government official should be supported by a technical committee that includes representation from the affected population, key government agencies, and the sponsoring agency.

Sources of Information

SA is not a single method but can incorporate various approaches and tools to obtain, verify, and analyze data. Validating data in the post-disaster situation may be a challenge but should not be neglected. Data-gathering issues include the following.

- The socio-cultural, historical, and political context of the project will influence the data that is gathered, and the tools used, as will the complexity of social structures and perspectives that need to be incorporated.
- The strengths and limitations of data-gathering tools should be evaluated with respect to their validity, efficiency, and social acceptability during the planning of the assessment.
- While the affected population is the principal subject of the SA, it may also be engaged in data-gathering, analysis, mapping, focus groups, or other activities, and should be represented in the technical committee.
- Given the difficulty of data gathering in post-disaster situations, the technical committee should strongly consider requesting that the consultant team (1) collect data in such a way that it can be used as the baseline for later project monitoring and evaluation, and (2) propose concrete indicators and benchmarks to be used in monitoring and evaluating the project.
- In addition to the initial SA (described here), ongoing SAs should be carried out simultaneously with the execution of the project.

Scope of the Social Assessment

After reaching agreement on the principal objectives and methodology, the consultants should familiarize themselves with the most current version of the post-disaster reconstruction policy, program, or project under consideration, if one has already been proposed, or otherwise with the broad goals of the reconstruction program. Based on this, the team will gather and analyze information on (1) the socio-cultural, institutional, historical, and political context where the project takes place; (2), the legal and regulatory context; and (3) the key social issues, including economic factors and income distribution, diversity and gender, the roles and behavior of community groups and affected stakeholders, the types of social participation, and any potential social risks. A detailed list of topics to be analyzed is shown in the table below. The relative weighting of these issues in the analysis depends on the project being considered and the context.

Further Guidance on the Social Assessment

Institutions, roles, and behavior. This component of the analysis should consider both formal and informal institutions, the political and administrative apparatus, and "rules of the game" at various levels of government, as well as the influence of private sector institutions, community, kin, and solidarity rules. Macro-institutional issues may also be relevant to the project, as well as an analysis of obstacles to equitable access to and benefit from institutions and their resources. The reasons for exclusions can include local customs, intergroup relations, formal and customary laws, or information and communication systems, and may be intentional or unintentional.

Social and economic diversity and gender. The information and analysis presented should be disaggregated by gender and income level, and vulnerabilities and their causes for each group should be identified. A special focus should be put on social equity impacts and on the distribution of impacts across the different identified social groups. Quantitative analysis should be accompanied by confidence intervals and significance levels. The following concepts should be kept in mind.

Topic	Elements to analyze
A. Institutions, roles, and behavior	1. Examine social groups' characteristics, intragroup and intergroup relationships, and the relationships of those groups with public and private institutions.
	2. Describe formal and informal behaviors, norms, and values that have been institutionalized through these relationships and how they affect the implementation of the project.
	3. Describe possible opportunities to influence behavior of such groups.
	4. Point out constraints or potentials among these institutions for the project's implementation.
	5. Summarize historical facts that are directly linked with the project framework and outcome range.
	6. Describe the political framework relevant to the project.
B. Legal and regulatory considerations	1. Review and summarize all national, local, and intermediate legislation and regulations pertinent to the project.
	2. Highlight in particular legislation and regulations that provide social assistance to poor and excluded groups.
C. Social and economic diversity and gender[2]	1. Describe the most significant social and cultural features that differentiate social groups in the project area.
	2. Examine how people are organized into different social groups, based on the ascribed status (ethnicity, clan, gender, locality, age, language, class, or other marker), achieved status, or chosen identity (ideology, education, citizen, political affiliation).
	3. Analyze the economic structure of the community and other factors that may influence local political decision making related to reconstruction, such as the allocation of assistance and public expenditures.
	4. Describe the assets and capabilities of diverse social groups.
	5. Analyze dynamic social and political power relations and their implications for the realization of the project.
	6. Explore current visible or underlying conflicts among the groups.
	7. Describe their different interests in the project and their level of influence.
D. Stakeholders	1. Identify and characterize the various stakeholders.[3]
	2. Explore the different stakeholder's interests, motivations, and incentives in the project.
	3. Describe the impacts the project will have on the different groups of stakeholders.
	4. Analyze their existing and lacking assets and capabilities, both material and intangible, and present them in a table.
E. Participation	1. Describe the local traditional systems of participation and its mechanisms of inclusion and exclusion, and evaluate its legitimacy to serve as project participation from.
	2. Based on the asset and capability table (see D, Stakeholders), explore opportunities and conditions for participation by stakeholders, particularly the poor and vulnerable, in the project process.
	3. Develop mechanisms to enhance marginalized groups' skills and encourage them to participate in the project.
	4. Develop communication strategies to inform stakeholders and a feedback mechanism to include stakeholder's reactions. The communication of information is a basic asset to be able to participate.
F. Social risks and vulnerability	1. Analyze all economic and social effects the project may have on the poor and excluded.
	2. Examine specific social risks[4] according to the different social groups identified, especially on vulnerable groups.
	3. Analyze the perceptions of the affected groups regarding vulnerability and social risk and compare this data with results from other activities.
	4. Identify the country risks caused by political instability; conflict; ethnic, religious, or social tensions; endemic corruption; etc.

- *Practical gender needs vs. strategic gender needs.* "'Practical gender needs'" are based on local traditional gender roles and responsibilities and focus on immediate practical needs, such as water, food, shelter, and health. In contrast, "strategic gender needs" analyze systemic factors that limit women's access to resources and benefits compared to men's. The analysis and comparison of these two types of needs may help facilitate a sustainable, long-term mitigation response.

- *Intrahousehold dynamics and relations.* It may be helpful to picture the household as a system that allocates resources among individuals, each of whom is supported by her or his own internal and external relations. In such a system, the modification of one part can affect the whole. Hence, a holistic understanding of the system is fundamental to estimate multiple social impacts of an external intervention.

Stakeholders. The stakeholder analysis should include the characteristics, interests, incentives, and mode of influence over the project, particularly elements that adversely affect the allocation of resources and control over the quality of design and implementation. Note that the degree of organization often affects the degree of visibility and the ability of groups to express and defend their interests. Vulnerable social groups are often not organized and for this reason need more support to be heard and included.

Participation. The development of communication strategies to share information and ensure the continuous flow of information contributes to participation. See ⌨ Chapter 3, Communication in Post-Disaster Reconstruction. Beside a communication strategy that reaches all stakeholders, the skills of vulnerable and marginalized groups may need to be enhanced to ensure their participation in the project. Procedures to involve stakeholders in monitoring and evaluation are important. Be aware that participation, while a fundamental element for project planning, implementation, and evaluation, does not guarantee the desired results.

Social risks and vulnerability. Make sure that the particularly vulnerable groups are identified, defining vulnerability beyond the traditional so that is includes groups that are socially stigmatized (such as battered women) or marginalized (people infected with HIV or suffering from AIDS). The analysis should examine the nature and roots of these vulnerabilities in the context of socioeconomic trends in the country or region.

Presentation of Findings and Recommendations

For each topic in the table above (and others the consultants may identify during the assessment), the consultants should provide a systematic summary of (1) their findings as they relate to the housing reconstruction policy, program, or project under consideration; and (2) the significant corresponding social impacts they have identified. The team should present short- and medium-term recommendations for improving the social outcomes or mitigating any adverse social impacts of the project. The recommendations should be grouped in the way that the consultants believe will make them the most understandable during the review process and, in the final

report, most useful for implementation. After an initial review by the technical committee and other stakeholders, as directed by the technical committee, the recommendations should be presented in a final report as a work plan that identifies both the sequence of activities and the party or parties responsible for carrying them out, focusing particularly on modifications in project design or social risk mitigation activities.

Expected Results and Outputs of the Social Assessment

The principal output is an in-depth SA for the policy, program, or project that will permit government and/or other agencies to mitigate any adverse social impacts or improve social outcomes by making adjustments in project design and designing a system for project monitoring. In the initial report, the consultants will present a strategy, plan, and schedule for the consultancy. The assessment itself should be presented in draft and final forms.

Time will usually be of the essence in carrying out this consultancy. The following schedule allows an SA to be completed in approximately 2 months. The following time intervals are ambitious, and, if necessary, can be adjusted, depending on the particular situation. Outputs will include:

1. an initial report, in which the consultants any recommendations for modification of the scope of work as well as a work plan and schedule for the presentation of outputs, presented within 7 days of the contract signing;
2. a draft report, presented within approximately 21 days of the acceptance of the initial report;
3. a final report, presented within 21 days of the receipt of comments on the draft report from the party or parties responsible for overseeing the assessment or 30 days of the presentation of the draft report, whichever is earlier.

The draft and final reports should be presented along with an executive summary or abbreviated version that can be widely circulated, in language(s) and a format that stakeholders can easily understand.

An effective review process will help guarantee the acceptance of the SA, and the consultants should take an active role in carrying it out, with assistance from government and the sponsor of the assessment. This may entail various meetings with government, the community, and other stakeholders; use of information technology; and/or other means to ensure wide distribution of the draft report and collection of feedback. Meetings may also be required once the report is finalized, to more widely disseminate the findings and recommendations.

Annex 2 Endnotes

1. World Bank, 2003, *Social Analysis Sourcebook: Incorporating Social Dimensions into Bank-Supported Projects* (Washington, DC: World Bank); John Twigg, 2007, *Tools for Mainstreaming Disaster Risk Reduction, Social Impact Assessment, Guidance Note 11* (Geneva: International Federation of Red Cross and Red Crescent Societies and the ProVention Consortium), http://www.proventionconsortium.org/themes/default/pdfs/tools_for_mainstreaming_GN11.pdf; *and* International Association of Impact Assessment, 2003, "Social Impact Assessment. International Principles," Special Publications Series No. 2, http://www.iaia.org/publicdocuments/special-publications/SP2.pdf.

2. According to the World Health Organization (WHO), "gender" refers to the socially constructed roles, behaviors, activities, and attributes that a given society considers appropriate for men and women. Gender is an important consideration in SA.

3. The term "stakeholder" includes the people affected by the project (beneficiaries, affected population) and people able to influence it (organizations, institutions). See also ⌨ Chapter 12, Community Organizing and Participation, for a discussion of this topic.

4. "Social risks" include country risks, political economy risks, institutional risks, exogenous risks, and vulnerability risks, among others.

Guiding Principles for Relocation

- An effective relocation plan is one that the affected population helps develop and views positively.
- Relocation is not an "either/or" decision; risk may be sufficiently reduced simply by reducing the population of a settlement, rather than by relocating it entirely.
- Relocation is not only about rehousing people, but also about reviving livelihoods and rebuilding the community, the environment, and social capital.
- It is better to create incentives that encourage people to relocate than to force them to leave.
- Relocation should take place as close to the original community as possible.
- The host community is part of the affected population and should be involved in planning.

**This Chapter Is
Especially Useful For:**
- Policy makers
- Lead disaster agency
- Agencies involved in reconstruction
- Affected communities

Introduction

Relocation is defined as a process whereby a community's housing, assets, and public infrastructure are rebuilt in another location. Relocation is sometimes perceived to be the best option after a disaster for one or more of the following reasons: (1) people have already been displaced by the disaster, (2) their current location is judged to be uninhabitable, or (3) relocation is considered the best option to reduce vulnerability to the risk of future disasters. In fact, relocation may be appropriate when the disaster is the result of site-specific vulnerabilities. Informal settlements in urban areas, for instance, are often located on sites where topography makes the site's vulnerabilities impossible to mitigate. In rural areas, settlements on fault times or in flood zones have vulnerabilities that may also be impossible to address.

However, relocation is often not the right solution: not all risks are site-specific and relocation itself entails numerous risks. Finding adequate sites for relocating disaster-affected communities can be an enormous challenge. Unsuitable new sites can lead to lost livelihoods, lost sense of community and social capital, cultural alienation, poverty, and people abandoning the new sites and returning to the location of their original community. The economic, social, and environmental costs of relocation should be carefully assessed before the decision to relocate is finalized, and other mitigation options should be considered. For instance, sometimes relocating only a portion of an at-risk community may be sufficient.

This chapter discusses the reasons for and against relocation of disaster-affected communities following a disaster, as well as the risks and risk mitigation strategies that can be used if relocation is necessary. It warns against choosing relocation out of organizational convenience without taking into consideration its potentially dramatic negative social consequences. This chapter is not about "resettlement" as defined by the World Bank and other international financial institutions (IFIs), nor is it a summary of IFI resettlement policies (which are discussed below). However, the approach recommended in this chapter is consistent in many ways with these policies.

Key Decisions

1. The **lead disaster agency** should coordinate with appropriate **government agencies, including local government**, to initiate an inclusive in-depth comparative analysis of disaster risk management (DRM) options that includes mitigation at the existing site.
2. As soon as relocation is raised as a serious post-disaster risk mitigation strategy, the **lead disaster agency** should initiate a process for defining the policy framework for relocation, the financing plan, the assistance strategy for those relocated, and the criteria for household selection and relocation site selection.
3. The **lead disaster agency**, in coordination with **local government**, should quantify the population subject to relocation through their joint participation in assessments that will provide these estimates.
4. **Local government** should carefully identify relocation sites, in the context of the post-disaster land use planning process, that offer the best potential to provide sustainable living and livelihood conditions to the relocated population.

5. **Agencies involved in reconstruction** should decide how to collaborate with government to establish common policies and criteria for relocation, and on the common procedures for applying them.
6. **Agencies involved in reconstruction** should decide and plan how their relocation projects will ensure the full restoration of livelihood and social conditions in the relocation site, including special attention to squatters and vulnerable groups.
7. **Populations subject to relocation** and **receiving communities** should demand that **agencies involved in reconstruction** give them a lead role in identifying sites and organizing relocation.
8. **Agencies involved in reconstruction** should decide how to organize and finance joint monitoring of relocation projects, and how to ensure that findings will be incorporated into ongoing projects.

Public Policies Related to Relocation

Public agencies at the national and local government levels in disaster-affected countries may have relocation or involuntary resettlement policies that apply in post-disaster situations or that can easily be adapted. Using them helps ensure that post-disaster relocation criteria and assistance schemes are consistent with other instances of relocation in the same country or state. If policies were established in connection with infrastructure projects, such as highway widening when squatters needed to be relocated from a public right-of-way, policy implementation may fall within the jurisdiction of the Ministry of Public Works or the Social Investment Fund.

At the local government level, resettlement or relocation policy may be established in connection with slum upgrading, local infrastructure projects, city development master plans, or DRM plans. The local agency with jurisdiction may be the planning department, the public works agency, or the agency responsible for environmental management. Policies intended to guide relocation from high-risk areas or to disperse illegal settlements may be readily applicable or may need to be modified to apply in a post-disaster situation. Such policies may include useful methodologies for selecting among mitigation options.

The World Bank safeguard policy on involuntary resettlement (Operational Policy 4.01), as well as those of many international and bilateral agencies and regional development banks, is designed to assist displaced persons in their efforts to improve or at least restore their income and standard of living after displacement; however, it may not apply in a post-disaster situation. (See 📖 Chapter 21, Safeguard Policies for World Bank Reconstruction Projects, for a description of how safeguard policies are applied in emergency [disaster] operations.) Resettlement policies are discussed later in this chapter.

International frameworks should be taken into consideration when the possibility of relocation arises, including the Pinheiro Principles on Housing and Property Restitution for Refugees and Displaced Persons.[1]

Whatever policy framework or frameworks are used to define the relocation policy, the policy and related procedures should be transparently and publicly reviewed with and communicated to the affected population throughout planning and implementation. See 📖 Chapter 3, Communication in Post-Disaster Reconstruction.

Technical Issues
The Typology of Reasons for Displacement

Disasters are only one cause of displacement, whether economic or physical. Others reasons for displacement that countries experience, and that countries may have policies and know-how to address, include the following:

- Development-induced involuntary resettlement,[2] including:
  - Relocation or loss of shelter
  - Loss of assets or access to assets
  - Loss of income sources or means of livelihood, whether or not the affected persons must move to another location
- Disaster-induced relocation
  - Voluntary
  - Involuntary
- Cyclical relocation, due to seasonal flooding, drought, or other factors
- Refugees from conflict

1 . Pinheiro Principles on Housing and Property Restitution for Refugees and Displaced Persons, http://www.cohre.org/store/attachments/Pinheiro%20Principles.pdf; *and* "World Bank Operational Policy 4.12: Involuntary Resettlement," http://web.worldbank.org/WBSITE/EXTERNAL/PROJECTS/EXTPOLICIES/EXTOPMANUAL/0,,contentMDK:20064610~pagePK:64141683~piPK:64141620~theSitePK:502184,00.html. The World Bank safeguard policies address the risks associated with involuntary resettlement. However, they are not always applied in post-disaster housing reconstruction projects.
2. World Bank OP/Bank Procedure (BP) 4.12, "Involuntary Resettlement," apply principally to these instances of resettlement.

This chapter focuses on disaster-induced relocation, whether voluntary and involuntary. Development-induced involuntary resettlement is discussed below.

Who Lives in Disaster-Prone Sites and Why

The urban poor in particular often inhabit hazardous areas because they can't afford to live elsewhere. The primary concern of people living in poverty is their immediate survival, which requires them to find affordable housing in close proximity to livelihood opportunities. For people with marginal incomes, even minor additional costs of rent, utilities, or transportation that might result from living in a safer location may be unaffordable. Safe and affordable sites are hard to find in areas where jobs are located, where land is likely to be scarce and prices higher. Poor urban dwellers often settle informally on public lands not suitable for development because of their inherent risk factors and then remain there for financial or political reasons until a disaster strikes.

DANIEL PITTET

Why Relocation Is Sometimes Necessary

Disasters will continue to displace people, often leaving no alternative but relocation. Relocation of vulnerable communities to physically safer places is often the best way to protect them from future disasters. Some locations are inherently unsafe, e.g., floodplains, unstable hillsides, and areas where soil is likely to liquefy as a result of seismic tremors. In particular, informal settlements of the urban poor are often located on highly vulnerable sites. In some cases, a disaster may have changed the topography, making a community's original site unsuitable for habitation. Finally, it may be too costly to provide safety to communities located in areas likely to be subject to future disasters. Risk-mapping is a tool that can provide data on the degree, probability, and characteristics of these risks. However, a relocation process that incorporates international lessons learned can prevent avoidable human suffering. In Aceh, Indonesia, following the 2004 Indian Ocean tsunami, changes in topography greatly complicated site selection for new housing, as described in the 🖼 case study, below.

Why Relocation Is Often Unsuccessful

Inadequacy of new sites. One of the chief reasons for relocation failure is underweighting the welfare of the population as a criterion for the selection of the relocation site. Inappropriate land may be chosen for a relocation project because it can be acquired quickly, is owned or controlled by government, or is easily accessible with topography that favors rapid construction. For similar reasons, people resettled to protect them from one risk (e.g., tsunamis) may find themselves exposed to new ones (e.g., risks to livelihood, high crime, lack of services). The 🖼 Disaster Risk Management section in Part 4, Technical References, describes the process for comparing risk mitigation options.

Distance from livelihoods and social networks. A lack of affordable land in areas close to sources of employment often necessitates relocation to peripheral areas where land is less expensive. Yet a key cause for unsustainable relocation solutions is the distance of the new site from vital resources (grazing land, food sources), relatives, social networks, livelihoods, and markets. In addition, bringing infrastructure and services to these remote areas may be extremely expensive, even when the land is cheap. The full cost analysis of new sites should include both infrastructure investment and the provision of services, such as public transportation. The 🖼 case study on the 2004 Indian Ocean tsunami in Sri Lanka, below, reveals how livelihoods can be affected when vendors relocate further away from markets.

Socio-culturally inappropriate settlement layouts. Housing design, layouts, and construction are often to blame for the rejection or failure of post-disaster relocation projects, in particular in rural areas. The following are frequently cited reasons for the abandonment of a new site by a resettled community.

- Settlements are designed using unfamiliar land use patterns that do not permit the clustering of kin and neighborhood groups vital to social cohesion in rural areas.

- There is insufficient space for tool sheds, livestock, and other agricultural needs, as well as poor soil conditions, along with lack of irrigation, tools, agricultural inputs, and livestock, making it is difficult to reestablish farm-based livelihoods in agricultural areas.
- Faulty house design and construction (such as the lack of thermal protection), limited plot dimensions, difficulty of extending and upgrading houses, and lack of space for domestic and livelihood activities.
- Poor access and lack of public transportation, particularly to markets and social facilities.
- Conflicts and competition with host or adjacent communities that do not receive any benefit from the relocation and lack structures for the governance of resources.
- Social conflicts caused by moving communities with different ethnic, religious, or social backgrounds into close proximity.
- Widows and female-headed households exposed to sexual and physical abuse.

Most of these risks also apply to reconstruction *in-situ* if the reconstruction plan entails land consolidation, changes in settlement layout, or introduction of new house designs and building technologies. A relocation plan (albeit abbreviated) may be needed even in these situations.

Lack of community participation. Consulting the people of a community, involving them in the selection and planning of a site, understanding their needs and values, and gaining insight from local experience and knowledge of the local environment can help reduce relocation risks. Importing outside labor to construct new settlements discourages community participation and deprives members of the community of employment opportunities. A lack of community participation can also hinder the development of a personal sense of ownership or responsibility for the home and settlement, which may lead to feelings of alienation and a prolonged dependency on external aid. The 📖 case study on the 2004 Indian Ocean tsunami reconstruction in India, below, describes a time-consuming, but successful, instance where communities took responsibility for selecting relocation sites.

Populations subject to relocation should demand that agencies give them a lead role in identifying and sites and organizing relocation.

Underbudgeting of relocation costs. Underestimating the cost of relocation is common and can undermine the entire process. Both hard costs (infrastructure, housing construction) and soft costs (facilitation, training, social assistance, temporary public services) should be estimated using conservative assumptions, and funded over a period of years, until communities fully adapt to their new location and livelihoods are reestablished. The estimates should include adequate provision for costs associated with assisting squatters or those without proof of land ownership and other land tenure issues. See the discussion of land tenure challenges in reconstruction in 📖 Chapter 7, Land Use and Physical Planning.

What Contributes to Successful Relocation?

Relocation of communities requires risk mitigation through well-planned and adequately financed programs that include such elements as land-for-land exchange, employment generation, ensured food security, improved access to health services, transportation to jobs, restoration of common properties, and support for community and economic development.

Relocation is more likely to be successful when:
- affected communities participate in critical relocation and implementation decisions (site selection, identification of basic needs, settlement planning, housing designs, and implementation);
- livelihoods are not site-specific and so are not disrupted;
- water, public transport, health services, markets, and schools are accessible and affordable;
- people are able to bring with them items of high emotional, spiritual, or cultural value (religious objects, salvaged building parts, statuary or other local landmarks);
- people belonging to the same community are resettled together to a new site;
- emotional, spiritual, and cultural attachment to the old site is not excessively high;
- housing designs, settlement layouts, natural habitat, and community facilities conform to a community's way of life;
- social, environmental, and hazard risk assessments confirm that risk cannot be mitigated in the old location, while the community can be assured of the suitability of the relocation site;
- communication with target groups is frequent and transparent, and mechanisms to resolve grievances are effective; and
- relocation and assistance to mitigate its economic impacts are adequately funded over a reasonable period of time.

Unjustified Relocation

Relocation to new sites is often decided for "practical reasons" that ignore risk management considerations and result in a massive waste of financial and natural resources. Examples include:

- Relocation to avoid rubble removal, simplify land tenure issues, or minimize the number of stakeholders "interfering" in the reconstruction project
- Relocation to reduce construction costs, without accounting for the cost of basic infrastructure and services, which can result in the building of houses or entire settlements that are later abandoned, sold by beneficiaries, or left unoccupied, due to the lack of services or costs to acquire them

Involuntary Resettlement Policies

Definition of involuntary resettlement. Resettlement is a term used to describe direct economic and social losses resulting from displacement caused by land taking or restriction of access to land, together with the consequent compensatory and remedial measures.[3] Resettlement activities in World Bank loans and projects are governed by the Safeguards Policy on Involuntary Resettlement, including Operational Policy (OP) and Bank Procedure (BP) 4.12.[4] The policy promotes the participation of displaced people in resettlement planning and implementation and prescribes compensation and other resettlement measures. Countries that borrow from the Bank often prepare resettlement plans; therefore, numerous examples are available.[5] For guidance on preparing a resettlement plan, see the annex to this chapter, 📖 How to Do It: Developing a Post-Disaster Resettlement Plan. Other international institutions, such as the Asian Development Bank[6] (ADB) and the Inter-American Development Bank,[7] have policies similar to OP/BP 4.12.

Resettlement policies may not apply following a disaster, because time does not allow for it or because the situation may not trigger the policy. 📖 Chapter 21, Safeguard Policies for World Bank Reconstruction Projects, describes how the World Bank's safeguard policies are applied in emergency (disaster) operations.

In resettlement policies, relocation is identified as one of several strategies to consider when either economic or physical displacement is taking place, generally as the result of public investment projects or other changes in land use. The ADB Involuntary Resettlement Policy includes the following matrix of types of losses from displacement and the mitigation measures that should be evaluated in resettlement plans to compensate for them.[8]

Type of loss	Mitigation measures
Loss of productive assets, including land, income, and livelihood	■ Compensation at replacement rates ■ Replacement for lost incomes and livelihoods ■ Income substitution and transfer costs during reestablishment plus income restoration measures in the case of lost livelihoods
Loss of housing, possibly entire community structures, systems, and services	■ Compensation for lost housing and associated assets at replacement rates ■ Relocation options, including relocation site development ■ Measures to restore living standards
Loss of other assets	■ Compensation at replacement rates ■ Replacement
Loss of community resources, habitat, cultural sites, and goods	■ Replacement ■ Compensation at replacement rates ■ Restoration measures

Comparing development-related and disaster-related displacement. Post-disaster relocation, like resettlement, may also be involuntary, and the same strategies used to reduce or avoid involuntary resettlement impacts can sometimes be applied in relocation. These include (1) mitigating the risks that are causing relocation to be evaluated as an option, using physical preventative or physical coping and adaptive measure (see 📖 Part 4, Technical References, Disaster

3. World Bank, 2004, *Involuntary Relocation Sourcebook: Planning and Implementation in Development Projects* (Washington, DC: World Bank), http://publications.worldbank.org/ecommerce/catalog/product?item_id=2444882.
4. World Bank OP/BP 4.12, 2001, "Involuntary Resettlement," http://go.worldbank.org/ZDIJXP7TQ0.
5. World Bank, "Resettlement Plans," http://go.worldbank.org/HRF9IQRLT0.
6. Asian Development Bank, 1995, "Involuntary Resettlement," http://www.adb.org/Documents/Policies/Involuntary_Resettlement/involuntary_resettlement.pdf.
7. Inter-American Development Bank, 1999, *Involuntary Resettlement in IDB Projects: Principles and Guidelines*, http://www.iadb.org/sds/doc/Ind-ADeruyttrePGIRPE.PDF.
8. ADB, 1998, *Handbook on Resettlement: A Guide to Good Practice* (Manila: ADB), http://www.adb.org/Documents/Handbooks/Resettlement/Handbook_on_Resettlement.pdf.

EARL KESSLER

Risk Management in Reconstruction, for more on disaster risk reduction options); (2) redesigning or replanning the physical site to accommodate all residents (internal relocation); and (3) providing incentives for residents to relocate themselves (voluntary relocation).

At the same time, the situation that confronts government, agencies, and households involved in infrastructure-related relocation is different from that encountered in a disaster-related resettlement, for a number of reasons that may affect the quality of the outcomes. In the case of a disaster, for instance, the land that is taken is often left vacant, rather than being transformed into something else, e.g., a roadway overpass, as it is with development-related resettlement, making it possible for the displaced population to return and making it necessary for local jurisdictions to prevent this from happening. There is ordinarily less time available to plan and implement a disaster-related relocation than there is a development-related resettlement program, which creates the risk that the full range of options may not be evaluated. If the property market has been affected by the disaster, voluntary resettlement may not be realistic, without the affected household moving a significant distance away from the area affected by the disaster. Properly planned resettlement may be a requirement of IFI financing for development projects, and, if so, technical and financial support are likely to be provided to assist in carrying it out, in contrast with post-disaster relocation, where neither of these factors may be present. Last, the population affected by a disaster may have been dispersed, making it more difficult to develop an approach to relocation that satisfies the entire community and keeps it intact, in contrast to a development-related resettlement project where there is time for participatory resettlement planning.

Risks and Challenges

1. Underestimation by decision makers of the social consequences of post-disaster relocation, in spite of the growing body of research that shows that it is rarely successful.
2. Loss of livelihoods, impoverishment, social and cultural alienation, loss of social coherence, increased morbidity, and loss of access to common property for the relocated community.
3. Conflicts and competition with hosting communities over scarce resources, such as land, food, fuel, water, and fodder for livestock.
4. Abandonment of relocation sites by relocated populations and return to areas where there may be inadequate provision for them or unsafe conditions. Failure of local officials to anticipate this event.
5. Insufficient consideration of the option of providing incentives to encourage voluntary relocation.
6. Government inaccurately reporting that relocation has taken place voluntarily in order to avoid the preparation of social and environmental impact assessments and relocation action plans.
7. Failure to recognize and mitigate risks of reconstruction projects in the same location that entail land consolidation, major demolition, and development of new settlement layouts.

Recommendations

1. Avoid relocation if at all possible. Especially avoid relocation to distant sites. Work hard to keep communities together.
2. If relocation is being considered, carry out a detailed participatory assessment of the environmental, social, and economic risks of relocation and of the cost of risk mitigation strategies for alternative sites.
3. Governments should not only avoid relocation in their own housing programs but should also regulate relocation in the reconstruction projects of nongovernmental agencies (private corporations and nongovernmental organizations [NGOs]), which often opt for relocation to gain visibility and for managerial convenience.
4. If relocation is unavoidable, involve the community in the decision-making processes by creating a community relocation committee, among other means.
5. Agencies should engage the services of qualified and experienced relocation specialists to design and implement relocation plans.
6. The technical, financial, and institutional feasibility of providing basic services such as water, electricity, health services, schools, markets, policing, and public transport in the relocation site must be demonstrated during project planning, and all arrangements put in place in advance of the relocation.
7. Use the relocation plan to carefully define, with the assistance of experts, how people will be assisted to restore their livelihood activities or develop alternative livelihoods in the relocation site.
8. Plan for the relocation of individual or collective cultural properties.
9. Assess and mitigate the impact of relocation on the hosting community, and be prepared to prevent social conflicts and problems of crime, delinquency, and secondary displacement.
10. Design, budget for, and implement measures to prevent the return of the relocated community or others to the site from which the relocation took place.
11. Be conservative when estimating the time a relocation program will take and the costs entailed.

Case Studies

1998 Hurricane Mitch, Honduras

The Consequences of Relocation without Prior Infrastructure Planning

As a camp-exit strategy, families living in temporary shelter camps in Tegucigalpa, Honduras, after Hurricane Mitch, were assisted with a voucher program that provided US$600 for the acquisition of a new house. This voucher program was to be combined with generous subsidies being offered by local and international NGOs in new relocation projects that they were building. The only affordable land available for the relocation projects was located in the Amarateca Valley, 35 kilometers from the center of Tegucigalpa. NGOs developed housing projects for more than 1,200 families who could contribute the voucher amount, and provided them with varying amounts of additional subsidies. However, these efforts were uncoordinated and poorly planned. The lack of planning was evident in the fact that at many of the sites there had been no arrangement with government and public utilities to provide infrastructure services (e.g., water, sewerage, electricity, and solid waste collection) and social services (e.g., schools, transportation, and health clinics) on a timely basis. Under pressure from the relocating families and from government (which was pushing to get families out of the shelters), relocation took place with improvised, temporary solutions (e.g., pit latrines and water supplied by tanker trucks). In some cases, individual housing projects included internal piped networks for water supply and sanitary sewerage, but the deep wells required for water supply and the facilities for wastewater treatment were not completed until years after the families had occupied the projects. As a result, there were sanitation and health hazards, defaults on house payments, loss of livelihood opportunities, disruption of social networks, and even social unrest and insecurity in the new settlements, all of which represented obstacles to developing a real sense of community. Ten years after the emergency, however, the Valley of Amarateca had attracted new employment opportunities ranging from textile factories, grain processors, and automobile parts assembly, attributed partly to the concentration of population in this location.

Source: Mario C. Flores, Habitat for Humanity International, 2009, personal communication, http://www.hfhi.org.

2004 Indian Ocean Tsunami, Sri Lanka

The Impact of Post-Tsunami Relocation on People's Livelihoods and Housing Choices

After the 2004 Indian Ocean tsunami, the government of Sri Lanka announced that no reconstruction would be allowed within a buffer zone, which varies from 100 to 200 meters, along the water. As a result, thousands of households had to be resettled. Research conducted in 2008 using a random sample of 211 households selected from 17 relocation sites in the Hambantota district of the country found that, while 96 percent of the households in the sample considered their new houses similar or superior in quality to their pre-tsunami houses, relocation generally had had an impact on their livelihoods. This was due to several factors, among them that in their pre-tsunami homes, many of the families had goats, cattle, and poultry; homestead gardens; and coconut trees (a staple food in Sri Lanka). They also enjoyed access to free fish. Livestock and poultry provided food security and constituted critical assets in case of financial emergencies. This changed in the relocation sites, where people were not able to keep the same number of animals. The number of animals owned by the sample households decreased from more than 6,400 before the tsunami to only 107 after the tsunami. People reported that they were consuming less fish, vegetables, and fruits than before the tsunami. Second, relocation led to a reduction in earning opportunities, in particular for women and the poor. The distance to markets from the relocation sites meant that the small incomes generated from micro-businesses in their homes, such as food processing, were now not sufficient to cover the transport expenses from their new homes to the market. As a result, there was a 59 percent decrease in the number of family members who were earning anything among the 211 households in the sample.

Reconstruction in Hambantota was unusual in that it produced more houses than were needed for the disaster-affected population, for several reasons. Being the home constituency of the country's president, it attracted generous resources from national and international NGOs; some families were not willing to relocate to new sites for reasons not picked up in the needs assessments; and delays in developing relocation sites led some families to purchase lands and construct their own houses using the housing grants before the abolition of the buffer zone policy. Also, because some people had not relinquished their pre-tsunami property, they were able to move back to their original housing sites after the buffer zone was reduced. For these reasons, some houses that were built outside the buffer zone by international NGOs for tsunami-affected communities in Hambantota were later given to non-affected households, for example, to people displaced by the construction of a new port. As of mid-2009, 63 percent of houses in the 17 relocation sites analyzed were occupied by people affected by the tsunami.

Officials involved in this reconstruction program in Sri Lanka have pointed out the importance of addressing the following issues in reconstruction: (1) the need to engage NGOs and to align their priorities with larger reconstruction program objectives; (2) the importance of clarity in public policies regarding relocation and occupation of environmentally sensitive areas, such as the buffer zone; and (3) how to simultaneously weigh and address the livelihood and housing reconstruction requirements of the same population.

Sources: World Habitat Research Centre and Centre for Environmental Studies, University of Peradeniya, 2009, "Preliminary Findings of an Ongoing Research Project on Post-Tsunami Resettlement and Livelihoods in Sri Lanka," http://www.worldhabitat.supsi.ch; *and* Narayanan Edadan, 2009, written communication.

2004 Indian Ocean Tsunami, Nagapattinam, India

Finding Land for Relocation through Community Participation

More than 30,000 families being suddenly rendered homeless is a nightmare under any circumstances. But in a backward district like Nagapattinam, India, it is a disaster—even worse when diverse cultures and livelihood systems are thrown into the mix. Although relocation from vulnerable coastal areas was deemed necessary after the 2004 Indian Ocean tsunami, moving fishing communities whose lifeline is the waterfront was not so easy. Relocation decisions needed to factor in safety, proximity to traditional livelihoods, and safeguarding the community cohesion that remains strong in traditional communities like fishers. The basic tenet of relocation decisions in Nagapattinam was that a hamlet—usually consisting of the same community—would be treated as an indivisible unit. While the decision to proceed in this manner was unanimous, two-thirds of Nagapattinam is below sea level and much available vacant land was considered inappropriate for housing, so the identification of suitable land took nearly six months. Ten teams of local administration officials searched geographically demarcated areas for appropriate land and initiated negotiations. However, no agreement could be finalized by the local administration until the community approved the land. On some occasions, as many as eight rounds of negotiations with the community were necessary before final approval was won. There were also cases where land was rejected by the community. In one case, where the land was away from the sea front, the local administration agreed to widen the backwater channel to allow boats to be brought to the site. In another case, the community objected to the proximity of the land to a cremation site, so a wall was built to separate the two. In a third case, prime property that had earlier belonged to Tata Steel Rolling Mills was handed over to the community when it was the only property the community could agree on. In all, 364 hectares were bought by the government of Tamil Nadu through negotiation with the land owners at a cost of US$5 million. The local administration's willingness to be sensitive to the communities' needs may have delayed relocation, but it ensured that citizens were satisfied with their decisions, and their basic right to a dignified life was preserved.

Source: C. V. Sankar, India National Disaster Management Authority, 2009, personal communication.

2008 Typhoon Frank, Iloilo City, Philippines

NGO Support for Relocation of Vulnerable People Using Low-Interest Loans

The flooding that resulted after Typhoon Frank lashed the Western Visayas region of the Philippines in June 2008 covered 80 percent of Iloilo City, a city of more than 400,000 people. The typhoon killed 24 people, damaged more than 6,000 houses, and affected 53,000 families.

While Typhoon Frank was obviously an extreme event, the urban poor located on Iloilo's river banks actually face flooding every year during the monsoon period. The Homeless People's Federation Philippines (HPFP), one of the biggest NGOs collaborating with the urban poor in the Philippines, had previously organized a city-wide network, the Iloilo City Urban Poor Network (ICUPN), consisting of three major Iloilo NGOs (HPFP; Iloilo City Urban Poor Federation, Inc. [ICUPFI]; and the Iloilo Federation of Community Associations). ICUPN had been working with local government units (LGUs) for some time to develop a flood control plan to address the exposure of poor households to the flooding problem. When implementation of the plan began after Typhoon Frank, land for relocation originally acquired by the LGUs in early 2000 was assigned to the Typhoon Frank-affected families. The land covered 16.2 hectares and was both in the city and within 6 km of where people had originally lived. Various organizations received land for reconstruction: HPFP received 1.5 hectares and constructed 172 housing units. The affected

families, selected by HPFP and ICUPN in collaboration with the communities, could choose from among three housing models, with prices between US$1,770 and US$3,650. The houses were purchased using low-interest loans (between 3 percent and 6 percent) from the Urban Poor Development Fund. A key factor in the success of the program was that, before the typhoon, the community had been organized into saving groups. These groups are now purchasing the land, and each family will receive its individual land title only after the loan is paid back. Families who cannot manage the loan payments can provide "sweat equity" during construction. While this approach has many positive aspects, one issue was the lack of infrastructure on the relocation sites when first settled. The aim is to complete the infrastructure incrementally over the next 3–5 years.

Source: Sonia Cadornigara, 2009, "Thinking City-Wide in Iloilo City, Philippines, Notes on a Visit to Iloilo City," HPFP (unpublished).

DANIEL PITTET

For access to additional resources and information on this topic, please visit the handbook Web site at www.housingreconstruction.org.

2004 Indian Ocean Tsunami, Aceh, Indonesia

Unsustainable In-Situ Reconstruction due to Topographic Changes

Aware of the undesirable social consequences of relocation, the Indonesian NGO Urban Poor Linkage Indonesia (UPLINK) successfully advocated against government's resettlement plan and for people's right to return to their native villages in Aceh, Indonesia, after the 2004 Indian Ocean tsunami. Yet the tsunami had caused significant changes in local topography that made in-situ reconstruction inadvisable in some cases. In several villages, houses were uninhabitable due to water intrusion, and significant land areas had also been lost to the sea. As a result, people returning to their villages were forced to build houses in former paddy fields, despite the fact that the land was too low

JENNIFER DUYNE BARENSTEIN

for homebuilding before the tsunami and was even more so after the topographic changes. The Rehabilitation and Reconstruction Agency of Aceh and Nias (Badan Rehabilitasi dan Rekonstruksi [BRR]) took some measures to mitigate this problem, but they often did not have the desired effect. For example, in Lam Awe, BRR built an embankment that ended up impeding the runoff of both storm water and sewage, due to the lack of drainage outlets or sluices. In other villages, land-filling activities carried by the BRR actually increased the vulnerability of some houses previously built by UPLINK. The photo, above, shows some results from these missteps. They illustrate the risks associated with the lack of land use planning and demonstrate that there are cases when post-disaster *in-situ* reconstruction is not the most appropriate approach, since the consequences can be as devastating as the disaster itself.

Source: Jennifer Duyne Barenstein, Methodius Kusumahadi, and Kamal Arif, 2007, "People-Driven Reconstruction and Rehabilitation in Aceh. A review of UPLINK's Concepts, Strategies and Achievement" (evaluation by World Habitat Research Centre under contract to Misereor).

Resources

Cernea, Michael. 1997. "The Risks and Reconstruction Model for Resettling Displaced Populations." *World Development* 25, no. 10: 1569–87. http://ideas.repec.org/a/eee/wdevel/v25y1997i10p1569-1587.html.

Cernea, Michael, ed. 1999. *The Economics of Involuntary Resettlement: Questions and Challenges.* Washington, DC: World Bank. http://www-wds.worldbank.org/external/default/WDSContentServer/WDSP/IB/1999/06/03/000094946_99040105542381/Rendered/PDF/multi_page.pdf.

Downing, Theodore E. 1996. "Mitigating Social Impoverishment when People Are Involuntarily Displaced." In *Understanding Impoverishment: The Consequences of Development-Induced Displacement.* Ed. Christopher McDowell, 33–48. Providence/Oxford: Berghahn Books.

International Finance Corporation (IFC). Environmental and Social Development Department. 2002. *Handbook for Preparing a Resettlement Action Plan.* Washington, DC: World Bank. http://www-wds.worldbank.org/external/default/WDSContentServer/WDSP/IB/2002/09/13/000094946_02090404022144/Rendered/PDF/multi0page.pdf.

Schacher, Tom. 2009. "Retrofitting: Some Basics." Presentation, SHA Construction Course, Walkringen. Swiss Agency for Development and Cooperation. http://www.sheltercentre.org/library/Retrofitting+some+basics. This presentation provides very clear technical advice on vulnerability assessment and decision making regarding retrofitting interventions.

World Bank. 2004. *Involuntary Resettlement Sourcebook: Planning and Implementation in Development Projects.* Washington, DC: World Bank.

World Bank, "Safeguard Policies." http://go.worldbank.org/WTA1ODE7T0. Particularly OP 4.12, Operational Policy on Involuntary Resettlement. http://go.worldbank.org/64J6NBJY90.

Systematic early planning should be used to identify the potential adverse impacts of resettlement and to mitigate them. The resettlement plan (also called a resettlement action plan) is a useful tool used by both the World Bank and the International Finance Corporation for planning resettlement. Rebuilding houses is a priority in resettlement. But a community is made not only of physical structures; it has social, economic, and cultural dimensions that are fundamental for its well-being and functioning. The resettlement plan can assist in addressing the entire scope of the resettlement impact. A summary of all World Bank safeguard policies is found in ⛫ Chapter 21, Safeguard Policies for World Bank Reconstruction Projects.

Objectives of a Resettlement Plan

Resettlement can have its benefits and its costs. The resettlement plan is used to identify ways to maximize improvements in the quality of life of the resettled community and to minimize and compensate for the costs. The general objective of the resettlement plan is to plan a resettlement process so that can be effectively carried out in a way that supports of the long-term development objectives of the affected population.[1]

The specific objectives of a resettlement plan are to operationalize resettlement by outlining eligibility criteria for the affected parties, analyzing and proposing appropriate levels of assistance, and helping program and schedule the activities that will take place during the resettlement process, as well as to detect and minimize possible adverse impacts and involve the population in designing and implementing the resettlement program.

Methodology for Preparing a Resettlement Plan

The resettlement plan process can use multiple tools and techniques, and can incorporate new and innovative techniques, as desired. Some recommendations on organizing the process follow.

- The development of a resettlement plan can take several months and should be overseen at the senior level by the manager of the reconstruction program or, preferably, a resettlement manager.
- To oversee development and implementation, it is useful that a resettlement unit be established, managed by the resettlement manager. The resettlement manager supervises staff and any consultants, oversees activities of the task force involved with planning resettlement activities, and ensures involvement of the community and coordination among all parties.
- Consultants with relevant expertise are often needed to help the task force conduct surveys and examine the complex social, environmental, economic, and physical dimensions of resettlement. Consultants can provide objective input to a process that may become conflictive.
- A resettlement task force should be created and may assist directly in the preparation of the plan (see Resettlement Preparation Activities and Potential Agencies Responsible, below, for a proposed breakdown of tasks). The task force should include representatives of the project sponsor, relevant government line and administrative departments, local governments, community organizations, and NGOs involved in support of resettlement, as well as representatives of the affected communities. The importance of direct involvement of implementing agencies in the development of the plan cannot be underestimated.
- A community resettlement committee should be created to articulate the interests and needs of the affected population and to facilitate the communication among the community, the consultants, and the resettlement task force.

Sources of Information

A combination of qualitative and quantitative methods should be used to obtain diverse information reflecting the complexity and multi-dimensionality of the relocation experience. Suggested tools include surveys, census, interviews, mapping, photographic documentation, and participatory data gathering, among others.

Resettlement Preparation Activities and Agencies Involved

Activity[2]	Actions	Agency involved
1. Relocation policy	Develop resettlement policy and minimum standards by considering national law, international agreements, and donor requirements.	National government Resettlement unit International humanitarian and financial institutions Local governments International and local NGOs Affected population
2. Census and socioeconomic surveys	Conduct detailed survey and data analysis.	Resettlement unit Local government officials NGOs Consultants for design of survey and analysis of survey data
3. Land acquisition assessment	Conduct detailed land survey of plots to be acquired and confirm ownership.	Resettlement unit Land registry office NGO (field verification)
4. Determination of eligibility criteria and resettlement entitlements	Determine legal obligations for compensation and resettlement. Agree on additional assistance for compensation and resettlement.	Project agency or resettlement unit Government agencies (legal, financial, technical, and administrative)
5. Consultations	Inform DP population. Discuss project area or route and extent of land acquisition. Discuss valuation and grievance procedures. Establish committees.	Resettlement unit NGOs
6. Feasibility study of resettlement sites	Determine viability of residential, commercial, and agricultural relocation sites.	Resettlement unit NGOs Relevant government agencies (land use planning, soils, urban development, water and sanitation, and so forth)
7. Feasibility of livelihood restoration measures	Determine the technical, economic, and financial feasibility of each proposed livelihood restoration strategy before it is included as an option to be made available to affected people.	Resettlement unit Relevant government agencies for livelihood restoration (planning, social departments) Labor agency Employment agency Welfare agencies Finance and microfinance organizations Consultants to conduct the economic feasibility studies of proposed strategies NGOs

  SAFER HOMES, STRONGER COMMUNITIES: A HANDBOOK FOR RECONSTRUCTING AFTER NATURAL DISASTERS

Various elements of the resettlement plan will require careful analysis, often by experts with experience in post-disaster resettlement. Some of the most important technical inputs are described in the following table. These descriptions can also be used in the development of consultant terms of reference.

Technical Inputs Needed for Resettlement Plan

Topic	Technical input
A. Identification of affected population and project impacts *This information will serve as the baseline to monitor and evaluate the impacts of the resettlement plan and the resettlement project*	1. Census that enumerates all affected people (including seasonal, migrant, and host populations) and registers them according to location. This census will be used to determine eligibility for resettlement assistance and to exclude the ineligible. 2. Thematic maps that identify population settlements, infrastructure, soil composition, natural vegetation areas, water resources, and land use patterns. 3. Inventory of lost and affected household, enterprise, and community assets, including land use/land capability, houses and associated structures, other private physical assets, private enterprises, common property resources, infrastructure, and cultural property. 4. Socioeconomic analysis of income sources and livelihood strategies to serve as the basis for developing livelihoods restoration program.
B. Resettlement policy development	1. Identification and analysis of minimum standards to be applied as required by government or funding sources. 2. Summary of local laws, decrees, policies, and regulations as they apply to resettlement and comparison with the minimum standards. 3. Development of policy and standards, and consultation with affected groups and stakeholders.
C. Determination of eligibility criteria and resettlement entitlements	1. Analysis of any compensation guidelines announced by government or project sponsors, and development of alternatives, including estimates of eligibility numbers, estimated cost, and delivery mechanisms. Common forms of compensation are land-for-land and cash. However, the post-disaster reconstruction assistance program may substitute for any resettlement compensation scheme. 2. Analysis and strategy for addressing difficulties in applying eligibility criteria, such as absence of legal title to land. An entitlement matrix can identify the losses classified according to land tenure situation (owner, renter, squatter, etc.) and the scope of any difficulties. Include disadvantaged groups, such as women, the elderly, the handicapped, or ethnic minorities, in this analysis. 3. Review of compensation guidelines with affected population and final proposal. 4. Announcement of method of compensation delivery.
D. Land acquisition assessment	1. Preparation of criteria for identification and analysis of sites that covers: ■ Quantity of land required ■ Location of land required ■ Use of land required ■ Estimated number of residential ■ Tenure status of present users ■ Presence of public or community infrastructure
E. Feasibility study of resettlement sites	1. Methodology for technical feasibility studies for resettlement sites (topographical, soil, irrigation, groundwater, land use planning, and public services issues). 2. Methodology to reach agreement on social acceptability of sites, which may require direct work with community to clarify criteria and establish decision-making processes.

For access to additional resources and information on this topic, please visit the handbook Web site at www.housingreconstruction.org.

Topic	Technical input
F. Design of livelihood restoration program	1. Analysis of any livelihood restoration strategies announced by government or project sponsors, or proposed by economic groups (such as farmers, fishers, tradespeople) and development of alternatives, if necessary, including estimates of eligibility numbers and estimated cost, and delivery mechanisms.
	2. Development of livelihood restoration plans, by subgroups, for major types of livelihood:
	■ Land-based livelihoods
	■ Wage-based livelihoods
	■ Enterprise-based livelihoods
	3. Analysis of need for special assistance for vulnerable and socially marginalized groups and those whose livelihood is especially affected by relocation.
	4. Identification of assistance that can be provided by specialized organizations (professional, trade, marketing chains) and means of coordination.
	5. Identification of livelihood assistance that may be needed in addition to compensation for lost assets (financial support, technical assistance, retraining). This may be combined with long-term efforts to overcome deep-rooted problems with economic ramifications, such as poverty or social discrimination.
G. Participation, consultation, and communication	1. Design of participation strategy for all phases of relocation program.
	2. Identification of stakeholders and process for consultation with them.
	3. Development of two-way communication strategy, to inform the affected population and to involve them in monitoring and providing feedback to executing agencies.
H. Grievance redress	1. Development of registration process.
	2. Establishment of policy and operational procedures to address grievances. This may include provision for civil courts procedures if other options fail.
	3. Communication plan for familiarizing population with grievance procedures.
I. Resettlement implementation	1. Identification of roles and responsibilities of public and private entities involved in implementation, including funding agencies of individual projects, local governments, NGOs, the affected population, and the task force and advisory group.
	2. Identification of needs for training, technical assistance, or institutional strengthening to improve the implementation of the resettlement plan.
	3. Development of and agreement on work plans for each group or entity, using the resettlement plan as the overall frame of reference.
	4. Agreement on coordination mechanisms to be used during project implementation.
	5. Identification of needs, funding, and terms of reference for consulting services needed during implementation, including those necessary to implement the monitoring plan.
J. Monitoring, evaluation, and completion audit	1. Development of a monitoring plan that covers inputs, process, outputs, and impacts. (See Note 1, below.)
	2. The following aspects of the resettlement plan should be monitored:
	■ The physical progress of resettlement activities
	■ The disbursement of compensation
	■ The effectiveness of public consultation and participation activities
	■ The sustainability of income restoration and development efforts
	3. Using census data and other information, development of the project baseline before implementation begins.
	4. Assurance that sufficient resources have been budgeted to monitor the affected population for an extended period post-resettlement and to carry out an ex post audit.
	5. Design of mechanisms to involve the affected population in monitoring and evaluation activities.
K. Project budget and financial procedures	1. Development of a program budget based on realistic assumptions about eligible population, per household assistance costs, program administration costs, and time to implement.
	2. Analysis of options for indexing financial assistance to mitigate effects of local currency fluctuation and price inflation.
	3. Establishment of a system that links project budget with the implementation schedule and that can monitor disbursements and disbursement patterns.
	4. Design and implementation of financial procedures to disburse funds to implementing agencies, communities, and/or households, depending on financial assistance strategy.

Monitoring Resettlement

Permanent monitoring identifies problems or potential conflicts early and allows adjustments on time. Monitoring should be carried out by an independent entity for a number of years beyond the completion of the resettlement plan to evaluate the long-term impacts. Three suggested components of the monitoring system are (1) performance monitoring—an internal management function to measure input indicators against proposed timetable (or milestones) and budget; (2) impact monitoring—to gauge the effectiveness of the resettlement plan and its implementation in responding to the affected communities needs; and (3) completion audit— to measure output indicators, such as productivity, gains, livelihood restoration, and development impact against baseline. This is undertaken when all resettlement plan activities are completed. Suggested information sources and indicators are shown below. For more guidance on post-disaster reconstruction monitoring and evaluation, see ⌨ Chapter 18, Monitoring and Evaluation.

Sources of Resettlement Monitoring Information

Activity	Source of information	Examples of indicators
Performance monitoring (inputs, process, and outputs)	Information from monthly or quarterly narrative	Public meetings held Census, inventories, assessment, interviews completed Grievance redress procedures in place and functioning Compensation payments disbursed Housing lots allocated, infrastructure completed Income restoration and development activities initiated Monitoring and evaluation reports submitted
Impact monitoring	Quarterly or semiannual quantitative and qualitative surveys Consultation of affected population regarding their experiences, if possible, to develop baseline indicators	*Quantitative* Education: primary school attendance Agriculture: average land/household, production Work: employment, wage, income Health: birth and death rate, infant mortality, incidence of diseases *Qualitative* Interviews Focus group discussions
Completion audit	External assessment based on performance and impact monitoring reports, independent surveys, and consultation with affected persons	*The same indicators are used as during the performance monitoring and impact monitoring, with a particular focus on surveys and consultations.*

Expected Outputs

The principal output of the resettlement planning process is a resettlement plan that is viewed positively by the affected population and is acceptable to other stakeholders. Acceptability by both groups will be a function of the level and quality of participation that has taken place during the development of the plan. The resettlement plan then serves as a guide during implementation. The resettlement plan must reflect the unique features of the project context, disaster scale, and institutional capacity, and must be open to modifications during implementation, as needs and priorities emerge

Annex Endnote

1. If the World Bank safeguards policy on resettlement (OP 4.12) applies in a post-disaster reconstruction project, a Policy Framework and a Process Framework will need to be prepared, in addition to a Resettlement Plan. This annex uses World Bank and IFC frameworks for resettlement to provide general guidance on good practice. Refer to World Bank, 2004, *Involuntary Resettlement Sourcebook: Planning and Implementation in Development Projects* (Washington, DC: World Bank), for extensive guidance on World Bank requirements.
2. World Bank, 2004, *Involuntary Resettlement Sourcebook: Planning and Implementation in Development Projects*, (Washington, DC: World Bank), 232.

Guiding Principles for Reconstruction Approaches

- Households begin reconstruction the day of the disaster and government—guided by its reconstruction policy—may have to play catch-up in order that households and builders conform to, or participate in, any proposed reconstruction approach.
- Communities and households must have a strong voice in determining the post-disaster reconstruction approaches and a central role in the reconstruction process.
- The reconstruction policy must address the needs of households in all categories of tenancy: owners, tenants, and those without legal status. More than one reconstruction approach will probably be employed.
- The building approaches adopted after disasters should be as similar as possible to those used in normal times for similar households and should be based on their capacities and aspirations.
- Building codes and standards for reconstruction should reflect local housing culture, climatic conditions, affordability, and building and maintenance capacities, and improve housing safety.
- Reconstruction should contribute to economic recovery and the restoration of local livelihoods.
- Good planning principles and environmental practices should be incorporated, whatever the reconstruction approach.

Introduction

Post-disaster housing reconstruction can be undertaken through different approaches, which vary principally in terms of a household's degree of control over the reconstruction process. The choice of the best reconstruction approach—or approaches—to be employed is context-specific and should take into consideration (1) reconstruction costs; (2) improvement in housing and community safety; (3) restoration of livelihoods; (4) political milieu; (5) cultural context; and (6) people's own goals for well-being, empowerment, and capacity. Consultation with the community and evaluation of requirements and capacities is critical before deciding on any reconstruction approach.

For analytical purposes, this chapter makes a distinction among five reconstruction approaches that may be pursued after a disaster. These approaches are not mutually exclusive and should be understood as fluid categories that are often found in combination. In addition to the construction of permanent houses, these approaches apply to projects of substantial repair and retrofitting and to transitional shelter. Considerations in deciding whether to formally incorporate support for transitional shelter in the reconstruction approach are discussed in Chapter 1, Early Recovery: The Context for Housing and Community Reconstruction.

- **Cash Approach:** Unconditional financial assistance is given without technical support.
- **Owner-Driven Reconstruction:** Conditional financial assistance is given, accompanied by regulations and technical support aimed at ensuring that houses are built back better.
- **Community-Driven Reconstruction:** Financial and/or material assistance is channeled through community organizations that are actively involved in decision making and in managing reconstruction.
- **Agency-Driven Reconstruction *in-Situ*:** Refers to an approach in which a governmental or nongovernmental agency hires a construction company to replace damaged houses in their predisaster location.
- **Agency-Driven Reconstruction in Relocated Site:** Refers to an approach in which a governmental or nongovernmental agency hires a construction company to build new houses in a new site.

The authors of this handbook advocate for what the World Bank and several other agencies have defined as owner-driven reconstruction, which has proven to be the most empowering, dignified, sustainable, and cost-effective reconstruction approach in many types of post-disaster situations. As one reconstruction expert aptly stated: "It is better to have 100,000 people each concerned

1. George Soraya, Lead Urban Specialist, World Bank, Jakarta, 2009, personal communication.

about one house than to have 100 people concerned about 100,000 houses."[1] Experience shows that empowering people to manage their own recovery and reconstruction, both individually and as a community, will be faster and more efficient, and will encourage people to use their creativity and to mobilize their own resources. If they are waiting for others to take care of them, they can become disempowered and may be more apt to complain and less likely to contribute. Of course, not all reconstruction situations will lend themselves to this approach, as explained in this chapter.

Key Decisions

1. **Government** should decide on the policy for housing and community reconstruction, based on the results of the damage and loss assessment, and in consultation with **the affected community** and the **lead disaster agency**. Important decisions include: the reconstruction approach or approaches to be employed; the financial contributions to be made by various parties, including households; mechanisms for coordination; and the administrative and project management procedures that all agencies will follow.

2. The **lead disaster agency** should determine, in consultation with **government** financial officials, the level of assistance that will be provided for transitional sheltering, repairing, retrofitting, and reconstruction, and on the system for delivering funds. **Government** may want to impose a maximum assistance level for nongovernmental agency projects to reduce competition among agencies. See 📖 Chapter 15, Mobilizing Financial Resources and Other Reconstruction Assistance.

3. **Agencies involved in reconstruction** should agree with **government** on performance benchmarks for all reconstruction approaches and on reporting procedures, and collaborate on establishing the baseline and the monitoring system.

4. **Affected communities** should decide which reconstruction approach or approaches are most suitable for them and collaborate with government in the selection process. They should also decide how they prefer to organize themselves during reconstruction and should have the right to select which agencies will assist them and to agree on the form of assistance. Depending on the community's political, social, and economic characteristics, organization of the community and collective decision making may require outside facilitation and support.

5. Whatever the approach, **local governments** must direct those aspects of reconstruction related to land use and physical planning and the regulation of construction. See 📖 Chapter 7, Land Use and Physical Planning.

Public Policies Related to Reconstruction Approaches

Unless government has a disaster management plan, there are unlikely to be public policies at either the national or local level that specifically address post-disaster reconstruction approaches. Yet there may be national or local housing sector programs that provide new housing to low-income people or subsidies for upgrading that can serve as a starting point for defining the post-disaster housing reconstruction approach.

Government should take an active role in setting the rules for and overseeing the activities of all agencies involved in reconstruction. It should provide the appropriate regulations and guidelines so that agencies conform to the following good planning and construction principles.

- Consistently apply good planning principles and conform to local development plans.
- Conform with local building codes and standards.
- Minimize environmental impacts in construction, site planning, and building design.
- Ensure community participation in all aspects of development, including those managed by outside agencies and private contractors.
- Maintain or improve the tenure status of households during the reconstruction process.

Public policies in other sectors may influence decisions on the reconstruction approach as well. Refer to 📖 Chapter 7, Land Use and Physical Planning; 📖 Chapter 9, Environmental Planning; and 📖 Chapter 11, Cultural Heritage Conservation, among others.

Technical Issues

The following are descriptions of five reconstruction approaches frequently used in post-disaster reconstruction, including a discussion of the advantages and disadvantages of each.

"In reconstruction, it is better to have 100,000 people each concerned about one house than to have 100 people concerned about 100,000 houses."

George Soraya, Lead Urban Specialist, World Bank

SAFER HOMES, STRONGER COMMUNITIES: A HANDBOOK FOR RECONSTRUCTING AFTER NATURAL DISASTERS

The Cash Approach (CA)

With this reconstruction approach, support for repair and reconstruction of damaged houses is provided exclusively by unconditional financial assistance. Any category of tenants, including squatters, may be entitled to and benefit from cash assistance, depending on the policy.

CA is appropriate for disasters that have a relatively limited impact and where housing damage was not caused by shortcomings in local construction practices. Emphasis with CA is on the distribution of financial assistance with minimal attention given to enabling measures. This approach may give affected people the choice to use the assistance based on their own priorities, which may not necessarily be housing. Some people may use the cash to migrate out of the disaster zone, for instance, if that is what they judge to be their best alternative.

Experiences with the Cash Approach
- After the 2004 floods in Santa Fe, Argentina, the World Bank supported a government CA program for housing repair and reconstruction.

Advantages	Disadvantages and risks	Recommendations
Most cost-effective, rapid delivery of aid to households.	May reproduce pre-disaster vulnerabilities.	Use CA only when damage is not severe and is not attributed to poor construction or poor building code enforcement.
Does not require complex delivery mechanisms.	No improvement of building skills.	
	No opportunity to introduce new building technologies.	
Assistance can be adjusted to household's income, family size, livelihoods, socio-cultural requirements, etc.	Vulnerable people may be unable to handle repair and reconstruction without assistance.	Ensure that housing labor and materials markets are functioning properly.
Does not discourage repair of houses or use of salvaged and local building materials.	Financial assistance may be used to meet other requirements while houses remain unrepaired.	
Best when local building capacity and financial support are adequate.	Risks of negative publicity if households use funds for questionable purposes.	
Families can employ cash according to their priorities.	May increase risk of corruption.	

Owner-Driven Reconstruction (ODR)

In an ODR program, people who lost their shelter are given some combination of cash, vouchers, and in-kind and technical assistance (TA) to repair or rebuild their houses. They may undertake the construction or repair work by themselves, by employing family labor, by employing a local contractor or local laborers, or by using some combination of these options. ODR is similar to the "aided self-help approach" that has been used extensively to provide housing assistance to the urban poor, particularly in Latin America.[2]

ODR is the most empowering and dignified approach for households, and it should be used whenever the conditions are right for it. The approach is viable for both house and apartment owners (in the latter case, the condominium association or cooperative society would manage construction), as well as for informal settlers, once their tenure is secured. In fact, the term "owner" in ODR refers as much to the ownership of the building process as to the ownership of the house. A common misunderstanding about ODR is that the owners will build their houses by themselves. Recent examples show that this is rarely the case because people tend to hire local contractors or laborers for at least part of the work. Thus, the key difference between this approach and agency-driven approaches is that contractors and paid laborers are accountable to the homeowner rather than to an external agency that may not be able to provide the intensive supervision and control that homeowners often can.

However, the risks of ODR need to be understood and addressed. ODR requires good oversight and governance, that is, a government capable of establishing and enforcing standards, and some agency (governmental or nongovernmental) to ensure the quality of construction. Where engineered building technologies are being used, or multifamily housing is being rebuilt, using ODR is more challenging, but not impossible. The oversight from supporting agencies or government will need to

2. Sultan Barakat, 2003, *Housing Reconstruction after Conflict and Disaster*, Humanitarian Policy Network Paper 43 (London: Overseas Development Institute), http://www.odihpn.org/report. asp?id=2577.

be more technical, and experienced contractors must be hired. Success lies in establishing a support system for homeowners appropriate to the local context, which may include:

- Training of tradespeople and homeowners
- Technical assistance and construction supervision and inspection
- Updating and enforcement of building codes and construction guidelines
- Mechanisms to regulate prices and facilitate access to building materials
- A system for providing financial assistance in installments as construction progresses

Experiences with ODR

- Formally adopted by the state government of Gujarat as its official reconstruction policy following the 2001 earthquake in Gujarat, India. Independent evaluations proved it produced high levels of satisfaction.[3]
- Used by the World Bank after the 2004 Indian Ocean tsunami in Thailand and Sri Lanka and after the 2005 North Pakistan earthquake. The Bank funded reconstruction and therefore was in a position to influence government reconstruction policy. In these cases, both official Bank documents and evaluations carried out by other agencies that pursued this approach confirm that this was the most successful housing assistance strategy.[4] Also see the; 🖳 case study on ODR in the North Pakistan earthquake reconstruction, below, and others in the case studies section of this chapter.

Advantages	Disadvantages and risks	Recommendations
Mobilizes households to take an active role in rebuilding, which speeds recovery from psychological trauma.	Without good standards and oversight, quality of construction may be poor, and pre-disaster vulnerabilities can be reproduced.	Establish a support system for homeowners that is responsive to local requirements.
Assistance can be adjusted to the needs of the household related to income, family size, livelihoods, socio-cultural requirements, etc.	Conversely, if building codes are too rigid and biased toward alien housing technologies, people can have trouble complying with requirements, even with oversight.	Ensure that assistance is equitable and sufficient to satisfy minimum housing standards. Establish a delivery mechanism for financial assistance that is easy to understand and access.
Consistent with normal incremental housing construction practices.		
Encourages repair of houses and use of salvaged and local building materials.	May be more difficult to implement in relocated communities and poor communities with no building experience (for example, urban squatters).	Ensure building codes are based on local building technologies and materials. Ensure adequate training for trades people and construction supervisors.
Tends to involve local building industry, thereby contributing to restoration of local economy and livelihoods.		Acknowledge housing rights and accommodate special needs of tenants, squatters, and the homeless.
Helps preserve community's cultural identity by ensuring continuity in local building tradition and architectural style.	Suitable for contractor-built multifamily and high-rise building reconstruction; however, skilled technical oversight is required.	Adjust the approach to reach geographically distant regions and socioeconomically disadvantaged people.
Allows people to "top up" housing assistance with their own savings and build a house reflecting their specific needs and aspirations.	Households of elderly and vulnerable groups will face difficulties managing reconstruction alone and may not reach milestones, making it impossible to receive second and subsequent disbursements.	Provide special attention and support to vulnerable groups (orphans, widows, the elderly, and the very poor).
Is less subject to disruptions caused by unstable political situation (for example, eastern provinces of Sri Lanka).		Adopt measures to prevent inflation and ensure access to quality construction materials.
Is viable for dispersed and remote settlements (for example, Pakistan, Gujarat).		Consider involving nongovernmental organizations (NGOs) as part of the enabling system.

Case Study: 2005 North Pakistan Earthquake, Pakistan

Flexibility in ODR Housing Reconstruction and Retrofitting

Following the North Pakistan Earthquake of 2005, the Pakistani government promoted ODR to rebuild some 400,000 houses. Under the lead of the Earthquake Reconstruction and Rehabilitation Authority (ERRA), a multitude of international NGOs joined this program. Homeowners were responsible for the reconstruction of their own houses, with technical assistance and financial support disbursed in tranches. Insufficient capacity in the field can slow down the pace of construction and increase the likelihood of substandard construction work. To prevent this, ERRA facilitated the mobilization of decentralized teams who could provide technical updates and on-site training to the scattered beneficiaries. ERRA also used field observations and field testing to decide whether to allow different construction techniques and developed retrofitting methods to

3. Jennifer Duyne Barenstein, 2006, "Housing Reconstruction in Post-Earthquake Gujarat: A Comparative Analysis," Humanitarian Practice Network Paper 54, Overseas Development Institute, http://www.odihpn.org/report.asp?ID=2782; and Abhiyan, 2005, *Coming Together: A Document on the Post-Earthquake Rehabilitation Efforts by Various Organisations Working in Kutch* (Bhuj: United Nations Development Programme/Abhiyan), http://openlibrary.org/b/OL3338629M/Coming_together.
4. See, for example, World Bank, 2009, "Implementation Completion and Results Report, Grants to the Democratic Socialist Republic of Sri Lanka for a Tsunami Emergency Recovery Program" and Yasemin Aysan, 2008, "External Evaluation of the Swiss Consortium's Cash for Repair and Reconstruction Project in Sri Lanka 2205-2008," study contracted by the Swiss Consortium of Swiss Solidarity, HEKS, Swiss Red Cross, and SDC, http://www.deza.admin.ch/ressources/resource_en_173148.pdf.

increase or maintain the seismic resistance of diverse housing styles. The approval of the local timber-frame construction style *Dhajji* was vital for the success of the reconstruction effort; statistical analysis indicates that, as compared to concrete block masonry, *Dhajji* houses are less costly and can be made acceptably seismic-resistant. Also, *Dhajji* construction techniques are easier for homeowners to understand, utilize, and adapt to local contexts, preferences, and resources. Three years after the earthquake, almost 300,000 seismic-resistant houses were nearing completion. An overarching factor in this success was the constructive way in which homeowners and those managing the implementation of the program were able to interact as the program was carried out.

Source: A. van Leersum, 2009, "Implementing Seismic Resistant Construction in Post-Disaster Settings: Insights from Owner-Driven Reconstruction in Pakistan" (MSc thesis, Eindhoven University of Technology). The opinions expressed are those of the author and do not necessarily reflect those of the involved organizations.

FEMA NEWS PHOTO

Community-Driven Reconstruction (CDR)

CDR entails varying degrees of organized community involvement in the project cycle, generally complemented by the assistance of an agency. The degree of control over reconstruction by the community in CDR projects varies between agencies and from project to project. The agency may take the lead, suggesting housing designs, technologies, and/or materials, and delivering construction inputs and training. The agency may also employ skilled and unskilled laborers from the community or facilitate the formation of construction committees. At the other extreme, the community may manage most of the reconstruction process and receive only the support of facilitators ("collective ODR"). In summary, CDR may involve one or more of the following roles for the community:

- Organization and planning of the entire reconstruction process, including housing and infrastructure
- Decisions regarding housing design and building materials
- Production of building materials such as bricks
- Distribution of building materials or other forms of housing assistance (e.g., cash and vouchers)
- Hands-on reconstruction
- Oversight of builders

Experiences with CDR

- Adopted by several national NGOs following the 2001 Gujarat, India, earthquake. The level of satisfaction was relatively high, but lower than for ODR houses.
- Used successfully as collective ODR following the 2006 Java earthquake in Indonesia. See the case study entitled Organizing Community-Based Resettlement and Reconstruction, in Chapter 12, Community Organizing and Participation.
- Adopted by the United Nations Centre for Human Settlements (UN-HABITAT), KfW, and Urban Poor Linkage Indonesia (UPLINK) in Aceh, Indonesia, following the 2004 Indian Ocean tsunami. Each of these agencies used a somewhat different interpretation of the approach. UPLINK gave people more choice in house designs, but community-based construction committees were given control over the purchase and distribution of building materials and over the mobilization of reconstruction labor. (In some cases, local contractors gained control of these committees.) KfW gave building materials and financial assistance directly to owners, but provided little choice over materials and designs.
- Used by the city of Ocotal, Nicaragua, to relocate and rehouse residents of displaced neighborhoods and highly vulnerable sites following Hurricane Mitch in 1998. Housing designs and building materials were proposed by a local architect, but receipt of a house was contingent on participation in construction of at least one family member. (See case study, below.)

For access to additional resources and information on this topic, please visit the handbook Web site at www.housingreconstruction.org.

Advantages	Disadvantages and risks	Recommendations
Useful where: ■ new building technologies, materials, or housing designs are being introduced; ■ agencies must bring in building materials; or ■ housing reconstruction is linked to community development activities. Can foster social cohesion when people from different communities work together to organize relocation and reconstruction. Has high levels of flexibility and accountability and provides control for owners over reconstruction. Access to construction materials more assured. Scale of project may contribute more strongly to reactivation of local economy.	Overheads may be high because of agency involvement. Agencies may leave little room for individual preferences by imposing standard designs and materials. Local contractors capture community construction committees that manage large amounts of resources. Real participation may be limited if: ■ consultation is only with community leaders whose views don't reflect those of the community; ■ processes are captured by local elites; ■ participation is perceived as excessively time-consuming; or ■ women's perspectives are not incorporated.	Require upfront community agreement on level and type of agency involvement. Ensure project staff is qualified to lead a participatory reconstruction process. Ensure community participation throughout the project cycle, site selection, settlement planning, and housing design. Avoid overruling community preferences and recognize the different needs and capacities of community members. Introduce governance mechanisms to prevent project resources from being diverted by local elites.

Case Study: 1998 Hurricane Mitch, Nicaragua

Successful CDR Project Built Social Capital

After Hurricane Mitch struck the town of Ocotal, Nicaragua, damaging 1,164 houses and destroying 328, the mayor initiated a CDR project for resettling the affected population as well as for households located in high-risk areas. The guiding principle was to prevent future disasters by protecting the people, while improving the social cohesion of the community. The social dynamics of the community were carefully analyzed and community participation was promoted. The reconstruction process was explained to the citizens in community meetings, and the damage and loss assessment was conducted to reflect the community's own priorities. Further, the new building site underwent an extensive planning process during which the proximity of the site to the future residents' income sources was analyzed, as were possibilities for the future growth of the community, an important consideration when rapid population growth is expected.

Culturally and environmentally appropriate house designs, including improved traditional building materials and techniques, were proposed by a local architect and presented to the community. Future residents discussed the design and could request modifications, which were incorporated when technically feasible. Access to a house was contingent on full participation in the construction by at least one family member. Because Ocotal constructed its own adobe factory, it created much-needed employment in an effort to reduce out-migration from the town. Beneficiaries were trained in hazard-resistant construction, including the modification of traditional adobe building practices. Participation in the joint construction work on the building site made it possible for residents-to-be to establish initial contacts with their new neighbors. People's pride and self-esteem increased as the project progressed, social cohesion was fostered, and a positive neighborhood identity was created. The Ocotal reconstruction project successfully incorporated prevention and built social capital, which has contributed to the sustainability of the project. In all, approximately 300 new homes have been built to date.

Sources: Esther Leemann, 2010, "Housing Reconstruction in Post-Mitch Nicaragua: Two Case Studies from the Communities of San Dionisio and Ocotal, " eds. DeMond S. Miller and Jason David Rivera, *Community Disaster Recovery and Resiliency: Exploring Global Opportunities and Challenges* (Auerbach Publications, forthcoming); and José Luis Rocha, 1999, "Ocotal: Urban Planning for People," *Envio digital* 218, http://www.envio.org.ni/articulo/2299.

Agency-Driven Reconstruction in-Situ (ADRIS)

In ADRIS, a governmental or nongovernmental agency hires one or more contractors to design and build the houses. Design, materials, and expertise are likely to be imported from outside the community. The community may or may not be consulted on certain aspects of the project, such as house designs. House owners may be asked to take over some building tasks, such as curing concrete. Whereas house owners may also hire contractors within the framework of ODR, the principal contractor is accountable to the agency and may be contracted through formal tendering procedures. A special case of ADRIS is when a public agency reconstructs government-owned housing, on public property.

Because ADRIS takes place on the owners' own land, it gives the homeowner some degree of control over quality, and sometimes the opportunity to participate in specific tasks. During construction, owners may be able to make suggestions to or modify the design. ADRIS eliminates the hurdle of land acquisition and generally allows the household to know where its house will be located. However, if housing designs are standardized or different from local designs, it may be difficult to fit the houses into pre-disaster settlement layouts or to modify them later. ADRIS, therefore, often results in similar or even worse outcomes than those of ADRRS, especially in the case of large-scale single-family reconstruction.

Experiences with ADRIS

- Many international NGOs and private companies "adopted" villages and used ADRIS to build houses after the 2001 Gujarat, India, earthquake, even though government adopted an ODR policy. These projects often became a mix of ADRIS and ADRRS in adjacent sites where the housing designs did not fit existing sites and individual households, humanitarian agencies, or local governments bought additional land for new construction. In some cases, contractors did not respect the heritage sites and spatial organization, and caused irreversible damage to historical villages.
- Many private voluntary organizations adopted ADRIS in Tamil Nadu, India, following the 2004 Indian Ocean tsunami. However, they required that the land be cleared of houses and vegetation before starting construction. As a result, hundreds of pre-tsunami houses that were culturally and climatically appropriate and easily repairable were demolished, and thousands of trees were felled, which negatively affected people's livelihoods and well-being.[5]

Advantages	Disadvantages and risks	Recommendations
Communities are not displaced.	A contractor's construction modes, designs, and settlement layouts are often not compatible with existing sites.	Avoid ADRIS if local building capacity is available.
People can be effectively involved in construction and monitoring.	Remaining built and natural environments may be considered an obstacle to reconstruction, leading to unnecessary house demolition and tree removal, causing high social and environmental impacts and conflicts.	If ADRIS is unavoidable, ensure community participation in choices regarding housing design, site layout, building materials, and construction.
New building technologies can be introduced.	Exogenous building technologies may be used that have negative environmental impacts and do not meet local requirements.	Ensure equitable distribution of project benefits with transparent allocation criteria based on social assessments, and monitor their application.
No land acquisition is required.	Community participation may be more difficult to incorporate or may be limited to community leaders, resulting in disproportionate benefits for elites.	Protect the heritage value of pre-disaster environment, both built and natural, including buildings and trees that survived the disaster.
	Construction quality is often poor due to inexperience of agency with oversight of housing construction, among other reasons.	Require contractors to use local building materials and designs.
	Contractors may encourage communities to demand additional benefits from government.	Hire a professional project manager or "clerk of the works" from the construction industry to supervise construction.
	Corruption and exploitation by contractors.	Establish social audit mechanisms to ensure local accountability. See 📖 Chapter 18, Monitoring and Evaluation, Annex 2, for a social audit methodology.
		Ensure quality control through an independent third-party audit. See 📖 Chapter 19, Mitigating the Risk of Corruption, Annex 2, for instructions on conducting a construction audit.

Agency-Driven Reconstruction in Relocated Site (ADRRS)

When using ADRRS, a governmental or nongovernmental agency contracts the construction of houses on a new site, generally with little or no involvement by the community or homeowners. The community, government, or agency supporting the reconstruction may purchase the land for the new settlement. Upon completion, the houses may be allotted through a lottery or using criteria defined by the community or the agency, or both. ADRRS, often justified as a risk-mitigation measure, may

5. Jennifer Duyne Barenstein, 2006, *Housing Reconstruction in Post-Earthquake Gujarat: A Comparative Analysis,* Humanitarian Policy Network Paper 54 (London: Overseas Development Institute), http://www.odihpn.org/report.asp?ID=2782.

be advisable when communities are being relocated. And agencies may favor ADRRS for the ease of constructing on a clear site without tenancy issues or other complications. ADRRS is used by public agencies to reconstruct government-owned housing in a relocated site, generally public land. However, for single-family homes, ADRRS can be problematic. It can lead to the construction of costly, inappropriate housing of poor quality and settlement arrangements that do not meet the socio-cultural and livelihood requirements of the people, causing severe economic consequences and low occupancy rates. The argument that ADRRS results in higher construction quality is rarely valid, because of poor supervision or the lack of qualified contractors. Moreover, finding an appropriate site can be a major challenge; failing to do so is, in fact, one of the principal reasons for dissatisfaction with this approach. The complexities of a decision to relocate are discussed in Chapter 5, To Relocate or Not to Relocate.

Experiences with ADRRS

- International NGOs and national private companies opted for ADRRS after the 2001 Gujarat, India, earthquake because of perceived organizational advantages and higher visibility, including naming rights to new settlements. Local elites were sometimes given incentives to sell this approach to local officials. By accepting these offers, people lost their access to government financial assistance. When they later found the designs, layouts, and construction quality to be subpar and refused to occupy these villages, they ended up having to liquidate their assets, such as land and livestock, so they could rebuild elsewhere. An independent study found that in villages that opted for ODR, housing conditions were considered better than before the earthquake and economic conditions unchanged, while in villages reconstructed with the ADRRS approach, a significant percentage of households reported high levels of indebtedness and worse economic conditions.[6]

- ADRRS has had positive results in urban contexts. Two examples are the city of Nagapattinam in Tamil Nadu, India, and Banda Aceh, Indonesia, after the 2004 Indian Ocean tsunami. In Banda Aceh, a Korean voluntary organization acquired land in a middle-class neighborhood for an urban housing project. Although the houses were small, high occupant satisfaction was attributed to housing design, good location, access to public services, and the fact that livelihoods were not site-dependent. See the case studies later in this chapter.

Advantages	Disadvantages and risks	Recommendations
Appropriate where pre-disaster settlements are located on hazardous sites.	Difficulties and delays in finding appropriate land.	Only adopt ADRRS if ODR is not possible on safety grounds.
May be faster and more cost-effective.	Negative socioeconomic impacts and disruption of livelihoods from relocation may cause occupancy rates to remain low.	Avoid this approach in rural areas, anywhere people can manage house construction on their own, and where livelihoods are very site-specific.
May allow pre-disaster housing problems to be addressed (for example, shortages, vulnerability, and poor housing conditions).	Poor site selection may cause negative environmental impacts or re–create vulnerability of original location.	Carefully assess relocation effects on livelihoods and provide mitigation measures.
More appropriate for dense urban settlements, rental housing, and complex building technologies (multistory construction).	Construction quality is often poor.	Identify beneficiaries and allot houses during the planning stage.
	Loss of local building culture and capacity.	
	Disruption of access to common property and to natural and cultural heritage sites.	Ensure community participation throughout the project cycle, site selection, settlement planning, and housing design.
Can contribute to heritage conservation by relocating from sensitive sites.	Settlement layout, housing designs, and building technologies can be alien to local communities and culturally inappropriate, particularly in rural areas.	Establish social audit mechanisms to ensure local accountability. See Chapter 18, Monitoring and Evaluation, Annex 2, for a social audit methodology.
Can address housing needs of various categories of the population simultaneously, depending on design of the settlement.	Repairs and extensions to houses built with exogenous building technologies may be unaffordable.	
	Contractors may encourage communities to demand additional benefits from government.	Ensure quality control through an independent third-party audit. See Chapter 19, Mitigating the Risk of Corruption, Annex 2, for instructions on conducting a construction audit.
	Lack of community participation or oversight may result in poor targeting, unequal distribution of houses, and elite capture.	Take into consideration socioeconomic and gender-specific requirements

6. Jennifer Duyne Barenstein, 2006, *Housing Reconstruction in Post-Earthquake Gujarat: A Comparative Analysis*, Humanitarian Policy Network Paper 54 (London: Overseas Development Institute), http://www.odihpn.org/report. asp?ID=2782.

Case Study: 2003 Bam Earthquake, Iran

Shift from ADRRS to ODR during Bam Earthquake Reconstruction

When the Housing Foundation of the Islamic Revolution (HF)-United Nations Development Programme (UNDP) joint housing reconstruction project started following the 2003 Bam earthquake, the government of Iran and the HF (the executing agency for the reconstruction) had not fully defined the reconstruction approach. For the first year of the project, the HF hired contractors to build housing units for the program's beneficiaries (129 female-headed households [FHHs]). But the poor performance and slow delivery by the contractors and their numerous claims for cost increases led the HF to shift after the first year to ODR with technical assistance.

The ODR approach followed several organized steps, namely, (1) submission of ownership documents or other verifiable proof of ownership in 1 of the 14 regional offices of the HF; (2) request for rubble removal from the property; (3) request for a demolition or leveling permit from the Bam Municipality; (4) delivery of a letter to the landowner by the HF office that introduced the landowner to the licensed consultancy firms that had established branches in the HF offices; (5) selection of a housing model from among those demonstrated by the private developers, contractors, UNDP, and international NGOs at the HF Technical and Engineering Site; (6) review and revision of the selected design with the consultancy firm until agreement on a final design; (7) receipt from the municipality of guidelines for engaging a contractor; (8) preparation of documentation for loans and grants from banks; (9) selection, negotiation, and contracting of a licensed contractor; and (10) commencement of construction. The beneficiaries received their first loan installment after the house foundation was complete. The shift to ODR resulted in more rapid reconstruction and higher satisfaction for the FHHs with the quality of the work.

Source: Victoria Kianpour, UNDP Iran, 2009, personal communication, http://www.undp.org.ir/.

Comparison of Reconstruction Approaches

Reconstruction approaches can be compared according to the degree of household control, the form of assistance, the role of the actors, and where the reconstruction takes place. The factors can be combined in many ways. The following table compares the five approaches discussed in this chapter.

Reconstruction approach	Degree of household control	Form of assistance		Role of actors			Location	
		Financial	Technical	Community	Agency	Contractor	*In-situ*	New site
Cash Approach	Very high	Cash only	None	None	None	Household may hire	Yes	No
Owner-Driven Reconstruction	High	Conditional cash transfer to household	TA/Training of household	None	Project oversight and training	Household may hire	Yes	No
Community-Driven Reconstruction	Medium to high	Transfer to household or community	TA/Training of community and household	Project organization and oversight	Project oversight and training	Community may hire	Yes	No
Agency-Driven Reconstruction *in-Situ*	Low to medium	Funds handled by agency	Limited or none	Limited	Management of project	Agency hires	Yes	No
Agency-Driven Reconstruction in Relocated Site	Low	Funds handled by agency	Limited or none	Limited	Management of project	Agency hires	No	Yes

Determining which reconstruction approach is preferable for an affected population—or even a subset of the population—is not a straightforward process. The disaster situation, and the conditions and preferences of households make each situation unique. This determination is also affected by the tenancy status of the household before the disaster and the desired tenancy status after reconstruction.

 For access to additional resources and information on this topic, please visit the handbook Web site at www.housingreconstruction.org.

However, some approaches may be more suitable to certain groups than others. The following table shows what may be the most suitable solutions for specific groups. It points out the importance of addressing the reconstruction requirements of owners who are landlords, since renters—a large proportion of the population in some countries, especially in urban areas—will be dependent on reconstruction by landlords. It is unlikely that a group of apartment dwellers (even if they were condominium or cooperative owners) would band together to reconstruct their units, particularly if reconstruction entailed relocation. However, this option is included here. More likely, they would liquidate their holdings and relocate elsewhere. The 📖 case study on the Gujarat earthquake, below, compares satisfaction levels of owner-occupiers with different reconstruction methods.

Tenancy categories of affected population	Suitable reconstruction approaches
1. House owner-occupant or house landlord	Any approach.
2. House tenant	If tenant can become a house owner-occupant during reconstruction, see #1. If tenant becomes an apartment owner-occupant, see #3. Otherwise house tenants are dependent on landlords to rebuild.
3. Apartment owner-occupant or apartment building landlord	Cash or ODR. CDR if owners as a group can function as a "community." Reconstruction of multi-family, engineered buildings will always involve contractors, but owners may not require help of agency.
4. Apartment tenant	If tenant can become a house owner-occupant during reconstruction, see #1. If tenant becomes apartment owner-occupant, see #3. Otherwise, apartment tenants are dependent on landlords to rebuild.
5. Land tenant (house owner)	With secure tenure, same as #1, house owner-occupant. Without secure tenure, same as squatter.
6. Occupant with no legal status (squatter)	If squatter can become a house owner-occupant during reconstruction, see #1. If squatter becomes an apartment owner-occupant, see #3. Otherwise, squatters are dependent on landlords to rebuild, or they remain without legal status.

Risks and Challenges

- Underestimating an affected community's capacity to rebuild its houses and, hence, opting for reconstruction by contractors.
- Allowing those who can provide reconstruction funding to impose the reconstruction scheme.
- Building houses that people refuse to occupy for reasons of location, materials, design, or loss of livelihood.
- Not providing households participating in ODR projects with adequate assistance, facilitation, and supervision, resulting in poor construction quality, price inflation for materials, and other problems.
- Failing to take advantage of reconstruction as an opportunity to reduce risk and to strengthen local building practices and construction capacities.
- Inadequate oversight of private construction companies, which results in higher costs or inferior quality of construction.
- Designing and building houses that do not meet the communities' cultural and individual requirements because of a lack of community participation in reconstruction planning.
- Local elites who hijack the project benefits because eligibility criteria and assistance schemes were poorly designed or not monitored during implementation.
- Pressure to overinvest in housing that leaves little or no funding for on-site investments such as infrastructure and restoration of natural habitat.
- Failing to provide sufficient technical assistance and facilitation to ensure that poorer households participating in ODR schemes reach construction milestones and obtain access to subsequent funding disbursements.
- In urban areas, adopting ODR without strengthening institutional capacity for land use planning, regulation, and building inspection, which can result in increased vulnerability.
- Neglecting the needs of tenant categories other than homeowners, e.g. owners of multiple family housing, tenants, landlords, and squatters.

Recommendations

1. When reconstruction is simple and mainly entails repair of damaged housing that is otherwise adequate, adopt CA; otherwise, whenever possible, adopt ODR.
2. Use CDR when community life and the local economy is disrupted by the disaster or relocation is required, or both.
3. Avoid ADRIS in rural areas and in places where the built environment and natural habitat are significantly intact.
4. If ADRRS is absolutely necessary, government should require community participation and establish simultaneous audit and oversight mechanisms.
5. Help communities rebuild their houses with facilitation and other appropriate enabling mechanisms identified through a social assessment that focuses on vulnerable households.
6. Ensure that reconstruction agencies take into consideration people's different housing needs, vulnerabilities, livelihoods, and family size in selecting reconstruction approaches and that socioeconomic factors and gender-related requirements are addressed.
7. Under every approach, ensure that construction methods embody good planning, risk reduction, and environmental principles.
8. Require community participation in all aspects of the process, even when outside agencies or the private sector are in the lead.

Case Studies

1999 Eje Cafetero Earthquake, Armenia, Colombia

Decentralization of the Rural Reconstruction Process using ODR

When an earthquake struck the coffee-growing region of Colombia in 1999, national authorities worried about the repercussions of the disaster on the coffee exports-based regional economy. The President of Colombia created FOREC, a national fund that was put in charge of managing the overall reconstruction program. FOREC, in turn, decentralized the reconstruction process by distributing responsibility among 32 NGOs, putting each one in charge of a small town or a sector of an affected city. Rural reconstruction was assigned to the Coffee Growers' Organizations (CGOs), a network of local, regional, and national committees represented internationally by the Coffee Growers' Federation. However, the mission of a CGO was promoting coffee production and exports, not building houses or infrastructure. Lacking the means to implement a housing program, the CGOs opted for a user- or owner-driven approach in which beneficiaries were give responsibility for designing, planning, procuring, and building their own projects. FORECAFE, a rural reconstruction fund created by the CGOs, was charged with controlling the quality of construction on individual projects and managing progress payments, which were disbursed based on approval of the use of the prior payments. More than 14,000 individual housing, infrastructure, income-generation, and community services projects were completed in less than 18 months, thanks to an effective system of coordination of information, financial control, and quality management. This post-disaster, user-driven reconstruction experience (one of the first in Latin America) demonstrated the benefits of transferring responsibility over design, planning, and management of reconstruction directly to the individual beneficiaries of that reconstruction.

Source: G. Lizarralde, C. Johnson, and C. Davidson, eds., 2009, *Rebuilding after Disasters: From Emergency to Sustainability* (London: Taylor and Francis), http://www.preventionweb.net/english/professional/publications/v.php?id=11329.

2005 Jammu and Kashmir Earthquake, India

Quality Transitional Shelter Built by ODR Gets Affected Population through the Winter

In October 2005, a massive earthquake hit the Jammu and Kashmir region of India, killing more than 1,000 people and injuring 6,300. The impact on housing in some communities was catastrophic. In Tangdhar region, for example, 5,393 of 6,300 houses collapsed and 266 were partially damaged. In addition, winter was fast approaching, threatening to block access roads to the affected area. In contrast to many post-disaster situations where temporary shelters are a makeshift solution for a few months, sometimes built with inappropriate materials, the Jammu and Kashmir government decided to provide robust interim shelters. A reconstruction policy was needed that reflected local needs, priorities, and climatic conditions, including a proposal for the interim shelter construction approach. The Jammu and Kashmir government analyzed such options as (1) government construction of houses, (2) contracting NGOs to construct housing, and (3) facilitating construction by households, as was done in Bhuj, India, after an earthquake hit that city. The option chosen was ODR, and enabling mechanisms were established, including providing cash assistance of Rs 30,000 (US$677) for those whose houses had fully collapsed (enough for a 200 sq. ft. shelter) and sending engineers to survey villages and to help communities with technical issues. The transitional shelter design chosen could be built in two days. Although access to construction material was facilitated, people were encouraged to use lumber from their old houses to prevent shortages in the spring when permanent reconstruction work would begin. To ensure completion of shelter construction before winter hit, an incentive of Rs 5,000 (US$112) was given to the families that finished their sheds before the end of November while respecting safety norms. The reconstruction policy and technical advice were communicated to communities using flyers in Urdu and English with easy-to-understand drawings. In the end, 15,000 shelters—90 percent of the total—were completed by the end of November. A crisis was averted, thanks to a combination of a practical transitional shelter strategy, a clear message, good incentives, and strong support by the state for ODR.

Sources: Kutch Nav Nirman Abhiyan, 2005, *An Owner Driven Interim Shelter Initiative in J & K. Report on Tangdhar Region,* http://www. kutchabhiyan.org/PDF/InterimShelter_Initiative_in_J&K.pdf.

2001 Gujarat Earthquake, India

Citizens' Satisfaction with Different Reconstruction Approaches

In 2004, an independent household survey compared citizens' satisfaction with different reconstruction approaches following the 2001 Gujarat, India, earthquake. The highest satisfaction was achieved with ODR with financial assistance and technical assistance from government, complemented by additional material assistance from local NGOs. All families whose houses were built using this model reported that their housing situation was better than before the earthquake. A second approach, government-supported ODR without NGO assistance, was almost as popular, with 93.3 percent of households reporting being fully satisfied. Relatively high levels of overall satisfaction (90.8 percent) were also reported under a third approach: local NGOs using CDR. Satisfaction decreased when houses were built by contractors. Only 71.8 percent of the people reported being satisfied with contractor-built houses built in-situ (equivalent to ADRIS). Contractors' profit imperative was held responsible for low construction quality. Only 22.8 percent of the people who received contractor-built houses in relocated sites (equivalent to ADRRS) reported being satisfied and only 3.5 percent considered the quality adequate. People complained about lack of participation, discrimination in favor of local elites, and disruption of family networks. Many people refused to move to new villages, and houses remained unoccupied. The study also showed that reconstruction by contractors was more costly and required more time than ODR.

Source: Jennifer Duyne Barenstein, 2006, "Housing Reconstruction in Post-Earthquake Gujarat: A Comparative Analysis," *Humanitarian Policy Network Paper 54* (London: Overseas Development Institute), http://www.odihpn.org/report.asp?id=2782.

Contractor-built houses in Gujarat

Owner-built houses in Gujarat

ALL PHOTOS: WHRC

Resources

Abhiyan. 2005. *Coming Together: A Document on the Post-Earthquake Rehabilitation Efforts by Various Organisations Working in Kutch.* Bhuj: UNDP/Abhiyan. http://openlibrary.org/b/OL3338629M/Coming_together.

Adams, L., and P. Harvey. 2006. *Cash for Shelter. Learning from Cash Responses to the Tsunami.* Humanitarian Policy Group. London: Overseas Development Institute. http://www.odi.org.uk/projects/details.asp?id=367&title=cash-voucher-responses-tsunami.

Ayesan, Yasemin. 2008. *External Evaluation of the Swiss Consortium's Cash for Repair and Reconstruction Project in Sri Lanka, 2005–2008.* Bern: Swiss Agency for Development Cooperation. http://www.sdc-cashprojects.ch/en/Home/Experiences/SDC_Cash_Transfer_Projects/Cash_for_Repair_and_Reconstruction_Sri_Lanka.

KUTCH NAV NIRMAN ABHIYAN

Barakat, Sultan. 2003. *Housing Reconstruction after Conflict and Disaster.* Humanitarian Policy Network Paper 43. London: Overseas Development Institute. http://www.odihpn.org/report.asp?id=2577.

Causton, A., and G. Saunders. 2006. "Responding to Shelter Needs in Post-Earthquake Pakistan: A Self-Help Approach." Humanitarian Policy Network Humanitarian Exchange 34. http://www.odihpn.org/report.asp?ID=2810.

Deutsche Gesellschaft für Technische Zusammenarbeit (GTZ), 2003, *Guidelines for Building Measures after Disasters and Conflicts* (Eschborn: GTZ), http://www.gtz.de/de/dokumente/en-gtz-building-guidelines.pdf.

Duyne Barenstein, Jennifer. 2006. "Challenges and Risks in Post-Tsunami Housing Reconstruction in Tamil Nadu." Humanitarian Policy Network Humanitarian Exchange 33. http://www.odihpn.org/report.asp?ID=2798.

Duyne Barenstein, Jennifer. 2006. *Housing Reconstruction in Post-Earthquake Gujarat: A Comparative Analysis.* Humanitarian Policy Network Paper 54. London: Overseas Development Institute. http://www.odihpn.org/report.asp?ID=2782.

Duyne Barenstein, Jennifer. 2008. "From Gujarat to Tamil Nadu: Owner-driven vs. Contractor-driven Housing Reconstruction in India." http://www.resorgs.org.nz/irec2008/Papers/Duyne.pdf.

National Disaster Management Agency, Pakistan (NDMA), 2007, *Earthquake 8/10, Learning from Pakistan's Experience* (Islamabad: NDMA), http://siteresources.worldbank.org/SOUTHASIAEXT/Resources/223546-1192413140459/4281804-1211943362217/1Earthquake.pdf.

PART 1

RECONSTRUCTION TASKS AND HOW TO UNDERTAKE THEM

SECTION 2

PLANNING RECONSTRUCTION

Guiding Principles

1 A good reconstruction policy helps reactivate communities and empowers people to rebuild their housing, their lives, and their livelihoods.

2 Reconstruction begins the day of the disaster.

3 Community members should be partners in policy making and leaders of local implementation.

4 Reconstruction policy and plans should be financially realistic but ambitious with respect to disaster risk reduction.

5 Institutions matter and coordination among them improves outcomes.

6 Reconstruction is an opportunity to plan for the future and to conserve the past.

7 Relocation disrupts lives and should be kept to a minimum.

8 Civil society and the private sector are important parts of the solution.

9 Assessment and monitoring can improve reconstruction outcomes.

10 To contribute to long-term development, reconstruction must be sustainable.

The last word: Every reconstruction project is unique.

Guiding Principles for Land Use and Physical Planning

- Laws, regulations, plans, and institutional frameworks should form the basis of reconstruction planning. If existing instruments are not realistic, or are contributing to informality, use the reconstruction process as an opportunity to improve them.
- The planning process should incorporate active collaboration among the reconstruction agencies, the affected community, the private sector, and other stakeholders, thereby engendering their ownership of the planning process.
- The planning process should respond to issues of land rights and titling and to discrepancies in the administration of land records, address the needs of informal occupiers of land, and work with them to identify viable alternatives.
- While addressing long-term development and DRR goals, land use and physical plans should still be flexible and offer choices, rather than become static "master plans."
- Land use and physical plans integrated with strategic planning can address reconstruction, DRR, and long-term development, yet be readily translated into action plans and investment proposals, including those that promote private investment.
- The planning process needs high-level support, active leadership from the government agencies that will actually implement the plans, and involvement from local communities.

This Chapter Is Especially Useful For:

- Lead disaster agency
- Local government officials
- Land use planners
- Agencies involved in reconstruction
- Project managers

Introduction

Land use planning is done to identify alternatives for land use and to select and adopt the best land use options. The main objective of land use planning is to allocate land uses to meet the economic and social needs of people while safeguarding future resources.

Land use and physical planning have an integrative function. Therefore, this chapter needs to be read along with 📖 Chapter 8, Infrastructure and Services Delivery; 📖 Chapter 9, Environmental Planning; 📖 Chapter 10, Housing Design and Construction Technology; and 📖 Chapter 11, Cultural Heritage Conservation. The issues dealt with in these chapters need to be addressed comprehensively during any meaningful planning process.

Land use and physical planning exercises provide a forum in which the interests of multiple stakeholders as well as the physical, social, and economic constraints on land uses can be debated and balanced in the post-disaster context. Specifically, post-disaster planning provides:

- tools and processes for organizing housing and infrastructure reconstruction in space and over time, addressing the impacts of the disaster and disaster risk reduction (DRR);
- a framework for stakeholders and elected representatives to relate reconstruction to longer-term mainstream development priorities; and
- an opportunity to modify policy, legislation, and regulations; strengthen institutions; and improve construction methods.

This chapter shows how land use and physical planning—complex processes with normative, institutional, and technical aspects—can be used to establish a coherent framework within which affected populations can permanently reestablish their housing, settlements, and livelihoods after a disaster.

Key Definitions

Land use planning is a public policy exercise that designates and regulates the use of land in order to improve a community's physical, economic, and social efficiency and well-being. By considering socioeconomic trends as well as physical and geographical features (such as topography and ecology), planning helps identify the preferred land uses that will support local development goals. The final outcome is allocation and zoning of land for specific uses, regulation of the intensity of use, and formulation of legal and administrative instruments that support the plan. A land use plan may be prepared for an urban area, a rural area, or a region encompassing both urban and rural areas.

Physical planning is a design exercise that uses the land use plan as a framework to propose the optimal physical infrastructure for a settlement or area, including infrastructure for public services, transport, economic activities, recreation, and environmental protection. A physical plan may be prepared for an urban area or a rural area. A physical plan for an urban region can have both rural and urban components, although the latter usually predominates. A physical plan at a regional scale can also deal with the provision of specific regional infrastructure, such as a regional road or a bulk water supply system.

Land use plans and physical plans are not necessarily mutually exclusive. It is common practice in many countries to prepare comprehensive development plans that address both land use zoning and the provision of physical infrastructure. Such an exercise is more meaningful if carried out in the context of a strategic planning process, whereby the proposals in the land use plan and the physical plan become part of a comprehensive development plan. While land use and physical plans are outcome-oriented, strategic plans are more process-oriented.

Key Decisions

1. The **lead disaster agency** should decide with **local government**, immediately after the disaster, how they will share responsibility and coordinate aspects of reconstruction related to local planning and land use, including decisions on land use changes and relocation. They should also decide whether technical assistance will be needed at the local level.
2. The **lead disaster agency** should determine how geographic, satellite photography, and other data useful for land use planning will be shared with and among all agencies involved in reconstruction, to save costs and improve planning outcomes. See 📖 Chapter 17, Information and Communications Technology in Reconstruction.
3. **Local government** should decide immediately whether its existing land use plans, regulations, and building codes are sufficient to manage the recovery and reconstruction or to what extent they need to be modified. Building code revision, if required, is time-consuming and needs to start immediately.
4. **Agencies involved in reconstruction** should establish a joint timeline for reconstruction early on that allows enough time for planning without impeding the reconstruction process, and should agree on a communications strategy with the public regarding land use issues. See 📖 Chapter 3, Communication in Post-Disaster Reconstruction.
5. **Local government** should participate in assessments or initiate studies to determine how existing land uses and construction technologies contributed to disaster impacts and to determine how regulations should be modified to reduce future disaster risk.
6. **Agencies involved in reconstruction, local government**, and **land administration agencies** should collectively decide during assessment whether relocation will be part of the reconstruction process, and whether land tenure issues will need to be addressed, so that preparation for these activities can begin immediately, due to their long lead time. See 📖 Chapter 5, To Relocate or Not to Relocate.
7. **Local government** should determine how it will manage (1) plan review and approval, (2) issuance of building permits, (3) contractor training, and (4) construction inspection. **Local government** should also determine whether local capacity and institutions are adequate to ensure safe rebuilding or what, if any, assistance will be needed.

Public Policies Related to Land Use and Physical Planning
Local and National Policies and Regulations

Public agencies at the national, state, and local government levels in disaster-affected countries may already have in place physical and land use policies and regulations, including special provisions to manage post-disaster planning. Implementation of the policy may fall within the jurisdiction of the Ministry of Public Works, the Ministry of Land, and/or the Ministry of Urban Development and Planning, at different levels in different departments. Most disaster-affected urban areas have physical and land use policies and regulations in place, generally under the jurisdiction of the local planning department or planning commission. Rural areas may have no plans or may be governed by regional or rural development plans overseen by a higher level of government, such as the state or province.

International Frameworks

Land use, land ownership, and land rights issues are addressed in a number of framework documents and instruments issued by international agencies. Some key agreements are included in the Resources section below.

Physical planning, because it is context-specific, is addressed less in international frameworks. However, numerous organizations, including the Oregon Natural Hazards Workgroup,[1] the Shelter Centre and the United Nations Office for the Coordination of Humanitarian Affairs (UN OCHA),[2] the United Nations Environment Programme (UNEP),[3] and the European Union's Sustainable Urban Development Unit[4], have prepared useful guides that address physical planning issues.

DANIEL PITTET

Technical Issues

Is There Time for Planning?

Planning as a process is inevitable in a reconstruction scenario, whether the decision is to just rebuild houses or to achieve comprehensive development resilient to future disasters. The key questions are: How detailed will the planning process be? Will the planning process enable reconstruction or become an impediment in itself? These questions are answered by defining the planning process. We may plan for simple housing layouts where land is available or we may take affected communities through a properly structured planning process, which results in communities expressing their needs and aspirations. While the latter takes more time, the time required can be managed by synchronizing outputs with the reconstruction process, as described below.

Is There Information for Planning?

Lack of information is frequently put forward as a reason to forgo planning, and the disaster may have worsened an already inadequate situation by destroying information or making access to it more difficult. But supplementary information can be readily mobilized using various means, including information technologies and participatory data-gathering methods. Lack of information should almost never be a reason not to plan. See 📖 Chapter 17, Information and Communications Technology in Reconstruction.

Is There Capacity for Planning?

Lack of institutional capacity often discourages decision makers from attempting a planning process. While a planning process usually presumes both the existence of a legal and institutional framework to mandate the process and the professional capacity to implement the prepared plans, these are not absolute prerequisites. Various pragmatic approaches have been tried out in post-disaster situations under less than ideal circumstances. 📖 Annex 2, How to Do It: Post-Disaster Planning Where Planning Law and Institutional Capacity Are Weak, summarizes these approaches.

Define a Planning Process to Guide Reconstruction

Need for a planning process. Keeping housing reconstruction in focus, there is a strong case for a planning process if (1) the impact of the disaster is intense and widespread; (2) there is a need to mitigate future risks from the same or other types of disasters; (3) displacement of the affected population is likely due to loss of land, land rights, titling issues, or high vulnerability of pre-disaster housing locations, or (4) the reconstruction tasks are complex. 📖 Annex 1 to this chapter, How to Do It: Undertaking a Comprehensive Planning Process, describes the key steps in carrying out a planning process.

Legal and institutional frameworks for planning. Land use and physical planning are usually governed by law and need to be carried out in conformity with such laws. Ad hoc reconstruction processes tend to result in legal tangles and titling issues later. At the same time, existing laws and

1. Oregon Natural Hazards Workgroup, Partners for Disaster Resistance & Resilience, 2007, "Post-Disaster Recovery Planning Forum: How-To Guide," http://nctr. pmel.noaa.gov/education/science/ docs/Reports/OR_Post-Disaster_ Guide_20070716_Draft.pdf.
2. United Nations Office for the Coordination of Humanitarian Affairs (UN OCHA) and Shelter Centre, 2010, *Shelter After Disaster: Strategies for Transitional Settlement and Reconstruction* (Geneva: UN OCHA), http://www.sheltercentre. org/library/Shelter+After+Disaster; *and* T. Corsellis and A. Vitale, 2005, Transitional Settlement: Displaced Populations (Oxford: Oxfam Publishing), http://www. sheltercentre.org/node/3128/ download/2179.
3. UNEP, 2008, "Environmental Needs Assessment in Post-Disaster Situations: A Practical Guide for Implementation," http:// www.humanitarianreform.org/ humanitarianreform/Portals/1/ cluster%20approach%20page/ clusters%20pages/Early%20R/ UNEP%20PDNA_pre-field%20 test%20draft.pdf.
4. European Commission, 2002, *Consultative Guidelines for Sustainable Urban Development Co-Operation towards Sustainable Urban Development: A Strategic Approach* (Luxembourg: European Commission), http://www.ucl. ac.uk/dpu-projects/drivers_urb_ change/official_docs/Tow_Sust_

institutions may be inadequate to respond to post-disaster needs, and therefore the need for updating should be assessed immediately following the disaster. This is especially important if the legal framework and instruments do not reflect the practicalities of how people build, and therefore contribute to the informal construction and land uses that may themselves have contributed to the scope of the disaster.

Functional scope and geographical jurisdiction. While a conventional planning process can address a wide range of concerns, post-disaster reconstruction planning should focus on topics that enable safe and sustainable community rebuilding. The geographical area for planning should also be clearly delineated.

Timeline for planning. The deliverables of the planning process should be in synch with reconstruction priorities. For example, identifying land for relocation and updating building codes are likely to be the most urgent priorities.

Define Principles to Guide Planning

No matter the scale of the planning exercise or whether it is used to update plans or develop new ones, good planning principles should guide the process. The best principles for planning are those that both professional planners and the community agree on and those that embody the larger development vision for the locality. Some principles to be considered are discussed below.[5]

At the town level

- **Plan for growth.** Land use planning may be done to reallocate existing land uses in the short term, but it generally has a longer, more prospective focus, especially in urban areas that are growing or whose growth may be affected by the disaster. Planning for the future relies on assumptions about population change and future demands for land and services. For this reason, developing reliable projections of population and use of services (such as highway traffic) is an essential early step in the land use planning process.
- **Restore connectivity.** Restoration of the social and economic linkages is important for revival of communities. Transportation, communication, and road networks must be priority items in reconstruction so that they can transport labor and material for reconstruction. Connectivity at the local level will make mobility easy for all means of transportation, including walking and bicycles, and will support livelihood activities.
- **Consolidate unused land.** Since reconstruction requires land, plans may be needed to consolidate land so that it can be made available for development. Unused public land and abandoned industrial land are two sources to be considered.
- **Improve energy efficiency and consider environmental impact.** Reconstruction presents excellent opportunities to promote housing designs, development patterns, and neighborhood layouts that lower energy consumption and encourage a lifestyle that has a low impact on the environment. Even landscaping and house orientation contribute to these goals and should be carefully planned.
- **Create development nodes.** Relocated and rehabilitated settlements should be planned so that they are attractive for investment and development. Providing quality public services and sites for services and other land uses desired by residents will help new settlements become vibrant communities.
- **Reconstruct strategic towns.** Economic centers that serve as growth engines should be reconstructed as early as possible. Strategic towns and cities absorb population, generate employment and nonwage economic opportunities, and provide social services to affected people. Even if such cities have the capacity to meet the needs of their own population and support other villages and towns, they may need assistance to plan reconstruction.[6]

At the site level

- **Integrate residential, ecological, and economic land uses.** Home-based businesses are the life blood of low-income communities. Ensure that neighborhood plans and housing designs provide adequate and appropriate space for these activities. A common tool for achieving this is mixed-use zoning that allows residential and certain commercial activities to be carried out from the same plot. The environmental setting and ecological footprint of a neighborhood also affect quality of life and should be priority concerns in site planning.
- **Avoid enclave development.** Relocating communities in enclaves isolates them socially, prevents economic integration, and brings a host of social problems. Therefore, while planning for relocation, careful integration into the fabric of the receiving settlement is essential.

5. World Bank, 2008, "Planning for Urban and Township Settlements After the Earthquake," Good Practice Notes, http://siteresources.worldbank.org/CHINAEXTN/Resources/318949-1217387111415/Urban_planing_en.pdf.
6. World Bank, 2008, "Planning for Urban and Township Settlements After the Earthquake," Good Practice Notes, http://siteresources.worldbank.org/CHINAEXTN/Resources/318949-1217387111415/Urban_planing_en.pdf.
7. Escape routes were found to be a high-priority infrastructure improvement for communities after the 2004 Indian Ocean tsunami. Improving escape routes may require the acquisition of land to expand rights-of-way. This can be accomplished through negotiation or through conventional compulsory land acquisition if a voluntary solution cannot be found.
8. Speech of Chinese Ambassador Zhou Wenzhong, Chinese Embassy in the United States, "Sichuan Earthquake: Relief, Recovery and Reconstruction," July 25, 2008, http://www.china-embassy.org/eng/sgxx/t478233.htm.

- **Plan for emergency access.** Any settlement plan must identify escape and evacuation routes and provide access for emergency vehicles and fire service engines. Those affected by the disaster may not return or occupy sites that don't have adequate escape routes.[7]

The Key Actors in a Planning Process and Their Roles

Level of government	Roles
Central or national government	■ Invoke federal/national law, only where the situation warrants. ■ Mobilize the relevant government agencies to undertake, commission, and supervise planning. ■ Provide funding or support for accessing international funding. ■ Provide specialized technical expertise if required. ■ Ensure public investments conform to plans and codes.
State or provincial government	■ Provide legal mandate for the plans. ■ Create the policy environment in which the plans are prepared. ■ Mobilize the relevant government agencies, including regional entities, to guide and support the planning process. ■ Provide technical expertise as required. ■ Provide funding or support for accessing funding. ■ If regional planning is required, carry out the planning process.
Local government	■ Carry out the planning process at the local level. ■ Create structures to enable meaningful community participation. ■ Be committed to implementing plans prepared with community participation. ■ Approve plans and establish the regulatory framework for implementation. ■ Carry out communications campaigns and training programs to ensure compliance with plans and codes. ■ Review and approve building plans, enforce building codes and land use regulations, carry out inspection, and administer sanctions.
Community (affected people as well as larger community)	■ Participate in the land use, physical, and strategic planning processes. ■ Develop a collective vision for the future of the community. ■ Arrive at consensus on policy issues that cut across communities. ■ Where relevant, prepare community-level detailed plans in conformity with larger policies. The ⬛ case study on the 2004 Indian Ocean tsunami in Aceh, Indonesia, below, describes how communities took the lead in remapping land parcels as the first step in a wide-ranging process to formalize land ownership.
Project facilitators (planners, nongovernmental organizations [NGOs], and other intermediaries)	■ Interpret government policies to set out the agenda for planning. ■ Educate the community on planning imperatives and the policy framework. ■ Interpret technical information and offer viable choices to government and communities to enable informed decision making. ■ Develop and carry out projects that comply with plans and codes.
Technical experts	■ Carry out technical investigations, data collection, and analysis to support planning. ■ Develop technical recommendations and options. ■ Assist with implementation of plans and codes.

The actual distribution of roles will depend on the existing legal and institutional frameworks as well as the actual capacity at the local government level. An interesting experiment in progress at the time of this writing is the pairing of cities in China following the 2008 Wenchuan earthquake. The cities and towns affected by the earthquake are being paired with those unaffected by the earthquake for financial, technical, and logistical assistance.[8]

Differentiating the Planning Process for Urban, Rural, and Regional Contexts

While the generic planning process, as well as the core issues that the process addresses, will remain largely similar in urban, rural, and regional contexts, there are differences that need to be recognized.

Urban areas. The planning process in urban areas tends to be more complex and prone to conflict and contestation. Land values are higher, property ownership is more complex, and flexibility to change land uses is often more limited. It is important to acknowledge that land use planning is going through a paradigm shift across the world. From an earlier, purist approach of exclusive zones for specific uses (e.g., residential, commercial), there is a shift toward appropriate mixes of compatible uses (e.g., residential with small businesses, institutional with offices). From an earlier approach of flat, low-density urban development, there is a shift toward more compact cities with variable density correlated with urban transport systems. Other distinctive characteristics of urban planning include the following.

- Developed or built-up areas predominate. Therefore, the land use plan needs to reflect and plan for diverse land uses.
- The demand for infrastructure will be higher (in both quantitative and qualitative terms) and the provision of infrastructure more complex and costly. Therefore, housing reconstruction must be closely coordinated with the development of infrastructure.
- Urban land use planning has an immediate and highly visible impact on urban land values. Therefore, a transparent approach to planning is essential.
- Urban areas are more likely to have agencies that undertake planning and regulation as well as professionals for design and supervision. Therefore, the approach to DRR is usually based on planning and regulation.
- Investments in urban settlements and infrastructure during reconstruction should contribute to already-established urban development goals.
- Development control and regulation systems are usually present in urban areas but tend to be flawed and complicated, creating high incentives for noncompliance. DRR initiatives therefore need to focus on simple and effective regulatory systems. The 📖 case study, below, on the 1985 Mexico City earthquake shows how the deterioration of apartment buildings caused by the lack of building code enforcement may have contributed to the disaster impact, and the cost of reconstruction.
- Stakeholder participation in urban areas is relatively difficult due to such factors as: the diversity of interests resulting in more conflict, higher sensitivity of residents to delays, and the volatility of public opinion in post-disaster situations. People are also more mobile and their free time is more limited. Urban residents often care more about their own space than about common space, since geography is only one basis for identity in an urban context.

Rural areas. In rural areas, the settlements and associated built-up areas form a relatively small part of the larger landscape. Land values are lower, and, while ownership and titling issues exist, they can often be resolved relatively easily through participation. The sense of ownership is higher in rural areas, and the social structure plays a major role in the dynamics of reconstruction. However, community participation is fully achievable in a rural context. Other features of rural planning include the following.

- Land use plans need to respond more significantly to natural features, such as geology, topography, hydrology, and ecology. The classification of uses within a settlement will assume less significance while in the larger landscape will reflect the diversity of uses in agriculture, animal husbandry, forestry, and other related activities.
- Institutional arrangements for regulating development are nonexistent in rural areas of most developing nations; there may be no designated planning agency whatsoever. The approach to DRR should be based on building awareness and training construction workers.
    - A land use plan in a rural area may not dramatically change land values, but can still have a significant impact on the sustainability of development.
    - Physical planning may be limited to a basic road network and essential services within the settlement. However, there may be planning required to support agriculture and other rural livelihoods.
    - Housing is usually designed and built by owners themselves or by local masons. It is important that building regulations are responsive to the local cultural context.

Regions. Regional plans become relevant if there are reconstruction requirements or vulnerability mitigation issues that are spread over large, geographically integrated areas. For example, if the road network in a large area has been damaged or if an entire floodplain has attracted high-risk land uses, a regional plan may be the appropriate vehicle. Coastal zones also may have special planning or regulatory regimes that govern set-backs and land uses across multiple jurisdictions. Other considerations with regional planning include the following.

DANIEL PITTET

- A regional land use plan will deal with macro-level issues, like locations of settlements, protection of forests, and management of coastal zones, river basins, and floodplains. However, such plans by themselves will not be enough to guide the post-disaster reconstruction process.
- Physical planning at the regional scale will primarily look at the facilities for regional infrastructure, such as regional roads, structures for watershed management, and bulk water pipelines.
- The institutional arrangements for regional planning can vary from state or provincial governments to special agencies set up to coordinate development in a particular zone. Their planning capability will vary.
- Regional plans are often developed with an economic focus. Their utility in a post-disaster context may be to connect disaster recovery to the economic goals set out in the plan.
- Regional plans have to be complemented by plans for the rural and urban areas within the region.

Note on planning terminology. A wide range of terms are in vogue for various kinds of plans: Vision Plan, Structure Plan, Outline Plan, Development Plan, Concept Plan, Master Plan, Strategic Plan, and so on. There are considerable overlaps and similarities in the contents of these differently named plans. Each country, state, or locality will have terminology that is accepted in that context. The following section describes a fairly generic process that captures most of the contents and formats that are used across the world. It is recommended that the process adopted in each context build on processes that are understood locally, while improving on them, adding whatever is missing in terms of content, format, or processes.

Planning in the Post-Disaster Context

When a planning process is contemplated, the first question asked is "What topics and issues should the plan address?" or "What are the components of a good post-disaster reconstruction plan?" This section describes what needs to be addressed in a comprehensive post-disaster reconstruction plan. The second question is "How do we go about it?" How to do planning is covered in 🖳 Annex 1, How to Do It: Undertaking a Comprehensive Planning Process. In situations where the institutional capacity is limited, realistic decisions will be required about how much planning can be achieved, which is dealt with in 🖳 Annex 2, How to Do It: Post-Disaster Planning Where Planning Law and Institutional Capacity Are Weak.

Generic Content of a Comprehensive Post-Disaster Reconstruction Plan

Land use

	In a non-disaster situation, a comprehensive plan would address land uses for all purposes, including transportation, governmental, industrial, commercial, and residential. After a disaster, the planning exercise may focus primarily on land for housing and infrastructure reconstruction, but should not ignore other land use requirements, especially any others that have been affected by the disaster. This component of the plan addresses the issues and questions listed below.
Housing needs assessment	How many houses have been destroyed or damaged? Is it safe to rebuild in the same location? Are there multi-dwelling buildings (apartments)? Are there tenancy, land rights, or titling issues? What is the housing need in different categories?
Assessment of land availability	If *in-situ* reconstruction is possible, can adequate DRR measures be implemented in available sites? If relocation is required, is there public land available? What are the criteria for choosing relocation sites? What are people's preferences in relocating? What are the underlying socioeconomic and political dynamics?
Land allocation planning	What is government policy on land for housing reconstruction and other purposes? Will housing reconstruction be plotted (single-family) development or apartment construction? What will be the process for acquiring and allocating land? What will be the policy on land allocation for social and physical infrastructure? Is there any need for land consolidation or land pooling?
Titling	What sort of tenure is to be granted to those who have been allotted land? How will the property rights documents be created and provided? How will the rights of women be protected? The outputs of this component will include (1) maps showing locations for housing reconstruction, (2) tentative or conceptual housing layouts (housing design is a separate activity), (3) housing project briefs with cost estimates, and (4) policy recommendations, if required.

Land use zoning and building codes

	Land use zoning is a systematic way of managing the nature and intensity of land use in a specific area. The output is a map (with an accompanying table) showing various zones where specific uses or a mix of uses may be permitted. This component should address such questions as: Is there a system for land use zoning? Is it adequate to address DRR requirements? To reduce risk while accommodating future growth, what type of land use zoning is required? What is the institutional mechanism for implementation of the zoning? Is it market friendly? User friendly? How will informal settlements of the urban and rural poor be integrated into the land use zoning? Land use zoning imperatives are different in rural and urban settings, as discussed later in this chapter.

Building codes and development regulations	This component relates to the design, construction, and performance of buildings. The issues that need to be addressed include: Is there a regulatory system in place? How effective is it? Are prevailing codes responsive to prevailing hazard risks? What codes need to be put in place? How would they relate to land use zoning? Do existing procedures for building permission need improvement? What is the architectural heritage of the region? How can building codes accommodate local traditions? Do local building techniques need enhancement for disaster resilience? How will the new building codes affect housing affordability? How will codes apply to informal settlements of the urban and rural poor? The typical output is a set of Building Codes, Building Bylaws, or Development Control Regulations (the rule book for building design and construction).
Guidelines and manuals	If time or institutional constraints make it unrealistic to update building codes and regulations in advance of reconstruction, an alternative is to produce advisory guidelines and manuals that can be used in reconstruction. These guidelines and manuals should be based on standards and codes from an area with similar building technologies and housing designs. There are risks associated with using standards that are inappropriately stringent or from areas with different building technologies. The promulgation of the guidelines should be accompanied by a social communications program, training of builders, and a strategy for overseeing reconstruction. See 🔖 Annex 2, How to Do It: Post-Disaster Planning Where Planning Law and Institutional Capacity Are Weak, for more guidance on this situation.

Physical plan

	Several key elements of physical planning are listed here. Planning may address them collectively, or each may be dealt with separately if the situation demands it.
Road layout	What is the existing road network in the settlement or region? Is it adequate for speedy evacuation and rescue in the event of a disaster? Are new road connections required to reduce risk and enhance preparedness? Are new roads required to provide connectivity to housing reconstruction locations? What is the extent of damage to roads? Are engineering improvements required? The output of this component will include road network maps and project briefs for road construction.
Plot layout	This relates to proposed housing reconstruction. While detailed design of housing layouts is a separate activity, at the planning stage it is important to prepare at least a conceptual layout of the proposed housing to ensure that the land allocation is adequate and that major issues have been addressed. The output is a set of plot layout plans.
Planning for infrastructure and services	This component deals with network alignments and land allocation for infrastructure services. The critical services include water supply, wastewater management, solid waste management, and storm-water management. Power supply and telecommunications networks may also be important. In all these cases, the existing systems need to be documented and proposed improvements need to be conceptually worked out to the extent that is required for assessing land-related issues. The output is a set of maps. Project formulation for infrastructure is a separate activity, but may be carried out concurrently or integrated with the planning process.
Planning for public buildings and social infrastructure	This component deals with allocation of land for facilities related to health, education, government, recreation, community development, and disaster shelters. In the planning process, the questions that need to be addressed are: What facilities existed pre-disaster? Should refuges be built? What is the extent of damage? Do any facilities need relocation? Were pre-disaster facilities adequate? What does the reconstruction policy envisage: restoration of pre-disaster levels or improvement? What is the land requirement? What facilities are required as part of new housing to be created? The output of this component is a set of maps showing locations of proposed facilities and project briefs for creating them. Planning infrastructure projects is covered in 🔖 Chapter 8, Infrastructure and Services Delivery.

Local economic development

	A comprehensive planning process needs to look at the economic base of the settlement/region and the need for interventions in the post-disaster situation. For example, if the disaster has destroyed livelihoods and economic diversification is a dire necessity, then the planning process needs to generate proposals for creating new job opportunities. In most cases, this will have a land allocation or land use zoning dimension. The output will consist of project briefs and, where relevant, maps showing land allocation.

Cultural heritage conservation

	Issues related to cultural heritage conservation are dealt with in detail in 🔖 Chapter 11, Cultural Heritage Conservation. In the planning process, conservation imperatives will find reflection in land use zoning, building regulations, and land allocation for cultural projects where relevant.

Implementation strategy

	Everything decided or developed in the planning process will remain wishful thinking if inadequate attention is paid to the strategy for implementation. While immediate post-disaster needs (usually "restoration") will find funding easily, for long-term recovery it may be necessary to develop strategies to generate funding from multiple sources. This section of a plan should bring together the "big picture" of the reconstruction process, define the implementation process, estimate overall funding requirements, and assign roles, responsibilities, and tasks.

Addressing Land and Property Rights and Land Titling Issues

In a post-disaster situation, the following issues related to land and property rights and titling typically emerge:

1. Determining land and property rights that existed before the disaster and the entitlement to land or housing assistance after the disaster
2. Addressing the situation of people with uncertain tenure rights in reconstruction policy making and reconstruction planning
3. Providing certainty of land title or expanded land rights in reconstruction to those affected by the disaster, irrespective of their pre-disaster situation.[9]

Lack of land tenure security is usually compounded by pre-disaster institutional weaknesses related to land and property rights, described below.

Pre-disaster dysfunctionalities. There is an inextricable connection between poorly functioning land systems and disasters.[10] Where tenure security is weak or land markets are not accessible by all groups, vulnerability is higher, disaster impacts are greater, and recovery is slower. Indicators of poorly functioning land systems include:

- a large number of settlements where occupants have extralegal or informal tenure[11];
- poor land governance, including outdated, incomplete, or erroneous land records and weak capacity in land administration institutions;
- lack of tenure security with all types of property rights;
- inferior land rights for women because of inheritance, marital law, or administrative practices; and
- highly unequal land distribution, including the inability of large sections of the urban population to afford formal land access in locations that are supportive of livelihood activities.

Before a disaster, weak land systems will cause populations to settle in high-risk areas and will leave these settlements beyond the reach of DRR measures. Occupants with weak tenure rights hesitate to evacuate, despite pre-disaster warnings, for fear of losing their land—and may instead lose their lives. Female-headed households, more apt to evacuate, face disproportionate risks of lost land or property from invaders or theft.

Post-disaster complexities. After a disaster, an influx of the poor and landless from other areas to take advantage of temporary emergency shelters can make closing these shelters difficult when the disaster subsides. Renters, squatters, and landless people who leave the area may have difficulty returning to their disaster-affected land, reestablishing their lease terms, or securing affordable rent. Legal owners may have lost records or have customary rights with no documentation. The number of tenancy categories may far surpass what is anticipated in the law.[12] None of these problems is solved easily or rapidly in most countries, even under the best conditions, and they will inevitably arise in housing reconstruction programs. Some potential solutions are summarized in the following table.

9. In Aceh, Indonesia, following the 2004 Indian Ocean tsunami, government announced that residents would have the "right to return," a policy that was subject to significant debate over its practical meaning for years afterward.
10. United Nations Centre for Human Settlements (UN-HABITAT), 2008, "Scoping Report: Addressing Land Issues after Natural Disasters," http://www.gltn.net/images/stories/downloads/utf-8nat_disaster_scoping_paper_jan_08.pdf.
11. Customary systems are not necessarily an indicator of poorly functioning land systems. In some locations, such as rural Africa, they may be the most appropriate system.
12. UN-HABITAT identified 31 different tenancy situations in the affected population in Peru following the 2007 Ica/Pisco earthquake. UN-HABITAT, the Department for International Development (DFID), and the Ministry of Housing, Construction and Sanitation, "Final Report, Land Ownership and Housing" ("Informe Final, Tenencia de la Tierra y la Vivienda"), 2008, Centro de Estudios y Promoción del Desarrollo.

Measures to Consider for Solving Post-Disaster Land Rights Problems

Common land rights problem	Potential solutions
Occupants of informal settlements whose occupancy is of uncertain tenure rights	The resolution of such a problem is inevitably linked to pre-existing government policies on slums and squatter settlements. However, the post-disaster situation usually offers a unique opportunity to provide these households with secure tenure as part of the reconstruction program, particularly if they are affected by the disaster.
Owners with documented land rights who have lost documentation in the disaster	Use community-based mechanisms or traditional authorities to validate claims. Provide new (or interim) documentation to landowners, including in previously undocumented areas.
Owners with rights to land, but no formal title or documentation to prove it or with property title but no land title	Provide technical assistance to the land administration agency to accelerate normal land administration procedures, including replacement of documents, formalization of unrecorded subdivisions, and transfers, especially those from one generation to another. Use duly authorized mobile "land administration teams" to expedite the process of consolidating legitimate claims to property. Provide assistance so that owners can negotiate clear title with the possessor of land. After the 2000-2001 Mozambique floods, the United Nations Human Settlements Programme (UN-HABITAT) helped the government analyze how poor land management had contributed to the disaster and identify projects to remedy weaknesses in the land management system, as discussed in the 📖 case study, below.

Common land rights problem	Potential solutions
Landless who need to be relocated, including disaster refugees	Acquire public land. A common solution is to use public land for relocation. However, availability is not sufficient justification; use should be preceded by a good site evaluation.
	Acquire private land: Market-based acquisition from landowners willing to sell.
	Government offers limited fiscal incentives to sellers (property tax rebate).
	Offer long-term rental assistance rather than land ownership.
	Use land pooling systems of the sort used in Kobe and Bhuj, India.
International agencies or individuals attempting to acquire legally titled land as part of reconstruction confronted by poor records and inefficient land administration procedures	Provide technical assistance to the land administration agency to accelerate normal land administration procedures.
	Find owners (public or private) willing to lease land to agencies either as a permanent or interim solution. Agency then follows through to establish freehold ownership of individual parcels, if called for.
	Provide financial assistance to families to cover legal costs, lower other transaction costs, or subsidize purchase price of land.
	Focus on procuring land rights and tenancy, rather than on securing formal land title for large numbers of those affected by the disaster.
Large parcel needing to be subdivided to provide individual title to those relocated	Provide technical assistance to the land administration agency to accelerate normal land administration procedures.
	Agency provides collective land rights to those relocated as a group as a permanent or temporary solution.
	Agency provides interim documentation until subdivision and titling are formalized.
	Consider rezoning the parcel(s) to allow subdivision into individual residential, mixed-use, commercial, and institutional (public use) plots. Nonresidential uses may help cover costs of residential land.
Land disputes between households affected by the disaster or with the receiving community, caused by relocation without formal titles or invasion	Avoid relocation programs that create questionable land rights at all costs.
	Negotiate with claimants to purchase land or land rights and resolve claims before relocation.
	Establish formal or community-based mediation mechanisms to resolve post-disaster land issues.
	Consider land consolidation and readjustment of parcels at the community level.
Inflated price of land needed for relocation	Use market valuation methods, validated with data from outside the area.
	Use competitive bidding process to establish prices.
	Use negotiated price to avoid eminent domain proceedings.
	Temporarily impose land price controls by government in the immediate post-disaster period.
Areas from which communities are removed for reasons of risk reduction are rapidly relocated	Zoning or other regulation should be established that forbids relocation in high-risk areas.
	Provide budgetary support for police or other enforcement body.
	Create community enforcement mechanisms.
	Secure abandoned areas or transform into an income-generating use that involves minimal built investment, e.g., urban agriculture and fish cultivation ponds, or into a recreational use, so that either private users or the public at large has an interest in monitoring and reporting relocation.
Governments considering land distribution to those displaced but fearing protests from other landless groups not affected by the disaster, or fearing influx of fraudulent claims	Announce assistance scheme and eligibility criteria after the census of affected population has taken place.
	Link assistance scheme or land distribution to existing social assistance schemes and policies.
	Use disaster to establish assistance or land distribution schemes that can be expanded to other groups post-disaster.
	Employ social communication tools to explain assistance strategy and rules.

For access to additional resources and information on this topic, please visit the handbook Web site at www.housingreconstruction.org.

118   SAFER HOMES, STRONGER COMMUNITIES: A HANDBOOK FOR RECONSTRUCTING AFTER NATURAL DISASTERS
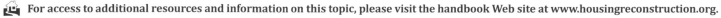

Other actions that should be taken during reconstruction include:

- focusing particularly on the housing, land, and property rights of women and children;
- conducting an early land-tenancy assessment and carrying out ongoing monitoring of the impact of measures to address land-rights issues as reconstruction progresses (see Annex 1, How to Do It: Conducting a Post-Disaster Housing Sector Assessment in 📖 Chapter 2, Assessing Damage and Setting Reconstruction Policy, for an example of an assessment methodology that can be used for this purpose); and
- ensuring that agencies or banks transfer land title to new owners once they occupy the house or loans are paid off.

The long-term land rights-related reforms that often are required and that may also be initiated during reconstruction include:

- long-term reforms to improve access to secure tenure and to improve land administration systems;
- the reduction and prevention of informal settlements in DRR strategies (see 📖 Part 4, Technical References, Disaster Risk Management in Reconstruction); and
- establishment of redundant and secure recordkeeping systems, and deployment of information and communications technology (ICT) data warehousing systems to digitize and protect records so that they are available after the next disaster (see 📖 Chapter 17, Information and Communications Technology in Reconstruction).

Promoting Disaster Risk Reduction in Reconstruction Planning

To avoid reproducing the vulnerabilities that contributed to the disaster, multiple actors need to be engaged in implementing and managing DRR.[13] The following actions should be taken to integrate DRR into the planning process.

- Include DRR as an integral element in every phase of planning for reconstruction.
- Strengthen both institutional and community capacity, and commit resources to ongoing DRR activities and training.
- Implement DRR training and information sharing that reaches all stakeholders.
- Develop model plans that incorporate risk reduction, safe construction techniques, siting, and building maintenance that integrate DRR into housing and settlement designs and maintenance.
- Identify how risks can be transferred through insurance and micro-insurance schemes, and facilitate the implementation of feasible measures.

Structuring Community Participation

Planning for settlements and livelihoods must be coordinated at the local level; therefore, the participation of the community is fundamental to ensure the sustainability of the reconstruction or relocation plan. See 📖 Chapter 12, Community Organizing and Participation, for a discussion of participatory methodologies. Conditions that contribute to the success of this approach include:

- an inclusive decision-making process that incorporates vulnerable populations, including women;
- a high level of interaction, cooperation, and partnerships among different stakeholders, including civil society, national and local governments, the private sector, the media, and national and international support agencies;
- support by local governments to establish a coherent framework, provide effective facilitation mechanisms, and ensure the necessary information is available (participatory data-gathering may be a useful tool); and
- established participatory activities in the community, or a pre-disaster consultative process with communities living in vulnerable locations to agree on preventive measures and mitigation options as groundwork for post-disaster collaboration, especially in communities affected by recurring disasters (a brief description of a community-based hazard mitigation planning process is found in 📖 Part 4, Technical References, Disaster Risk Management in Reconstruction).

The 📖 case study, below, on the reconstruction project undertaken by the South Indian Federation of Fishermen Societies (SIFFS) in Tamil Nadu following the 2004 Indian Ocean tsunami describes how homeowners carried out a geographic and socioeconomic mapping of reconstruction settlements that included a study of the use of outdoor common space.

We may plan for simplified housing layouts where land is available or we may take affected communities through a properly structured planning process, which results in communities expressing their needs and aspirations.

13. International Strategy for Disaster Reduction (ISDR), 2006, "Words Into Action: Implementing the Hyogo Framework for Action Document for Consultation," http://www.preventionweb. net/english/professional/ publications/v.php?id=594, p. 81; *and* ProVentium Consortium, 2007, "Construction Design, Building Standards and Site Selection, Guidance," Note 12, http://www.sheltercentre.org/ sites/default/files/ProVention_ ToolsForMainstreaming_GN12.pdf.

Risks and Challenges

- Initial damage and loss assessments inadequately incorporate land issues.
- Data needed for planning—for example, land records and GIS data—have been lost or are inadequate.
- The land administration system is weak or flawed and, as a result, the information base related to land is inadequate for proper planning.
- The institutional framework for planning and regulation is weak or flawed, making the planning process difficult and possibly resulting in redundant regulation.
- Ad hoc planning processes create a disconnection between reconstruction and existing land use or physical plans.
- Delays in the planning process and/or institutional and community disagreement over land use and site location delay reconstruction.
- A variety of obstacles prevent acquisition or legalization of land for relocation, reconstruction, and infrastructure rights-of-way.
- A lack of coordination between infrastructure rehabilitation and housing reconstruction.
- Local cultural, social, and economic life is ignored in planning reconstruction.
- The opportunity to improve urban land markets, and to expand land rights and access to land, is missed in post-disaster planning.
- The legality of reconstruction investments is later challenged because of unresolved land ownership claims or violations of land use regulations.

Recommendations

1. Assess the need for a post-disaster planning process as soon as the emergency is over. If the legal and institutional framework is weak, consider using the alternative methods described in ⌂ Annex 2 of this chapter.
2. Delineate the geographical area for planning and identify the functional jurisdiction of the plan.
3. Assess the existing legal and institutional framework for planning. Build on existing systems, customizing and improving them for post-disaster reconstruction.
4. Structure the planning process in a comprehensive manner covering land use planning, physical planning, and strategic planning, addressing all the issues relevant to reconstruction, DRR, and, to the extent required, mainstream development objectives.
5. Translate plans into project briefs, including both capital investments for reconstruction and supportive measures, such as awareness building and institutional development.
6. Carry out a stakeholder mapping, and structure systematic and meaningful community participation at all stages of planning—from mapping and data collection to analysis and proposal formulation.
7. Assess the situation with respect to land rights and titling early on. Include in reconstruction planning specific measures to respond to the needs of vulnerable groups, including informal and illegal occupiers of land.
8. Use technologies and tools for planning that are appropriate, considering resource and time constraints.
9. Include DRR as an integral element in every phase of planning for reconstruction.
10. When relocation is unavoidable, integrate relocation areas into the existing fabric of receiving towns and cities.

Case Studies

2004 Indian Ocean Tsunami, Aceh, Indonesia

Consolidating Distressed Communities through Local Services and Community-Driven Land Adjudication

The Indian Ocean tsunami of December 26, 2004, killed more than 150,000 persons in Aceh and left many communities in trauma. The tsunami wiped out not only houses, but also residential property boundaries of villages near the coast and civil records in many local sub-district offices. With support from a nongovernmental organization, through YIPD and Forum Bangun Aceh (FBA), an initiative was launched to involve people in rebuilding their settlements. One activity was the repair and recovery of destroyed local sub-district offices. Sub-district offices were refurbished and reorganized so that they could once again provide citizens with services, such as issuing temporary ID cards (as many were lost) and other identification documents. These offices also provided information on available supplies and recovery programs. As a result of this effort, these locations became regular meeting points for the survivors and helped sub-district staff develop lists of survivor families by sub-village.

Another early initiative was community-driven land adjudication (or community land mapping). The initial activity took place in three sub-districts in the city of Banda Aceh and was funded by the United States Agency for International Development (USAID) through YIPD. This was eventually expanded to more than 400 villages across the Aceh province, supported by the Australian Agency for International Development (AusAID) through Local Governance Infrastructure for Communities in Aceh (LOGICA) and by the Reconstruction of Aceh Land Administration System (RALAS) Project with US$28.5 million from the Multi-Donor Fund for Aceh and Nias. Community members carried out a land inventory process using guidelines developed by the national Land Administration Agency (LAA). Volunteers were trained to identify landmarks and to produce drawings. The map of parcel boundaries, once completed, was signed by a family member and owners of neighboring parcels (left, right, front, back). The LAA validated the community's findings regarding ownership and boundaries, using land records from before the tsunami and parcel measurements, and then secured community agreement. The ruling on the land parcels was published and registration of titles was offered free of charge. The goal was to formalize 600,000 land titles. By December 2008, 211,839 parcel surveys had been completed, 126,107 land titles were registered, and 112,460 had been distributed. The land management system was computerized, and joint titling for married couples was introduced. This activity set the stage for reconstruction in these disaster-affected regions.

Sources: Azwar Hassan, Forum Bangun Aceh, 2009, personal communication, http://www.forumbangunaceh.org/ *and* Multi Donor Fund for Aceh and Nias, 2008, "Investing in Institutions: Sustaining Reconstruction and Economic Recovery. Progress Report of the Multi Donor Fund for Aceh and Nias," http://www.multidonorfund.org/documents/2008-12-18-MDF%20Progress%20Report_V.pdf.

2000–2001 Floods, Mozambique

Development of a Technical and Legal Framework to Address Land Issues

During 2000 and 2001, flooding in Mozambique affected 2.5 million people and left 200,000 people homeless. The floods magnified poor land management practices that had been in place prior to the disaster. While some disaster-affected areas had land use plans, and measures to mitigate erosion and landslides, they often were not followed. Land records and equipment were destroyed by the floods. Insecurity of land and housing tenure meant that some people affected by the flood refused to leave low-lying lands. Other concerns included the property rights of people affected by the flood who did not return to their previous locations and of those settled in new areas, particularly women. (Land rights in Mozambique are acquired through occupation or authorization of statutory use rights. Family law recognizes equal land rights for women; however, in practice, access can be limited.) An assessment by UN-HABITAT noted that the allocation of housing and development of new settlements after the flooding lacked adequate technical and legal backing, presaging the emergence of land disputes between affected households and those already living in relocation areas. The UN recommended that irregular allocations and unlawful occupations of land be remedied as quickly as possible. UN-HABITAT and the government of Mozambique have since implemented a portfolio of projects intended to:

- improve the capabilities of municipalities and the Directorate of Geography and Cadastre;
- rehabilitate the offices of institutions involved in land registration;
- prepare maps of rural and urban settlements affected by the floods;
- delineate and register properties in new settlements; and
- review the legal and institutional framework governing rights to land.

Source: UN-HABITAT, n.d., "Scoping Report: Addressing Land Issues after Natural Disasters," http://www.gltn.net/images/stories/downloads/utf-8nat_disaster_scoping_paper_jan_08.pdf.

DANIEL PITTET

2004 Indian Ocean Tsunami, Tamil Nadu, India
SIFFS-Built Houses and Habitat with Owners in Nagapattinam

The South Indian Federation of Fishermen Societies (SIFFS) undertook reconstruction of 1,380 houses in the villages of Chinnangudi, Karantheru, Puduppalayam, and Tarangambadi in the Nagapattinam district of Tamil Nadu, India, after the 2004 Indian Ocean tsunami. For SIFFS, the reconstruction project was about more than the provision of four walls and a roof. Instead, it aimed to create an organic community, taking people's social and cultural needs into account while planning the houses.

SIFFS adopted an ambitious plan to customize every house for its owner, with the motto "1,380 houses in 1,380 designs." With the correct approach to layout, planning, design, and construction, combined with a positive mix of a scientific approach and cultural sensitivity, SIFFS was confident that it could achieve its goals. SIFFS house owners were involved in every step of the process, from design through construction. Some important elements of the project were (1) habitat mapping: a geographic and socioeconomic mapping of the four settlements that included the study of common space as well as space usage within the houses; (2) a mass contact program: to inform people of mapping results and to seek their opinions on designs; (3) design development: layouts were presented and discussed before settlement plans were finalized; (4) model houses: seven prototype house designs were selected using a participatory process and "real" model houses were constructed to help families make their choice; (5) plot allotment: with the help of local authorities and traditional village panchayats, plots were allocated to each family prior to construction; (6) family design meetings: used to finalize the design before construction; and (7) cluster committees: representatives from each cluster of houses were trained to monitor the construction process. There has been nearly 100 percent occupancy of these houses by their owners since they were completed, and families have already invested their own money to expand or beautify the houses. The goal of building organic habitats—not just houses—was substantially achieved.

Sources: C. V. Sankar, India National Disaster Management Authority, 2009, personal communication; *and* SIFFS, 2008, "SIFFS Hands Over Houses in Marthandanthurai," http://www.siffs.org/Index.aspx?Page=NewsContent.aspx&FirstNews=SIFFS&NewsId=54.

1985 Mexico City Earthquake, Mexico
Improvement of the Tenure Situation of Urban Low-Income Families after Reconstruction

An earthquake with a magnitude of 8.1 on the Richter scale hit Mexico City on the morning of September 19, 1985, killing approximately 10,000 people, leaving 250,000 homeless, and damaging the houses of another 900,000. The disaster impact was magnified by the high concentration of population and economic activity in the affected area and by the deteriorated condition of the buildings, which was attributed to the city's rent-control policies.

Before the earthquake, the Mexican government had made a public commitment to improve low-income housing. Government used the post-earthquake situation as an opportunity to speed up the low-income housing improvement initiative. The Cabinet approved an in-situ reconstruction approach and passed a decree expropriating some 5,500 rental properties that were either damaged or in dangerous condition. Landlords offered little resistance, since most of the properties had ceased to be profitable. This radical measure had positive consequences for the affected families, protecting them from eviction and allowing them later to become homeowners.

The housing reconstruction program cost US$392 million and was partially financed by a World Bank loan. (The US$400 million disaster recovery loan for Mexico's earthquake reconstruction is still one of the largest disaster loans ever made by the Bank.) This program required a relatively large subsidy and has been criticized for its somewhat arbitrary targeting, since earthquake-affected families located outside the zone where properties were confiscated were excluded from the reconstruction program. At the same time, the speed with which the reconstruction took place, in an extremely complex environment, sets it apart from many other reconstruction projects of this scale.

Sources: Sosa Rodriguez and Fabiola Sagrario, n.d., "Mexico City Reconstruction after the 1985 Earthquake," Earthquakes and Megacities Initiative, http://emi.pdc.org/soundpractices/Mexico_City/SP2_Mx_1985_Reconstruction_Process.pdf; World Bank, 2001, "Bank Lending for Reconstruction: The Mexico City Earthquake," http://lnweb90.worldbank.org/oed/oeddoclib.nsf/DocUNIDViewForJavaSearch/9C4EA21BB92 73C74852567F5005D8566; *and* Aseem Inam, 1999, "Institutions, Routines, and Crisis. Post-Earthquake Housing Recovery in Mexico City and Los Angeles," Cities 16(6):391-407.

SAFER HOMES, STRONGER COMMUNITIES: A HANDBOOK FOR RECONSTRUCTING AFTER NATURAL DISASTERS

2001 Gujarat Earthquake, India
Two-Stage Planning Process with Effective Policy for Relocation in Bhuj City

In the aftermath of the 2001 earthquake in Gujarat, India, the state of Gujarat's comprehensive reconstruction program covered urban and rural housing reconstruction and local/regional infrastructure development. Housing reconstruction was principally owner-driven. Government played a facilitating role, providing land and infrastructure. Village councils undertook physical planning for rural housing. For four of the severely affected urban areas, including the city of Bhuj, government commissioned detailed land use planning and physical planning exercises, leading to the preparation of development plans at the city level and town planning schemes at the micro level. New development control regulations were also framed for each city.

In the densely populated, densely built-up "old city" area of Bhuj (and two other towns), land readjustment projects known as Town Planning Schemes were carried out. Neighborhoods were reconfigured where buildings collapsed, creating a vastly safer street network and built form.

This map shows the layout of a neighborhood in the old city of Bhuj before the land readjustment exercise. The streets were narrow, with dead ends and bottlenecks. Few buildings were standing (the plots that are shaded). During the quake, buildings had collapsed on the streets, preventing people from escaping and later hampering rescue.

This map shows the layout of the same neighborhood after the land readjustment exercise. The streets have been made wider, continuous, and much safer. The plots where buildings collapsed have been reorganized. Final plots, somewhat smaller than the original plots, have been allocated in roughly the same location as the original ones, retaining overall community configurations.

While the detailed planning for in-situ reconstruction was in process, government prepared three relocation sites with fully serviced plot layouts for voluntary relocation. This greatly reduced the resistance to improvements in the old city area. About 4,000 households relocated voluntarily to these relocation sites.

Sources: B. R. Balachandran and Purvi Patel, 2006, "Reconstruction of Bhuj: A Case of Post-Disaster Urban Planning," *The ICFAI Journal of Urban Policy,* Vol. 1, No. 1, pp. 24-34, http://ssrn.com/abstract=1093589; *and* B. R. Balachandran, 2006, "The Reconstruction of Bhuj," in *The Role of Local Government in Reducing the Risk of Disasters,* , Katalin Demeter, Ekin Erkan, and Ayse Güner, eds. (Washington, DC: World Bank), http://go.worldbank.org/5K1D5BO8V0.

B. R. BALACHANDRAN (2)

Resources

Brown, O., A. Crawford, and A. Hammill. 2006. *Natural Disasters and Resource Rights Building: Resilience, Rebuilding Lives.* Winnipeg, Canada: International Institute for Sustainable Development (IISD). http://www.iisd.org/pdf/2006/tas_natres_disasters.pdf.

Centre on Housing Rights and Evictions. n.d. *The Pinheiro Principles: United Nations Principles for Housing and Property Restitution for Refugees and Displaced Persons.* Geneva: COHRE. http://www.cohre.org/store/attachments/Pinheiro%20Principles.pdf.

Food and Agriculture Organization (FAO). 1996. *Guidelines for Land Use Planning.* Rome: FAO. http://www.fao.org/docrep/t0715e/t0715e00.HTM.

The Global Land Tool Network (GLTN) developed a series of guidelines and experiences not explicitly directed to disasters but applicable in such cases. http://www.gltn.net/images/stories/downloads/utf-8nat_disaster_scoping_paper_jan_08.pdf.

IISD. 2006. Land use and land titled cases, experiences and policies, especially in Indonesia and Sri Lanka are explored in "Natural Disasters and Resource Rights: Building Resilience, Rebuilding Lives," a report issued by IISD. http://www.iisd.org/pdf/2006/tas_natres_disasters.pdf.

International Strategy for Disaster Reduction (ISDR). 2006. "Words Into Action: Implementing the Hyogo Framework for Action Document for Consultation." http://www.preventionweb.net/english/professional/publications/v.php?id=594.

UN-HABITAT. n.d. "Scoping Report: Addressing Land Issues after Natural Disasters." Covers major literature and presents a series of cases studies. http://www.gltn.net/images/stories/downloads/utf-8nat_disaster_scoping_paper_jan_08.pdf.

UN Inter-Agency Standing Committee (IASC). 2006. *Operational Guidelines on Human Rights Protection in Situations of Natural Disasters, with Particular Reference to the Persons Who Are Internally Displaced (Guidelines on Human Rights and Natural Disasters).* Washington, DC: Brookings-Bern Project on Internal Displacement, 18.

World Bank. "Safeguard Policies." http://web.worldbank.org/WBSITE/EXTERNAL/PROJECTS/EXTPOLICIES/EXTSAFEPOL/0,,menuPK:584441~pagePK:64168427~piPK:64168435~theSitePK:584435,00.html.

 For access to additional resources and information on this topic, please visit the handbook Web site at www.housingreconstruction.org.

In this guide, comprehensive planning is defined as a planning process that incorporates land use, physical planning, and strategic planning. This annex outlines a generic process for developing a comprehensive plan, along with its key activities. The specific planning exercise needs to be adjusted to the requirements of the situation.

Activities, Outputs, and Deliverables

Key activities of a planning process. The planning activities and outputs described in this annex are generic descriptions that may vary by country or locality. The common factor is that these activities and outputs are the critical inputs to produce the deliverables of the planning process described below. As with the plan's content, the right strategy for the planning process is to build on processes that are already understood in the local context, customizing and improving these processes for the purpose of guiding reconstruction. The outputs that can be expected for each activity are described in the table below. These outputs are the building blocks for the preparation of the plan. By themselves, these outputs don't qualify as deliverables in a planning process.

Activity	Considerations	Output
Delineate geographical and functional jurisdiction of the plan	The first step in the planning process is to delineate the geographical area for which planning is to be done and to determine which components need to be addressed by the plan. This decision should take into consideration the impact of the disaster, issues emerging with respect to future risks, and the existing legal and institutional framework.	Base map showing geographical jurisdiction and report providing the rationale for the geographical area and a preliminary listing of plan components, along with the rationale for those components
Carry out stakeholder mapping	The primary stakeholders for reconstruction planning are those affected by the disaster and the agencies involved in reconstruction, including government and nongovernment agencies. Other stakeholders include the larger community, businesses, line departments not directly involved in reconstruction, and those who can contribute to reconstruction (resource groups, institutions, and individuals). Those affected by the disaster are unlikely to be a homogenous group. The diversity within this group needs to be recognized, particularly with respect to vulnerable groups. The negotiation of interests among stakeholders is a key function of planning. Stakeholders also add value to the planning process through their knowledge of the local context and their skills.	Inventory and brief description of stakeholder groups
Prepare maps and collect data	Prepare a base map of the area to be planned and collect demographic information. Then, for each of the components of the plan, prepare a set of maps showing the pre-disaster and post-disaster situations and collect base data. It is quite common that reliable maps and data are not available. There are various technologies, such as remote sensing, and methodologies, such as participatory mapping and rapid appraisals, that can be used to supplement available information.	Maps capturing the existing situation Reports compiling and presenting data on the existing situation A summary of the maps and data presented in a manner that the stakeholders can understand and use
Undertake participatory strategic planning	The planning process should involve stakeholders in a structured manner. There are two broad stages in this process.	
Analyze existing situation and articulate vision	Using the mapping and data collection that has been carried out, undertake a participatory analysis of the existing situation. Various tools can be used such as a SWOT (Strengths, Weaknesses, Opportunities, and Threats) analysis. Studies such as land suitability analysis and risk assessment provide the basis for land use planning. From these analyses, generate a collective, overarching vision for the community, settlement, or region that stakeholders agree upon.	SWOT Analysis Report for each component of the plan Vision Statement

Activity	Considerations	Output
Formulate objectives, strategies, and projects	This should be done based on the analysis above and the vision. ■ Identify plan components and, for each component, articulate specific objectives. ■ To support these objectives, formulate strategies that build on strengths, overcome weaknesses, take advantage of opportunities, and ward off threats. ■ Translate these strategies into implementable projects. For example, an objective in land use zoning may be to minimize flood impact on housing. The strategy to achieve this may be to make public land in safe locations available for housing. This strategy may be translated into (1) reservations on public land in the land use plan, (2) formulation of a housing project for relocation of households from at-risk locations, and (3) regulations on construction in low-lying areas.	Comprehensive Strategic Plan compiling all components of the plan
Approve, publish, and implement the plan	The success of the plan depends on the measures that are taken to ensure its full implementation.	
Approve the plan	The plan with all its components should be officially sanctioned under the provisions of the relevant planning legislation. Of utmost importance is the publication of revised building codes.	Legal notification issued
Publish the plan	The plan documents, including all relevant maps and reports, should be made publicly available using various media. The entire set of documents should be made available on Web sites. Hard copies of the full set should be available for purchase and hard copies of a short and user-friendly summary may be made available for free.	Plan documents available in various forms
Assign institutional responsibility	Prepare a detailed work plan for the implementation and monitoring of the plan. Identify appropriate institutions for undertaking the various tasks, assess their capacity, officially designate roles, and commission capacity building activities where required. The work plan should include a social communication effort to notify and educate residents about the plan.	Work plan with allocation of roles and responsibilities Monitoring plan
Develop and implement regulations	Development and implementation of regulations and enforcement mechanisms and sanctions will ensure full enforcement.	Regulations

Deliverables of a planning process. All the outputs of the planning process are combined to create four broad categories of deliverables, as shown below.

These deliverables are the documents (including maps) that are used to create the legal and regulatory framework for development. The deliverables listed here are generic. The legal terms for these deliverables will vary by country or locality.

Deliverable	Description
Land use plan and regulations	This is a map or a set of maps with supporting documents that show and describe (1) proposed housing reconstruction locations with basic details, (2) the nature and intensity of land uses permitted in different zones in the planning area, (3) areas reserved for particular uses, (4) areas where development is restricted, and (5) guidelines for the design and construction of buildings. It is common practice to also show any proposed road network and other transportation networks on the same maps. Supporting documents include draft regulations and/or ordinances that must be approved legislatively in order for the plan to be put in force and implemented. The regulations should include sanctions that penalize noncompliance with the plan.
Physical plan	This is a map or set of maps that show the proposed layout of the road network, alignments of various other infrastructure networks, and locations of major facilities.
Compendium of project briefs	This is a list of projects, each one with a brief description, cost estimate, and implementation strategy. Projects may include capital investments as well as other interventions, such as public awareness or capacity building. Reconstruction happens through project formulation and implementation. The planning process is the means of establishing the ground rules for development (or reconstruction) and organizing projects within a systematic framework.
Implementation plan	This a detailed work plan that assigns roles and responsibilities, describes the design of programs for capacity building where required, and a explains the strategy for monitoring plan implementation.

Tools and Methods for Planning

The following are some examples of tools and methods that can be employed in the planning process described above.

Tools for participatory planning. Genuine involvement of stakeholders is critical for preparing plans that capture local knowledge and that address local aspirations and concerns. Examples for such participation are available from experiences in many disasters.

- **Participatory mapping.** In a post-disaster situation, good maps are usually not available. In such situations, it is possible to start with available maps and enhance the information content by involving local people. For example, in Indonesia, post-tsunami planning exercises have used public participation to establish pre-tsunami land holding patterns and also to reorganize those patterns in response to post-tsunami needs.
- **Participatory strategic planning.** SWOT analysis is an effective tool for strategic planning and can be used in a participatory mode as well. In the city planning process in Bhuj, India that was carried out after the Gujarat earthquake citizens conducted a SWOT analysis of their city with respect to various plan components, and came up with strategies for reconstruction and management of the city's growth. See 📖 case study, below, on the Bhuj planning process.

Tools for risk mapping. These tools provide a sound basis for decisions related to relocation and infrastructure development.

- **Risk analysis.** An all-hazards risk assessment (or risk analysis) is a determination of the nature and extent of risk developed by analyzing all potential hazards and evaluating existing conditions of vulnerability that could pose a potential threat or harm to people, property, livelihoods, and the environment on which they depend. The risk analysis shows vulnerabilities in a particular location and quantifies the potential impact of a disaster on a community. A detailed methodology for risk analysis is included in 📖 Part 4, Technical References, Disaster Risk Management in Reconstruction.
- **Land suitability assessment.** Land suitability mapping uses multiple parameters, such as topography, ecology, demography, and infrastructure availability, which are assessed and weighted to determine suitability of land for specific purposes.

Technologies for mapping and spatial analysis. The choice of technology depends on the resources available as well as the time frame.

- **Total Station survey.** This is a survey method that uses computerized survey equipment (called Total Stations) and is by far the most accurate method for creating a topographical and cadastral map, but requires considerable time. Where land values are high and variations in topography play an important role in vulnerability (for example, hilly areas), this is an appropriate method.

- **Remote sensing.** Creating maps from satellite images and aerial photographs falls in the category of remote sensing. There are various kinds of satellite images and they have varying resolution. To illustrate, for planning in rural or regional contexts, False Color Composite images can be used to map land use patterns, while in urban areas, high-resolution monochrome images can be used to identify physical features like buildings and roads. Stereoscopic images can be used to get topographical information and create digital terrain models. Mapping using remote sensing is a relatively fast and less expensive option when high levels of accuracy are not required.
- **GIS data.** GIS is a generic term for software platforms that can combine spatial information (maps) and alphanumeric data in a seamless manner and that use sophisticated algorithms to perform a range of analyses. GIS can be used in post-disaster reconstruction planning for simple functions, e.g., the production of thematic maps depicting impact, to more complex functions, e.g., vulnerability and land suitability analysis, to highly complex functions, e.g., simulation of future disaster scenarios. See 📖 Chapter 17, Information and Communications Technology in Reconstruction.

This chapter may give the impression that land use and physical planning processes require powerful laws, institutions, and resources. While that may be the ideal situation, this annex addresses the frequently-encountered situation of a disaster occurring where the post-disaster institutional capacity for planning is very limited. In such contexts, consideration of planning usually precipitates three primary issues and related concerns. It should be used jointly with Annex 1.

Issue	Related concern
Legality	
What is the validity of a land use/physical plan that has no legal or statutory backing?	Who will approve and regulate the plan? How will compliance be ensured? Where there is reorganization of land involved, how will legal title be assured?
Capacity	
How can the leadership, technical expertise, and financial resources be mobilized in the absence of installed institutional capacity?	Planning a housing site in an ad hoc manner is not a big challenge, but can a whole city or a large portion of one be planned using a "rapid planning exercise"?
Implementation	
How should the activities in the land use and physical planning process be prioritized? How will decisions be implemented?	Are there some activities that are more critical than others? Once the planning exercise is completed, who implements it?

Even though the need for post-disaster planning in these situations is common, there is no standard approach to conducting it. However, there are a number of strategies that have been used around the world.

Create a Mandate for Planning

The strategies that can be deployed to create a mandate for land use and physical planning include amending existing legislation, introducing new legislation (using national or international models), and creating a mandate through ordinances and government orders.

Experience in Japan. Following the 2005 Hanshin earthquake in Japan, the Japanese government enacted a "Special Act for Disaster Afflicted Urban Areas," with special provisions for urban planning, including the creation of neighborhood committees for land readjustment projects.[1]

Experience in China. Following the 2008 earthquake in Wenchuan, China, government's national implementation guidelines established three categories of land uses for national land available for reconstruction: (1) suitable for reconstruction, (2) appropriate reconstruction, and (3) ecological restoration. The guidelines also defined parameters, such as the priority in which services should be restored and the types of land uses that should be expanded in which areas. Within this framework, specific decisions were delegated to provinces and municipalities.[2]

Mobilize Institutional Resources

The reconstruction policy should address how the resources for local planning will be made available at the local level, if capacity at this level is at all in doubt. The strategy may need to be quite specific. For instance, if planning capacity is adequate, but the procedures for eminent domain are weak, national or state government may need to intercede in that specific area. The options for mobilizing institutional resources are numerous. They include:

- accessing planning expertise from another level of government;
- bringing in a planning agency from a comparable location inside the country;
- getting technical support from an international agency and/or comparable location outside country (international exchange programs like the City Links program of the International City/County Management Association [ICMA] or other pairing programs sponsored by international nongovernmental organizations (NGOs) or bilateral agencies;
- creating a platform for collaboration between all the players in the reconstruction process and carrying out a collaborative planning exercise;
- borrowing "planners" from another sector in the same country or the private sector (e.g., economic development, health services, corporate strategy); and
- government officials from one or more jurisdictions acting as planners (mayor's office, cabinet).

In each of these cases, the technical resources within government can be supplemented by procuring services from the private sector or seeking skilled volunteers.

Experiences in India. To provide assistance to the cities of Cuddalore and Nagapattinam following the 2004 Indian Ocean tsunami, ICMA, with support from the U.S. Agency for International Development (USAID), paired these two Indian cities with three cities in Florida that are at continual risk of hurricanes. Together, they created maps for the Indian cities with detailed data layers of features, such as public infrastructure systems and facilities, land uses, and relevant building structures. The maps were used to develop evacuation plans and a flood mitigation program, among other purposes.

After the Gujarat earthquake of 2001, the Indian state government of Gujarat created special Area Development Authorities in four affected towns and brought in planners using state resources to manage a process where private sector professionals were hired to provide planning services.

In carrying out rural reconstruction work in Kutch, a collaborative platform was created by Kutch Nav Nirman Abhiyan, a collective of more than 20 NGOs in Kutch. This platform was used to undertake various planning exercises at a regional scale. See the case study on Kutch Nav Nirman Abhiyan in ⛁ Chapter 14, International, National, and Local Partnerships in Reconstruction. Under the guidance of planners experienced with the methodology, participatory planning exercises can be organized and carried out in a matter of two to three weeks.

Experience in China. The Chinese experience following the 2008 Wenchuan earthquake (referred to elsewhere in this chapter) entailed pairing more capable cities unaffected by the earthquake with earthquake-affected cities. The assisting cities were asked to make both technical and financial contributions. In this model, a city with better institutional capacity can help to coordinate planning, draft guidelines, and formulate regulations, based on its own planning framework, while also compensating temporarily for any deficiencies in staffing or local capacity caused by the disaster itself.

Prioritize Planning Activities

When there is a scarcity of institutional resources, it becomes important to prioritize the critical elements of the land use planning process. Listed below are the most important land use and physical planning activities, along with suggestions on how they can be handled expeditiously.

Critical planning activities	Suggestions to expedite activity
Identifying land for housing reconstruction	
The selection and allocation of land based on sound principles and with due consideration of disaster vulnerability issues is a fundamental requirement. It is also important to ensure that clear title be given to the beneficiaries.	Use mostly alternative sources: secondary sources, such as a regional disaster risk reduction agency; satellite or GIS data gathered for the disaster assessment; maps created for other purposes, such as a nearby environmental study; or new local maps developed through a participatory process.
Revising land use zoning	Consider grandfathering existing land uses, altogether or in low-risk zones, to reduce approvals of reconstruction exceptions to existing zoning laws.
While it may not be essential to carry out a comprehensive land use zoning exercise, it is very important to assess disaster risk and vulnerability and to formulate development control regulations that respond to it.	Seek the leadership of an experienced planner or planners deployed for a short period of time, potentially with support from humanitarian or development agencies, if local expertise is not available.
Physical planning	
Planning the housing layout in a neighborhood is a priority task that may be able to be done quickly without professional planners.	Gather data and maps from government planners preparing for reconstruction of major infrastructure may have GIS data, maps, and/or satellite images that can be extended into local areas or made available directly to local planners.
If the disaster impact is widespread, and reconstruction entails extensive infrastructure, public facilities, relocation, and connectivity issues, then physical planning is required and the deployment of a professional planning team by one of the means described above is necessary.	Useful data for planning are increasingly available publicly. For information on this topic, see ⛁ Chapter 17, Information and Communications Technology in Reconstruction.
	Seek the leadership of an experienced planner or planners deployed for a short period of time, potentially with support from humanitarian or development agencies, if local expertise is not available.
Improving construction methods	
A full revision of building codes, which may be the perfect solution, is rarely feasible in a post-disaster content. Instead, guidelines and manuals for the reconstruction process, tailored to the requirements of the specific post-disaster situation, should be prepared and people should be trained to use them.	Guidelines and manuals are the minimum acceptable activity. See ⛁ Chapter 10, Housing Design and Construction Technology, and ⛁ Chapter 16, Training Requirements in Reconstruction, for guidance on developing guidelines.
	Agencies involved in reconstruction may have experience and materials that can be adapted. Ask agencies to develop and propose common reconstruction standards and procedures for adoption by the local jurisdiction.

Annex 2 Endnotes
1. Supporters Network for Community Development, 1999, "Key Terminology in Restoration from Hanshin Earthquake Disaster," http://www.gakugei-pub.jp/kobo/key_e/index.htm#Men1005.
2. National Development and Reform Committee, 2008, "The Overall Planning for Post-Wenchuan Earthquake Restoration and Reconstruction," http://en.ndrc.gov.cn/policyrelease/P020081010622006749250.pdf.
3. ICMA, "City Links: Global Problems, Local Solutions, 2003-2008," http://www.icma.org/citylinks. Assistance through City Links can be accessed through local USAID offices.
4. Brian Hoyer, 2009, "Lessons from the Sichuan Earthquake," Humanitarian Exchange Magazine, No. 43, http://www.odihpn.org/report.asp?id=3008.

8 INFRASTRUCTURE AND SERVICES DELIVERY

Guiding Principles for Infrastructure and Services Delivery

- Infrastructure restoration and housing reconstruction rarely progress on the same schedule after a disaster. Those planning housing reconstruction must make sure there are plans in place both for infrastructure reconstruction and for interim services, if necessary.
- Successful infrastructure reconstruction requires extensive coordination on many fronts: with planners, households, and multiple agencies involved in housing reconstruction, among others.
- Project developers should build infrastructure that conforms to planning and regulatory requirements, or help bring these instruments up to date if not adequate. Minimum technical standards are needed for retrofitting and reconstruction of infrastructure that incorporate disaster risk reduction (DRR) and sustainable development objectives.
- All four types of DRR measures—policy and planning measures, physical preventative measures, physical coping and adaptive measures, and community capacity building measures—are relevant to infrastructure and should be utilized in retrofitting and reconstruction.
- Entities that will operate and maintain infrastructure facilities—which may include the community—should be directly involved in infrastructure project planning and implementation.
- Involving built environment experts, such as architects and engineers, in project development increases the chance that upgraded standards are incorporated into local infrastructure projects over the long term.

This Chapter Is Especially Useful For:

- Lead disaster agency
- Local government officials
- Local service providers
- Agencies involved in reconstruction
- Project managers

Introduction

Post-disaster infrastructure restoration happens in stages. During the disaster response, the focus is on stabilizing systems and preventing secondary damage (e.g., fires from gas leaks or contamination from sewage plants). Soon thereafter, attention shifts to repairing lifeline infrastructure and networks such as roadway connections and basic communications. During reconstruction, restoring permanent infrastructure services, including those in residential neighborhoods, is the priority. However, restoration of full services may not happen right away.

Infrastructure reconstruction requires planning and coordination on several fronts. Reading
🕮 Chapter 7, Land Use and Physical Planning; 🕮 Chapter 9, Environmental Planning; and 🕮 Disaster Risk Management in Reconstruction in Part 4, Technical References, along with this chapter will provide a more comprehensive understanding of the issues affecting infrastructure. Depending on the severity of the disaster, infrastructure reconstruction is likely take many years longer than the reconstruction of housing, so interim solutions and full reconstruction must both be planned. Multiple agencies are likely to be involved, and it is crucial to coordinate decisions among them.

The approach to housing reconstruction affects how infrastructure reconstruction is managed; infrastructure construction at a large-scale new housing site will be different from that in a location where owner-driven in-situ reconstruction is planned. In all cases, housing should be designed to accommodate public services, even if the services are not available at the time of housing reconstruction.

The definition of infrastructure in this chapter includes lifeline systems and related local public services. The importance to community reconstruction of the rehabilitation of public facilities such as public buildings and meeting spaces, and of educational and health facilities, is also touched on.

Key Decisions

1. The **lead disaster agency** should work with **affected communities** and **local government** immediately after the disaster to assess the state of infrastructure systems and the capability of local service providers to restore both lifeline services and full infrastructure services, and to identify the assistance required to do so.
2. The **lead disaster agency** and **local government** should work with **communities** to prioritize the public services needed to restore community life and to agree on the division of labor to restore facilities between government and the community.

3. The **lead disaster agency** should establish and publicize the infrastructure standards all agencies involved in reconstruction should meet.

4. The **lead disaster agency** should collaborate with **local service providers** to estimate the cost of, and raise and channel the resources needed for, the restoration of local services and facilities and for infrastructure reconstruction.

5. **Agencies involved in reconstruction** should decide how to ensure the provision of interim and permanent infrastructure to reconstruction sites, especially if they do not expect to provide it themselves.

6. **Agencies involved in reconstruction** should decide how to support **local service providers** to build back better and conform to standards established by national, regional, or local governments for any infrastructure they finance.

7. **Local service providers** should work with government and other funders to ensure they have adequate resources to build back better and restore services in an economically sustainable manner. This may entail reviewing local tariffs or service fee schemes.

8. **Communities** should insist that **agencies involved in reconstruction** provide them a lead role in planning and implementing projects related to services for which they will be responsible.

Public Policies Related to Infrastructure and Services Delivery

Countries with emergency management or DRR plans will be much better prepared to plan and implement infrastructure and service restoration, since protecting infrastructure and restoring services are generally among the chief concerns of these plans.

Post-disaster projects for reconstruction, rehabilitation, and retrofitting of infrastructure should be aligned with the country's and the locality's overall development vision, particularly with respect to long-term development and land use plans, the allocation of institutional roles, and the standards for infrastructure improvement. While not strictly considered "infrastructure," various public facilities are essential for communities and should also be restored early on, and the comments in this chapter apply to these investments as well. Community facilities include schools, clinics, refuges, buildings for local government administration, and meeting spaces. Schools and clinics contribute to the resumption of normal life by providing space for social services. Local government buildings and meeting spaces allow local public services to resume and facilitate community planning and the reestablishment of local governance.

To the extent possible, reconstruction should be carried out using standard procedures and in accordance with local plans and regulatory requirements. For instance, plans should conform to local building codes, or existing intergovernmental fiscal channels may be used to transfer funds. Infrastructure investments must be guided by environmental policies as well, due to the potential for extremely negative environmental impacts from both the construction and the operation of poorly planned infrastructure systems. Where the standards or the legal and regulatory frameworks are inadequate, they may need to be updated before reconstruction begins.

Because land use and development are generally governed by local land use planning agencies, local government should be involved in decisions regarding new land uses and acquisition, and should coordinate the acquisition of rights-of-way for infrastructure, especially if eminent domain procedures are involved. Land acquisition for infrastructure can be a long and contentious process, and lack of site control poses a significant construction risk. See 📖 Chapter 7, Land Use and Physical Planning.

The agencies that should be involved in decisions regarding infrastructure redevelopment, rehabilitation, or retrofitting are a combination of the national ministries responsible for the regulation and financing of the systems being restored (e.g., ministries of water and sanitation, roads, transport, environment, and power) and—equally important—local service providers: local government and community-based entities responsibility for local infrastructure investment, maintenance, and service provision. Local service providers may include departments of water and sanitation, transportation, environment, solid waste management, environment, and community-based water service providers or other community organizations. The sustainability of service provision after a disaster depends largely on the commitment and capability of local entities responsible for service provision.

DANIEL PITTET

Technical Issues

Types of Damage to Infrastructure from Disasters

The magnitude of damage to infrastructure depends on the hazard type, its intensity, and the ex ante preparedness. The following graphic shows relative magnitudes of common impacts by disaster type.[1]

	Earthquake	Volcano	Landslide	Hurricane	Flood	Drought
Structural damage to system infrastructure	●	○	●	●	●	○
Rupture of mains and pipes	●	○	●	◐	●	○
Obstructions in intake points, intake screens, treatment plants, and transmission pipes	○	●	◐	◐	●	○
Pathogenic contamination and chemical pollution of water supply	◐	●	○	●	●	○
Water shortages	◐	◐	○	○	○	●
Disruption of power, communications, and road system	●	○	◐	●	◐	◐
Shortage of personnel	●	◐	◐	◐	◐	○
Lack of equipment, spare parts, and materials	●	○	◐	●	●	○

Legend:
● Severe effect
◐ Moderate effect
○ Minimal effect

Moving from Recovery to Reconstruction

Two-pronged approach. To reconcile the need to act quickly while still allowing time for design and consultation, the negative impacts of a disaster should be contained and the lifeline infrastructure should be rehabilitated during the recovery period, while the planning and design of long-term infrastructure reconstruction begins. This approach responds to critical service needs and demonstrates visible efforts, while allowing lead time for land use planning, consultations, infrastructure design, land acquisition, and procurement. Government should avoid the temptation to shortchange infrastructure planning and design to take advantage of the availability of relief funds.[2] The 📖 case study on infrastructure reconstruction following the 2001 Gujarat earthquake, below, describes some of the common reasons why infrastructure projects often take longer than originally anticipated.

Long lead times for infrastructure reconstruction mean that housing reconstruction is likely to take place before infrastructure is fully restored or reconstructed; therefore, short-term interventions may need to address the availability of basic services and safety of households in communities where reconstruction is taking place. The use of transitional shelter, which allows households to resettle on their own land while rebuilding, will make this especially crucial.[3] International standards, such as the Sphere Standards, can be used to define minimum standards for various basic services and shelter.[4] The Inter-Agency Network for Education in Emergencies (INEE) defines minimum standards for public education.[5]

Greater Priority / Least Effort

Least Priority / Greatest Effort

- **Essential Facilities** (police and fire stations, schools, hospitals, emergency operations centers)
- **User-Specified Facilities** (government buildings, historical landmarks, stadiums)
- **Transportation Lifeline Systems** (road segments, bridges)
- **Hazardous Materials Facilities** (storage, industrial labs)
- **High-Potential Loss** (dams, power plants, military bases)
- **Utility Lifeline Systems** (power lines, sewers, and water mains)
- **General Building Stock** (number of buildings, occupancy, and construction classification)

Source: Federal Emergency Management Agency (FEMA), 2004, *Using HAZUS-MH for Risk Assessment* (FEMA 433), http://www.fema.gov/plan/prevent/hazus/dl_fema433.shtm.

1. Pan-American Health Organization (PAHO), 2002, "Emergencies and Disasters in Drinking Water Supply and Sewerage Systems: Guidelines for Effective Response," http://www.reliefweb.int/rw/lib.nsf/db900sid/LGEL-5S6BNE/$file/paho-sew-02.pdf?openelement.
2. Sisira Jayasuriya and Peter McCawley, 2008, "Reconstruction after a Major Disaster: Lessons from the Post-Tsunami Experience in Indonesia, Sri Lanka, and Thailand" (Working Paper 125, Asian Development Bank Institute), http://www.adbi.org/working-paper/2008/12/15/2766.reconstruction.post.tsunami.experience/.
3. See 📖 Chapter 1, Early Recovery: The Context for Housing and Community Reconstruction, for a discussion of transitional shelter.
4. The Sphere Project, 2004, *Humanitarian Charter and Minimum Standards in Disaster Response* (Geneva: The Sphere Project), http://www.sphereproject.org. Updated standards are due to be issued in late 2010.
5. INEE, 2004, *Minimum Standards for Education in Emergencies, Chronic Crisis and Early Reconstruction* (Geneva and New York: INEE), http://www.ineesite.org/.

Infrastructure Interventions Relevant to Housing and Community Reconstruction

Short-term interventions	Medium- to long-term interventions
Electric power systems	
Give priority to functions that support other lifelines, such as treatment and pumping of water.	Incorporate DRR mechanisms in reconstructed systems and facilities.
	Provide power for households and community facilities and for pumping water and running generators and tools used in reconstruction.
	Consider alternative energy generation options in housing and community building design and community planning.
	Develop a DRR plan for electric power installations.
Transport systems	
Prioritize access to critical facilities, such as hospitals, emergency centers, and fire stations.	Incorporate DRR mechanisms in reconstructed systems and facilities.
Initial rehabilitation of roads should support housing reconstruction, especially transport of materials to disaster site. Consider modest early repairs and more permanent reconstruction later on.	Provide housing site access and egress, including access by emergency vehicles for delivery of construction materials.
	Retrofit and upgrade to improved codes and standards.
	Design roadway systems for sites to encourage walking and bicycling.
	Plan for public transit access.
	Develop a DRR plan for the transport sector.
Water systems	
Water loss increases health and fire hazards, and causes loss of cooling systems for telecommunications and computers.	Incorporate DRR mechanisms in reconstructed systems and facilities.
Strengthen and support structures.	Test for availability and quality of potable water before selecting relocation sites.
Provide alternative domestic water supply until systems are restored.	Provide water for reconstruction purposes, such as mixing concrete.
Repair, clean, and disinfect wells, boreholes, water storage tanks, and tankers.	Provide water for households.
	Consider meter installation during rehabilitation of system.
Improve leak detection. Monitor water quality.	Develop a DRR plan for all water installations and facilities.
Rehabilitate water distribution and treatment works.	
Educate population on point-of-use treatment of drinking water.	
Sewerage system and storm-water runoff	
System loss causes untreated sewage discharge into water bodies or increased environmental and health hazards.	Incorporate DRR mechanisms in reconstructed systems and facilities.
Provide emergency sanitation systems.	Improve shut-off and diversion systems. Segregate combined overflow systems.
	Consider small-scale sewage treatment options.
Prevent defecation in areas likely to contaminate food chain or water supplies.	Design site for rainwater capture for landscaping and other non-potable purposes.
Educate population on hygiene.	Use permeable paving materials to maximize infiltration of water.
	Consider incorporating cisterns in site designs for collection of rainwater.
	Develop a DRR plan for all sewerage and storm-water installations and facilities.
Solid waste	
Unmanaged waste can pollute and obstruct water sources and provide breeding grounds for insects and vermin.	Develop integrated solid waste management plan if none exists.
Develop systems and designate sites for domestic, industrial, construction, hospital, and hazardous waste management, including recycling of disaster debris. See 📖 Chapter 9, Environmental Planning, Annex 1, How to Do It: Developing a Disaster Debris Management Plan, for advice on debris management.	Maintain interim facilities until normal operations resume, and maintain debris and construction waste recycling until reconstruction tapers off. See a case study about an ambitious debris recycling program following the 1994 earthquake in Northridge, California, in 📖 Chapter 9, Environmental Planning.
	Reestablish normal solid water management services as soon as possible.
	Incorporate recycling and composting services in solid waste management plan.

Short-term interventions	Medium- to long-term interventions
Public buildings (health facilities, schools, and police and fire stations)	
Social consequences and compromised health and safety result from the lack of these facilities. Prioritize restoration of power supply, transportation access, and water supply.	Incorporate DRR mechanisms in reconstructed buildings. Prioritize school reconstruction to minimize disruption to school, and therefore family, life. Construct community meeting spaces or incorporate community space in other early public building reconstruction projects. Restore public facilities to improved construction and service standards. Design new public buildings with energy efficiency and multiple uses in mind. Develop a DRR plan for all public buildings.

Sources: FEMA, 1995, *Plan for Developing and Adopting Seismic Design Guidelines and Standards for Lifelines,* FEMA Publication 271 (Washington, DC: FEMA), http://www.fema.gov/library/viewRecord.do?id=1528; FEMA, 2004, *Using HAZUS-MH for Risk Assessment,* FEMA Publication 433 (Washington, DC: FEMA), http://www.fema.gov/plan/prevent/hazus/dl_fema433.shtm; PAHO, 2002, "Emergencies and Disasters in Drinking Water Supply and Sewerage Systems: Guidelines for Effective Response," http://www.reliefweb.int/rw/lib.nsf/db900sid/LGEL-5E2DJV/$file/paho-guide-1998.pdf?openelement; *and* World Health Organization (WHO), 2005, "Technical Guidance Notes for Emergencies, Nos. 1, 2, 3, 4, 5, 6, 9, 10, 11, 12, 13, and 14," http://www.who.int/water_sanitation_health/hygiene/envsan/technotes/en/index.html.

A DRR-Oriented Infrastructure Project Development Sequence

DRR measures can be categorized as policy and planning measures, physical preventative measures, physical coping and/or adaptive measures, and community capacity building, all of which are of paramount importance during post-disaster infrastructure reconstruction.[6] DRR is a top priority in infrastructure reconstruction.[7] Not only should infrastructure facilities be built so that the risk of future damage from disasters is reduced, but the infrastructure itself—such as a system for storm-water runoff—can provide protection from the impacts of disasters. With respect to the phases of DRR, the most relevant to infrastructure are:

- **Mitigation:** structural (physical) or non-structural (e.g., land use planning, public education) measures undertaken to minimize the adverse impact of potential natural hazard events; and
- **Rehabilitation and reconstruction:** measures undertaken in the aftermath of a disaster to restore normal activities and restore physical infrastructure and services, respectively.

The table below provides information on the stages of infrastructure reconstruction and some key considerations to be taken into account during that reconstruction effort. Part 4, Technical References, 🔧 Disaster Risk Management in Reconstruction includes a methodology for risk assessment useful in infrastructure planning and provides information on sources of hazard and vulnerability data.

6. Department for International Development (DFID), 2005, *Natural Disaster and Disaster Risk Reduction Measures: A Desk Review of Costs and Benefits,* Draft Final Report (London: ERM), http://www.preventionweb.net/english/professional/publications/v.php?id=1071.

7. Based on Charlotte Benson and John Twigg with Tiziana Rossetto, 2007, "Guidance Note 12: Construction Design, Building Standards and Site Selection," *Tools for Mainstreaming Disaster Risk Reduction,* International Federation of Red Cross and Red Crescent Societies and ProVention Consortium, http://www.proventionconsortium.org/themes/default/pdfs/tools_for_mainstreaming_GN12.pdf.

Stage	Key considerations
Damage and loss assessment and project prioritization	Locate or conduct an inventory of infrastructure assets and remaining capacity and a preliminary assessment of reconstruction and resource requirement from the post-disaster damage and loss assessment. When prioritizing projects, inventoried assets can be categorized and tasks sequenced taking into consideration priority and effort involved. Use economic and social criteria to evaluate costs and benefits of projects. Infrastructure planning, design, and construction must be coordinated with the plan for housing reconstruction to ensure the availability of basic services and sanitary conditions in such settlements. The timing of housing reconstruction affects the prioritization process for infrastructure.
Define roles and responsibilities	Clearly define the roles and responsibilities of the various individuals, agencies, and organizations involved in the hazard risk assessment; the design and siting of appropriately hazard-resilient infrastructure; the enforcement of design; and the quality control of construction, operation, and maintenance, while ensuring that local service providers have a lead role. Ensure local governments are given the lead when these issues fall under their jurisdiction. Provide assistance if local capacity is a constraint. Coordinate with other development or relief organizations working in the area to avoid duplication of research on hazard-proof construction and to promote a harmonized use of hazard-proof construction standards. Set up a system of consultation and collaboration with engineers, academics, local government, and the affected community. Ensure that engineers and other infrastructure service providers participate fully in the design of projects, so that they contribute more than just building or supplying to order.

Stage	Key considerations
Hazard assessment	Assess the frequency and dimension of all potential sources of natural hazards (geological, meteorological, or hydrological) in the area and determine the most likely hazard scenarios for consideration in the infrastructure design.
	Ideally, development organizations working in the country should have already analyzed some aspects of disaster risk. Make this information public and use it in planning.
	Existing academic studies and hazard maps may provide information for the hazard evaluation. However, depending on the prevalent hazards and the site, it may also be necessary to conduct site-specific risk analysis or micro-zonation studies.
	Local secondary disaster effects (e.g., landslides from excessive rain or ground shaking) should be anticipated and considered.
Review of legislation and good practice	Assess existing codes of practice for hazard resistance and determine whether they are adequate for use in infrastructure reconstruction.
	▪ If this review has already been conducted at the national level by a development organization or by a local research/academic body, draw on whatever information is relevant to the specific project context.
	▪ If there is no existing review, conduct research on existing codes of practice for hazard resistance, which might include the following.
	▪ Investigate the history of the code development and level of hazard inclusion.
	▪ Analyze the performance of buildings and infrastructure designed to the codes during past hazard events.
	▪ Compare loading and design criteria to building codes developed for countries with similar hazards and neighboring countries with similar construction practices.
	▪ Review international good practices, building codes, and design guidelines appropriate to the identified hazards, and assess their applicability.
Review of construction methodologies and local capacity	Identify the normal local construction practices for the relevant type of infrastructure. A rapid assessment may be made in the case of new construction. A more detailed analysis is required in a retrofitting project.
	Weaknesses in structures and in the vulnerability of infrastructure to the identified natural hazards must be assessed. This may include a study of the rate of degradation of the structure and its materials over time to assess resilience against projected hazards.
	Determine the strengths and durability of materials in existing infrastructure or proposed.
	Identify those who will carry out the design and construction (engineered, non-engineered, self-built, or contractor-built) and ensure their ability to comply with codes.
	Assess program management and administration capacity and strengthen it with training or outside expertise.
	Assess local construction practices, their resistance to the determined hazards, and the level of risk this poses.
Set hazard safety objectives	Establish clear and measurable objectives for hazard safety, based on the level of risk that can be supported by the affected public and government agencies. Take into account development agency accountability issues.
	Consider different performance objectives for critical facilities and infrastructure, factoring in the potential impact on the users or clients who would be negatively affected to varying extents by loss of service.
Site selection	The site for development will typically be defined by local government based on availability, land use plans, and economic criteria. The suitability of these sites needs to be assessed.
	Any hazard assessments carried out in previous stages should be considered.
	Determine whether additional works are required to render the site viable for development or whether land use should be restricted to reduce vulnerability to natural hazards.
	Consider whether resiting to a location of reduced risk is an option.
	Topographical features and landscape can be used to reduce the impact of potential natural hazards (e.g., to minimize flood risk or modify wind speed and direction).
	Land swaps might be a potential solution in collaboration with local government, but make sure that environmental protection is taken into consideration.
	Project cost estimates should plan for possible land acquisition.
Technology selection	In evaluating infrastructure technology options, evaluate the following.
	▪ Consider the financial and operational capacity of the entity responsible for service provision.
	▪ Assess capital investment and operation and maintenance (O&M) costs over the life of the project.
	▪ Review the availability of parts and supplies over the life of the project.
	▪ Consider rebuilding zoned and decentralized infrastructure systems, which may be more resistant to system failures.

Stage	Key considerations
Design and procurement	Design a sustainable and socially acceptable strengthening or construction solution that satisfies the DRR objectives.
	Consider limitations of finance, construction skills, and material availability.
	In a rehabilitation project, take into account potential disruption to normal activity.
	Ensure that the environmental and social impacts of the proposed solution are acceptable.
	Ensure (through testing and research) that the proposed solution will in fact yield the performance objectives established for the project.
	Develop a procurement strategy that provides overall value during the entire life of the service or facility.
	Apply "build back better" principles, even if they have not been translated into specific codes or standards.
	Assess the competency of contractors and ensure adequate site supervision.
	Address training needs for the implementation of the proposed solution (e.g., on-the-job training included in the implementation stage).
	Develop building codes and guidelines that account for local hazard conditions, building material characteristics, and construction skills and quality, and ensure that:
	■ building codes cover retrofitted facilities;
	■ standards are coordinated with respective ministries and local planning departments;
	■ streamline permissions and permits;
	■ work with government to streamline repair permits and demolition procedures; and
	■ enhance technical and human capacity, if necessary, to ensure speed in reviewing and issuing construction permits.
Construction	The quality of any post-disaster construction must not compromise the design intent. Establish procedures for multidisciplinary inspection and check against specifications of works throughout the building process in the following ways:
	■ Test materials and check adherence to design guidelines.
	■ Ensure implementation of the quality assurance systems.
Operation and maintenance	Require that guidelines for O&M be provided by the builder so that the design level of hazard resilience can be maintained.
	Institute measures to ensure adequate human capacity for O&M of constructed facilities and management of ongoing risk management activities.
	Define procedures for the approval of structural alterations carried out during the life of the facilities.
	Set up structures for funding O&M and risk management activities, including cost recovery mechanisms.
Evaluation	Assess the adequacy of the restored infrastructure system and the success of the project as a whole. This assessment should include evaluation of:
	■ functionality, social acceptability, and sustainability;
	■ project cost with respect to potential benefits of hazard-proof design in future events, skills provided to builders, and new construction guidelines introduced; and
	■ reporting of infrastructure performance under any hazard events that have occurred.
	Lessons learned regarding strengthening hazard resilience should be summarized, shared, and drawn on for future projects.

Other Considerations in Infrastructure Reconstruction

Local institutional capacity. Since infrastructure systems and housing are interdependent and often fall under multiple geographic and administrative jurisdictions, both public and private, coordination across sectors and among agencies in constructing and rehabilitating infrastructure is necessary. Reconstruction speed and quality depends on pre-disaster conditions, such as the state of the infrastructure, record keeping, data management, and institutional capacity. Infrastructure assessments should analyze each of these issues. The capacity of local government, communities (if their role is operational), and the consulting and construction sector is particularly relevant. The involvement of academic, professional, and licensing bodies will help ensure that architects, engineers, and builders correctly apply appropriate codes and construction techniques during reconstruction and in the future. To ensure long-term sustainability and economic development, reconstruction should emphasize the use of local resources (technical, financial, operational).

If there is a risk of reconstruction approvals for infrastructure becoming a bottleneck, consider helping local government in setting up a "single window" where environmental and engineering studies, site plans, and building plans can be approved simultaneously, and building permits issued in a single location. Partnerships with the private sector and other nongovernmental partners may help support this aspect of implementation.

In the agencies that will take over management of new facilities, training, staffing, and other institutional strengthening needs should be identified and funded. If local communities will operate or maintain the infrastructure, they should be trained as well. Local agencies should not be strapped with new infrastructure they cannot afford to operate. Assistance with analyzing not just design and construction, but also the financial feasibility of operating new services must be provided. This may entail designing new tariffs or other cost recovery strategies. The 🏚 case study on relocation after Hurricane Mitch in 1998 in Honduras, below, illustrates some of the risks of underfunding new infrastructure services.

A fast pace of reconstruction can result in cost escalations; a slower reconstruction, while conducive to managing costs, results in losses from delays in service provision. A balance needs to be struck between these two considerations.

Public notifications and consultations. Local legislation may require stakeholder participation in siting, planning, and land acquisition processes, even though the community may not be involved in operation of the services. Participation may need to be accelerated in a post-disaster environment, but this should not be done in a way that compromises the intent of these processes. Acceleration may be successfully achieved through an enhanced communications and outreach effort. See 🏚 Chapter 3, Communication in Post-Disaster Reconstruction, for guidance on developing a communications plan.

Urban infrastructure development. Infrastructure reconstruction in urban areas can be more challenging due to the higher population and the built environment densities, more complex infrastructure technologies and materials, the need for temporary relocation or interruption of services, and the complexities of the social structure, including diverse income levels. When considering an infrastructure project, local service providers, as well as agencies responsible for regulation in the sector and those responsible for local urban planning, should be given the lead in identifying the best approach to rehabilitation or reconstruction. See 🏚 Chapter 7, Land Use and Physical Planning, for more discussion of reconstruction issues in urban versus rural contexts. Social assessment is a tool to help identify and plan for social issues in reconstruction. For guidance on social assessment, see 🏚 Chapter 4, Who Gets a House? The Social Dimension of Housing Reconstruction, Annex 2, How to Do It: Conducting a Post-Disaster Social Assessment.

Managing logistics and cost overruns. The following are some options for government and agencies involved in reconstruction to reduce bottlenecks and manage cost overruns in infrastructure reconstruction:

■ In establishing equipment, material, and supply requirements, make specific plans for procuring items that require a long lead time and could create supply bottlenecks.

■ Expect and budget for cost increases due to an increased demand for material and labor. Increases will be a function of the size and pace of the reconstruction relative to the national economy and the supply capacity.

■ If a fast pace of reconstruction will result in unacceptable cost escalations, analyze whether losses from delays in full service provision from slower reconstruction are acceptable and identify interim service provision options. The phasing and pace of reconstruction should strike a balance between costs and benefits.

■ If faster reconstruction is a priority, build price incentives into construction contracts, without sacrificing the quality of workmanship.

■ Ensure timely approval of plans, issuance of permits, and inspections, so that these procedures never hold up construction.

■ Use the design/build approach, but only when the expertise and capacity exist to properly oversee it.

■ Facilitate material imports and clearances. See 🏚 Chapter 15, Mobilizing Financial Resources and Other Reconstruction Assistance, Annex 1, Deciding Whether to Procure and Distribute Construction Material, for advice on this option.

The 🏚 case study on infrastructure construction in a relocation site in Sri Lanka following the 2004 Indian Ocean tsunami, below, shows how reconstruction agencies adapted to a government policy that infrastructure would only be provided once housing construction was complete.

 For access to additional resources and information on this topic, please visit the handbook Web site at www.housingreconstruction.org.

Risks and Challenges

- Failure to improve the disaster resiliency of rehabilitated and reconstructed infrastructure systems.
- Restored infrastructure systems that are later unaffordable to the users and not properly maintained.
- Setting unrealistic reconstruction time frames in response to local political and social pressures, or collapsing reconstruction schedules in an attempt to avoid having donated funds diverted elsewhere.
- Not adequately coordinating infrastructure and housing reconstruction so that residents live for years without proper services.
- Underbudgeting program management and administration costs, which, in a post-disaster environment, can cost more than twice as much as those in regular projects.[8]
- Time and cost overruns due to limited project management capacity and increased demand for resources in local markets.
- Environmental damage from improperly planned or engineered infrastructure projects.
- Failure to involve the local service providers in planning and executing infrastructure reconstruction programs.

Recommendations

1. Government should enforce measures to ensure that infrastructure planning and reconstruction is closely coordinated with housing reconstruction, using a broad definition of infrastructure to include community facilities.
2. From the first day, support local service providers, such as local government and the community, in the planning and implementation of infrastructure projects, or at a minimum involve them in these efforts.
3. Plan in advance for activities that require long lead times, especially land acquisition and public consultations.
4. Develop realistic reconstruction schedules and service delivery strategies that take into consideration the fact that infrastructure reconstruction can take much longer than housing reconstruction.
5. Apply the "build back better" principle to infrastructure reconstruction, both in terms of its resilience to hazards and its environmental sustainability.
6. Provide a reserve for material and labor cost increases, because these costs will grow in proportion to the speed of the reconstruction effort. A contingency of at least 20 percent is realistic.
7. Make generous provisions for project management and for construction management and quality control, recognizing that the post-disaster environment will be more complex and that there is a risk that the work will be of lower quality than in normal conditions.
8. Use local technical resources in infrastructure planning and design, risk reduction, and construction.
9. Plan and budget for the human capacity development needed for the O&M of reconstructed infrastructure facilities.

Case Studies

1998 Hurricane Mitch, Honduras

Relocation without Infrastructure

Hurricane Mitch had a major impact on the housing situation in the city of Choluteca, located along the Choluteca River in southern Honduras. More than 25 high-density neighborhoods located within the natural floodplain of the Choluteca River were completely washed away, displacing approximately 3,000 families. Although many of these houses in these neighborhoods had running water and electricity, most families did not have clear title to their land. Site selection for reconstruction was based on the availability of a large parcel of land with clear title rather than on its suitability for the creation of a sustainable community. *Banco de Occidente* sold land that it owned 15 kilometers from the city of Choluteca for monthly payments per lot of approximately US$100 over 10 years. The settlement, later called *Nueva Choluteca* or *Limon de la Circa*, consisted of 2,154 lots laid out with little consideration of urban design, transportation needs, or environmental impact. The layout made the provision of infrastructure—water, sewage, electricity, drains, and communications—potentially more expensive, although in fact no provision was made for any of these services at the time. Nongovernmental organizations (NGOs), including Caritas, Atlas Logistique, Iglesia de Cristo, International Organization for Migration, and CECI, participated in

8. World Bank, 2008, "World Bank Good Practice Note on Overall Reconstruction: Design, Implementation and Management," http://siteresources.worldbank.org/CHINAEXTN/Resources/318949-1217387111415/Overall_Reconstruction_en.pdf.

the reconstruction effort. In 2001, only 42 percent of the houses were occupied by their owners. The rest were rented, transferred to non-owners (friends and family), or not occupied. One-tenth of the houses were in poor condition, and the neighborhood was considered very dangerous. Poor-quality housing, continued lack of infrastructure, increased segregation of residents, lack of employment (one study estimated the national unemployment rate at 68 percent immediately after Mitch), high rates of crime, and public health problems were all evident This project demonstrates that relocation at this scale generates a tremendous need for public services, including electricity, water, sewage, and storm-water and solid waste management, as well as social necessities, such as employment, health centers, and schools. International agencies must ensure not only that there is local capability to build infrastructure and provide services to such settlements over the long term, but even more that these projects contribute to the longer-term development of the community where they are built.

Sources: Priya Ranganath, 2009, personal communication; Priya Ranganath, 2000, "Mitigation and the Consequences of International Aid in Post-Disaster Reconstruction" (working paper, McGill University), http://www.colorado.edu/hazards/publications/wp/wp103/wp103.html#casestudy; *and* Gonzalo Lizarralde and Marie France Boucher, n.d., "Learning from Post-Disaster Reconstruction for Pre-Disaster Planning," Groupe de Recherche IF, http://www.grif.umontreal.ca/pages/papers2004/Paper%20-%20Lizzaralde%20G%20&%20Boucher%20M%20F.pdf.

DANIEL PITTET

2001 Gujarat Earthquake, India

Causes for Delays in Rebuilding Urban Infrastructure

The Gujarat Urban Development Company (GUDC) was responsible for reconstruction of urban infrastructure in 14 towns following the 2001 earthquake. Government decided to go beyond replacing lost capital stock and planned the urban reconstruction program in a holistic manner. The program included both *in-situ* reconstruction and relocation. Development plans were prepared for the four most severely damaged towns using a 20-year horizon. Development codes were amended to incorporate national codes for seismic and cyclone safety. GUDC eventually awarded 89 contracts worth US$80.7 million, using financing from the Asian Development Bank (ADB). The four most severely damaged towns received new infrastructure, while damaged infrastructure in 10 less-affected towns was selectively upgraded. A 3-year implementation period was originally projected for the Gujarat reconstruction, yet the scope of the effort necessitated extension of the reconstruction program to six years. The ADB's completion report attributed the delays to common post-disaster factors, including (1) the time required for acquisition of easements and removal of encroachments; (2) delayed contract awards due to multiple agency approvals; (3) selected contractors who were not able to meet requirements; (4) fraudulent bank guarantees presented by contractors; (5) the inability of suppliers to honor supply commitments due to price increases; (6) frequent transfers of implementing agency officials; and (7) late receipt of funds by the Gujarat State Disaster Management Authority, causing delayed payments to contractors.

Source: Asian Development Bank, 2008, "Gujarat Earthquake and Reconstruction Project, Completion Report," http://www.adb.org/Documents/PCRs/IND/35068-IND-PCR.pdf.

2004 Indian Ocean Tsunami, Mandana, Sri Lanka

Coordinating the Timing of Housing Construction with Infrastructure Provision

In September 2005, after the devastating 2004 Indian Ocean tsunami, Habitat for Humanity-Sri Lanka (HFHSL) began construction on a 196-house community on unimproved land in Mandana, five kilometers inland from Thirrukkovil, on the southeast coast of Sri Lanka. The beneficiaries were selected by government authorities from tsunami-affected families that had previously lived within the high-tide line and that were not allowed to rebuild in the coastal buffer zone. Land for relocation was provided by government. Although HFHSL had originally planned to quickly provide small, 150 sq. ft. "core houses" with at least a permanent structure at a relatively low cost, the Sri Lankan government mandated a 500 sq. ft. minimum for houses built on government-donated land, in an effort to improve the quality of life of those relocated.

The first 96 families moved into new permanent houses in February 2006. Although toilets, wiring, and plumbing were in place, there was no electric service, piped water and sewage, public transportation, or graded road. Officials had stated upfront that houses must be completed before other services would be provided, but the lack of infrastructure presented challenges to the families who moved in, as well as for the construction crews.

As an interim measure, HFHSL negotiated with other NGOs to provide wells, water tanks, and water delivery; these services were continued and expanded to meet the needs of homeowners. Generators used to run cement mixers also provided limited emergency power. As homes were completed and more families moved into

EARL KESSLER

the community, HFHSL and the Sri Lankan management of chemical company BASF (the corporate sponsor of the project) joined with the homeowners' new community association to press for provision of full services. Three years later, formal electric services, piped water, and septic tanks were in place and supporting a growing community. In all, HFHSL built 2,049 housing units to support post-tsunami reconstruction in Sri Lanka.

Source: Kathryn Reid, Habitat for Humanity International, 2009, personal communication, http://www.hfhi.org.

Resources

Benson, Charlotte, John Twigg with Tiziana Rossetto. 2007. "Tools for Mainstreaming Disaster Risk Reduction. Guidance Note 12: Construction Design, Building Standards and Site Selection." International Federation of Red Cross and Red Crescent Societies, ProVention Consortium. http://www.proventionconsortium.org/themes/default/pdfs/tools_for_mainstreaming_GN12.pdf.

CDMP. 1997. "Basic Minimum Standards for Retrofitting." Organization of American States and USAID's Unit of Sustsainable Development and Environment. http://www.oas.org/CDMP/document/minstds/minstds.htm.

CDMP. 2006. *Hazard-resistant Construction.* Organization of American States and USAID's Unit of Sustainable Development and Environment. http://www.oas.org/CDMP/safebldg.htm.

Jayasuriya, Sisira and Peter McCawley. 2008. "Reconstruction after a Major Disaster: Lessons from the Post-Tsunami Experience in Indonesia, Sri Lanka, and Thailand." Working Paper No. 125. ADB Institute. http://www.adbi.org/files/2008.12.15.wp125.reconstruction.post.tsunami.experience.pdf.

PAHO. 2002. "Emergencies and Disasters in Drinking Water Supply and Sewerage Systems: Guidelines for Effective Response." Pan American Health Organization. http://www.reliefweb.int/rw/lib.nsf/db900sid/LGEL-5S6BNE/$file/paho-sew-02.pdf?openelement.

Sphere Project. 2004. *Humanitarian Charter and Minimum Standards in Disaster Response.* Geneva: Sphere Project. http://www.sphereproject.org/.

WHO. 2005. "Technical Notes for Emergencies." World Health Organization. http://www.who.int/water_sanitation_health/hygiene/envsan/technotes/en/index.html.

World Bank. 2008. "World Bank Good Practice Note: General Considerations for Infrastructure Planning." http://siteresources.worldbank.org/CHINAEXTN/Resources/318949-1217387111415/Infrastructure_Planning_en.pdf.

World Bank. 2008. "World Bank Good Practice Note on Overall Reconstruction: Design, Implementation and Management." http://siteresources.worldbank.org/CHINAEXTN/Resources/318949-1217387111415/Overall_Reconstruction_en.pdf.

 For access to additional resources and information on this topic, please visit the handbook Web site at www.housingreconstruction.org.

Guiding Principles for Environmental Planning

- During reconstruction, there are two principal environmental concerns: restoring damage to the environment from a disaster and minimizing the environmental impact of the reconstruction process itself.
- Site planning in new settlements should be governed by ecological concerns.
- Construction methods, building designs, and choice of materials all have an environmental impact; they should be based on local practices while being eco-friendly.
- Disaster debris is a valuable resource that should be reused during reconstruction whenever possible. However, materials that can be harmful to workers or the environment, such as asbestos or toxic substances, must be managed carefully.

This Chapter Is Especially Useful For:
- Lead disaster agency
- Environmental specialists
- Local officials
- Project managers

Introduction

Disasters almost always have negative environmental impacts, ranging from damage to ecosystems to the production of vast quantities of waste. Post-disaster reconstruction can either be an opportunity to address these impacts and long-standing environmental problems in the disaster location or it can cause a second wave of damage. The choice is up to decision makers responsible for assessment, planning, and implementation of reconstruction programs. Assessment allows the disaster's environmental impacts to be identified and priority areas for corrective action to be determined. Physical and environmental planning present opportunities to analyze and rebalance the relationship between the built environment and the natural environment. And in implementation, actions can be taken that aid environmental recovery, mitigate the impacts of the reconstruction itself, and promote long-term sustainable development goals.

The scope of "environmental issues" is broad and encompasses built, social, and economic and ecological aspects, and each of these affects those who live where the disaster took place. This chapter focuses principally on critical ecological and built environment issues related to housing demolition and reconstruction. It attempts to persuade those involved in reconstruction that restoration of the environment should be one of their highest priorities. To that end, it covers environmental impact assessments, relocation, waste management, ecological planning of new settlements, environmental needs of habitat, and environmental assessment of housing reconstruction.

Key Decisions

1. **National and local governments** must decide on the legal framework for environmental management to be applied during reconstruction and on a division of labor that will ensure its successful implementation.
2. **Government** should decide immediately which agency will be in charge of post-disaster debris management and that agency should plan and coordinate the debris management program in a way that reduces risk, facilitates recovery and reconstruction, and disposes of debris in a cost-effective and environmentally sound manner, while keeping disposal of reusable or salable materials to a minimum.
3. The **lead environmental agency** must decide how to provide environmental guidance to all institutions active in reconstruction, keep this information updated, and monitor reconstruction implementation. It must also decide what incentives and sanctions will be employed.
4. **Land use planning and environmental institutions** need to agree on the mechanisms to ensure that post-disaster environmental planning and management activities are integrated with land use and site planning, as well as on how these local activities will be coordinated with the lead disaster agency.
5. In a consultative manner, **government** should define how **local community and civil society organizations** can contribute to environmental protection during reconstruction and on coordination mechanisms among the organizations and with government. These organizations can participate in local debris management, assessments, reconstruction monitoring, technical assistance and project implementation.

Some Environment-Related Consequences of Common and Recurrent Natural Disasters

Type of Disaster	Associated Environmental Impact
Hurricane/cyclone/typhoon	■ Loss of vegetation cover and wildlife habitat ■ Inland flooding ■ Mudslides and soil erosion ■ Saltwater intrusion to underground freshwater reservoirs ■ Soil contamination from saline water ■ Damage to offshore coral reefs and natural coastal defense mechanisms ■ Waste (some of which may be hazardous) and debris accumulation ■ Secondary impacts by temporarily displaced people ■ Impacts associated with demolition, reconstruction, and repair to damaged infrastructure (e.g., deforestation, quarrying, waste pollution)
Tsunami	■ Groundwater pollution through sewage overflow ■ Saline incursion and sewage contamination of groundwater reservoirs ■ Loss of productive fisheries and coastal forest or plantations ■ Destruction of coral reefs and natural coastal defense mechanisms ■ Coastal erosion or deposition of sediment on beaches or small islands ■ Marine pollution from back flow of wave surge ■ Soil contamination ■ Loss of crops and seed banks ■ Waste accumulation—additional waste disposal sites required ■ Secondary impacts by temporarily displaced people ■ Impacts associated with demolition, reconstruction, and repair to damaged infrastructure (e.g., deforestation, quarrying, waste pollution)
Earthquake	■ Loss of productive systems (e.g., agriculture) ■ Damage to natural landscapes and vegetation ■ Possible mass flooding if dam infrastructure is weakened or destroyed ■ Waste accumulation—additional waste disposal sites required ■ Secondary impacts by temporarily displaced people ■ Impacts associated with demolition, reconstruction, and repair to damaged infrastructure (e.g., deforestation, quarrying, waste pollution) ■ Damaged infrastructure as a possible secondary environmental threat (e.g., leakage from fuel storage facilities) ■ Release of hazardous materials from industries, medical facilities, and nuclear plants
Flood	■ Groundwater pollution through sewage overflow ■ Loss of crops, trees, livestock, and livelihood security ■ Excessive siltation that may affect certain fish stocks ■ River bank damage from erosion ■ Water and soil contamination from fertilizers and/or industrial chemicals ■ Secondary impacts by temporarily displaced people ■ Sedimentation in floodplains or close to river banks
Volcanic Eruption	■ Loss of productive landscape and crops buried by ash and pumice ■ Forest fires as a result of molten lava ■ Secondary impacts by temporarily displaced people ■ Loss of wildlife following gas release ■ Secondary flooding should rivers or valleys be blocked by lava flow ■ Damaged infrastructure as a possible secondary environmental threat (e.g., leakage from fuel storage facilities) ■ Impacts associated with demolition, reconstruction, and repair to damaged infrastructure (e.g., deforestation, quarrying, waste pollution)
Landslide	■ Damaged infrastructure as a possible secondary environmental threat (e.g., leakage from fuel storage facilities) ■ Secondary impacts by temporarily displaced people ■ Impacts associated with demolition, reconstruction, and repair to damaged infrastructure (e.g., deforestation, quarrying, waste pollution)

Source: United Nations Environment Programme (UNEP), 2008, *Environmental Needs Assessment in Post-Disaster Situations: A Practical Guide for Implementation* (Nairobi: UNEP), http://www.humanitarianreform.org/humanitarianreform/Portals/1/cluster%20approach%20page/clusters%20pages/Early%20R/UNEP%20PDNA_pre-field%20test%20draft.pdf.

 For access to additional resources and information on this topic, please visit the handbook Web site at www.housingreconstruction.org.

6. The **lead environmental agency** must decide on and implement mechanisms that ensure that trees, groundwater, and other natural resources and other local environmental assets will be protected on a site-specific and regional basis during demolition and reconstruction. Community and advocacy organizations can play an important role in this.

7. **Local authorities** need to establish measures to ensure that decision points, such as the approval of site plans and the issuance of demolition and building permits, are used to ensure compliance with the environmental guidelines. These are opportunities to address such issues as the integration of infrastructure development with housing reconstruction and the use of local and eco-friendly materials and designs.

8. **National and local governments** should define any technical assistance requirements related to implementing post-disaster environmental management systems, norms, and procedures, and identify a point person to raise the necessary funding and to manage procurement. International agencies, including the World Bank, can frequently be of assistance.

Public Policies Related to Environmental Planning

National and local environmental law and regulations should be applied in reconstruction, although additional guidance may be needed to address the unique post-disaster situation. The national environmental ministry and local governmental environmental agency should be involved early and should participate in assessments. The World Bank will apply its environmental safeguards, as explained in ᴸ Chapter 20, World Bank Response to Crises and Emergencies, and ᴸ Chapter 21, Safeguard Policies for World Bank Reconstruction Projects. Policy guidance should be widely accessible to different actors, including all government agencies, the private sector, international agencies, NGOs, and local communities. If existing legal and regulatory instruments require updating, or strengthening, donors and other sources should finance technical assistance to develop reconstruction environmental policy guidelines that address the issues discussed in this chapter. Government should consider updating its environmental policies as part of its disaster risk reduction program so that the country is prepared to apply the policies in the event of a disaster. The objective is to provide environmental guidelines that balance environmental protection with the need to support reconstruction. The lead agency should also designate a group of experts to provide advice on specific cases and exceptions and to propose modifications to the policy as reconstruction experience is gained. The ᴸ case study on the 1999 Armenia post-earthquake reconstruction, below, describes how Colombia designed a comprehensive environmental management plan.

Technical Issues

The following paragraphs discuss in detail some of the technical issues related to environmental planning and provide examples of how these issues applied to real-world situations. Case studies involving some of these issues are found later in this chapter.

Rapid Environmental Impact Assessment

Governments, international aid agencies, NGOs, and communities use rapid environmental impact assessments (REAs) as the key starting point after any disaster. An REA needs to be conducted within 120 days of the event.[1] There are standards manuals and guidelines for REA on organization-level assessments, community-level assessments, consolidations, and analyses. Personnel required for an REA include specialists on disaster relief and environmental impact assessments (EIAs). Community REAs can be conducted by NGOs and field practitioners.[2] During the early recovery phase, UNEP recommends the use of the Environmental Needs Assessment (ENA) methodology.[3] More detailed environmental studies may also be needed to analyze the particular issues of environmental impact at the relevant scale. For instance, groundwater contamination may need to be evaluated for the entire watershed, or the availability of local natural resources used in housing construction, such as lumber or stone, may need to be evaluated at the national or regional level. At the end of the housing reconstruction process, an integrated environmental assessment should be part of the project evaluation.

In Aceh, Indonesia, after the 2004 tsunami, the following 10 priority areas for environmental management in the recovery process were identified: (1) contaminated groundwater; (2) sanitation; (3) lost livelihood; (4) lack of coordination in relief or recovery response during the emergency response phase; (5) shelter and related domestic needs; (6) enhanced roles identified for local governance and the role of communities in environmental management; (7) volume of (mixed) waste; (8) uncertain land tenure for tsunami survivors; (9) strengthening

1. Charles Kelly, 2005, *Guidelines for Rapid Environmental Impact Assessment in Disasters* (Geneva: CARE International), http://www.reliefweb.int/rw/lib.nsf/db900SID/EVOD-6FCH52?OpenDocument.
2. Ministry of the Environment Republic of Indonesia, 2005, *Rapid Environmental Impact Assessment, Banda Aceh, Sumatra* (Jakarta: Republic of Indonesia), http://www.humanitarianinfo.org/sumatra/reference/assessments/doc/gov/GoI-EnvironmentalImpactAssessment-050405.pdf.
3. UNEP, 2008, *Environmental Needs Assessment in Post-Disaster Situations: A Practical Guide for Implementation* (Nairobi: UNEP), http://www.humanitarianreform.org/humanitarianreform/Portals/1/cluster%20approach%20page/clusters%20pages/Early%20R/UNEP%20PDNA_pre-field%20test%20draft.pdf .

of local government to overcome the loss of infrastructure, staff, and resources; and (10) increase of capacity to direct and absorb relief assistance for sustainable development. After the 2008 earthquake in Wenchuan, China, the government reconstruction policy promoted the reuse of waste and encouraged improving the environmental sustainability of industrial plants rehabilitated after the earthquake, including those producing construction materials using recycled inputs, as described in the case study, below.

Post-Disaster Waste Management

Post-disaster waste management is one of the most crucial and urgent issues following a disaster. Different types of waste are produced in urban and rural areas. Much of the waste from rural housing (stone, adobe or mud brick, and wood) can be recycled, while that from urban areas needs proper separation, collection, and treatment. In urban areas, asbestos and electrical appliances are a potential source of hazardous waste; therefore, proper separation and treatment of these wastes is required. Rubble and debris represent resources that have value in reconstruction; however, they can also represent a risk for communities and should be analyzed and handled with care. In case of water-related disasters, a large amount of biological waste is produced and needs to be treated properly. See Annex 1, How to Do It: Developing a Disaster Debris Management Plan, in this chapter. Also see text box "Managing Asbestos in Housing and Community Reconstruction" later in this chapter.

Typhoon Tokage, in the city of Toyooka, Japan (2004), produced disaster waste that was 1.5 times the annual waste production in the city. It took significant time and financial resources to process the waste in order to start the reconstruction process. Information and communications technology (ICT) tools and systems can be deployed. Catalogue and communicate availability of recycled materials to facilitate local economic activity. The case study on the 1994 Northridge earthquake, below, discusses how the city of Northridge, California, recycled more than 50 percent of all disaster debris.

In-Situ Construction versus Relocation

The decision to relocate or build *in-situ* has environmental consequences. Likewise, the amount and nature of waste produced in a disaster often influences decisions about the reconstruction process. The environmental consequences of the *in-situ* versus relocate decision should be discussed with community members, government, and multilateral and bilateral donors. Local environmental guidelines should be consulted as well.

After the 2004 Indian Ocean tsunami, many settlements in Aceh, Indonesia had to be relocated 2-3 kilometers inland because of water logging and disaster debris, thereby causing challenges to the livelihoods of fishing communities. Some tsunami-affected countries like Sri Lanka imposed strict limits based on the Coastal Regulatory Zone Act. See Chapter 5, To Relocate or Not to Relocate, for more information and a case study on this issue.

Ecological Planning of New Settlements

New housing settlements are often sited in areas with rich ecological resources and biodiversity, without evaluating the ecological footprint of the project, creating both new risks and an environmental conservation challenge. If the environmental assessment used for site selection is not properly conducted, relocation may create new risks. After a coastal hazard (like a typhoon or tsunami), the new settlement may be developed on mountain slopes. Yet the higher ground may have a high landslide risk. Therefore, proper ecological analysis and hazard mapping is required before selecting new settlements after a disaster. This is particularly important for fragile ecosystems, such as small islands and mountainous areas with higher biodiversity. Protection of natural habitat should be a priority after a disaster, including mangroves and nesting grounds of birds, along with architectural heritage, such as structures, since both contribute to the cultural, psychological, and economic recovery of the community. The case study on the Indian Ocean tsunami reconstruction in Tamil Nadu, India, below, shows how the protection of trees was not fully considered in planning housing reconstruction.

Green and Clean Recovery and Reconstruction

Rural housing styles have evolved in harmony with local cultural and climatic conditions. Vernacular designs and techniques are often optimal because of their cost-effectiveness, local availability, and minimal environmental impact. There is increasing support for using local, environment-friendly housing materials in reconstruction (e.g., stone, mud brick, wood, and slate), especially in rural areas. False perceptions about environmental impacts can discourage the use of local materials (e.g., the

"Without the trees the village is not alive. It is another village, not our village anymore."

4. UNEP, 2005, *Environmental Management and Disaster Preparedness: Lessons learnt from the Tokage Typhoon* (Geneva: UNEP), http://www.unep.or.jp/ietc/wcdr/unep-tokage-report.pdf.
5. Sphere Humanitarian Charter and Minimum Standards in Disaster Response, http://www.sphereproject.org/.

ban on timber products in Aceh in the initial stage of the post-tsunami reconstruction). This makes reconstruction more difficult for homeowners who may be unfamiliar with new building materials and construction methods. Materials and design should be selected using environmental and climate change-oriented criteria, such as energy use, greenhouse gas emissions, the sustainability of production chains, the use of water, and the potential for recycling and reuse. See  Chapter 10, Housing Design and Construction Technology, for background on these issues.

Need for Basic Environmental Services

Lack of basic infrastructure such as water, sanitation, and waste management can cause serious environmental and environmental health problems and can lead to low occupancy rates of new and reconstructed housing. Sphere standards, which establish minimum health, sanitation, water supply, and housing standards for humanitarian operations, can be useful as a frame of reference in reconstruction.[5] See Chapter 8, Infrastructure and Services Delivery, for detailed guidance on post-disaster infrastructure restoration.

Tools for Environmental Planning

Community participation is absolutely critical at each stage of environmental planning and assessment. Public hearings, held to inform the community of environmental assessments and planned actions, can bring together all stakeholders, including project proponents, environmental agencies, NGOs, citizens, and project-affected persons.

The tools outlined below aim to apply core principles of building local capacity of communities to prevent and mitigate disasters, create partnerships among stakeholders, share and exchange information, and develop learning and decision-making tools to address disaster impacts. All tools incorporate common elements, such as assessment, stakeholder involvement mechanisms, and monitoring.[6]

Assessment Tools

Rapid Environmental Impact Assessment. Helps identify and prioritize likely environmental impacts in natural disaster conditions. A qualitative assessment approach is used to rank issues and identify follow-up actions.[7]
Environmental (or Ecological) Risk Assessment. Evaluates the adverse effects that human activities and pollutants have on the plants and animals in an ecosystem, and identifies impacts on human, ecological, and ecosystem health.[8]
Environmental Impact Assessment. Involves analysis of baseline environment, identification and evaluation of impacts, and mitigation measures to remedy adverse effects of natural and man-made disasters. See Annex 2, How to Do It: Carrying Out Environmental Impact Assessment and Environmental Monitoring of Reconstruction Projects, for guidance on carrying out an EIA.
Strategic Environmental Assessment. Evaluates the consequences of plans, policies, and programs on the natural environment using a systematic approach, taking into account social and economic considerations.[9]

Planning Tools

Eco and Hazard Mapping (EHM). Serves as a simple systematic and visual tool that aids in post-disaster reconstruction planning by using maps and plans of cities, neighborhoods, and buildings. The mapping process involves multi-stakeholder participation. Participants mark all environmental aspects, hazards, and risks on plans and maps that contribute to the formulation of post-disaster recovery plans.
Environmental Profiling. Provides planning and management options based on a study of development setting, environmental setting, and disaster setting of a city or village. The development setting studies the socioeconomic structure, institutional structure, and environmental resources. Environmental setting studies the natural and built environment in detail. Disaster setting provides an analysis of hazards and vulnerability faced by communities.[10]

Implementation Tools

Environmental Management System. Used as a problem-solving and problem-identification tool based on the concept of continual improvement. EMS forms the core of the international environmental standard ISO 14001. The EMS adopts the Plan-Do-Check-Act cycle to develop environmental policies; frame the EMS; and implement, review, and revise performance.[11]
Environmental Management Plan. An Environmental Plan (EMP) is used to monitor the impacts and mitigation measures agreed to in the EIA of a specific project. See Annex 2, How to Do It: Carrying Out Environmental Impact Assessment and Environmental Monitoring of Reconstruction Projects, for guidance on carrying out an EIA and implementing an EMP.

DANIEL PITTET

6. United Nations Centre for Human Settlements (UN-HABITAT) and UNEP, 1999, *The SCP Source Book Series, V 5, Institutionalising the Environmental Planning and Management* (EPM) Process (Nairobi: UNCHS and UNEP), http://www.unhabitat.org/pmss/getPage.asp?page=bookView&book=1652.
7. Charles Kelly,2005, "Guidelines for Rapid Environmental Impact Assessment in Disasters,"CARE International, http://www.reliefweb.int/rw/lib.nsf/db900SID/EVOD-6FCH52?OpenDocument.
8. See U.S. Environmental Protection Agency, National Center for Environmental Assessment, "Ecological Risk Assessment," http://cfpub.epa.gov/ncea/cfm/ecologic.cfm.
9. World Bank, Environment, "Strategic Environmental Assessment Toolkit," http://go.worldbank.org/XIVZ1WF880; *and* Organisation for Economic Co-operation and Development Development Co-operation Directorate, Strategic Environmental Assessment Network, "Applying SEA: Good Practice Guidance for Development Co-operation," http://www.seataskteam.net/guidance.php.
10. UN-HABITAT and UNEP, 1998, *The SCP Source Book Series, Volume 1: Preparing the SCP Environmental Profile* (Nairobi: UN-HABITAT and UNEP), http://www.unhabitat.org/pmss/getPage.asp?page=bookView&book=1427.
11. International Organization for Standardization, "ISO 14000 Essentials," http://www.iso.org/iso/iso_catalogue/management_standards/iso_9000_iso_14000/iso_14000_essentials.htm.

Managing Asbestos in Housing and Community Reconstruction

What Is Asbestos and Where Is It Found?

Asbestos is the name given to a number of naturally occurring fibrous minerals with high tensile strength, the ability to be woven, and resistance to heat and most chemicals. Because of these properties, asbestos fibers have been used in a wide range of manufactured goods and construction materials, including roofing shingles, ceiling and floor tiles, paper and cement products, textiles, and coatings. In-place management dictates having a building management program to minimize release of asbestos fibers into the air and to ensure that when asbestos fibers are released, either accidentally or intentionally, proper control and cleanup procedures are implemented. However, in a disaster, there is a likelihood that construction debris—especially debris from engineered buildings—may include asbestos-containing materials (ACMs), making it necessary to develop abatement procedures as part of the debris management program. Under normal circumstances, abatement entails removal of asbestos before building demolition; however, after a disaster this may not be possible.

Managing Asbestos Health Effects

Exposure to airborne friable asbestos may result in a potential health risk, because people breathing the air may breathe in asbestos fibers. Fibers embedded in lung tissue over time may cause serious lung diseases, including asbestosis, lung cancer, and mesothelioma. Disease symptoms may take several years to develop following exposure. Continued exposure can increase the amount of fibers that remain in the lungs. Exposure to asbestos increases your risk of developing lung disease. That risk is made worse by smoking.

Good practice is to minimize the health risks associated with ACMs by avoiding their use in new construction and renovation, including disaster relief and reconstruction, and, if installed ACMs are encountered, by using internationally recognized standards and best practices to mitigate their impact. In reconstruction, demolition, and removal of damaged housing and infrastructure construction materials, asbestos hazards should be identified and a risk management plan adopted as part of the EMP that includes disposal techniques and end-of-life sites.

How Asbestos Is Detected

ACMs are mixtures of individual asbestos fibers and binding material. The asbestos content of manufactured items ranges from 1 percent to 100 percent. Asbestos fibers cannot be seen without a special microscope. Analysis by an accredited testing laboratory is the only way to know for certain whether a material contains asbestos. Workers should be protected from asbestos exposure even in the sampling process.

Disposal of Asbestos

Asbestos waste or debris should not be burned since the fibers can be released; it should be disposed of at an approved disposal site. Laws should require (1) safe methods to contain asbestos waste (wet, double-bagged), (2) procedures for hauling waste, (3) disposal of ACM in an authorized landfill, and (4) formal record keeping of asbestos waste disposal. Landfilling is the environmentally preferred method of asbestos disposal because asbestos fibers are immobilized by soil. Asbestos cannot be safely incinerated or chemically treated for disposal.

Information on Asbestos Regulation

Because the health risks associated with exposure to asbestos are now widely recognized, global health and worker organizations, research institutes, and some governments have enacted bans on the commercial use of asbestos, and they urge the enforcement of national standards to protect the health of workers, their families, and communities exposed to asbestos through an International Convention. Information on these standards and emerging legal frameworks are available from the sources below.

The International Ban Asbestos Secretariat (IBAS), http://ibasecretariat.org/. IBAS keeps track of national asbestos bans.

International Finance Corporation, 2007, "Environmental, Health, and Safety Guidelines," http://www.ifc.org/ifcext/sustainability.nsf/Content/EnvironmentalGuidelines.

World Health Organization, 2006, "Elimination of Asbestos-Related Diseases," http://www.who.int/occupational_health/publications/asbestosrelateddisease/en/index.html.

World Bank Group, 2009, "Good Practice Note on Asbestos: Occupational and Community Health Issues," http://siteresources.worldbank.org/EXTPOPS/Resources/AsbestosGuidanceNoteFinal.pdf.

Sources: World Bank Group, 2009, "Good Practice Note on Asbestos: Occupational and Community Health Issues," http://siteresources.worldbank.org/EXTPOPS/Resources/AsbestosGuidanceNoteFinal.pdf; *and* US Environmental Protection Agency, "Asbestos," http://www.epa.gov/asbestos/.

Risks and Challenges

- Ignoring environmental issues in any phase of reconstruction and not involving environmental experts in decision making at the policy and programmatic level.
- Delays in conducting the environmental assessment increase environmental risks created by the disaster.
- Dangerous or hazardous rubble and debris (such as toxic or ignitable substances, asbestos, explosives, collapsing buildings) are not handled with caution, with negative effects on communities and the environment
- Damage to infrastructure leads to secondary impacts like fire and floods before problems are identified and addressed.
- Political and institutional factors, rather than community and environmental priorities, drive site-selection decisions.
- Poor planning permanently destroys environmental assets, such as endangered habitats, coastal sand dunes, and mangroves.
- Infrastructure and site development negatively affect groundwater quality and quantity.
- Social and cultural assets are destroyed because of ad hoc development planning.
- Community participation in environmental decision making is downplayed because of political and commercial interests.
- Local building practices are combined in an unsafe way with practices promoted by external actors.
- Commercial interests influence material and technology selection, with negative ramifications on the environment and community.

Recommendations

1. Include government staff and consultants in the environmental assessment teams so that they acquire firsthand knowledge of environmental issues in the affected area and can identify how incentives for environmentally sustainable reconstruction can be incorporated in the reconstruction policy.
2. Identify the legal framework for environmental management to be applied in reconstruction early on, how it will be implemented and by whom, and how it will be monitored and evaluated.
3. Mobilize the post-disaster debris management effort immediately after the disaster, carrying out a rapid planning exercise if a debris management plan was not in place before the disaster.
4. Ensure that the environmental requirements for reconstruction are effectively and continually communicated to all agencies participating in the reconstruction program.
5. In developing the reconstruction policy, government, UN shelter cluster partners, and environmental organizations should work together to minimize the environmental impact and maximize the local sustainability of the building materials and practices to be used.
6. Use the environmental review process to evaluate the ecological footprint of a relocation site or in-situ reconstruction project and to select the site, develop mitigation measures for the project and its construction, and adjust project parameters.
7. Plan new settlements or the rehabilitation of existing systems so that sanitation and other basic infrastructure are provided as early as possible to ensure healthy environmental conditions for new residents.

Limitations

- Environmental issues are not restricted to the disciplinary boundary of environmental management. In a post-disaster context, environmental issues also deserve consideration when making decisions regarding, among other things, financial management, technical and engineering aspects of housing reconstruction (safer design), material availability, accessibility, cost, and time.
- Environmental issues tend to become a lower priority when measured against the desire to speed up the reconstruction. Respecting the existing environmental policy framework of the country and documenting and mapping environmental hazards and assets may help rebalance these considerations. In the long run, wise environmental decisions will pay off.

 For access to additional resources and information on this topic, please visit the handbook Web site at www.housingreconstruction.org.

Case Studies

1999 *Eje Cafetero* Earthquake, Armenia, Colombia

An Integrated Response to Post-Disaster Environmental Management

The devastating earthquake in Armenia, Colombia, in January 1999, left 1,230 people dead and 200,000 affected, and damaged or destroyed 80,000 homes. Given the economic importance of this agricultural and coffee-growing region, recovery of the environment was immediately identified as one of the most critical concerns. The President not only declared an economic and social state of emergency, but—for the first time in Colombian history—declared an ecological state of emergency in the affected region. This action, together with the creation by the President of the Fund for the Reconstruction of the Coffee Zone (*Fondo para la Reconstrucción del Eje Cafetero*, or *FOREC*), which was charged with integrated reconstruction of the zone, ensured that the environmental dimensions of the disaster would be prominent in the reconstruction plan. The reconstruction strategy was also designed to respect and further national environmental strategies and laws, while promoting the sustainable economic development of the region. The importance of the environment was also reflected in the degree of central government involvement in this aspect of the recovery process, not fully delegating responsibility to NGOs and local governments, as was done with most other aspects of the reconstruction program. A broad range of activities were developed to promote environmental goals: (1) careful management of almost 4 million cubic meters of debris, (2) development of environmental guiding principles, and of environmental guidelines for land use planning and reconstruction, (3) formulation of integrated land use plans that incorporated environmental management and disaster prevention, (4) investment in new infrastructure for ecotourism in the Nevados Park, (5) new environmental regulations for the mining industry, and (6) stabilization of critical mountain slopes. In addition, a sustainable management was implemented during reconstruction for *guadua* (a type of bamboo used in the region as a construction material). As a result, 1,045 hectares of culms were planted to compensate for over-exploitation in an effort to reduce avoid soil erosion, improve air and water quality, and contribute to improving the quality of life in the region.

Sources: Ana de Campos, 2009, personal communication *and* FOREC, El Ministerio del Medio Ambiente, las corporaciones autónomas regionales del Valle del Cauca (CVC), Quindío (CRQ), Risaralda (CARDER), Caldas (CORPOCALDAS), Tolima (CORTOLIMA), el Instituto de Hidrología, Meteorología y Estudios Ambientales - IDEAM, el Instituto de Investigación e Información Geocientífica Minero –Ambiental y Nuclear - INGEOMINAS, el CORPES de Occidente, 2002, *Plan de Manejo Ambiental para la Reconstrucción del Eje Cafetero. Informe Final de Gestión y Resultados,* Armenia.

2004 Indian Ocean Tsunami, Tamil Nadu, India

Neglecting the Importance of Trees for Livelihoods and Thermal Comfort

"Without the trees the village is not alive. It is another village, not our village anymore."

Some of the housing reconstruction projects in the aftermath of the 2004 Indian Ocean tsunami in Tamil Nadu, India, should have more carefully considered the space around the house and the surrounding vegetation as equally important aspects of the inhabited space. People now live in houses of a new design, built with foreign materials, in a strange settlement layout, and sometimes in a new location, without any trees. In fact, the loss of trees is described in this region as one of the worst consequences of contractor-built reconstruction. In several villages, contractors refused to start any reconstruction work before the ground was completely cleared of houses, trees, and other vegetation. In one village, people estimated that 800–1,200 trees were cut down, demonstrating a lack of understanding of the importance and the central role of trees in these communities. Tree products are abundantly used in every home as food, fodder, and firewood, and to fabricate utensils—and are also a valued source of income. Trees also have cultural importance: trees are connected to notions of health, protection, beauty, and sacredness. In a tropical climate where temperatures exceed 40°C most of the year, the importance of shade cannot be overemphasized. Areas with trees demarcate locations where people sit together, talk, and play, in short, where social life takes place. Even though the sites were specifically chosen by the communities themselves and the cutting of trees was probably inevitable, the fact remains that the tree cover is gone with serious adverse effects. The demolition of trees resulting from reconstruction projects risks causing a long-term detrimental impact on the social networks, livelihoods, and general well-being of the village community.

DANIEL PITTET

Source: Jasmin Naimi-Gasser, 2009, "The socio-cultural impact of post-tsunami housing reconstruction programs on fishing communities in Tamil Nadu, India: An ethnographic case study" (thesis, University of Zurich); *and* C. V. Sankar, 2009, written communication.

2008 Wenchuan Earthquake, China

Using Waste as a Resource to Create an Environment-Friendly Society

Following the 2008 Wenchuan earthquake in China, some people proposed that the concept of a circular economy be applied in reconstruction. The idea was to use the resources available for reconstruction, including debris from the earthquake, in the most efficient and productive way possible. It also translated into a focus on industrial rebuilding for industries that could contribute to the circular economy in the long term and on the way in which industrial activities would be carried out once rehabilitated, seeking to reduce energy consumption; improve the conservation of water, land, and materials; and reduce their impact on the surrounding communities. The policy mentions emission reduction of high energy-consuming enterprises and promotion of cleaning production technology. Lastly, it encourages the recycling of construction waste, industrial solid waste, and coal gangue to develop environmental friendly construction materials. These activities both conserve resources and protect the environment, which, in turn, promote the community's economic, social, and environmental development in a way that is healthier, integrated, and sustainable.

Source: People's Republic of China, National Development and Reform Committee (NDRC), 2008, "The Overall Planning for Post-Wenchuan Earthquake Restoration and Reconstruction," http://en.ndrc.gov.cn/policyrelease/P020081010622006749250.pdf.

1994 Northridge Earthquake, California

Acting Quickly to Recycle Debris after a Major Urban Earthquake

On January 17, 1994, residents of the Los Angeles region of southern California were awakened by a 6.7 magnitude earthquake that proved to be the most costly earthquake in United States history. Fifty-seven people died, more than 9,000 were injured, and more than 20,000 were displaced. Surprisingly, the city of Los Angeles did not have a disaster debris management plan in place, but quickly developed procedures afterward. City officials updated an existing list of licensed, insured debris removal contractors and asked them to attend an orientation and to sign hastily drafted contracts for debris removal. At first, contracts were only two pages long and covered one week of work, but the contracts ultimately grew to 22 pages, each contractor was assigned a grid of streets to clear, and the work

FEMA NEWS PHOTO

periods were extended. These early contracts allowed the city to begin removing debris quickly. Yet recycling was not included until two months after the date of the disaster, due to a dispute about whether the costs would be eligible for Federal reimbursement. Once recycling was approved, the city developed contract terms that rewarded haulers for source-separated materials while working with businesses to develop processing for mixed debris. The city also provided training and financial incentives to haulers. Most of the materials collected were recyclable; wood, metal, dirt, concrete and asphalt, and red clay brick were separated. After four months, the city was recycling about 50 percent of the debris collected each week. A year later, the city was recycling more than 86 percent of the debris, totaling more than 1.5 million tons. City inspectors (pulled from other assignments) monitored the contractors. By the end of the program, the city had recycled almost 56 percent of all materials from the earthquake for less than the cost of disposal, a total that would have been much higher had the city implemented recycling from the beginning of recovery. To prepare for the possibility of future disasters, Los Angeles later issued a request for proposals for a contingency contract for various disaster waste management activities, including the use of sites in the event of a natural disaster.

Sources: U.S. Environmental Protection Agency, "Wastes - Resource Conservation - Reduce, Reuse, Recycle - Construction & Demolition Materials," http://www.epa.gov/osw/conserve/rrr/imr/cdm/pubs/disaster.htm#la *and* U.S. Geological Survey, "USGS Response to an Urban Earthquake: Northridge '94," http://pubs.usgs.gov/of/1996/ofr-96-0263/,

2004 Indian Ocean Tsunami, Sri Lanka
Ecological Planning of Settlements to Address Waste Management

After the 2004 Indian Ocean tsunami in Sri Lanka, waste management became an additional challenge to the problem of dealing with the regular waste generated by the growing population. There was a need to address the waste generated by the changing consumption patterns of the tsunami-affected people, many of whom were housed in transitional shelters. Many new housing schemes, settlements, and townships were developing in numerous, dispersed locations, and in these locations there was inadequate space and capacity to tackle this problem. Therefore, it was important to ensure that local authorities were provided the resources and capacity to manage the impacts of these settlements on the waste stream, to avoid waste management becoming a major issue when these settlements were occupied. New ecological plans were developed in many cases, with the assistance of outside experts.

Sources: Satoh Tomoko, 2007, *Study on Evolution of Planning and Responses to Water-Related Disaster in Japan, and Its Application to Indian Ocean Tsunami Case in Sri Lanka* (master thesis, Kyoto University); *and* Aat van der Wel, Valentin Post, 2007, "Solid Waste Management in Sri Lanka: Policy & Strategy," http://www.waste.nl/page/1554.

Resources

Humanitarian Reform in Action. "Mainstreaming the Environment into Humanitarian Response." http://oneresponse.info/crosscutting/environment/publicdocuments/ERM_%20Final%20 Report_08%2011%2007.pdf.

Inter-Agency Technical Committee of the Forum of Ministers of the Environment of Latin America and the Caribbean. 2000. "Panorama of the Environmental Impact of Disasters in Latin America and the Caribbean." Report given at the 12th Forum of Ministers of the Environment of Latin America and the Caribbean, Bridgetown, Barbados, March 2–7. http://www.gdrc.org/uem/disasters/disenvi/ Panorama-Envi-Impact.pdf.

Kelly, Charles. 2005. *Guidelines for Rapid Environmental Impact Assessment in Disasters.* Geneva: CARE International. http://www.reliefweb.int/rw/lib.nsf/db900SID/EVOD-6FCH52?OpenDocument.

Sphere Project. 2000. "Humanitarian Charter and Minimum Standards in Disaster Response." http:// www.sphereproject.org/component/option,com_docman/task,cat_view/gid,17/Itemid,203/ lang,english/.

UNEP. 2005. *After the Tsunami: Rapid Environmental Assessment.* Geneva: UNEP. http://www.unep. org/tsunami/tsunami_rpt.asp.

UNEP. 2005. *Environmental Management and Disaster Preparedness: Lessons learnt from the Tokage Typhoon.* Geneva: UNEP. http://www.unep.or.jp/ietc/wcdr/unep-tokage-report.pdf.

World Bank. 1999. "OP/BP 4.01. Environmental Assessment." *Operational Manual.* Washington, DC: World Bank. http://go.worldbank.org/9MIMAQUHN0.

World Bank. 2007. "OP/BP 8.00. Rapid Response to Crises and Emergencies." *Operational Manual.* Washington, DC: World Bank. http://go.worldbank.org/ILPIIVUFN0.

 For access to additional resources and information on this topic, please visit the handbook Web site at www.housingreconstruction.org.

152 SAFER HOMES, STRONGER COMMUNITIES: A HANDBOOK FOR RECONSTRUCTING AFTER NATURAL DISASTERS

Natural disasters can generate tremendous quantities of debris. After a disaster, some institution must immediately take the lead to develop and direct a plan for collecting and managing disaster debris. Failure to do so will increase the secondary risks for the affected community and will delay reconstruction. If disasters are anticipated, a disaster debris management plan should be in place that lays out the roles and responsibilities of different agencies, a plan of action, and the mechanisms for coordination. While this sort of planning is becoming more common, especially in countries with strong local governments, it is more likely that both the preparation and the execution of the debris management plan will be done immediately after the disaster strikes, sometimes by an inexperienced lead agency. This section provides basic guidance on how institutions can collaborate to manage post-disaster debris. It does not assume pre-planning has been done and therefore covers planning as well as some important topics to consider in each component. It is based on a range of publicly available documents.[1]

Phases of Disaster Debris Management

Post-disaster debris management typically occurs in two overlapping phases: initial clearance and long-term removal, management, and processing. The overall plan should address both.

Phase 1. Initial clearance of debris. Debris clearance will be the primary debris management activity during the first few days. During this phase, debris is cleared from power lines and key roadways to restore transportation, emergency access, and utility services as quickly as possible. Households and businesses will set debris at the side of the road, for later collection. Various agencies may be available to provide assistance, including the national guard or military, utility companies, local and state police, and public works and highway agencies. Coordination among them will be required. This phase will last approximately 10 days.

Phase 2. Long-term removal, management, and processing of debris. Following initial clearance, debris management generally shifts to local public agencies, and becomes more complex. It will include removing, collecting, processing, and disposing of debris, including all debris in public areas, as well as debris set out by residents for collection. The rules for handling institutional, commercial, and industrial waste must be part of the plan. This phase may last up to one year.

Components of a Disaster Debris Management Plan

Disaster debris may be viewed as pure waste or as a resource. Disaster debris may be viewed as pure waste or as a resource. The reality is somewhere in between; some portion is a usable resource and some portion must be disposed of. The goals of post-disaster debris management are to reduce risk, facilitate the recovery and reconstruction efforts, and dispose of debris efficiently and in a cost-effective and environmentally sound manner, while keeping final disposal of reusable or salable materials to a minimum.

The management plan must cover collection of waste and a hierarchy of waste disposal options that usually includes: reuse, reduction, recycling, composting, combustion, and land-filling. The plan should also include strong monitoring and regulatory mechanisms, such as controls to prevent and sanction illegal dumping by both households and businesses, a very common occurrence in many countries. The demands of post-disaster debris management may mean that normal operating procedures have to be rapidly expanded or strengthened, even in communities with well-run solid waste management systems. This could include locating additional debris staging and storage areas, contracting out services normally performed "in-house," and/or finding ways to reuse or market debris materials. A comprehensive disaster debris management plan should include the following activities.

Disaster Debris Management Plan Activities

Activity	Considerations
A. Define requirements and management approach	
1. Define roles and responsibilities (national/ local government, public/ private entities, households, and institutions).	Pre-planning of roles and responsibilities significantly speeds the start-up of debris management. The default lead should be the local government, even if other actors are principally responsible for Phase 1 activities. Actors who are likely to be involved include utility companies (water and power), local police, national guard or military, public works and highway agencies, local government, local emergency management agency, the private sector (e.g., contractors and property owners), institutions, households, community and civil society organizations, and volunteers.
2. Identify debris types and forecast amounts.	Take the time to categorize the waste stream in order to properly design the management strategy. Data from prior disasters, sampling, and estimation tools can be used. Identify any toxic or hazardous substances in the debris, such as fiberglass or asbestos. See **Note 1: Identify debris types and forecast amounts**, below.
3. Identify applicable national and local environmental regulations to be followed.	The disaster will already have caused significant environmental damage. Don't compound the problem by ignoring environmental law in the handling the disaster debris. See section on Public Policies Related to Environmental Planning, in this chapter.

Activity	Considerations
4. Inventory current operational, regulatory, and financial capacity and requirements for debris management, including equipment and administrative needs, establish debris-tracking mechanisms.	Identify public and/or private local resources that are available to assist with debris collection and management. Identify local or national contractors that own heavy equipment needed for debris removal and collection, such as bulldozers, dump trucks, skid steer loaders, front end loaders, and logging trucks, and that can provide skilled operators to run the equipment. Analyze the financial resources available for debris management and develop a financial plan, which may include taxes, user fees, donations, and resources from a higher level of government. Debris management costs often exceed estimates and can undermine the financial stability of local agencies. External support is usually required. Tracking should be by weight, volume, and type of debris, and will be useful for control and later reimbursement of costs.
5. Identify activities to be contracted out and agree on contracting approach.	Common areas for contracting include (1) collection, (2) recycling, (3) DMS operation, (4) hazardous waste management, and (5) monitoring of all of the above. Ideally, contracts will have been pre-arranged, companies pre-qualified, and/or contract scope, terms and prices pre-defined. If not, identify contracting mechanisms and procurement rules to be used and agree on areas for contracting and contract types. Types include (1) time and materials (good early on; likely to be more expensive used long-term), (2) unit price (useful when quantities are hard to define), and (3) lump sum (if scope of work is clearly defined). Due to the opportunities for revenue generation and livelihood, consider a preference for community groups or other civil society organizations to carry out contracted debris management services, assuming they demonstrate competency and ability to manage any risks associated with materials handling.
6. Select debris management sites (DMS).	Identify an environmentally safe site between 10 and 50 acres, with good egress and ingress. See **Note 2: Selecting debris management sites**, below.
7. Identify DMS management approach.	DMS management may be done by the public agency or contracted out. Good management will permit the site to be closed and returned to its original use within a reasonable time. A pre-negotiated contract allows a quicker set-up of the DMS and better prices than what might be offered after the disaster. Key contract requirements include (1) provision of a pre-approved site (optional), (2) documentation of all costs and monitoring and auditing of all activities to guard against fraudulent cost claims or diversion of materials, and (3) compliance with all applicable legal requirements including environmental laws.
8. Establish a monitoring and regulatory system.	Good practice dictates the contracting of monitoring, particularly for any contracted services. Private contract terms, waste management behavior of households and businesses, DMS management, and the environmental impact of the plan are some critical areas to monitor.[2] Ensure that regulations and contracts allow violations to be adequately sanctioned.
9. Develop a communications plan.	Communications regarding the Disaster Debris Management Plan must be effective for the audiences for which they are intended. What's important is what people hear, not just what is said, so consultation with target groups regarding the messages should take place before and as communications take place. For guidance on communications in reconstruction settings, see ⛴ Chapter 3, Communication in Post-Disaster Reconstruction.
10. Plan for DMS closure.	Closure of the DMS should be the goal once the post-disaster debris stream returns to manageable volumes and normal composition. If site management is contracted out, the contract should include benchmarks and financial incentives to evaluate and facilitate closure.

B. Develop the debris removal and disposal strategy

Activity	Considerations
1. Design a debris collection system.	Collection options may include one or more the following: (1) curbside collection using existing solid waste and recycling system; (2) additional clearance and collection routes run by agency staff or additional contractors, potentially including specialized contractors to handle volume or for certain types of debris (e.g., hazardous waste, white goods, electronics, or vehicles); and (3) drop-off and exchange locations for debris and recyclables.
2. Establish hazardous materials categories and procedures for hazardous materials and medical waste identification and handling.	Separation is a critical aspect of hazardous waste management. Procedures are directly influenced by the findings of activity A3, above. These wastes are often regulated at the national level. If no regulation applies, refer to international guidelines. See **Note 1: Identify debris types and forecast amounts**, below.
3. Create incentives to encourage household reduction and reuse of waste.	Use the communications plan to promote reuse of building materials and on-site reduction, such as guidelines for salvaging of household items. Do not encourage practices that expose residents to toxins or mold. Take measures to prevent illegal scavenging and resale of private property. Consider a financial incentive so households or community organizations clear local debris.

Activity	Considerations
4. Maximize recycling; identify recycling options and procedures.	Assist recycling systems to scale up if necessary. Allow scavenging of recyclable materials to reduce the waste stream. Offer small businesses access to raw materials, such as trees for saw mills, at reduced or no cost. Publicize safe reuse methods for different types of waste and promote their use. Provide testing if any safety issues exist. Ensure reconstruction guidelines are clear on use of recycled materials to avoid inappropriate and unsafe reuse.
5. Analyze waste-to-energy options.	Unless existing waste-to-energy plants are in operation, this option is unlikely to be employed. Best practice is to have pre-negotiated contracts and prices by type of waste.
6. Identify disposal options and procedures.	Make sure public and private entities understand the range of options and that procedures are widely publicized. Create a hierarchy of disposal options that reduces the waste stream at the source and minimizes the costs and environmental impacts of disposal.
7. Evaluate the open burning option and establish rules.	The risks of burning include fires, pollution from particulate matter, and release of hazardous materials. Establish procedures based on existing rules on burning waste. If post-disaster procedures diverge from existing rules, publicize them as temporary and limit their scope. Requiring permits is an option, but may be difficult to manage in post-disaster circumstances.
8. Investigate options for sale of materials.	Existing commercial markets for sale of glass, metals, wood, and other recyclables of value should respond to the increased materials stream created by the disaster, although temporary storage may be necessary to allow market to "catch up." Promote to potential users options for reuse of materials, such as the use of crushed concrete and glass for roads. Ensure that users are experienced materials handlers and do not expose others to harm.
9. Establish guidelines and secure locations for preservation of historical materials.	Local museums or historical societies may need help with storage and may be able to provide guidelines for the handling or storage of these materials. Monitor informal markets to ensure historical assets are not being scavenged or sold.

Note 1: Identify debris types and forecast amounts.

The categories of waste that will have to be handled after a disaster include the following.

Vegetative Waste: Typically one of the largest volume debris streams. Much can be diverted as lumber, chipping for mulch, composting, or fuel.

Construction and Demolition (C&D) Debris: Large amounts are produced in most disaster events. May be possible to divert by reprocessing for construction, such as crushing concrete for aggregate and reusing brick and stone. Some paving materials, such as asphalt blacktop, can be recycled for road repair. If C&D debris contains asbestos, it must be managed separately and safety practices and personal protective equipment must be used by workers to minimize exposure. Asbestos-containing materials should not be burned. Governments should have regulations or procedures for asbestos removal, handling, and disposal personnel and permits. In their absence, an effort might be made to use international standards, such as those of the USEPA. However, these may be difficult to implement under time pressure and without an adequate institutional framework.[3] See "Managing Asbestos in Housing and Community Reconstruction" text box in this chapter.

Bulky Waste: Material such as carpet, furniture, and mattresses. Usually must be sent for disposal.

Appliances and Electronics: Should be collected separately and component materials recycled.

Vehicles and Boats: Should be inventoried by vehicle identification number (VIN) or license plate number, and held for a reasonable time for reclaiming or insurance purposes, then recycled/crushed, using normal environmental safeguards.

Trash: Household trash volume will decline if people are displaced and will increase if they return and dispose of damaged household items. Household collection service may need to be increased at that time.

Soils and Sediments: High rainfall and flooding can produce large quantities of soil and sediments. These may be contaminated, containing bacteria or toxins; testing is advised. Workers around flood waters and sediments may require safety practices and personal protective equipment to minimize exposure.

Business and Household Hazardous Waste: Manage and dispose of these wastes separately. If the normal household hazardous waste collection system is good, simply ramp it up; otherwise, disposal procedures should be established and communicated and a qualified contractor hired to oversee them. Businesses should be responsible for managing their own hazardous wastes, if adequate systems are in place, although small business hazardous waste may be handled with household waste. If systems are inadequate, government will have to establish arrangements for handling these materials, including industrial chemicals and other industrial inputs and wastes, paints, solvents, underground storage tanks, etc. If tracking systems exist for hazardous wastes, do not let them lapse in the post-disaster environment. Consider a special charge for this service if it will not unduly discourage responsible handling by producers, since disposal costs may be high. **Putrescible wastes:** This includes fruits, vegetables, meats, dairy products, and other produce from grocery stores, restaurants, institutions, and residences. It can also include animal carcasses. These rot or decay quickly and should be segregated accordingly and quickly managed. Some putrescible wastes can be composted or rendered. More information about composting food and other putrescible wastes can be found at USEPA's Food Waste Recovery Hierarchy Web site.[4]

Infectious/Medical Waste: In certain disasters, there will likely be large amounts of infectious and medical waste, as well as human bodies. These materials require special handling and management, and a major effort to keep them separate from other trash. National standards should exist; if they don't, procedures should be quickly set up based on international guidelines.[5] Workers exposed to this material should wear personal protective equipment to protect against infectious agents. Incineration of these wastes is often the best disposal solution.

Forecasting debris quantities. Models that can be used for forecasting debris quantities for a local area include the United States Army Corps of Engineers (USACE) Hurricane Debris Prediction Model.[6] The calculation and its parameters are as follows:

Q = H (C) (V) (B) (S)   where:

Q = estimated debris total generated in cubic yards
(Note: The predicted accuracy of the model is ±30%)

H = number of households, or population/3 (average household size is 3)

C = hurricane category factor (cat1 = 2, cat2 = 8, cat3 = 26, cat 4 = 50, cat5 = 80)

V = density of vegetation (1.1 for light, 1.3 for medium, 1.5 for heavy)

B = percentage of commercial structures (1.0 for light, 1.2 for medium, 1.3 for heavy)

S = precipitation factor (1.0 for none to light, 1.3 for medium to heavy)

Note 2: Selecting debris management sites

If no site has been identified for debris disposal before the disaster, use GIS information or land records to identify a large open space, generally between 10 and 50 acres, depending on results of debris stream analysis. If properly managed, the site can be closed or returned to its prior use once all materials are disposed of. At a minimum, the following site characteristics, should be considered when selecting the DMS: (1) publicly owned land; (2) good ingress and egress with room for scale, (3) relatively flat topography; (4) location near final disposal sites to reduce hauling distances, if possible; (5) can accommodate separation and reduction of types of debris and capacity for debris operations, such as chipping, grinding, crushing, burning, and recycling; (6) minimal effect on residential neighborhoods, educational facilities, or health care facilities; (7) no impact on environmentally sensitive areas, such as wetlands, endangered species, rare ecosystems, or other areas with environmental restrictions or on historic or archaeological sites.

Before being put into use, the DMS should be equipped with (1) fencing surrounding the site; (2) a scale and/or other means of registering weights and quantities; (3) signage and security measures to limit unauthorized access; (4) fire control equipment; storm-water controls to prevent discharge of contaminated runoff into water bodies; (5) controls to prevent migration of dust, wood chips, or other debris from both haulers and handling of debris on the site; (6) clearly marked sorting, staging, and processing areas for all categories of waste; and (8) monitors to correctly identify and segregate waste types.

Annex 1 Endnotes

1 . State of Connecticut, 2008, "Disaster Debris Management Plan," September 2008 (Annex to the State Natural Disaster Plan, 2006), State of Connecticut Department of Environmental Protection, http://www.ct.gov/dep/lib/dep/waste_management_ and_disposal/debris_management/final_ddmp_plan_september_2008_(pdf).pdf; U.S. Environmental Protection Agency (USEPA), 2008, "Planning for Natural Disaster Debris Guidance," USEPA, Office of Solid Waste and Emergency Response, http://www. epa.gov/osw/conserve/rrr/imr/cdm/pubs/pndd.pdf; California Waste Management Board, "Disaster Preparedness and Response," http://www.ciwmb.ca.gov/Disaster/ Links.htm and Integrated Waste Management Disaster Plan, http://www.ciwmb. ca.gov/Disaster/DisasterPlan/; USEPA, "Disaster Debris," http://www.epa.gov/ epawaste/conserve/rrr/imr/cdm/debris.htm; *and* Federal Emergency Management Agency, 2007, "Public Assistance Debris Management Guide," FEMA-325, FEMA, http:// www.fema.gov/pdf/government/grant/pa/demagde.pdf.

2 . Numerous sample contracts for post-disaster debris management and monitoring are available on the Internet, for example: http://iaemeuropa.terapad.com/ resources/8959/assets/documents/SAMPLE%20DEBRIS%20MANAGEMENT%20 PLAN.pdf; http://www.barkerlemar.com/organicmanagement/resources_loader. aspx?ID=57; http://www.nctcog.dst.tx.us/envir/SEELT/disposal/DDM/docs/TAB_I_ Debris_Monitoring_Scope_of_Services.pdf; and http://sema.dps.mo.gov/Debris%20 Management%20&%20Public%20Assistance/Example%20Locals%20Tonnage%20 Debris%20Contract.pdf.

3. USEPA, "Asbestos in Demolition and Renovation," http://yosemite.epa.gov/R10/OWCM. NSF/webpage/Asbestos+in+Demolition+and+Renovation.

4. USEPA, http://www.epa.gov/epawaste/conserve/materials/organics/index.htm.

5. California Integrated Waste Management Board, 2007, "Receipt of Medical Waste at Solid Waste Facilities and Operations," http://www.ciwmb.ca.gov/publications/ facilities/23206006.pdf.

6. U.S. Army Corps of Engineers Hurricane Debris Estimating Model, http://www.gema. state.ga.us/ohsgemaweb.nsf/1b4bb75d6ce841c88525711100558b9d/f715ec607d3bd dc6852571e30055c99a/$FILE/Appendix%20A.pdf.

How to Do It: Carrying Out Environmental Impact Assessment and Environmental Monitoring of Reconstruction Projects

Conducting an Environmental Impact Assessment

Environmental impact assessment (EIA) is the process of identifying, predicting, evaluating, and identifying and selecting options for mitigating the biophysical, social, and other relevant effects of proposals for development projects prior to finalizing project designs and commitments. EIAs are required in some form and for some types of projects in nearly all countries, although the specifics of what is required vary.

Frameworks for Environmental Impact Assessment

Each country has its own environmental assessment requirements that are applied at the project level, although there may be pressure to suspend them in a post-disaster environment. Environmental ministries generally promulgate and oversee environmental regulations under environmental laws, the implementation of which is sometimes delegated to lower levels of government. Reconstruction policy should define the environmental framework to be applied in reconstruction. The World Bank also defines what it requires in the projects it finances, but this will generally not replace local environmental review requirements (although the Bank may in some cases accept country procedures as a substitute for its own). See 📖 Chapter 21, Safeguard Policies for World Bank Reconstruction Projects, for a description of World Bank requirements.

The content and organization of the framework for environmental management varies from one country to another and from one region to another.

- In China, the Environmental Impact Assessment Law requires an EIA prior to project construction. However, if a developer ignores this requirement, the only penalty is that the Environmental Protection Bureau may require the developer to do a make-up environmental assessment. This lack of enforcement has resulted in a significant percentage of projects not completing EIAs prior to construction. However, China's State Environmental Protection Administration has used the legislation to halt projects, including three hydro-power plants under the Three Gorges Project Company in 2004.
- In India, the Ministry of Environment and Forests of India has been involved in promoting the EIA process. The main national laws are the Water Act (1974), the Indian Wildlife (Protection) Act (1972), the Air (Prevention and Control of Pollution) Act (1981), and the Environment (Protection) Act (1986). The responsible body is the Central Pollution Control Board.[1]
- The European Union provides separate guidelines for environmental assessment that is undertaken for individual projects, such as a dam, motorway, airport, or factory ("Environmental Impact Assessment") or for plans, programs, and policies ("Strategic Environmental Assessment").[2]

- International standards may also be called for or required in certain situations, such as the International Organization for Standardization (ISO) 14000[3] or the Convention on Environmental Impact Assessment in a Transboundary Context (Espoo Convention).[4]

Objectives of Environmental Impact Assessments

- To ensure that environmental considerations are explicitly addressed and incorporated into the development decision-making process
- To anticipate and avoid, minimize, or offset the adverse significant biophysical, social, and other relevant effects of development proposals
- To protect the productivity and capacity of natural systems and the ecological processes that maintain their functions
- To promote development that is sustainable and optimizes resource use and management opportunities

EIA Principles and Scope

The EIA process should be applied[5]:

- as early as possible in decision making and throughout the life cycle of the proposed activity;
- to all development projects that may cause potentially significant effects;
- to biophysical impacts and relevant socioeconomic factors, including health, culture, gender, lifestyle, age, and cumulative effects consistent with the concept and principles of sustainable development;
- to provide for the involvement and input of communities and industries affected by a project, as well as the interested public; and
- in accordance with internationally agreed measures and activities.

The environmental resources that may be affected by a project will vary by sector.[6] Many environmental agencies develop checklists or guidelines that apply to projects in specific sectors. In housing and community reconstruction, environmental impacts may result from (1) demolition, (2) site preparation and development, (3) building and infrastructure construction, and (4) occupancy of the site once developed. A general list of the resources to be evaluated includes the following.[7]

(i) Physical Resources
  - Atmosphere (e.g., air quality and climate)
  - Topography and soils
  - Surface water
  - Groundwater
  - Geology/seismology

(ii) Ecological Resources
- Fisheries
- Aquatic biology
- Wildlife
- Forests
- Rare or endangered species
- Protected areas
- Coastal resources

(iii) Economic Development
- Industries
- Infrastructure facilities (e.g., water supply, sewerage, flood control)
- Transportation (e.g., roads, harbors, airports, navigation)
- Land use (e.g., dedicated area uses)
- Power sources and transmission
- Agricultural development, mineral development, and tourism facilities

(iv) Social and Cultural Resources
- Population and communities (e.g., numbers, locations, composition, employment)
- Health facilities
- Education facilities
- Socioeconomic conditions (e.g., community structure, family structure, social well-being)
- Physical or cultural heritage
- Current use of lands and resources for traditional purposes by indigenous peoples
- Structures or sites of historical, archeological, paleontological, or architectural significance

EIA processes generally provide for the following steps or elements.

Elements of an Environmental Impact Assessment Process[8]

Screening	To determine whether or not a proposal should be subject to EIA and, if so, at what level of detail.
Scoping study or initial assessment	To identify the issues and impacts that are likely to be important and to establish terms of reference for EIA or other environmental assessment.
Examination of alternatives	To establish the preferred or most environmentally sound and benign option for achieving proposal objectives.
Impact analysis	To identify and predict the likely environmental, social, and other related effects of the proposal. In most environmental policy frameworks, projects are categorized at this stage by their potential environmental impact (Category A, B, or C), and this category determines the scope and content of the EIA or other environmental assessment that is required.
Mitigation and impact management	To establish the measures that are necessary to avoid, minimize, or offset predicted adverse impacts and, where appropriate, to incorporate these into an environmental management plan or system.
Evaluation of significance	To determine the relative importance and acceptability of residual impacts (i.e., impacts that cannot be mitigated).
Preparation of environmental impact statement (EIS) or report statement	To document clearly and impartially impacts of the proposal, the proposed measures for mitigation, the significance of effects, and the concerns of the interested public and the communities affected by the proposal.
Review of the EIS	To determine whether the report meets its terms of reference, provides a satisfactory assessment of the proposal(s), and contains the information required for decision making.
Decision making	To approve or reject the proposal and, if approved, to establish the terms and conditions for its implementation.
Follow-up	To ensure that the terms and condition of approval are met; to monitor the impacts of development and the effectiveness of mitigation measures; to strengthen future EIA applications and mitigation measures; and, where required, to undertake environmental audit and process evaluation to optimize environmental management.

Monitoring, evaluation, and management plan indicators should be designed so that they contribute to local, national, and global monitoring of the state of the environment and sustainable development.

Initial Environmental Assessment
The initial assessment (IA) is an important tool for incorporating environmental concerns at the time of initial project planning. It should be carried out as early as the project planning stage as part of feasibility so that it can ensure that the project will be environmentally feasible. The IA is conducted if the project is likely to have minor or limited impacts, which can easily be predicted and evaluated and for which mitigation measures are prescribed easily. The IA is also used to confirm whether a more extensive EIA is required.

The IA study should provide the following information:
- General environmental settings of the project area, including baseline data
- Potential impacts of the project and the characteristics of the impacts, magnitude, distribution; who will be the affected group; and the duration of the impacts
- Potential mitigation measures to minimize the impact, including mitigation costs
- The best alternative project with the potential for greatest benefit at least cost in terms of financial, social, and environment; it is not always necessary to change location of the project, but it can be changed in project design or project management
- Information for formulating a management and monitoring plan

If the IA determines that an full EIA is required, the assessment is conducted in more detail, focusing on the issues identified in the initial assessment. Mitigation measures are then defined, depending on the findings of the EIA.

The environmental assessment should analyze not only the impact of the project and their corresponding mitigation measures, but also the potential impact and mitigation measures for the construction activities, including traffic impacts, air pollution, noise pollution, and management of runoff or other potential contamination from the construction activities.

Mitigation Plan

The EIA should identify feasible and cost-effective measures that may reduce potentially significant adverse environmental impacts to acceptable levels. The plan includes compensatory measures if mitigation measures are not feasible, cost-effective, or sufficient. Specifically, the EIA should:

- identify and summarize all anticipated significant adverse environmental impacts (including those involving indigenous people or involuntary resettlement);
- describe—with technical details—each mitigation measure, including the type of impact to which it relates and the conditions under which it is required (e.g., continuously or in the event of contingencies), together with designs, equipment descriptions, and operating procedures, as appropriate;
- estimate any potential environmental impacts of these measures; and
- provide linkage with any other mitigation plans (e.g., for involuntary resettlement, indigenous peoples, or cultural property) required for the project.

Outline of Environmental Impact Assessment Report

A. Introduction
B. Description of the Project
C. Description of the Environment
D. Potential Environmental Impacts and Mitigation Measures
E. Institutional Requirements and Environmental Monitoring Plan
F. Public Consultation and Information Disclosure
G. Findings and Recommendation
H. Conclusions

Developing an Environmental Monitoring Plan[9]

A project's EMP consists of the set of mitigation, monitoring, and institutional measures to be taken during implementation and operation to eliminate adverse environmental and social impacts, offset them, or reduce them to acceptable levels. The plan also includes the actions needed to implement these measures. The content of the management plan is based on the results of the EIA, on the project design documents, and on any other regulations that apply. Another important objective of the EMP is to ensure that the mitigation measures and monitoring requirements approved during the environmental review are actually carried out in subsequent stages of the project.

An EMP for a housing or infrastructure reconstruction project should address the impact of the project on:

- the environment;
- the existing surrounding communities; and
- those who will take up residence at the site.

If a project is being built in phases, there may need to be EMPs for different phases, or the EMP may need to be updated as the project progresses.

To support timely and effective implementation of environmental project components and mitigation measures, the EMP draws on the EIA to:

- identify the principal and alternative responses to potentially adverse impacts;
- determine requirements for ensuring that those responses are made effectively and in a timely manner; and
- describe the means for meeting those requirements.

An EMP for a construction project should include the components and subcomponents described below.[10]

Environmental Management Structure and Procedures

The EMP should describe the following.

- The organization chart of the project management and the management responsibilities and lines of authority, including those for environmental management; if necessary, the EMP should recommend the hiring of outside consultants or other measures to strengthen the environmental management capacity of project management, such as the training of staff, in order to ensure implementation of EIA recommendations
- The permits and licenses that will be acquired for the project and assign responsibility for compliance with any conditions
- The measures that will be taken on the site to manage potential environmental impacts on any of the resources identified in the EIA
- The measures that will be taken on the site to manage environmental impacts from demolition and construction, such as noise, water, and air pollution
- Procedures for dealing with accidents or other unexpected environmental events that affect any of the resources analyzed in the EIA or with unexpected resources or contaminants found on the site during demolition and construction

Monitoring and Auditing

Environmental monitoring and auditing during project implementation provide information about key environmental aspects of the project, particularly the environmental impacts of the project and the effectiveness of mitigation measures. Such information enables the project sponsor to evaluate the success of mitigation as part of project supervision and allows corrective action to be taken when needed.

For access to additional resources and information on this topic, please visit the handbook Web site at www.housingreconstruction.org.

The EMP identifies monitoring objectives and specifies the type of monitoring, with linkages to the impacts and the mitigation measures identified in the EIA, including:

- a specific description, and technical details, of monitoring measures, including the parameters to be measured, methods to be used, sampling locations, frequency of measurements, detection limits (where appropriate), and definition of thresholds that will signal the need for corrective actions; and
- monitoring and reporting procedures to provide early detection of conditions that necessitate particular mitigation measures and information on the progress and results of mitigation.

Implementation Schedule and Cost Estimates

With respect to implementation, the EMP should provide:

- a description of the works to be undertaken as part of the project;
- an implementation schedule for the works;
- a schedule of environmental management measures that will be carried out as part of the project, showing phasing and coordination with overall project implementation plans; and
- capital and recurrent cost estimates and sources of funds for implementing the EMP.

The costs of implementing the EMP should be incorporated into the total project cost estimate to ensure that they are provided for as part of project financing.

Annex 2 Endnotes

1. India, Ministry of Environment and Forests, "Role of EIC in Environmental Impact Assessment India," http://www.eicinformation.org/internal.asp?id=14&type=normal&title=Environmental+Impact+Assessment.
2. European Union, "Environmental Assessment," http://ec.europa.eu/environment/eia/home.htm.
3. International Organization for Standardization, "ISO 14000 Essentials," http://www.iso.org/iso/iso_catalogue/management_standards/iso_9000_iso_14000/iso_14000_essentials.htm.
4. United Nations Economic Commission for Europe, "Convention on Environmental Impact Assessment in a Transboundary Context," http://www.unece.org/env/eia/welcome.html.
5. International Association for Impact Assessment, 1999, "Principles of Environmental Impact Assessment Best Practice," http://www.iaia.org/publicdocuments/special-publications/Principles%20of%20IA_web.pdf.
6. For example, the U.S. National Park Service has guidelines for the assessment of potential sources of environmental liability associated with real property. U.S. National Park Service, 1999, "Pre-Acquisition Environmental Site Assessment Guidance Manual," http://www.nps.gov/policy/DOrders/ESAGuidance.pdf.
7. Asian Development Bank, "Content and Format: Initial Environmental Examination (IEE)," http://www.adb.org/documents/Guidelines/Environmental_Assessment/Content_Format_Initial_Environmental_Examination.pdf.
8. International Association for Impact Assessment, 1999, "Principles of Environmental Impact Assessment Best Practice," http://www.iaia.org/publicdocuments/special-publications/Principles%20of%20IA_web.pdf.
9. World Bank, 1999, "Operational Policy 4.01, Annex C: Environmental Management Plan," http://go.worldbank.org/B06520UI80.
10. Red Tree, 2009, "Chapter 4: Outline Construction Environmental Management Plan," http://www.redtreellp.com/downloads/Masterplan%20Book/chapter%204bvii.pdf.

10 HOUSING DESIGN AND CONSTRUCTION TECHNOLOGY

Guiding Principles for Housing Design and Construction Technology

- The housing designs and construction technologies (HDCTs) used in reconstruction (there may be several) should be selected by taking into consideration local building practices, desired standards, culture, and economic and climatic conditions.
- The HDCTs used in reconstruction may affect prices and supply in the building materials market; interventions may be needed.
- Local expertise is invaluable in selecting HDCTs, but if changes are needed to improve resilience, builders should be supported by training, and their expertise augmented by global knowledge and best practices.
- A structure's entire life span, from construction through maintenance to eventual demolition or reuse, should be considered in evaluating the suitability of technology options.
- Repairing and retrofitting partially damaged houses are legitimate alternatives to full reconstruction but deserve similar attention and assistance to improve their resilience.

This Chapter Is Especially Useful For:

- Lead disaster agency
- Local officials
- Agencies involved in reconstruction
- Project architects, engineers, and chartered surveyors
- Project managers
- Affected communities

Introduction

When a disaster affects housing, there are important choices to be made in the rebuilding effort related to the design and construction technology to be employed and whether to repair or retrofit housing as opposed to demolishing it. These choices must take into account environmental, economic, social, institutional, and technical factors. The size and scale of the project as well as the geographic concentration of the affected area also play a significant role in the decision-making process. Ignoring these factors or making the wrong decisions about them can significantly affect whether or not stakeholders are satisfied with the reconstruction and whether or not the resulting housing solution is sustainable.

This chapter is particularly relevant for stakeholders responsible for the design, construction, and retrofitting of houses. It covers the three principal subjects related to efforts to rebuild housing after a disaster: design, construction technology, and the decision whether to repair or retrofit versus demolish. The chapter provides guidance to help practitioners make decisions that result in the most appropriate solutions. All considerations and recommendations developed here are relevant for both urban and rural contexts, but they need to be adapted to the context to ensure the appropriateness of a given building technology in the urban or rural environment.

Key Decisions

1. The **lead disaster agency** should select and engage a multidisciplinary team of experts, which may include experts from outside the country, to analyze the disaster impact on common HDCTs and help select the HDCTs to be used in reconstruction.
2. The **lead disaster agency**, having decided on HDCTs for reconstruction, must ensure that they are fully integrated into the reconstruction policy, including the housing financial assistance scheme, and must determine how to ensure that the norms and standards are uniform across the disaster area.
3. The **lead disaster agency** must decide the conditions under which repairing or retrofitting will be promoted as an alternative to full reconstruction.
4. The **lead disaster agency** should decide on and implement a range of mechanisms to fully involve **local governments, local communities**, and the **building industry** in decision making regarding HDCTs and in implementation of reconstruction.
5. **Agencies involved in reconstruction** should decide how to conform to the HDCT standards set by the **lead disaster agency**, including those for repairing and retrofitting of partially damaged houses, if it is agreed that they are appropriate approaches.

6. **Agencies involved in reconstruction** should decide jointly how the choice of HDCTs affects the need for training and should cooperate to ensure that quality training is available.
7. **Agencies involved in reconstruction** should decide, while planning their programs, how to lower the environmental impact of reconstruction.
8. **Agencies involved in reconstruction** should decide how to manage the impact of design and technology on building materials market, if necessary.

Public Policies related to Housing Design and Construction Technology

Building codes, if they exist, are the principal public policy instrument that governs choices regarding HDCTs. Countries that have recently updated building codes may have incorporated into them emerging policy objectives, such as energy efficiency, reduction of the environmental impact of building materials and construction technologies, or use of universal design that makes buildings accessible to those with differential abilities (see box). Where building codes do not exist, or are not adequate, they can potentially be updated for purposes of carrying out reconstruction, although the time required for designing, consulting with the public on, approving, and developing regulations for implementation of building codes can easily hold up reconstruction schedules.

A more practical approach may be to establish standards and guidelines for safety, comfort, and environmental impact for use during the reconstruction and repair program, to adjust them as reconstruction and repair work proceeds, and to use them as the basis for establishing or updating the building codes once the reconstruction program is completed. Whether the decision is to update building codes or to develop standards and guidelines, it is critical to involve building industry professionals, such as architects, engineers, builders, and chartered surveyors when developing specifications and codes.[1]

Universal Design

Universal design entails the design of products and environments to be usable by all people, to the greatest extent possible, without the need for adaptation or specialized design. The intent of universal design is to simplify life for everyone by making products, communications, and the built environment more usable by as many people as possible at little or no extra cost. Universal design benefits people of all ages and abilities.

Principle One: Equitable Use. The design is useful and marketable to people with diverse abilities.

Principle Two: Flexibility in Use. The design accommodates a wide range of individual preferences and abilities.

Principle Three: Simple and Intuitive Use. Use of the design is easy to understand, regardless of the user's experience, knowledge, language skills, or current concentration level.

Principle Four: Perceptible Information. The design communicates necessary information effectively to the user, regardless of ambient conditions or the user's sensory abilities.

Principle Five: Tolerance for Error. The design minimizes hazards and the adverse consequences of accidental or unintended actions.

Principle Six: Low Physical Effort. The design can be used efficiently and comfortably and with a minimum of fatigue.

Principle Seven: Size and Space for Approach and Use. Appropriate size and space is provided for approach, reach, manipulation, and use regardless of user's body size, posture, or mobility.

Source: The Center for Universal Design, 1997, *The Principles of Universal Design,* Version 2.0. Raleigh, NC: North Carolina State University. Copyright © 1997 NC State University, The Center for Universal Design.

Technical Issues and Recommendations: Housing Design

Housing design involves the form, dimensions, orientation, natural lighting, ventilation, and spatial organization of dwellings. There is no "ready-made" solution for housing design in reconstruction. Careful and contextualized integration of many issues determine whether or not a rebuilt house's stakeholders, most importantly, its inhabitants, are satisfied. The table below contains several of the issues involved in housing design, how the issue is relevant, and recommendations for designing the most suitable option.

1. See Royal Institution of Chartered Surveyors (RICS), 2009, *The Built Environment Professions in Disaster Risk Reduction and Response Guide* (London: University of Winchester), http://www.rics.org/site/scripts/download_info.aspx?downloadID=829&fileID=991.
2. Gujarat State Disaster Management Authority (GSDMA), 2003, *Guidelines for Construction of Compressed Stabilized Earthen Wall Buildings* (Gujarat: GSDMA), http://www.gsdma.org.
3. See International Code Council, "Codes and Standards," http://www.iccsafe.org/cs/.
4. Paul Gut and Dieter Ackerknecht, 1993, *Climate Responsive Buildings, Appropriate Building Construction in Tropical and Subtropical Regions* (St. Gallen: SKAT), http://www.nzdl.org/fast-cgi-bin/library?e=d-00000-00---off-0envl--00-0----0-10-0---0direct-10---4-----stt--0-1l-11-en-50---20-about-climate+responsive+building--00-0-1-00-0-0-11-1-0utfZz-8-10&cl=search&d=HASH7fb3fd71d302d3efdfe64e&gc=1.
5. Roland Stulz and Kiran Mukerji, 1993, *Appropriate Building Materials, A Catalogue of Potential Solutions (Revised)* (St. Gallen: SKAT), http://nzdl.sadl.uleth.ca/cgi-bin/library?e=d-00000-00---off-0cdl--00-0----0-10-0---0---0direct-10---4------0-1l--11-en-50---20-about--00-0-1-00-0-0-11-1-0utfZz-8-00&cl=CL2.1&d=HASH51495f31 4e8d35f51533d4.2&gc=1.

Issue	Relevance	Recommendations
Town, settlement, territory, land, planning	Planning criteria determine position, size, function, form, and materials of the house and the relation between buildings and infrastructure.	Modify, improve, or obtain an exemption for elements of the proposed plan that hinder implementation of sustainable housing solutions.
Policies, guidelines, building codes, standards, strategies	Existing documentation may not provide appropriate instructions.	Identify and suggest possible improvements (hazards, environmental impact, socio-cultural aspects, flexibility, etc.). Propose guidelines and standards for new alternative technologies that provide more appropriate solutions,[2] not only for use in the reconstruction period, but covering the needs of further long-term housing development.[3]
Infrastructure and community services	Water supply, drainage, treatment, sanitation, access roads, energy supply, communication systems, and community services directly influence housing design.	Ensure housing design is consistent with infrastructure plan so that all necessary services are provided (either in the community or in the individual house) and are not redundant. Examples: sanitation systems provide for local and/or community treatment of sewage; kitchen design accommodates available energy source for cooking.
Beneficiaries' needs, social structure, culture, livelihoods, aspirations	Social structure determines spatial organization and size; culture affects forms, function, and aesthetics; livelihoods dictate spatial organization, morphology, size, land use; community's aspirations determine the "housing standard."	Ensure intense community participation in the design and decision-making process (house size, morphology, spatial organization, functions, form, position on the plot). Example: houses without verandas or shading areas in hot climates affect the social structure by not providing gathering places for social interaction.
Climatic conditions	Indoor conditions must be within the human comfort zone, which varies according to population's culture, apparel, and activities. The main function of a house with respect to climate is to protect against and take advantage of the climatic conditions.	Design the house and landscape to take advantage of the climate and reduce the demand for operating energy: sun/shadow exposure, solar shading, thermal insulation, passive solar energy, solar hot water, photovoltaic electricity, rain water collection, wind ventilation system, etc. Consider biodiversity enhancement as a tool for improving the local climatic conditions. Example: trees are essential for improving indoor and outdoor conditions in hot climates and can help reduce the impact of wind, soil erosion, and solar radiation.[4]
Need for flexibility, modular design, expandability, incremental housing	As a family grows, the needs of space and functions change; a house needs to adapt to these changes. Housing and public buildings should be accessible to all (see box, above, on universal design).	Incorporating flexibility, modular design, and expandability in the housing design and concept will make those operations easier and cheaper to carry out when necessary. Incremental housing provides a basic house structure, allowing the users to complete it according to their will and means. Universal design principles reduce the barriers to use and movement by the handicapped and elderly.
Environmental impact	Worldwide, the housing sector has a huge environmental impact, contributing substantially to the deterioration of the local environment and natural resources. See 📖 Chapter 9, Environmental Planning, for a detailed discussion of environmental issues in reconstruction.	Study vernacular architecture and tradition; they are the best reference for developing new designs that lessen environmental impact. Assess environmental impact over the entire life span of a house. Employ basic rules for low environmental impact design: land use that respects and safeguards the soil and biodiversity; simple and reasonable design and size limits that minimize the quantity of building materials and the house's energy requirements; and use of building materials with low environmental impact.[5] In regions under water stress, incorporate rainwater-harvesting systems.
Cost	The entire life span of the house, not just the construction phase, determines the true cost of a design option; higher initial construction cost may lower the life span cost.	Consider the cost of upkeep as well as initial investment. Include materials transport cost. Use an appropriate factor to discount future costs. Design a house that facilitates future expansion (or reduction); it will reduce modification costs. Limit the needs of operating energy through the design; heating and cooling costs may force inhabitants to forego comfort.

Issue	Relevance	Recommendations
Exposure to risks and hazards	Improving a house's physical resistance to hazards is an essential element of risk reduction and disaster preparedness.	Limit a house's vulnerability to hazards through its design elements, especially form, dimension, morphology, and detailing.[6] Identification and analysis of a house's vulnerability should be observed so that improved structures can be designed. Consider not only the risk of the particular disaster, but the risks from other possible hazards.
Available construction technologies and building materials	Housing design may be influenced by the construction technology and materials and vice versa. The ⛏ case study on the 2003 Bam earthquake reconstruction, below, explains how demonstration buildings were used at an exhibition site to show locally appropriate construction technologies and materials.	When possible and appropriate, use traditional technologies. They often provide the most appropriate solutions by integrating costs, climate, culture, and technical capacity. When possible and appropriate, adapt traditional solutions by integrating modern technologies. Assess and factor into the design the availability of local material and manpower, especially after a large-scale disaster. In many cases, reuse and recycling of debris can be an alternative material source; however, measures may be needed to store, sort, and reprocess rubble.
Relation with the built heritage	A house's form, size, and construction material has a visual impact on the environment, and its relation with nearby historical and vernacular elements affects an area's overall architectural quality.	Observe and carefully consider the existing built environment in designing new dwellings; incorporate its context into the design.

Technical Issues and Recommendations: Construction Technology

6. Andrew Chalesson, 2008, *Seismic Design for Architects* (Oxford: Architectural Press), http://www.elsevier.com/wps/find/bookdescription.cws_home/716362/description#description.

Construction technology involves the choice of building materials and the technique and means used to erect a house. As with the housing design process, cautious consideration of contextual conditions is crucial to developing appropriate construction technologies. In addition, any selected technology must be constantly reviewed and, if necessary, upgraded during the construction process. The following criteria can be used to compare various construction technologies and identify the most suitable technology options.

Issue	Relevance	Recommendations
Policies, guidelines, building codes, standards, strategies	Existing documentation may not provide appropriate instructions. The ⛏ case study from Kenya on building code reform, below, describes how the building code was updated to reflect popular housing construction practices and designs.	Identify and suggest possible improvements (hazards, environmental impact, socio-cultural aspects, flexibility, etc.). Propose guidelines, standards, and building codes for new alternative technologies that provide more appropriate solutions.[7] Once the guidelines to be followed have been agreed on, use them as a tool to unequivocally determine which technical solutions can be applied and which cannot. Carefully and systematically monitor compliance with guidelines and standards.
Housing design	Housing design influences the choice of construction technology and materials.	Ensure that the physical characteristics and limits of a particular technology are coherent with the design. Example: the size of a room can determine the choice of the roofing technology; a big room may not allow for the use of locally available wood for the roof.
Availability of construction materials	Indigenous materials—unlike those imported from outside—support the local economy and livelihoods.	To the greatest extent possible, use indigenous materials, unless the scale of the disaster, the origin of the materials, and/or the available transportation hinder access to local materials. Use the materials from demolished houses as much as possible.[8]
Costs: materials technology	Local and abundant construction materials reduces transportation costs and limits price inflation of alien materials. Technology easily adopted by local builders limits the expensive involvement of external skilled manpower or contractors. Local technologies and materials that are durable and inexpensive to maintain reduce long-term maintenance costs.	Reduce both immediate and long-term costs by using local materials and technologies that are abundant, easily adopted, affordable, durable, and easily maintained). Save costs by using materials and technologies that can be easily dismantled, demolished, and recycled.[9] Establish the cost/benefit and comparative analysis of building materials considering the quantitative needs and availability, with consideration of possible inflation risks.

⚒ ⚒ *SAFER HOMES, STRONGER COMMUNITIES: A HANDBOOK FOR RECONSTRUCTING AFTER NATURAL DISASTERS*

Issue	Relevance	Recommendations
Exposure to risks/hazards	Engineers and architects are not always trained in the use of recent developments in the engineering of hazard-resistant structures. Contextual conditions guide the choice of appropriate solutions.	Mitigate risks by merging modern technology components with traditional construction practices and improving existing traditional practices. Carefully adapt unfamiliar solutions to the contextual conditions of every situation.[10]
Construction speed	Shortages of materials and manpower can drastically slow reconstruction.	Provide training to increase the number of skilled builders. Use a large number of people to construct houses rather than a few specialists, because more can be built concurrently. An owner-driven reconstruction approach, combined with the promotion of upgraded indigenous technology, generally allows for faster reconstruction. See Chapter 6, Reconstruction Approaches, for more information.
Climatic conditions, indoor comfort, operating energy needs	Thermal transmission, thermal storage, and vapor diffusion of materials play a large role in determining a house's thermal comfort and its energy consumption.	Select building materials by considering their impact on indoor comfort to ensure an appropriate climate-responsive house.
Socio-cultural appropriateness, acceptance	Technology and building materials influence a community's way of life. People may want modern imported technologies because of the social status they confer rather than improved traditional solutions that are more appropriate and far cheaper to maintain.	Help communities make appropriate decisions by demonstrating how to analyze advantages and disadvantages of materials and technologies and their relevance to the social and cultural context. Combinations of materials could be a good option to enhance the acceptance of alternative technologies (e.g., one room [core house] made of brick and concrete and the rest made of bamboo, as experimented with in Bihar).
Environmental impact (including transportation, maintenance, and demolition and recycling possibilities)	Certain technologies and materials can substantially contribute to the deterioration of both the local and the global environment and natural resources. See Chapter 9, Environmental Planning, for a detailed discussion of environmental issues in reconstruction.	Whenever possible, use: ■ locally available, low-energy-consumption building materials, especially those produced with renewable energy sources; ■ materials from sustainable production chains (e.g., avoid use of timber from savage deforestation); ■ non-toxic materials; ■ materials easily dismantled (and recyclable as building materials or energy sources); and ■ in regions under water stress, materials that require minimum amounts of water (including the curing, drying, and maintenance processes).[11]
Availability and capacity of local skills	The quality of construction depends on manpower skills. Skilled manpower from other regions is likely to migrate to the disaster area.	Address a skilled manpower shortage with proper training, management, and monitoring. Technical instruction should be provided to builders at all skill levels. Construction quality must be monitored and documented through systematic quality control procedures. See Chapter 16, Training Requirements in Reconstruction.
Opportunities for participation and livelihoods	Traditional methods and materials are generally easier for local people to implement and replicate. The feasibility of community participation in the reconstruction phase is largely determined by the technology being applied. When local artisans understand what the problem is, they can often devise appropriate solutions.	Train and monitor local laborers regarding new components, such as earthquake-resistance features and imported technologies. Use model houses to teach improved technologies. Devise simple measures to test resistance in the field. Assimilate new technologies in a community with long-range measures that ensure their replicability beyond the reconstruction period.

7. GSDMA, 2003, *Guidelines for Construction of Compressed Stabilized Earthen Wall Buildings* (Gujarat: GSDMA), http://www.gsdma.org; and National Information Centre of Earthquake Engineering, "IITK-GSDMA Project on Review of Seismic Codes & Preparation of Commentary and Handbooks," http://www.nicee.org/IITK-GSDMA_Codes.php.
8. See Chapter 9, Environmental Planning.
9. Roland Stulz and Kiran Mukerji, 1993, *Appropriate Building Materials, A Catalogue of Potential Solutions (Revised)* (St. Gallen: SKAT), http://nzdl.sadl.uleth.ca/cgi-bin/library?e=d-00000 00---off-0cdl--00-0----0-10-0---0---0direct-10---4-------0-1l--11-en-50---20-about---00-0-1-00-0-0-11-1-0utfZz-8-00&cl=CL2.1&d=HASH51495f314e8d35f51533d4.2&gc=1.

10. Rajendra Desai, 2008, "Case Studies of Seismic Retrofitting – Latur to Kashmir and Lessons Learnt," National Centre for Peoples' Action in Disaster Preparedness (NCPDP), http://www.ncpdpindia.org/Retrofitting%20Case%20Studies.htm; http://www.ncpdpindia.org/images/03%20RETROFITTING%20LESSONS%20LEARNT%20LATUR%20TO%20KASHMIR.pdf; and other sources listed in Resources section.
11. R Roland Stulz and Kiran Mukerji, 1993, *Appropriate Building Materials, A Catalogue of Potential Solutions (Revised)* (St. Gallen: SKAT), http://nzdl.sadl.uleth.ca/cgi-bin/library?e=d-00000 00---off-0cdl--00-0----0-10-0---0---0direct-10---4-------0-1l--11-en-50---20-about---00-0-1-00-0-0-11-1-0utfZz-8-00&cl=CL2.1&d=HASH51495f314e8d35f51533d4.2&gc=1.

Lesson Learned related to Construction Technology

Availability of skills. Following the 2001 earthquake in Gujarat, India, there were initially insufficient local mason skills to properly use hollow interlocking compressed stabilized earth blocks, which resulted in slow construction. But news of employment opportunities spread fast, and trained artisans from other parts of the country came to Gujarat to provide their services. The same happened following the 1999 earthquake in Uttarakhand, India: bricks from the plains soon arrived, as did masons from Bihar state. Likewise, following the 2005 earthquake in Kashmir, masons and laborers from Bihar played a pivotal role in speedy reconstruction.[12]

Participation and livelihoods. After the 2005 earthquake in Kashmir, as part of the National Centre for People's Action in Disaster Preparedness (NCPDP) project, sponsored by the Aga Khan Development Network, local building systems, architecture, lifestyle, and preferences formed the basis of the reconstruction design. Local artisans played a large role in the development of the reconstruction technology. Feedback from local women and from local master artisans was incorporated, and technical guidelines from government were translated into practical guidance and gradually improved to ensure replicability and affordability.

The Importance of Technical Guidelines. Technical guidelines on HDCTs are essential references in the process of housing reconstruction and retrofitting. They should provide guidance on standards and codes to be respected; damage assessment; structural safety related to various risks; construction techniques, means, and procedures; building materials and quality; and professional skills needed for successful implementation to take place. To be effective, it is crucial that technical guidelines are appropriate to the given context. If unavailable, they must be developed. If available but not suitable to the current post-disaster situation, they must be modified. Guidelines must be incorporated as an integral part of the training curricula for builders and used as a working tool throughout the reconstruction phase, including during inspection and monitoring.

The Debate Concerning the Promotion of Vernacular Technologies

Vernacular technologies are often appropriate solutions in terms of cost, environmental impact, climate, and cultural and architectural suitability, and should generally be given priority. However, these technologies are not always optimal due to such concerns as their vulnerability to hazards and durability, and often need to be improved through the introduction of modern technology or components. There is considerable debate in the development community concerning the promotion of vernacular technologies in reconstruction. Agencies should ensure that a reputable organization has tested the hazard resilience of a particular technology, and that any recommended improvements or retrofitting approaches are incorporated in housing designs, before financing a large-scale program to repair vernacular buildings. (For some of the resources available from organizations working to merge modern and vernacular technologies to produce more appropriate solutions, see the annex to this chapter.) Vernacular building technologies were approved for use by the Earthquake Reconstruction and Rehabilitation Authority (ERRA) following the 2005 North Pakistan earthquake, as discussed in the ⌨ case study, below.

Technical Issues and Recommendations: Repair/Retrofit versus Demolition

In reconstruction efforts, repairing and retrofitting a house may make more sense than demolishing and rebuilding it. Many practitioners and policy makers think that programs designed to repair and/ or retrofit housing are difficult to design and implement. However, such programs can save many partially damaged houses, often with excellent results. Properly designed and monitored projects for repairing and retrofitting houses can drastically improve the reconstruction process in terms of cost, environmental impact, speed, supply of resources, community participation and satisfaction, recovery of psychological well-being, and heritage conservation. In addition, structures that are vulnerable but not damaged by the disaster can have their vulnerabilities addressed as part of the program. Comprehensive reconstruction programs should have a component for repair and retrofit that addresses similar technical issues as those considered for reconstruction, including layout, infrastructure, and building technology and materials. Detailed guidelines and training should be made available for ensuring efficient and safe retrofitting; if guidelines do not exist, they may have to be developed for a particular context. Listed below are some of the issues that should be addressed in a repair/retrofit program.

Properly designed and monitored projects for repairing and retrofitting houses can drastically improve the reconstruction process in terms of cost, environmental impact, speed, supply of resources, community participation and satisfaction, recovery of psychological well-being, and heritage conservation.

12. Rajendra Desai, 2009, personal communication.

Issue	Relevance
Relocation	The repair or retrofit option is moot if a house must be relocated.
Damage level	Before a decision can be made about whether the repair/retrofit option is appropriate, the level of damage to the house, to the neighborhood, and to the infrastructure, as well as the related risks to residents, must be considered.
Cost of the repair or retrofit option versus reconstruction	To be justifiable, the total cost of the repair or retrofit option should generally be lower than that of demolition and reconstruction.
Willingness and capacity of people to repair or retrofit their houses	It is essential that the local population participate in the discussion about repairing and retrofitting. People do not always perceive repairing or retrofitting as viable or desirable options. Without local support, a project to repair or retrofit may even encounter passionate objections. Communications, public outreach, and training are all crucial elements of a successful repair and retrofit program, as they are for reconstruction.
Architectural, historical, cultural, and socioeconomic value of damaged houses	If a particular house has a high architectural, historical, cultural, or socioeconomic value, substantial efforts to overcome any cost or technical difficulties and prevent it from being demolished may be justified. The owner may be offered extra financial or technical assistance if the house is considered part of community heritage, in order to encourage preservation of the property.

The Vulnerabilities of Houses and How to Reduce Them

Disasters affect house structure in a variety of ways. Consequently, the technical solutions for constructing and repairing affected buildings have to respond to the type of disaster and have to take into consideration the building technology and materials being used. This principle applies to risk reduction in both new and retrofitted houses. A variety of technical materials on this topic are listed below in the Resources section. See 📖 Chapter 16, Training Requirements in Reconstruction, for instructions on designing a training program for builders. For a list of organizations working to improve the risk of vernacular buildings, see the annex to this chapter. The figure above depicts the range of vulnerabilities associated with housing, taken from the *Manual on Hazard-Resistant Construction in India*. The example shown here is applicable to both engineered and non-engineered. building types. Training builders and supervisors to understand these vulnerabilities is crucial.

Vulnerability of Non-Engineered Buildings against Earthquake, Cyclone, and Flood Hazards

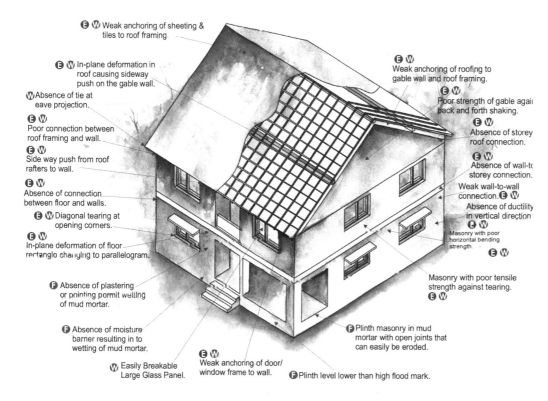

(E) (W) Weak anchoring of sheeting & tiles to roof framing.

(E) (W) In-plane deformation in roof causing sideway push on the gable wall.

(W) Absence of tie at eave projection.

(E) (W) Poor connection between roof framing and wall.

(E) (W) Side way push from roof rafters to wall.

(E) (W) Absence of connection between floor and walls.

(E) (W) Diagonal tearing at opening corners.

(E) (W) In-plane deformation of floor rectangle changing to parallelogram.

(F) Absence of plastering or pointing permit wetting of mud mortar.

(F) Absence of moisture barrier resulting in to wetting of mud mortar.

(W) Easily Breakable Large Glass Panel.

(E) (W) Weak anchoring of door/window frame to wall.

(F) Plinth level lower than high flood mark.

(E) (W) Weak anchoring of roofing to gable wall and roof framing.

(E) (W) Poor strength of gable again back and forth shaking.

(E) (W) Absence of storey roof connection.

(E) (W) Absence of wall-to storey connection.

Weak wall-to-wall connection. (E) (W)

Absence of ductility in vertical direction (E) (W)

Masonry with poor horizontal bending strength. (E) (W)

Masonry with poor tensile strength against tearing. (E) (W)

(F) Plinth masonry in mud mortar with open joints that can easily be eroded.

Key
(E) Earthquake
(W) Wind/Cyclone
(F) Flood/Rain

Source: United Nations Development Programme (UNDP) India and NCPDP, 2008, *Manual on Hazard-Resistant Construction in India: For Reducing Vulnerability in Buildings Built without Engineers* (Gujarat: UNDP India and NCPDP), http://www.ncpdpindia.org/Manual_on_Hazard_Resistant_Construction_in_India.htm.

Risks and Challenges

- The lack of local knowledge about appropriate housing design and current construction practices.
- Specialized expertise to inform the choice of building technology is not available.
- Building materials and skilled labor are in short supply, leading to inflated prices.
- Poor construction quality results in structures that are vulnerable, fragile, and expensive to maintain.
- Imported building technologies and materials require more energy to produce comfortable indoor conditions, leading to increased costs and negative environmental impact.
- Building technologies are poorly adapted to risks in the environment in which they are located, or adapted to only one of numerous risks (e.g., they provide wind protection, but are vulnerable to earthquakes).
- New housing designs or building technologies are incompatible with local traditions or with the local population's willingness to change.
- Design and construction contribute to local and global environmental damage.
- Demolition of reparable houses results in loss of cultural identity and heritage, slower psycho-social recovery, adverse environmental impacts, and extended time for reconstruction.
- Improperly designed or implemented repair and retrofitting projects damage the architectural integrity and quality of a house.
- Donors are not willing to finance non-standard reconstruction approaches, such as repair and retrofitting.[13]
- Building codes and regulation prohibit the use of local building technologies or do not adequately incorporate the use of local materials and practices.

Case Studies

2003 Bam Earthquake, Iran

System for Classification of Housing Damage Level

After the 2003 earthquake in Bam, Iran, reconstruction planning was based on a rating of housing damage that had two components: "damage to residential area" and "damage to residential unit."

Damage to residential area. This rating took into consideration issues related to geology, such as soil stability, land condition, and the percentage of the village that was damaged (ratio of damaged housing units to total housing units).

Damage to residential unit. This rating took into account the condition of the damaged units, the type of damage, and the type of technical expertise that might be needed to develop the reconstruction plan. The following diagram demonstrates the rating procedure.

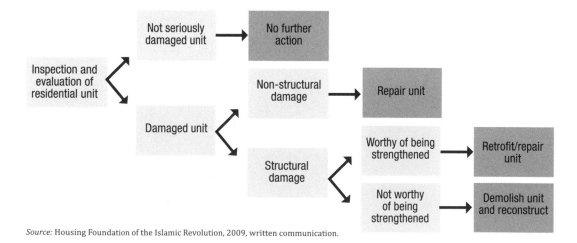

13. NCPDP, "Good Practices Review: Latur, Kashmir, Kutch, Uttrakhand," http://www.ncpdpindia.org.

Source: Housing Foundation of the Islamic Revolution, 2009, written communication.

2003 Bam Earthquake, Iran

Using Demonstration Buildings and Local Professionals to Improve Reconstruction Outcomes

Only few months after the 2003 earthquake in Bam, Iran, the HF established the Engineering and Technical Services Exhibition Site in Bam, where more than 500 housing designs, techniques, and building materials were publicly exhibited, enabling people to choose the housing model they preferred. Also at the site, modular and real-scale model buildings demonstrated environment-friendly, cost-effective, and locally appropriate shelter models. The structure and material of the model buildings were tested and licensed by the Building and Housing Research Center of the Iran Ministry of Housing.

Quake-proof model buildings built by trained local workers under the UNDP-backed project were cost-effective and consistent with the cultural identity of the ancient city of Bam.

The High Council of Bam Architecture mobilized and trained engineering, architectural, and technical firms to provide professional services to people who selected housing designs and construction techniques for their homes. Under the HF-UNDP joint program, engineers and architects organized a series of consultations to ensure that the views of the program's female-headed households (FHHs) were incorporated into the construction and design of the units (see photo). CRATerre-EAG, a French construction research center, provided training for builders on earthquake-resilient construction. In addition, 20 skilled local masons and other builders were trained on quake-resilient traditional building techniques. In the exhibition site, trainees demonstrated the features of earthquake-resilient construction using small model buildings they built themselves. The trainers were able to transfer their knowledge and skills to masons from Bam and elsewhere in the country. A recovery loan of US$220 million from the World Bank helped HF procure building materials in large quantities. The success of the UNDP-backed training encouraged government and CRATerre to later replicate these experiences in other earthquake-prone areas of the country.

Source: Victoria Kianpour, UNDP Iran, 2009, personal communication, http://www.undp.org.ir/.

Design consultations among engineers, architects and planners, and FHHs, 2005.

⌂ **For access to additional resources and information on this topic, please visit the handbook Web site at www.housingreconstruction.org.**

1995 Building Code Revision, Kenya
Stakeholder Participation in Legislative Reform

As a consequence of the significant urban population growth in the last several decades in Kenya, informal housing settlements became ubiquitous. (In fact, more than half of the population of Kenya's capital, Nairobi, currently lives in informal housing.) Kenya's building codes, like those in all countries, were intended to protect people from poor construction and to reduce vulnerability to disasters. However, in the 1980s, many of Kenya's building codes still dated from the colonial era, and progressive building—the most common form of home construction in the country—was not even permitted by the codes. As a result, the minimum acceptable house according to the code was beyond the means of poor and even middle-income families. The effect was to discourage investment in housing and limit the types of construction that lenders could finance.

In 1990, the Intermediate Technology Development Group (ITDG, now Practical Action) initiated a participatory effort to modify the codes to reflect local building practices while still encouraging safe building practices. With the participation of stakeholders from the construction field, "Code 95" was developed and, in 1995, was approved by Parliament. Code 95 is performance-based and permits the use of innovative and popular materials, alternative building technologies, outdoor cooking areas, and pit latrines. A Department for International Development (DFID)-funded pilot project in the city of Nakuru, managed by ITDG, demonstrated that housing construction under the new code reduced costs by at least 30 percent and helped trigger an upsurge in residential construction. Understanding the importance of building codes in the housing supply chain, the business community, housing finance institutions, developers, and materials manufacturers have supported the implementation of Code 95. Although updating the code was a first step toward improving building safety and increasing disaster resilience, other challenges, such as simplifying approval processes and establishing incentives to fully apply the building codes, must still be tackled.

Sources: Saad Yahya et al., 2001, *Double Standards, Single Purpose: Reforming Housing Regulations to Reduce Poverty* (London: ITDG Publishing); Mohini Malhotra, ed., 2002, "The Enabling Environment for Housing Microfinance in Kenya," Cities Alliance Shelter Finance for the Poor Series, http://www.citiesalliance.org/citiesalliancehomepage.nsf/Attachments/kenya/$File/kenyafinal.pdf; *and* Shailesh Kataria and Saad Yahya, Royal Institution of Chartered Surveyors (RICS), 2009, personal communication.

DANIEL PITTET

2005 Northern Pakistan Earthquake, Kashmir
Earthquake-Resistant Dhajji Dewari Technology

During the 2005 earthquake in Kashmir, buildings constructed using several traditional methods held up much better than did many "modern" structures. Two types of construction that did well were those that used the techniques known as *Taq* (timber-laced masonry bearing wall) and *Dhajji Dewari* (complete timber frame with a wythe of masonry forming panels within the frame). Although there were many cracks in the masonry infill, most of these structures did not collapse, thereby preventing the loss of life. During rural owner-driven reconstruction, the use of the *Dhajji Dewari* technology was promoted and facilitated. It was rapidly adopted by local communities and it provided a high level of satisfaction among beneficiaries. Not only do these construction techniques stand up well in earthquakes (when properly constructed), but they also make use of local materials (wood, stones, mud), have low environmental impact, are economical, and are part of the housing culture and know-how. Other vernacular construction techniques that performed well in the earthquake included wooden log houses; construction that employed the use of well-laid masonry with through-stones and well-designed arches; and buildings with trusses, tongue and groove joinery, and balconies resting on projecting wooden joists. Traditional building technology has been approved by the government of Pakistan's ERRA on the basis of the observation of existing structures.

Sources: Maggie Stephenson, United Nations Centre for Human Settlements (UN-HABITAT), 2009, personal communication; *and* Rohit Jigyasu, 2009, personal communication.

Resources

ADPC. 2005. "Handbook on Design and Construction of Housing for Flood-Prone Rural Areas of Bangladesh." Dhaka: ADPC. http://www.adpc.net/AUDMP/library/housinghandbook/handbook_complete-b.pdf. Focuses solutions for various construction technologies exposed to flooding.

Applied Technology Council (ATC). http://www.atcouncil.org/. ATC provides a range of materials on the hazard resistance of building technologies, building retrofitting techniques, and methodologies for post-disaster building inspections.

Arya, A. S. et al. 2004. *Guidelines for Earthquake Resistant Non-Engineered Construction.* Kanpur: National Information Center of Earthquake Engineering. http://www.nicee.org/IAEE_English.php. Information specifically on earthquake-exposed structures.

Benson, Charlotte, John Twigg with Tiziana Rossetto. 2007. "Tools for Mainstreaming Disaster Risk Reduction. Guidance Note 12: Construction Design, Building Standards and Site Selection." International Federation of Red Cross and Red Crescent Societies, ProVention Consortium. http://www.proventionconsortium.org/themes/default/pdfs/tools_for_mainstreaming_GN12.pdf.

Blondet, Marcial, Gladys Villa Garcia M., and Svetlana Brzev. 2003. *Earthquake-Resistant Construction of Adobe Buildings: A Tutorial.* Oakland: Earthquake Engineering Research Institute. http://www.preventionweb.net/english/professional/trainings-events/edu-materials/v.php?id=7354.

CDMP. 2001. "Hazard-Resistant Construction." Caribbean Disaster Mitigation Project. http://www.oas.org/CDMP/document/papers/parker94.htm.

Earthquake Engineering Research Institute (EERI)/International Association of Seismology and Physics of the Earth's Interior (IASPEI). 2006. "International norm for seismic safety programs." Draft paper of the working group of the International Association for Earthquake Engineering (IAEE) and the International Association of Seismology and Physics of the Earth's Interior. http://www.iitk.ac.in/nicee/skj/Norms_for_Seismic_Safety_Programs-2-23-06.pdf.

Minke, Gernot. 2001. *Construction Manual for Earthquake-Resistant Houses Built of Earth.* Eshborn: Building Advisory Service and Information Network at GTZ GmbH. http://www.basin.info/publications/books/ManualMinke.pdf.

Papanikolaou, Aikaterini and Fabio Taucer. 2004. "Review of Non-Engineered Houses in Latin America with References to Building Practices and Self Construction Projects." European Commission Joint Research Center. http://elsa.jrc.ec.europa.eu/showdoc.php?object_id=26.

Patel, Dinesh Bhudia, Devraj Bhanderi Patel, and Khimji Pindoria. 2001. "Repair and strengthening guide for earthquake-damaged low-rise domestic buildings in Gujarat, India." Gujarat Relief Engineering Advice Team (GREAT). http://awas.up.nic.in/linkfile/Disaster/Retrofitting%20Low%20rise%20houses.pdf.

RICS. 2009. *The Built Environment Professions in Disaster Risk Reduction and Response Guide.* London: University of Westminster. http://www.rics.org/site/scripts/download_info.aspx?downloadID=829&fileID=991.

Schneider, Claudia et al. 2007. *After the Tsunami: Sustainable Building Guidelines for South East Asia.* Nairobi: United Nations Environment Programme. http://www.preventionweb.net/english/professional/publications/v.php?id=1594.

Szakats, Gregory A. J. 2006. *Improving the Earthquake Resistance of Small Buildings, Houses and Community Infrastructure.* Wellington, NZ: AC Consulting Group Limited. http://www.preventionweb.net/english/professional/publications/v.php?id=1390.

Twigg, John. 2006. "Technology, Post-Disaster Housing Reconstruction and Livelihood Security." http://www.practicalaction.org.

UN-HABITAT. 2003–2005. Building *Materials and Construction Technologies: Annotated UN-HABITAT Bibliography.* Nairobi: UN-HABITAT. http://www.unhabitat.org/pmss/getPage.asp?page=bookView&book=1087.

UNDP India. 2008. "Manual on Hazard-Resistant Construction in India." Gujarat: UNDP India and NCPDP. http://data.undp.org.in/dmweb/pub/Manual-Hazard-Resistant-Construction-in-India.pdf. Includes illustrated practical solutions covering earthquake, cyclone, and flood situations for various technologies.

Technology/Project	Country/Region	Organizations/Links	Remarks
"Bunga" houses built with compressed stabilized earth blocks; earthquake-resistant structures derived from traditional houses of cylindrical shape	India, Gujarat State, Kutch District	Hunnarshala Foundation for Building Technology and Innovations, Bhuj, India, http://hunnar.org	Governmental approval through GSDMA, 2003, *Guidelines for Construction of Compressed Stabilized Earthen Wall Buildings* (Gujarat State Disaster Management Authority)
Manual on hazard-resistant construction in India	India	UNDP India and Government of India, Ahmedabad, India, http://www.ncpdpindia.org	For reducing vulnerability in buildings without engineers; focuses on construction and retrofitting of masonry buildings
Manual for restoration and retrofitting of rural structures in Kashmir	Pakistan, Kashmir India, Jammu	UNDP India and NCPDP, Ahmedabad, India, http://www.ncpdpindia.org	For reducing vulnerability of existing structures in earthquake-affected areas
Guidelines for earthquake-resistant construction of non-engineered rural and suburban masonry houses in cement sand mortar in earthquake-affected areas	Pakistan	ERRA, Government of Pakistan, http://www.erra.gov.pk	Practical guidance for non-engineered structures, covering site-selection issues, planning, construction, and retrofitting measures of various housing elements
Model bamboo house	Ecuador, Guayaquil	International Network for Bamboo and Rattan, http://www.inbar.int	Demonstration and comparison of 10 different technologies based on the use of bamboo for walling systems that have been developed to improve the quality and reduce the cost
Earth-based building materials and technologies	India	Auroville Earth Institute, Tamil Nadu, India, http://www.earth-auroville.com	Development, training programs, publications, and realization of numerous constructions using earth as building material and integrating modern technology for improving structural safety
Construcción de casas saludables y sismorresistentes de Adobe Reforzado con geomallas	Peru	Pontificia Universidad Católica del Perú, http://www.pucp.edu.pe	Technology based on the use of mud blocks walls reinforced with plastic nets for improving seismic resistance
Mitigation measures for post-hurricane reconstruction	Honduras	Centre des Etudes et Coopération International, http://www.ceci.ca	Technical improvement of dwellings for reducing the vulnerability to hurricane and flood
Collection of information and publications on seismic resistance of various building technologies	Various	World Housing Encyclopedia, an EERI and IAEE initiative, http://www.world-housing.net	

Guiding Principles for Cultural Heritage Conservation

- Cultural heritage conservation helps a community not only protect economically valuable physical assets, but also preserve its practices, history, and environment, and a sense of continuity and identity.
- Cultural property may be more at risk from the secondary effects of a disaster than from the disaster itself, therefore quick action will be needed.
- Built vernacular heritage offers a record of a society's continuous adaptation to social and environmental challenges, including extreme events, such as past disasters. This record can often be drawn on to design mitigation strategies for new construction or retrofitting.
- Communities should prioritize which cultural assets to preserve, considering both cultural meaning and livelihood implications, although reaching a consensus may be difficult.
- Cultural heritage conservation plans are best designed before a disaster, but, in their absence, heritage authorities can and should collaborate to develop effective post-disaster heritage conservation strategies.
- Because vernacular cultural properties are sometimes capable of withstanding local climate conditions, they may serve as safe havens where surrounding communities can temporarily relocate.

This Chapter Is Especially Useful For:

- Lead disaster agency
- Cultural heritage specialists
- Local officials
- Affected communities
- Project managers

Introduction

Once restricted to monuments, archeological sites, and movable heritage collections, the definition of cultural heritage now includes historic urban areas, vernacular heritage, cultural landscapes (tangible heritage, which include natural and cultural sites), and even living dimensions of heritage and all aspects of the physical and spiritual relationship between human societies and their environment (intangible heritage).

The World Bank uses a broad definition of physical cultural resources: "Movable or immovable objects, sites, structures, groups of structures, and natural features and landscapes that have archeological, paleontological, historical, architectural, religious, aesthetic, or other cultural significance."[1] The World Bank also recognizes that "physical cultural resources are important as sources of valuable scientific and historical information, as assets for economic and social development, and as integral parts of a people's cultural identity and practices."[2]

This chapter addresses the importance of protecting the cultural heritage of communities, especially traditional housing, which should be an integral part of any post-disaster recovery program.

Key Decisions

1. Immediately after a disaster, **government** should mobilize the **lead agency for post-disaster heritage conservation**, if one is already designated, or if not, appoint one to address damage to resources of national significance and to assist local communities.
2. The **lead agency for heritage conservation** should collaborate with the **lead disaster agency** and **local governments** to ensure cultural resources are considered in post-disaster damage and loss assessments.
3. **Communities** in collaboration with **local government** and the **lead agency for heritage conservation** should identify and prioritize cultural resources that require conservation during recovery and reconstruction and document the condition of these resources.
4. **Communities** in collaboration with **local government** and the **lead agency for heritage conservation** should decide whether adequate instruments or plans are in place to address post-disaster cultural heritage risks. If so, they should be activated. If not, stakeholders should work together to carry out the cultural heritage planning.

1. World Bank, 2006, "Operational Policy 4.11, Physical Cultural Resources," http://go.worldbank.org/IHM9G1FOO0.
2. World Bank, 2006, "Operational Policy 4.11, Physical Cultural Resources," http://go.worldbank.org/IHM9G1FOO0.
3. See Resources section for names of organizations that provide assistance.

5. The **lead agency for heritage conservation** should decide whether available local resources are adequate to address the post-disaster cultural heritage risks that have been identified. If not, it should identify and mobilize outside financial and technical assistance.[3]

6. **Churches, tribal organizations**, and other **guardians of cultural resources** should ensure that their resources are included in post-disaster assessments and should request assistance in conserving them, if required.

7. **Communities being relocated** and **receiving communities** should demand that the conservation of cultural resources be a consideration in resettlement planning, site selection, and relocation plans.

Public Policies related to Cultural Heritage Conservation

Local planning departments and local disaster management agencies are responsible for the implementation of the instruments mentioned in this chapter (disaster management plans and urban development plans, for example). They should be involved when heritage conservation issues arise in a post-disaster situation, as should historical societies involved in protection of the affected cultural assets, academic institutions involved in heritage research, and local government arts and cultural agencies.

DANIEL PITTET

Heritage conservation may be guided by national-level policies and by public agencies, such as the Iranian Cultural Heritage Organization, or quasi-public entities, such as the Indonesian Heritage Trust. The Swiss system is considered an international good practice for integrated disaster management planning. The Swiss Federal Office for Civil Protection, which provides aid in the event of a disaster and protection from armed conflict, includes a heritage section. The office mandates that localities provide legislative and administrative support to safeguard heritage and that they make specific financial contributions.[4] Entities of this nature should also be involved when cultural properties are affected by a disaster.

At the international level, the 2005 Kyoto Declaration on the Protection of Cultural Properties, Historic Areas, and Their Settings from Loss in Disasters established a framework for work on the preservation of cultural properties and historic areas.[5] The United Nations Educational, Scientific and Cultural Organization (UNESCO), the International Centre for the Study of the Preservation and Restoration of Cultural Property (ICCROM), and the International Council of Monuments and Sites (ICOMOS) are closely involved in the implementation of the Kyoto Declaration, including working to reduce disaster risk at World Heritage sites.[6] These agencies are often active in post-disaster situations and may provide technical assistance to public officials and owners of heritage assets.

Technical Issues
Disaster Preparation for Cultural and Natural Heritage Properties

Ideally, awareness about the socioeconomic value of cultural heritage and measures to protect it are established in "normal" times. This way, risks to cultural heritage and the related losses of livelihoods, cultural identity, and social cohesion can be mitigated before disaster strikes. In this scenario, the concern after a disaster is only with implementation. Cultural heritage risks can be addressed by various means, including the instruments listed below:

- Disaster risk management plans that incorporate cultural heritage consideration
- Culturally sensitive land use and spatial plans
- Raising the cultural sensitivity of disaster management authorities, the families, and other users that occupy heritage properties
- Systematic documentation of cultural heritage
- Regular maintenance and monitoring for risk reduction of heritage properties
- Post-disaster response and recovery programs that are consistent with management plans for heritage sites

But even if these measures are not in place, post-disaster reconstruction is an opportunity to "build back better," even for heritage properties, without compromising on their value. Reforms can be

4. June Taboroff, n.d., "Natural Disasters and Urban Cultural Heritage: A Reassessment," in *Building Safer Cities: The Future of Disaster Risk* (Washington, DC: World Bank), http://www.preventionweb.net/files/638_8681.pdf.

5. Adopted at the Kyoto International Symposium, "Towards the Protection of Cultural Properties and Historic Urban Areas from Disaster," January 16, 2005. See also "Recommendations from the United Nations Educational, Scientific and Cultural Organization/ International Centre for the Study of the Preservation and Restoration of Cultural Property/Agency for Cultural Affairs of Japan Thematic Session on Cultural Heritage Risk Management Kobe, 2005," http://www.unisdr.org/wcdr/thematic-sessions/thematic-reports/report-session-3-3.pdf.

6. UNESCO, "Natural and Environmental Disasters: UNESCO's Role and Contribution," http://portal.unesco.org/en/ev.php-URL_ID=31605&URL_DO=DO_TOPIC&URL_SECTION=201.html.

made and measures taken to reduce risks to cultural heritage from normal development and from future disasters. An example of preventive conservation is described in the 📖 case study, below, on cultural heritage in Georgia.

Conserving Cultural and Natural Heritage following a Disaster

Coordinating disaster management with heritage authorities. Lack of coordination between disaster management and heritage authorities often causes much of the damage to heritage within the framework of emergency operations and reconstruction programs. This can be avoided through an immediate cooperation between disaster management and heritage authorities following a disaster. (The first 48 hours following a disaster are considered very important to avoid irremediable losses to cultural heritage sites.) Natural and cultural heritage sites may be affected by the location of temporary camps for displaced populations that place increased pressure on related resources. It is therefore important to consult with and involve representatives of heritage agencies in planning reconstruction.

A multidisciplinary approach to damage and assessments. Damage assessment teams need to be multidisciplinary and include the expertise of heritage and conservation experts, including archeologists, conservation architects, seismologists, engineers, and social anthropologists. As a rule, the damage assessment should be carried on as a comprehensive exercise, avoiding separate assessments, because an integrated assessment allows timely identification of priorities. However, depending on the context, the nature of local heritage assets, and the type of damage, separate damage assessments—including detailed inspection of the building fabric—may have to be undertaken for cultural heritage buildings and sites. Temporary works may also be needed, including strutting and shoring walls, temporary roofing, underpinning, and protection of integral works of art/cultural property (e.g., carvings, murals). The World Bank financed an innovative project in Yunnan, China, that combined rules for earthquake-resilient construction with historic preservation regulations, as described in the 📖 case study, below.

Recognizing the value of built vernacular heritage. Vernacular housing and building practices often offer an affordable, environmentally sustainable, aesthetic and culturally appropriate response to people's sheltering needs. Their value, however, is often not recognized. While post-disaster reconstruction can be an opportunity to upgrade a community's housing condition, it should not result in the systematic demolition of vernacular houses and their surrounding habitat. Such practices can be avoided through culturally sensitive planning that recognizes the functional and aesthetic value of the vernacular.

Creating incentives for the conservation of vernacular housing. If new houses are provided for free without timely, adequate support for repair and retrofitting, reconstruction policies may directly encourage the demolition of undamaged or partially damaged vernacular houses. Reconstruction policies often give priority to the construction of new houses. Even though repair and retrofitting programs can at times be initiated almost immediately and at lower cost, they are often given marginal attention as housing strategies or incorporated only at a later stage.

Developing building codes compatible with vernacular building practices. Historic or vernacular buildings should not be condemned, destroyed, or stripped of their beneficial use simply because they do not or cannot comply with building codes for new construction. Building codes are important to ensure safety of new construction and repairs. However, the same codes may not be appropriate for historic buildings. Land use and site plans, building guidelines, and codes being developed for post-disaster reconstruction must reflect local building designs, culture, technologies, skills, and materials. The lead agency may need international technical assistance with this aspect of code revision.

Harmonizing new housing and settlements with local cultural and natural heritage. It is important that new construction be built in harmony with local building culture and settlement layouts, especially when building new houses within or near existing historical or vernacular settlements. If reconstruction entails relocation, the heritage value of a new site needs to be assessed so that irreversible losses can be mitigated or avoided altogether. The need for making tradeoffs in reconstruction is illustrated by the 📖 case study on the 1993 Latur earthquake reconstruction, below.

Providing storage for movable heritage properties. Storage facilities allow communities to store salvaged materials with heritage value and use them later during reconstruction, helping

Repairing and strengthening heritage buildings may be necessary elements of a post-disaster reconstruction program. When local craftspeople are given a significant role in restoration activities, conserving cultural heritage can also help restore local livelihoods.

ensure much-needed cultural continuity after a disaster. Without adequate inventory and storage facilities, movable heritage properties with high cultural and emotional value for their owners or the community may be subject to looting and further damage.

Using authentic materials and skills for repairing and retrofitting heritage buildings.
Repairing and strengthening heritage buildings may be necessary elements of a post-disaster reconstruction program. Ideally, repairs should have no impact on the heritage value, authenticity, or integrity of a building and its surroundings. However, in cases where this is not possible, the impact should be minimal and reversible and the work should reflect recommended international practices. Using local skills and materials may be the best way to achieve these aims. If traditional craftspeople are given a significant role in restoration activities, conserving cultural heritage can also help restore local livelihoods. The ⌨ case study on the 2003 Bam earthquake reconstruction, below, describes the challenges of rebuilding using traditional materials.

Ensuring community participation. The cultural heritage significance of a place or element may be very localized. Even within a community, there may be variations in the spiritual and emotional importance attributed to specific sites or elements. Accordingly, effective protection of cultural heritage can be achieved only through wide community participation in recovery and reconstruction planning. This participatory planning should focus both on cultural importance and on the cultural and livelihood activities that depend on the conservation of these properties.

Risks and Challenges

- Cultural heritage is affected by primary risks, that is, direct damage from the natural disaster.
- Cultural heritage is also threatened by secondary risks that arise during recovery and reconstruction, including:
    - rescue and relief measures that are carried out with no regard to heritage value of damaged areas (e.g., water damage from fire fighting and debris removal with no regard to heritage value);
    - looting of heritage buildings; and
    - reuse of cultural and natural heritage resources as fuel, food, and reconstruction materials.
- Infrastructure repair or replacement (e.g., road widening) disregards or encroaches upon cultural assets.
- Temporary camps are sited without regard to cultural heritage concerns.
- Illegal and uncontrolled relocation and reconstruction spoil heritage landscapes or damage other assets
- Financial assistance policies encourage demolition of heritage buildings.
- Authenticity and integrity may be lost because of inadequate repair and retrofitting measures.

Recommendations

1. Coordinate disaster management with heritage authorities beginning in the first 48 hours following a disaster to avoid irremediable losses to cultural heritage sites.
2. Make sure that temporary camps for displaced populations are not located so that they create risks to heritage sites or properties.
3. Incorporate heritage and conservation experts in housing damage assessment teams or conduct specific assessments of cultural heritage housing and community resources.
4. Determine whether temporary works, such as strutting and shoring walls or temporary roofing, are needed to protect cultural properties or specific components (e.g. carvings, murals).
5. In post-disaster reconstruction, avoid the systematic demolition of vernacular houses and their surrounding habitat in an attempt to upgrade a community's housing condition.
6. Create incentives for the conservation of vernacular housing, or consider declaring historic properties community property if the owners are not able or willing to save them.
7. Develop building guidelines and codes that are compatible with vernacular building practices.
8. Harmonize designs and building materials of new housing and settlements with local cultural and natural heritage.
9. Provide storage facilities for movable heritage properties so that they are not looted, sold, or removed from the community.
10. Use authentic materials and skills in repairing and retrofitting heritage buildings.
11. Ensure community participation in decisions regarding heritage conservation, and realize that the cultural and spiritual importance of heritage sites and properties may be very location-specific.

Case Studies
2003 Bam Earthquake, Iran
How Reconstruction Affected the Architectural Landscape

ARAD MOJTAHEDI

The newly rebuilt city of Bam, Iran, retains few features of the architectural fabric of the old city that existed before the 2003 earthquake. In the course of the city's reconstruction, the overall landscape was significantly altered. What is especially notable to the city's residents is the loss of the harmony between the beautiful, commonly used, and climatologically appropriate mud-brick houses and other physical structures and the city's extensively damaged citadel (Arg-e-Bam), a harmony that had endured for centuries. The changes were largely due to pressure to speed up reconstruction by using pre-made steel frame structures and conventional building materials. Other factors that contributed to the changes included (1) fears about the safety of old mud-brick construction techniques, (2) the lack of skills to apply old construction techniques in a way that ensured risk reduction, (3) the lack of an approved national building code or guidelines for promoting the mud-brick construction (the Iranian 2,800 building codes and national building regulations discourage mud-brick construction), and (4) the slow pace of construction with mud-brick techniques compared to conventional building techniques and the perception of higher costs. In fact, if government subsidies for production and transport of conventional building materials, such as cement, steel, and bricks, were removed, the traditional technique would have emerged as the more cost-effective choice. Since the earthquake, CRATerre-EAG (the same French construction research center that cooperated in the Housing Foundation of the Islamic Revolution/United Nations Development Programme demonstration project in Bam) has mobilized resources from the European Community to assist the Building and Housing Research Center of the Ministry of Housing and Urban Development of Iran in reincorporating the mud-brick construction techniques into Iran's building codes.

Source: Victoria Kianpour, UNDP, 2009, personal communication, http://www.undp.org.ir/.

1993 Latur Earthquake, Maharashtra, India
Traditional Housing and Settlement Patterns after Reconstruction

As part of the rehabilitation program following the 1993 earthquake in Marathwada, India, more than 52 villages were relocated and reconstructed, using a layout and a construction technology selected on the basis of earthquake safety. The large number of deaths in Latur villages was attributed to the traditional mud and stone houses with common walls that had grown incrementally over time, encroaching on common spaces and restricting streets to narrow lanes. The villages also consisted largely of caste clusters: all dalits lived together, and higher caste members lived in the center. When large Gaddi houses collapsed, they often fell on the huts and small houses with common walls. The chaotic growth made it nearly impossible to construct in-situ due to the excessive cost to remove the vast quantities of debris needed and the difficulty of getting all to agree on the demolition of common walls and reestablishment of property boundaries.

Traditional settlements in this area were characterized by a hierarchy of public and private open spaces used for various activities and clusters of housing of distinct types. In the new relocations, village sites had wide streets forming a grid pattern; housing with no common walls that could lead to a house collapsing and destroying other neighbors' houses; and well-defined, unencroachable common spaces and streets. The new settlements also broke the caste clusters.

However, the spatial plans of these reconstructed villages and the new house designs were a significant departure from the local population's traditional way of life. The new settlement was more spread out and provided limited spaces for traditional activities, especially those of artisans. While town planners perceived that the plan would encourage the development of backward rural areas, it took a number of years for people to make the necessary modifications so that the houses and settlements better suited their lifestyle. As one official involved noted, there are no easy answers in post-disaster planning. While reconstruction cannot correct all existing and underlying social imbalances, it should address the most important concerns.

Sources: Rohit Jigyasu, 2002, "Reducing Disaster Vulnerability through Local Knowledge and Capacity" (PhD thesis, Trondheim: Norwegian University of Science and Technology), http://ntnu.diva-portal.org/smash/record.jsf?searchId=1&pid=diva2:123824; *and* Praveen Pardeshi, 2009, written communication.

1998, Cultural Heritage Project, Georgia
Preventive Conservation of Historical Buildings and Traditions

Preventive conservation and maintenance play an important role in protecting historical buildings and artifacts from the ravages of natural disasters. In 1998, to promote economic growth through development of the tourism industry, a World Bank cultural heritage project was launched in Georgia to rehabilitate historic sites and revitalize cultural traditions. Because Georgia is prone to seismic activity, preventive conservation was included in the project through an Emergency Rehabilitation Program. In fact, toward the end of project implementation, two earthquakes shook Georgia's capital, Tbilisi.

The project provided US$1 million to prevent the loss and damage to cultural heritage throughout Georgia, and was implemented through a selection process run by the Georgian Cultural Revival Board. The selection committee and beneficiary groups received assistance from the Fund for the Preservation of Culture Heritage of Georgia. The project received proposals for stabilizing buildings, archiving old manuscripts, and recording traditional songs and dances, and ultimately funded the protection of more than 100 cultural and historic treasures. It is noteworthy that several projects were implemented jointly by different ethnic groups, meaning that the project may have helped strengthen social cohesion and foster a sense of national identity in these areas.

The project also sponsored conservation of historic buildings in Tbilisi's Old Town. According to the World Bank's evaluation, the project contributed to the revitalization and economic development in the city's historic core, and media coverage of the project increased the public's interest in preserving Georgia's varied and rich cultural heritage.

Source: World Bank, 2004, "Implementation Completion and Results Report, Cultural Heritage Project, Georgia," http://go.worldbank.org/UU6JPWJ7X0.

© FIONA STARR UNESCO WORLD HERITAGE

1996 Lijiang Earthquake, Yunnan, China
Post-Earthquake Conservation in a World Heritage City

The 1996 earthquake in Yunnan, China, killed 200 people and injured 14,000 more. Approximately 186,000 houses collapsed, and 300,000 people were forced out of their damaged homes. There was widespread destruction of dwellings; businesses; schools; hospitals; and water, power, and transportation systems. There was also significant damage to the Old City of Lijiang's historic homes, bridges, paving, and infrastructure. (The city was later designated a UNESCO World Heritage site.) Here, the traditional construction technique of loosely attaching mud-brick walls to timber frames allowed the frames to shake without collapsing. However, the walls collapsed. Residents' low income levels and dislocation made rebuilding a daunting task.

Using a loan from the World Bank, the Lijiang Country Construction Bureau (CCB) provided grants for home repair and guidelines on reconstruction techniques that emphasized earthquake-resilient materials and techniques. Within a few weeks of the earthquake, CCB issued the "Design and Construction Technical Requirements for Houses in Lijiang Prefecture." These guidelines explained the materials and reinforcing techniques that should be used, which included vertical and horizontal reinforcement poles; netting walls; and fired, hollow-cement brick instead of sun-dried mud-brick. In support of existing historic preservation regulations, residents were also advised against using nontraditional materials or visibly contemporary building techniques. A village committee appraised the damage to each house and households received grant funds for purchase of materials—US$95, US$120, or US$300, depending on the degree of damage. In addition to the grant program, residents used a mutual self-help approach in which families organized to repair one house before moving on to the next. CCB staff reported that the amount of private money put into the housing reconstruction was often 5–10 times the amount of the grant.

Source: Geoffrey Read and Katrinka Ebbe, 2001, "Post-Earthquake Reconstruction and Urban Heritage Conservation in Lijiang," in *Historic Cities and Sacred Sites: Cultural Roots for Urban Futures*, Ismail Serageldin, Ephim Shluger, and Joan Martin-Brown, eds. (Washington, DC: World Bank), http://go.worldbank.org/GW737LG7U1.

Resources
Organizations

Heritage Emergency National Task Force (HENTF). "Resources for Response and Recovery." http://www.heritagepreservation.org/PROGRAMS/TASKFER.HTM. Sponsored by the nonprofit Heritage Preservation and the U.S. Federal Emergency Management Agency (FEMA), HENTF is a partnership of 40 U.S. service organizations and federal agencies. The Web site includes information on useful tools for post-disaster assessment of heritage resources and on locating professional help for post-disaster heritage conservation.

International Council for Monuments and Sites (ICOMOS). http://www.icomos.org/.

International Centre for the Study of Preservation and Restoration of Cultural Property (ICCROM). http://www.iccrom.org.

Research Center for Disaster Mitigation of Urban Cultural Heritage. Ritsumeiken University. http://www.rits-dmuch.jp/en/index.html.

United Nations Educational, Scientific and Cultural Organization (UNESCO) Culture. "Emergency Situations." http://portal.unesco.org/culture/en/ev.php-URL_ID=34329&URL_DO=DO_TOPIC&URL_SECTION=201.html.

UNESCO World Heritage. "Rapid Response Facility." http://whc.unesco.org/en/activities/578/. The Rapid Response Facility provides timely resources to address threats and emergencies affecting Natural World Heritage Sites and surrounding areas of influence.

Documents

Duyne Barenstein, J., and D. Pittet. 2007. "Post-Disaster Housing Reconstruction. Current Trends and Sustainable Alternatives for Tsunami-Affected Communities in Coastal Tamil Nadu." Lausanne: EPFL. *Point Sud.* http://www.isaac.supsi.ch/isaac/Gestione%20edifici/Informazione/post-disaster%20housing%20reconstruction.pdf.

Feilden, B. M. 1987. *Between Two Earthquakes: Cultural Property in Seismic Zones.* Rome: ICCROM. http://www.amazon.com/Between-Two-Earthquakes-Cultural-Property/dp/089236128X.

ICOMOS. 1999. "Charter on the Built Vernacular Heritage." Ratified by the ICOMOS Twelfth General Assembly. Mexico. http://www.international.icomos.org/charters/vernacular_e.htm.

Jigyasu, Rohit. 2006. "Integrated Framework for Cultural Heritage Risk Management." *Disasters and Development 1*, no. 1. http://www.radixonline.org/resources/jigyasu-iccrom.doc.

Matthews, Graham. 2007. "Disaster Management in the Cultural Heritage Sector: A Perspective of International Activity from the United Kingdom: Lessons and Messages." Presentation at World Library and Information Congress: 73rd International Federation of Library Associations (IFLA) General Conference and Council, August 19-23, 2007, Durban, South Africa. Contains an excellent bibliography of disaster-related cultural heritage resources. http://ifla.queenslibrary.org/IV/ifla73/papers/140-Matthews-en.pdf.

Stovel, H. 1998. *Risk Preparedness: A Management Manual for World Cultural Heritage.* Rome: ICCROM. http://www.iccrom.org/pdf/ICCROM_17_RiskPreparedness_en.pdf.

UNESCO/ICCROM/ICOMOS/UNESCO World Heritage Centre. Forthcoming. "How to Develop Disaster Risk Management Plans for World Heritage Properties: A Resource Manual." Paris: UNESCO World Heritage Resource Manual Series.

UNESCO World Heritage. "Reducing Disasters Risks at World Heritage Properties." http://whc.unesco.org/en/disaster-risk-reduction#rrf.

World Bank. 2008. "Risk Preparedness for Cultural Heritage." Good Practice Notes. http://siteresources.worldbank.org/CHINAEXTN/Resources/318949-1217387111415/Cultural_Heritage_en.pdf.

World Conference on Disaster Reduction. 2005. *Kobe Report Draft. Report of Session 3.3, Thematic Cluster 3.* Geneva: Cultural Heritage Risk Management. http://www.unisdr.org/wcdr/thematic-sessions/thematic reports/report-session-3-3.pdf.

World Heritage Centre for the International Disaster Reduction Conference (IDRC). 2006. "Integrating Traditional Knowledge Systems and Concern for Cultural and Natural Heritage into Risk Management Strategies." Conference Proceedings from the special session organized by ICCROM and IDRC, August 31, 2006. Davos, Switzerland. http://whc.unesco.org/uploads/events/documents/event-538-1.pdf.

RECONSTRUCTION TASKS AND HOW TO UNDERTAKE THEM

PROJECT IMPLEMENTATION

Guiding Principles

1. A good reconstruction policy helps reactivate communities and empowers people to rebuild their housing, their lives, and their livelihoods.
2. Reconstruction begins the day of the disaster.
3. Community members should be partners in policy making and leaders of local implementation.
4. Reconstruction policy and plans should be financially realistic but ambitious with respect to disaster risk reduction.
5. Institutions matter and coordination among them improves outcomes.
6. Reconstruction is an opportunity to plan for the future and to conserve the past.
7. Relocation disrupts lives and should be kept to a minimum.
8. Civil society and the private sector are important parts of the solution.
9. Assessment and monitoring can improve reconstruction outcomes.
10. To contribute to long-term development, reconstruction must be sustainable.

The last word: Every reconstruction project is unique.

Project Implementation

12 COMMUNITY ORGANIZING AND PARTICIPATION

Guiding Principles for Community Organization and Participation

- Reconstruction begins at the community level. A good reconstruction strategy engages communities and helps people work together to rebuild their housing, their lives, and their livelihoods.
- Community-based approaches require a somewhat different programming flow that begins with mobilizing social groups and communities and having the community conduct its own assessment.
- A very strong commitment and leadership from the top are needed to implement a bottom-up approach, because pressure is strong in an emergency to provide rapid, top-town, autocratic solutions.
- "The community" is not a monolith, but a complex organism with many alliances and subgroups. The community needs to be engaged in order to identify concerns, goals, and abilities, but there may not be consensus on these items.
- The scale at which community engagement is most effective may be quite small, for example, as few as 10 families.
- Engagement of the community may bring out different preferences and expectations, so agencies involved in reconstruction must be open to altering their preconceived vision of the reconstruction process.
- Numerous methods exist for community participation, but they need to be adapted to the context, and nearly all require facilitation and other forms of support.
- Transparency and effective communication are essential to maintaining engagement and credibility with the community and within the community during the reconstruction process.
- The reconstruction approach may affect the type and level of direct participation in reconstruction.

Introduction

Community participation is seen by some as a way for stakeholders to **influence** development by contributing to project design, influencing public choices, and holding public institutions accountable for the goods and services they provide.[1] Some view participation as the **direct engagement** of affected populations in the project cycle—assessment, design, implementation, monitoring, and evaluation—in a variety of forms. Still others consider participation an operating philosophy that puts affected populations **at the heart of** humanitarian and development activities as social actors with insights, competencies, energy, and ideas of their own.[2]

Community engagement has numerous benefits and is critical in every stage of post-disaster recovery and reconstruction. This chapter encourages agencies involved in reconstruction to offer affected communities a range of options for involvement in reconstruction.[3] It addresses the organization of affected communities and participation by individuals, communities, and community-based organizations (CBOs). (See Chapter 14, International, National, and Local Partnerships in Reconstruction, for a typology of civil society institutions that participate in reconstruction.)

Key Decisions

1. The **lead disaster agency** should work with affected communities, local government, and agencies involved in reconstruction to define the role of communities in planning and managing reconstruction. The agreements that emerge from this dialogue should be an integral part of the reconstruction policy.
2. **Affected communities** should decide how they will organize themselves to participate in the reconstruction effort.
3. **Agencies involved in reconstruction** should decide how they will support and empower communities to play the roles they have agreed to take on, and how two-way communication with communities will be established and maintained throughout reconstruction.

This Chapter Is Especially Useful For:
- Policy makers
- Lead disaster agency
- Local officials
- Project managers
- Affected communities

1. World Bank, 1996, *The World Bank Participation Sourcebook* (Washington, DC: The International Bank for Reconstruction and Development/ World Bank), http://www. worldbank.org/wbi/sourcebook/ sbhome.htm.
2. Active Learning Network for Accountability and Performance in Humanitarian (ALNAP), 2003, *Participation by Crisis-Affected Populations in Humanitarian Action: A Handbook for Practitioners* (London: Overseas Development Institute), http://www.alnap.org/resources/guides/participation.aspx.
3. Opportunities for participation are also discussed in other chapters of this handbook.

4. **Local government** should define its role(s) in supporting reconstruction at the community level, in consultation with affected communities, the lead disaster agency, and agencies involved in reconstruction.
5. **Agencies involved in reconstruction** should decide with communities how to monitor and evaluate the involvement of the community in reconstruction to ensure that agreements regarding role(s) and responsibilities are fulfilled on all sides. This monitoring should take place at both the community level and at the national level for the overall reconstruction program.

Public Policies Related to Community Participation

Incorporating participation by the community in reconstruction projects funded by local and international nongovernmental organizations (NGOs) is largely voluntary, yet the commitment to participation is generally quite high. However, the level of this commitment may vary in projects sponsored by the public sector. In their decentralization framework, planning law, or local government ordinances, some countries require community participation and information disclosure for publicly supported projects. This participation may include anything from public hearings on project budgets to comment periods on procurement documents to "sweat equity" contributions in community infrastructure projects. Compliance with these laws may at times be pro forma; this is even more likely to be the case if government is operating on an emergency footing and fears that projects will be delayed. Under these circumstances, the pressure—even from the communities themselves—may be to act quickly and to impose top-down, technocratic solutions.

Experience is increasingly demonstrating that an emergency is the time to expand, rather than reduce, participation, even if there is no formal policy framework for participation in place. By including properly structured community participation mechanisms, physical outcomes and

DANIEL PITTET

the quality of oversight can actually be improved, especially when large sums of money are involved. Local and international NGOs and other agencies involved in reconstruction can help operationalize these mechanisms. Or local government may be able to establish the guidelines and coordinate community participation in reconstruction. The role of local government is especially critical when local land use decisions and infrastructure reconstruction are involved.

If government is reticent to take a decentralized, participatory approach, agencies with experience in participatory approaches to reconstruction may want to present concrete examples of where these efforts have been successful in an effort to advocate for this approach on behalf of communities. A number of examples of good practice are included in the case studies below.

Technical Issues
Types of Participation

As shown in the table below, forms of community involvement differ in terms of the extent of citizen involvement in decision making and with respect to the desired outcomes.

Type of participation	Role of affected population	Level of control
Local initiatives	Conceives, initiates, and runs project independently; agency participates in the community's projects.	High
Interactive	Participates in the analysis of needs and in program conception, and has decision-making powers.	
Through the supply of materials, cash, or labor	Supplies materials and/or labor needed to operationalize an intervention or co-finances it. Helps decide how these inputs are used.	
Through material incentives	Supplies materials and/or labor needed to operationalize an intervention. Receives cash or in-kind payment from agency.	
By consultation	Asked for its perspective on a given subject but has no decision-making powers.	
Through the supply of information	Provides information to agency in response to questions but has no influence over the process.	
Passive	Informed of what is going to happen or what has occurred.	Low

Source: Active Learning Network for Accountability and Performance in Humanitarian (ALNAP), 2003, *Participation by Crisis-Affected Populations in Humanitarian Action: A Handbook for Practitioners (London: Overseas Development Institute)*, http://www.alnap.org/resources/guides/participation.aspx.

The Purposes of Participation

As the World Bank sees it, participation allows stakeholders to collaboratively carry out a number of activities in the program cycle, including the following[4]:

- Analyzing: Identifying the strengths and weaknesses of existing policies and service and support systems
- Setting objectives: Deciding and articulating what is needed
- Creating strategy: Deciding, in pragmatic terms, directions, priorities, and institutional responsibilities
- Formulating tactics: Developing or overseeing the development of project policies, specifications, blueprints, budgets, and technologies needed to move from the present to the future
- Monitoring: Conducting social assessments or other forms of monitoring of project expenditures and outputs

Other agencies have more expansive views of participation. Participation is known to have outcomes that are social in nature: empowering individuals, increasing local capacity, strengthening democratic processes, and giving voice to marginalized groups. Another set of benefits has to do with program effectiveness and leverage: creating a sense of ownership, improving program quality, mobilizing resources, and stimulating community involvement in execution.

Community Participation in Reconstruction

Since communities know the most about their own local environment, culture, vulnerabilities, requirements, and building techniques, reconstruction should be planned by them or, at a minimum, under their direction. However, a true community-based approach requires a different programming flow, one that begins not with assessment, but with mobilization of social groups and communities, which is then followed by a community-based assessment. This mobilization may be done by the community on its own initiative or as a response to signals from government about how reconstruction will be undertaken. Alternatively, agencies involved in reconstruction, including national and local NGOs, or local governments may initiate the mobilization process.

This mobilization may be more or less difficult, depending on the impact of the disaster and the nature of the pre-existing organization of the community. Communication with the community is a critical element of a successful participatory process. See 📖 Chapter 3, Communication in Post-Disaster Reconstruction, for extensive guidance on post-disaster communication.

The 📖 case study on the 2006 Java earthquake reconstruction, below, demonstrates how the reconstruction of thousands of housing units was managed by communities.

A very strong commitment and leadership from the top will be needed to implement a bottom-up approach, because pressure is strong in an emergency to provide rapid, top-town, autocratic solutions.

Conventionally trained planners may need to adjust their thinking in order to successfully participate in this type of reconstruction project. Also, because the success of this type of approach depends on community decision making, assistance may be needed to restart institutional mechanisms for consensual decision making and to establish or reestablish other governance structures

How Communities Participate in Reconstruction

Reconstruction activity	Opportunities for community participation in reconstruction
Assessment	Conduct: ■ housing assessment and census ■ community-led needs assessments ■ local environmental assessments ■ mapping of affected area and changes ■ stakeholder analysis
Planning and design	Prioritize and plan projects Carry out participatory site planning and site evaluations Identify targeting criteria and qualify households Participate in training (DRR and construction methods) Assist with grievance procedures
Project development and implementation	Carry out and/or oversee: ■ housing reconstruction, including housing of vulnerable households ■ infrastructure reconstruction ■ reconstruction of public facilities (schools, community buildings, and clinics) Manage: ■ financial disbursements ■ community warehouses
Monitoring and evaluation	Supervise construction Participate in monitoring and social audit committees Conduct participatory evaluations See the annex to 📖 Chapter 18, Monitoring and Evaluation, for guidance on conducting a social audit of a reconstruction project

An exemplary case of community participation in post-disaster planning is described below in the 📖 case study on the 2003 Bam earthquake reconstruction.

The reconstruction approach. The housing reconstruction approach will affect the level and type of community participation. 📖 Chapter 6, Reconstruction Approaches, discusses five approaches to reconstruction. Of those five, the Cash Approach (CA), Owner-Driven Reconstruction (ODR), and Community-Driven Reconstruction (CDR) offer the greatest opportunity for direct involvement in housing reconstruction. Owners have some limited opportunities for involvement in a Agency-Driven Reconstruction *In-Situ* (ADRIS) project; however, Agency-Driven Reconstruction in Relocated Site (ADRRS) largely excludes an owner from any role in the rebuilding effort. Housing reconstruction and infrastructure reconstruction offer different opportunities for community involvement that should be coordinated, but identified, planned, and managed separately[5]

Training and facilitation. Training and facilitation are key ingredients of a participatory approach to reconstruction. Communities need training that supports their particular role(s). Training in housing reconstruction methods is important if community members are acting as builders or overseeing housing reconstruction (see 📖 Chapter 16, Training Requirements in Reconstruction). If supervision of infrastructure projects is a community responsibility, some members will need training to understand plans and specifications. Facilitation is different from training, but is also critically important. Facilitation involves activities that help the community reestablish their decision-making processes, develop and implement plans, get access to resources, resolve conflicts, etc. Finding, training, and keeping good community facilitators are absolutely critical roles for government and agencies involved in participatory community-based reconstruction. Expect turnover in the ranks of facilitators, since it is a demanding job and requires establishing a rapport with the specific community. The experience with the use of facilitators in the Yogyakarta

4. World Bank, 1996. *The World Bank Participation Sourcebook* (Washington, DC: The International Bank for Reconstruction and Development/World Bank), http://www.worldbank.org/wbi/sourcebook/sbhome.htm.

earthquake reconstruction is summarized in ⬚ Annex 1 of
this chapter entitled How to Do It: Establishing a Community
Facilitation System for Post-Disaster Housing and Community
Reconstruction.

The Institutional Context

The context can enable or constrain participation, depending on
factors such as the enabling environment for participation; the
constraints created by the culture, including the culture of the
agency involved in reconstruction; and the community's prior
organization.

DANIEL PITTET

The enabling environment. The term "enabling environment" as
used here means the rules and regulations, both national and local,
that provide the freedom and incentives for people to participate.
Examining the legal framework within which affected beneficiaries and communities operate will
identify any legal constraints that must be addressed to permit genuine participation. There are at
least three important considerations: whether the community has access to information, whether
the community has the right to organize and enter into contractual agreements, and the project
approach taken by agencies involved in reconstruction.[6]

Reconstruction agency constraints. The participatory process can be affected by constraints
emanating from agencies involved in reconstruction, including enormous time pressure and political
pressure to resolve the housing problem and create on-the-ground results; a lack of commitment,
skills, or capacity to conduct participatory reconstruction activities; operating with a short-term
emergency mind-set rather than a development perspective; and an inability to make a long-term
commitment to a community because of the nature of the agency's programs. Agencies involved
in reconstruction may also have a limited understanding of the context, especially if it is complex
or changing rapidly, and may therefore be reticent to make plans with the community when the
outcomes are unpredictable.[7] A committee of agency and affected community representatives
could be created specifically to monitor the quality of participation in reconstruction and to
address community grievances related to this issue. Community participation can take time, but
time is also lost if opposition to projects arises because the community was excluded. In the 1993
Latur earthquake reconstruction, government–recognizing the limits of its capacity to manage
participation–appointed two respected nongovernmental organizations to assist them, as described
in the ⬚ case study, below.

Organization of the community. A community's organization can be invisible to outsiders, but
tools such as community assessments and institutional mapping can help reveal it and any effects
it may have on a proposed project.[8] A range of organizations with various degrees of formal
structure is already operating in any given community, performing a variety of functions, including
channeling community demands.[9] Planning intervention without understanding this reality not only
is disrespectful of the community, but also can create conflicts and lead to unexpected delays or even
rejection of the project. The sponsors of any new initiative—even if it is just a single project that
seeks the community's participation—need to decide how the project will relate to the community
as it is already organized. As early as possible, an analysis should be carried out of the community's
characteristics, including its organizational structure and its capabilities. The methodology
described in ⬚ Annex 2 to this chapter, How to Do It: Developing a Community Participation Profile,
will provide input needed to make decisions about the demand and starting point for community
participation.

The existing organizational structure may be based on wealth, political party, caste, culture, or
power relationships, among other things. Self-appointed spokespeople for the community and
organizations that claim to be representative of local community needs and aspirations, including
national NGOs, may not be seen as such by members of the community. The role of women will
need to be carefully considered in planning reconstruction, and women's organizations may have
an important role to play. The ⬚ case study on the 1992 Pakistan floods describes how a local NGO,
PATTAN, worked to broaden women's role in reconstruction.

5. United Nations Human Settlements
Programme (UN-HABITAT),
Regional Office for Asia and
the Pacific, n.d., "Community
Contracts," http://www.fukuoka.
unhabitat.org/event/docs/
EVN_081216172311.pdf.

Formalizing community involvement. For community-based reconstruction, community contracts are a tool developed by the United Nations Human Settlements Programme (UN-HABITAT) to establish the terms of community involvement.[10] The formalization of the involvement of NGOs is addressed in 🖺 Chapter 14, International, National, and Local Partnerships in Reconstruction.

Participation Strategy and Tools

While the participation strategy may be best refined during the participatory process itself, agencies involved in reconstruction may want to define for themselves the basic parameters before the process is set in motion. A participation strategy defines why participation is called for, proposes who will be involved, and defines the objectives. It also defines the purpose of the participatory activities, which participation approach is most suitable, the tools and methods to be used, whether community members will be engaged directly or through existing organizational structures, and which, if any, partnering agencies will be involved.

It is not necessary to create participatory processes; over the years, organizations have systematized myriad instruments and methodologies that can be adapted to the context in which the participation will take place. The table below contains examples.[11]

Tools for Facilitating Community Participation

Contextual analysis	Understanding stakeholders	Identifying assets and vulnerabilities	Defining needs, demands, and projects
▪ Interviews with key informants ▪ Storytelling ▪ Focus groups ▪ Timelines ▪ Mapping damage, risks, land uses ▪ Activity or climatic calendars ▪ Community mapping	▪ Socio-anthropological analysis ▪ Participatory stakeholder analysis ▪ Interaction diagrams ▪ Venn diagrams ▪ Proximity-distance analysis ▪ Wealth ranking	▪ Capacity and vulnerability analysis ▪ Proportional piling ▪ Institutional analysis ▪ Cultural asset inventories	▪ Surveys ▪ Hearings ▪ Participatory planning ▪ Design charts ▪ Participant observation ▪ Preference ranking ▪ Information centers and fairs

Who Are the Stakeholders?

Stakeholder involvement is context specific; thus, who needs to or is willing to be involved varies from one project to another. The figure on the next page graphically depicts the connections among a common set of project stakeholders. The table below uses a hypothetical project (helping a community avoid relocation by implementing structural measures to reduce risk) to show common stakeholder categories.

6. World Bank, 1996, *The World Bank Participation Sourcebook* (Washington, DC: The International Bank for Reconstruction and Development/World Bank), http://www.worldbank.org/wbi/sourcebook/sbhome.htm.
7. ALNAP, 2003, *Participation by Crisis-Affected Populations in Humanitarian Action: A Handbook for Practitioners* (London: Overseas Development Institute), http://www.alnap.org/resources/guides/participation.aspx.
8. Jeremy Holland, 2007, *Tools for Institutional, Political, and Social Analysis of Policy Reform. A Sourcebook for Development Practitioners* (Washington, DC: World Bank), http://siteresources.worldbank.org/EXTTOPPSISOU/Resources/1424002-1185304794278/TIPs_Sourcebook_English.pdf.
9. World Bank, 1996, *The World Bank Participation Sourcebook* (Washington, DC: The International Bank for Reconstruction and Development/World Bank), http://www.worldbank.org/wbi/sourcebook/sbhome.htm.

Type of stakeholder	Example
Those who might be affected (positively or negatively) by the project	Homeowners who prefer to relocate the community versus homeowners who prefer the existing site
The "voiceless" for whom special efforts may have to be made	Squatters who risk being relocated if structural disaster risk reduction investments are built
The representatives of those likely to be affected	Existing community group that has managed the response
Those who have formal responsibility related to the project	Government risk management agency or local planning department
Those who can mobilize for or against the project	Unaffected communities that were already awaiting assistance now delayed by the disaster-related project
Those who can make the project more effective by participating or less effective by not participating	Another NGO working on a related issue in the same community
Those who can contribute financial and technical resources	Microfinance institution or governmental agency
Those whose behavior has to change for the effort to succeed	Government agency already planning the community's relocation
Those who must collaborate for the project to succeed	Landowner who will need to sell land where structural measures will be built

International Aid Organizations

International Non-Governmental Organizations

Local committees

Affected Populations

Individuals and families

National Aid Organizations

Local NGOs

CBOs

Government Institutions

Source: ALNAP, 2003, *Participation by Crisis-Affected Populations in Humanitarian Action: A Handbook for Practitioners* (London: Overseas Development Institute), http://www.alnap.org/resources/guides/participation.aspx.

The level of power, interests, and resources of each stakeholder will affect that stakeholder's ability to collaborate. Therefore, an environment needs to be created in which stakeholders can participate and interact as equals. Consensus-building is not always easy; specific measures may need to be taken to promote negotiation and resolve disputes.

Stakeholders of a project may not all have equal status, because they have different "stakes" in project outcomes. For instance, the head of a household that may be relocated has more invested in the outcome of a relocation project than the representative of the local planning department, although both are considered stakeholders.

The Unintended Consequences of Participation

Participation empowers communities; however, the outcomes of that participation can be unpredictable. The participatory process may give rise to new actors or interests or may create conflicts between organizations that had previously worked together harmoniously. Guiding the participation process includes making sure that people's expectations are realistic, especially if they believe that large amounts of funding are available. At the same time, an agency may observe a multiplier effect from its support of a participatory project, as the community realizes its capabilities and new ideas for activities and projects emerge.

The organization and facilitation of community participation should not be done on a purely *ad hoc* basis. Trained facilitators and other experts in community participation should be part of the management team for any project that entails participation. See the text box entitled "The Role of Facilitators in Empowering Community Reconstruction," in Chapter 6, Reconstruction Approaches, for guidance on this topic.

Risks and Challenges

- Government forgoing genuine participation, due to political and social pressures to show that the reconstruction process is advancing.
- Lack of support by the community for the reconstruction project because of limited involvement of stakeholders, particularly the affected community, in planning and design.
- Failing to understand the complexity of community involvement and believing that "the community" is a unified, organized body.
- Ignoring how the community is already organized when introducing participatory activities.
- Underestimating the time and cost of genuine participatory processes.
- Conducting poorly organized opinion surveys and believing that the responses to those surveys are representative of the community.

10. United Nations Human Settlements Programme (UN-HABITAT), Regional Office for Asia and the Pacific, n.d., "Community Contracts," http://www.fukuoka.unhabitat.org/event/docs/EVN_081216172311.pdf.
11. ALNAP, 2003, *Participation by Crisis-Affected Populations in Humanitarian Action: A Handbook for Practitioners* (London: Overseas Development Institute), http://www.alnap.org/resources/guides/participation.aspx.

- Failing to find or develop facilitators and trainers who understand and believe in the community-based approach.
- Rejecting established models of community organization—or alternatively blindly adopting models from other countries or contexts—without evaluating how they should or could be adapted to the specific conditions of the locality in question.
- Thinking that all community organizations are democratic and representative, or forgetting that they have their own agendas.
- Confusing the role of national NGOs with that of genuine CBOs.
- Agencies believing that they are being participatory by establishing a relationship with one specific local organization or spokesperson.

Recommendations

1. Analyze the community's capacity and preferences for participation by working with the community to carry out a Community Participation Profile early in the reconstruction process.
2. Work with the community to reach agreement not only on how it will organize itself, but also on activities and outcomes, i.e., the reconstruction priorities, projects, and goals.
3. Find the right scale for community involvement, which may be smaller than expected.
4. Provide the facilitation and support to make the community an effective actor in reconstruction, and involve the community in monitoring the quality of this support. There will be turnover in the ranks of facilitators, so providing the community with proper support is a continuous process.
5. Consider creating a monitoring mechanism with representation from both the agencies involved in reconstruction and the community, specifically to monitor the quality of community involvement.
6. Do not hesitate to demand good governance and accountability from the community, especially if funding is involved.
7. Do not disempower existing community initiatives by introducing new and unfamiliar organizational structures that compete; find ways to combine forces.
8. Consider using existing tools that foster participation, but make sure that they are adapted to the project and context.
9. Understand that stakeholder identification is one of the most important steps in a participatory process; use participatory methods to identify and engage stakeholders.
10. Understand that community participation can have unintended consequences. Maintain a constructive relationship with participants, and look for opportunities to support additional activities that spin off from the original participatory process.

Case Studies
2006 Java Earthquake, Indonesia
Organizing Community-Based Resettlement and Reconstruction

Somewhat hidden from the world by the ongoing flurry of Aceh tsunami recovery, the 2006 Java earthquake with a magnitude of 6.3 on the Richter scale was nevertheless an enormously destructive event. Over 350,000 residential units were lost and 5,760 persons were killed, most from the collapse of non-engineered masonry structures. Using lessons learned from the tsunami experience and resources from the ongoing Urban Poverty Project (UPP), the Indonesian government was able to respond quickly and efficiently. Facilitators were recruited and villages elected boards of trustees, which later were instrumental in organizing community meetings and supervising implementation. Key activities included (1) identifying beneficiaries and prioritizing the most vulnerable; (2) establishing housing groups of 10-15 families, who chose their leaders and a treasurer; (3) developing detailed plans to use the construction grants for each group; (4) opening group bank accounts; and (5) obtaining approval of plans, disbursement in tranches, and group procurement, construction, and bookkeeping. Training was provided to community members and local workers to ensure earthquake-resistant construction. Later, the community developed plans to rebuild village infrastructure and facilities, with a particular focus on disaster resilience. Communities conducted self-surveys, prepared thematic maps, analyzed needs and disaster risks, agreed on priority programs, and established procedures for operations and maintenance. Grants for infrastructure were also disbursed in tranches through the selected bank as work progressed. An adequate understanding of rules and a sense of ownership by the community were essential to ensuring good targeting and plans, accountability, and social control of implementation. The involvement of women increased accountability and enhanced the appropriateness of technical solutions. The role of facilitators was crucial, as they both ensured effective communication and adaptability of the program to local situations as well as compliance with program principles. In all 6,480 core houses were funded by a World Bank loan under UPP, and another 15,153 units were funded by the multi-donor Java Reconstruction Fund. This approach to reconstruction became the model for the much larger government-financed rehabilitation and reconstruction program, under which about 200,000 houses were rebuilt in Java.

Sources: Sri Probo Sudarmo, World Bank, 2009, personal communication; *and* World Bank, 2007, *Community-Based Settlement Rehabilitation and Reconstruction Project for Central and West Java and Yogyakarta Special Region*, project documents, http://web.worldbank.org/external/projects/main?pagePK=64283627&piPK=73230&theSitePK=226309&menuPK=287103&Projectid=P103457.

2003 Bam Earthquake, Iran

Community Participation in Developing the Structure Plan (2015) for the City of Bam

After the 2003 earthquake in Bam, Iran, a national strategy for housing reconstruction was published. For urban areas, the strategy featured (1) provision of interim or transitional shelters on existing vacant lots, including the distribution of prefabricated units to address housing needs for a 2-year period; and (2) provision of permanent shelter after preparation of a detailed city master plan and the approval of technologies and legal and procedural mechanisms for reconstruction.

VICTORIA KIANPOUR

The provision of interim shelter in the city of Bam gave government time to revise the existing city plan before beginning reconstruction. The most recent Bam City Master Plan had been developed by a consulting firm and approved by the High Council of Architecture and Urban Development of the Ministry of Housing and Urban Development in the year prior to the earthquake. However, the disaster raised significant new issues, so the same consulting firm was brought back to update the plan. A comprehensive survey sought inputs from local authorities, implementing agencies, community leaders, NGOs, women, youth, and children.

In April 2004, the Housing Foundation of the Islamic Revolution-United Nations Development Programme (UNDP) joint housing project organized a technical consultation in which UNDP; the United Nations Children's Fund (UNICEF); the United Nations Educational, Scientific and Cultural Organization (UNESCO); the World Health Organization (WHO); the United Nations Industrial Development Organization (UNIDO); and other UN agencies provided technical assistance and capacity building for the participatory city micro-planning process, to explain such concepts as child-friendly and healthy cities and to discuss the socioeconomic aspects of city planning. The final Structure Plan specifically addressed the need to respect the traditional architecture and urban design of the city and villages, to protect buffer zones, to minimize relocation, and to minimize expropriation through reuse of land. This plan formed the basis for subsequent detailed planning of 11 priority reconstruction areas in the city of Bam. To reduce the chance of excessive uniformity, each area had a different planning team. The modified plan and detailed plans were ratified by the High Council in October 2004, 10 months after the earthquake. Subsequently, the Ministry of Housing and Urban Development, with support from UNDP, published the results of the consultative process and the Bam Housing Typology.

Sources: Victoria Kianpour, UNDP Iran, 2009, personal communication *and* World Bank, 2009, "Planning for Urban and Township Settlements after the Earthquake," http://siteresources.worldbank.org/CHINAEXTN/Resources/318949-1217387111415/Urban_planing_en.pdf.

1993 Latur Earthquake, Maharashtra, India

Community Participation in the Maharashtra Emergency Earthquake Rehabilitation Program

With the help of the World Bank, the government of Maharashtra, India, developed the Maharashtra Emergency Earthquake Rehabilitation Program (MEERP), which institutionalized community participation and ensured that beneficiaries were formally consulted at all stages of the post-earthquake program. Every village created a local committee headed by the *sarpanch* (the head of the village council), and its subcommittees included women and disadvantaged groups. Consultative committees were also proposed at the level of the *taluka* (an administrative unit that includes several villages) and the district. To ensure the village-level committees interacted with the project management unit at all levels, government took an innovative step and appointed two respected community organizations to carry out the process, the Tata Institute of Social Sciences and the Society for Promotion of Resource Area Centre.

Source: Rohit Jigyasu, 2002, "Reducing Disaster Vulnerability through Local Knowledge and Capacity" (PhD thesis, Trondheim: Norwegian University of Science and Technology), http://ntnu.diva-portal.org/smash/record.jsf?searchId=1&pid=diva2:123824.

1992 Floods, Pakistan
Grassroots NGO Introduces Measures to Engage Women in Housing Reconstruction

Northern Pakistan's catastrophic floods in 1992 were attributed to large-scale deforestation in mountainous watersheds, and led eventually to government imposing a ban on commercial harvesting of forests. After the floods, PATTAN, a local NGO, introduced a number of measures that specifically addressed women's issues in the disaster recovery process. Female relief workers were engaged to assess the needs of women after the floods and to involve them in the planning, implementation, and rehabilitation activities. Local women were registered as heads of their households to help ensure efficient distribution of relief food. Village women's organizations were established (in parallel with men's groups) to articulate women's needs and to take responsibility for community development. These groups also provided a forum for discussing women's views regarding the design and layout of new houses. As a result, women became actively involved in reconstruction activities. Later, women were made responsible for collecting money to repay loan installments on the houses. Some women also participated in construction, traditionally a male activity. Perhaps most important, PATTAN introduced the concept that married couples should own houses jointly.

Source: World Bank, Food and Agriculture Organization, and International Fund for Agricultural Development, 2008, *Gender in Agriculture Sourcebook* (Washington, DC: World Bank), http://publications.worldbank.org/ecommerce/catalog/product?item_id=8612687.

Resources

Abarquez, Imelda, and Zubair Murshed. 2004. *Community-Based Disaster Risk Management: Field Practitioners' Handbook.* Bangkok: ADPC, UNESCAP, and DIPECHO. http://www.adpc.net/pdr-sea/publications/12Handbk.pdf.

Active Learning Network for Accountability and Performance in Humanitarian Action (ALNAP). 2003. *Participation by Crisis-Affected Populations in Humanitarian Action: A Handbook for Practitioners.* London: Overseas Development Institute. http://www.alnap.org/resources/guides/participation.aspx.

Davidson, C. H. et al. 2006. "Truths and Myths about Community Participation in Post-Disaster Housing Projects." *Habitat International.* Volume 31, Issue 1, March 2007, pp. 100–115. http://www.cbr.tulane.edu/PDFs/davidsonetal2006.pdf.

Dercon, Bruno, and Marco Kusumawijaya. 2007. "Two Years of Settlement Recovery in Aceh and Nias. What Should the Planners have Learned?" (Paper for 43rd International Society of City and Regional Planners Congress). http://www.isocarp.net/Data/case_studies/952.pdf.

Office of the United Nations High Commissioner for Refugees. 2006. *The UNHCR Tool for Participatory Assessment in Operations.* Geneva: United Nations. http://www.unhcr.org/publ/PUBL/450e963f2.html.

UN-HABITAT. Government of Indonesia, Provincial Government of Aceh, and the United Nations Development Programme. 2009. "Post-Tsunami Aceh-Nias Settlement and Housing Recovery Review."

UN-HABITAT. 2007. "People's Process in Aceh and Nias (Indonesia), Manuals and Training Guidelines." (In English and Bahasa). http://www.unhabitat-indonesia.org/publication/index.htm#film.

Volume 1	Orientation and Information:	http://www.unhabitat-indonesia.org/files/book-153.pdf
Volume 2	Community Action Planning and Village Mapping	http://www.unhabitat-indonesia.org/files/book-1407.pdf
Volume 3	Detailed Technical Planning for Housing and Infrastructure	http://www.unhabitat-indonesia.org/files/book-1417.pdf
Volume 4	Housing and Infrastructure Implementation	http://www.unhabitat-indonesia.org/files/book-1420.zip
Volume 5	Completion of Reconstruction Works	http://www.unhabitat-indonesia.org/files/book-1421.pdf
Volume 6	Monitoring, Evaluation and Controls	http://www.unhabitat-indonesia.org/files/book-225.pdf
Volume 7	Socialization and Public Awareness Campaign	http://www.unhabitat-indonesia.org/files/book-226.pdf
Volume 8	Training and Capacity Building	http://www.unhabitat-indonesia.org/files/book-229.pdf
Volume 9	Complaints Handling	http://www.unhabitat-indonesia.org/files/book-231.pdf

UN-HABITAT. Regional Office for Asia and the Pacific. n.d.. "Community Contracts." http://www.fukuoka.unhabitat.org/event/docs/EVN_081216172311.pdf.

World Bank. 1996. *Participation Sourcebook.* Washington, DC: The International Bank for Reconstruction and Development/World Bank). http://www.worldbank.org/wbi/sourcebook/sbhome.htm.

World Health Organization. 2002. "Community Participation in Local Health and Sustainable Development: Approaches and Techniques." http://www.euro.who.int/document/e78652.pdf.

How to Do It: Establishing a Community Facilitation System for Post-Disaster Housing and Community Reconstruction

The reconstruction process following the 2006 Yogyakarta and Central Java, Indonesia, earthquake demonstrated the effectiveness of a community-based approach to reconstruction. More than 150,000 houses were reconstructed in the first year following the earthquake and, by the second anniversary of the earthquake, a total of 275,000 houses had been built using a community-based model.

However, this reconstruction model entails establishing a community facilitation system. The facilitation system depends on the recruitment, training, and deployment of community facilitators. Finding enough quality facilitators and getting them into affected communities quickly allows reconstruction to be scaled up and gives people certainty about how reconstruction will proceed and what their role in reconstruction will be.

This certainty is considered to be an important factor in the satisfaction of the population with the program.

This section describes the approach to the use of community facilitators in the Community-Based Settlement Rehabilitation and Reconstruction Project funded by the Java Reconstruction Fund in Yogyakarta/Central Java reconstruction from 2005–2007, as well as previously in the 2004 Indian Ocean tsunami reconstruction in Aceh. While the participatory methodologies were similar for all reconstruction, in this program, each community also received a pool of funds to help finance infrastructure improvements that contributed to risk reduction in the community. More than 15,000 housing units were financed by this program in Yogyakarta and Central Java.

Key Features of the Facilitation System

Feature	Explanation
Recruitment	Facilitators were chosen from people who had qualifications in one of the following areas: engineering or construction, finance, and community development or organizing. All facilitators needed to have practical skills, as well as the ability to work with communities to empower them to carry out their role in reconstruction and to manage community expectations.
	The selection process for facilitators was managed by an outside consultant, and included:
	■ written application
	■ interview
	■ psychological testing
	The psychological testing component was carried out by the psychology departments of local universities, under the supervision of the recruitment consultant.
	Because community-based projects were a major source of post-disaster construction financing, the compensation offered to facilitators reflected no more than the market rate for their level of training and experience so that the hiring of community facilitators would not contribute to a post-disaster escalation of salaries in the market.
Training	Candidates who passed the recruitment process received approximately three weeks of training in two components.
	Basic. All candidates received the same basic training, during which time they were still being evaluated and following which they had to pass an examination. The trainers explained the facilitation process and the "people skills" that were required. Facilitators were taught that the building of houses is the entry point that gives them the opportunity to organize the community, but that the process they were facilitating is about community mobilization and empowerment, not just housing construction.
	Technical. Each facilitator that passed the basic training was then assigned to one of three roles—community development, technical (construction), or finance—for additional training. In this component of training, they received instruction on training community members in the procedures of the project. For instance, finance facilitators were taught how to train community members to manage project finances.
Assignment of Facilitators	Facilitators were organized into teams of nine people, consisting of two community development facilitators, two engineering facilitators, one finance facilitator, and four construction inspectors (called building controllers). This team provided support to a community of approximately 275 households over a period of six months.

Feature	Explanation
Oversight	Oversight was provided through weekly visits of financial, community development, and technical experts to each project, where they identified problems specific to a particular community, as well as general problems within the program. When general problems were identified, facilitators were called together for additional training or problem solving. Facilitators' log books were reviewed by the experts during their visits. Facilitators were evaluated on the quality of the results in the community, and their salary could be held back if project standards and milestones were not met.
	Monitoring was an essential element, and provided detailed information on the progress of every house and follow-up on any complaints, all of which was managed on a Web site, accessible to the public, which was designed specifically for this purpose.
Community Leadership	The success of this model depends on the involvement of the community. Every aspect of the project is run by the community. The facilitators simply make the community more effective in carrying out its roles. The roles of the community members include (1) prioritizing, building, and overseeing infrastructure projects; (2) managing their own housing reconstruction; (3) managing project finances; (4) handling complaints; (5) selecting beneficiaries through a participatory process; and (6) leading collective action when it is required.

Costs and Benefits of the Facilitation Model

This model is considered very cost-effective, since, for the housing component, only about 5 percent of program costs were spent on the facilitation process. (The percentage varied on the infrastructure component because it was smaller and projects were more diverse.)

However, the community-based reconstruction model used in Yogyakarta is not without its challenges. Some of the principal challenges that must be overcome to make this model work well are following:

1. Finding the right people to be facilitators
2. Thinking big. That is, scaling up the process quickly enough, so that communities have certainty within a short time about what is going to happen
3. Preventing facilitators from being hired away by other agencies once they are trained
4. Bureaucratic bottlenecks that slow down disbursements to communities and payment of facilitator salaries and operating costs

The Bank is working to document and systematize its experience with this model. Even without completion of this documentation, a community-based model for reconstruction has become the official reconstruction model of the Indonesian government, and is expected to be applied—with some minor improvements, based on the Aceh and Yogyakarta experiences—in the reconstruction following the 2009 Padang earthquake. One measure that has been suggested to make the model even more effective is the training of a nationwide cadre of facilitators who could be quickly mobilized, thus reducing the need to conduct the recruitment and training on short notice following a disaster.

Source: World Bank Indonesia team, 2009, personal communication.

The Community Participation Profile (CPP) serves as a reference for program development for government and agencies involved in reconstruction alike. It assists agencies in making judgments about the feasibility of and the starting point for community participation. It can also help the community to define its own requirements. The key questions that the CPP answers are:

1. Is there is a viable community structure in place that can establish priorities and respond to the most needy and vulnerable?
2. Will the community need help to manage its finances and provide oversight over community resources?
3. Are systems in place to ensure transparency and accountability at the community level?

4. What resources and skills are available in the community to contribute to reconstruction or other aspects of recovery (skilled and unskilled labor, building materials, land, and wealth)?
5. What are the attitudes toward and demand for participation in the reconstruction process?
6. What training will be required for the community to be successful in carrying out its responsibilities?

Not surprisingly, consultation with and involvement of a community are the best approaches for gathering the information needed to develop a CPP. The entire community does not need to participate; however, those who do should be relatively representative of the affected community that may participate in reconstruction.

Steps in Designing a Community Participation Profile

Define approach and objectives

Use town hall meetings and working groups. Engage with the community from the start of a reconstruction program to establish a transparent working relationship. Prepare and review an agenda for working sessions that includes identifying expected outcomes from the meetings. Working groups should include a broad range of community members. If good representation of all groups is difficult to accomplish, hold additional sessions to gather other views. Elements to discuss with the community are presented below.

1. Expected outputs	Explain and validate what is expected from the process, including:
	■ a description of the community's population;
	■ an organigram of the community's political structure;
	■ a list of the community's skills and resources;
	■ a statement of the community's commitment to participate and any conditions that may limit or enhance participation;
	■ a history of past experience in disaster reconstruction; and
	■ priorities and needs in the reconstruction program, such as retrofitting, relocation, and livelihood activation.
2. Facilitation	Identify facilitators with local language capacity from the community or outside to manage the consultations.
3. Review	Organize community review sessions to verify and validate results of the consultations.

Data collection

Divide the community participants into working groups. Collect data and conduct analysis using existing data and new data collected by working groups.

1. Population	Population and demographics of the community, and impact of the disaster on these characteristics
2. Education	Literacy by gender, age, economic group
3. Cultural aspects	Languages, religions, and customs that enhance or limit participation, such as women's ability to participate in meetings, segregated sessions, and preferences for community gatherings
4. Resources/skills (human, technical, financial)	Community structure, how representative, inclusive, and participatory it is
	Experience in managing funds, designing and implementing activities, monitoring and evaluation
	Nature of any system for community financial contribution
	Nature of household economic activities

5. Current community responsibilities	Community responsibilities, such as school maintenance, service provision, etc.
6. Community political structure	How local governance structures are put in place (election, appointment, etc.)
	Whether there is a traditional structure in addition to the political-administrative system
7. Reconstruction program	Tap into the housing condition assessment, among other sources, to identify what will need to be accomplished in reconstruction
8. Attitudes, demands, and expectations	Map the range of attitudes, expectations, and level of interest of various subgroups of the community toward participation in the reconstruction program.

Validate data

Organize community review sessions to verify and validate results of the data collection and consultations.

Present findings and agree on action plan

The findings of the community self-assessment will help determine the assistance that will be necessary to support a participatory approach to reconstruction. This activity should be done jointly with agencies offering to assist with reconstruction in the community, since it will help them understand what will be necessary to successfully carry out participatory reconstruction. After the findings are reviewed, the following issues must be agreed upon.

1. Organizational proposal	How to structure participation, such as size of groups, internal organization, etc. Roles and responsibilities of individuals, families, local government, etc.
2. Governance structure	How to organize decision making, measures for transparency and accountability, monitoring, sanctions, etc.
3. Need for institutional strengthening	Requirements to improve governance, transparency, financial administration of groups
4. Need for facilitation	Proposal for scope and nature of facilitation function
5. Need for training	Training activities for individuals, their focus, desired results, institutions that can provide the training, sources of funding
6. Need for outreach	Outreach activities that will be necessary to include the larger community in the reconstruction program
7. Budget	Budget for community participation activities

For access to additional resources and information on this topic, please visit the handbook Web site at www.housingreconstruction.org.

13 INSTITUTIONAL OPTIONS FOR RECONSTRUCTION MANAGEMENT

Guiding Principles for Reconstruction Management

- Government should lead the effort to define reconstruction policy and should coordinate its implementation. These policy decisions must be properly communicated to the public.
- Best practice is to establish a reconstruction policy and an institutional response structure, including one for housing and community reconstruction, before a disaster.
- The institutional arrangements for managing reconstruction should reflect reconstruction policy. The agency put in charge should be provided with a mandate, a workable structure, and a flexible operational plan.
- The reconstruction agency, even if it is new or temporary, must work closely with existing line ministries and other public agencies to provide efficient and effective post-disaster reconstruction.
- Mechanisms are needed to coordinate the actions and funding of local, national, and international agencies involved in reconstruction and to ensure that information is shared among them.
- Funding must be allocated equitably and should stay within agreed limits. Broad controls and good monitoring of all sources minimize corruption.

This Chapter Is Especially Useful For:
- Policy makers
- Lead disaster agency
- Agencies involved in reconstruction

Introduction

The management of the recovery and reconstruction process following a major disaster presents massive and often unprecedented challenges to any country, especially those with limited or no prior experience with such situations. Where post-disaster recovery planning that anticipates institutional requirements exists, early recovery is likely to go more smoothly. Where plans are not in place, governments may need advice on designing an appropriate organizational structure to manage reconstruction. In either case, assistance may be needed to put the reconstruction management arrangements in place and to establish an effective system of coordination among governmental and nongovernmental entities. This assistance may be provided by humanitarian agencies. See 📖 Chapter 1, Early Recovery: The Context for Housing and Community Reconstruction. Coordination is particularly important for housing and community reconstruction, since a large number of organizations are often involved.

This chapter analyzes organizational options for the management of post-disaster housing reconstruction in the context of the larger disaster management institutional framework. For a discussion of partnerships and the role of civil society, see 📖 Chapter 14, International, National, and Local Partnerships in Reconstruction.

Key Decisions

1. **National and local governments** should decide on their respective reconstruction approach before a disaster, by defining policies and designing at least the general outlines of the institutional structure of the reconstruction agency.
2. Based on the results of the initial assessment, in which local government must participate, the **lead disaster agency** and **local government** need to make specific decisions about how housing and community reconstruction will be managed.
3. **Agencies involved in reconstruction** should prioritize and decide how to incorporate the strengthening of governmental capacity into their post-disaster assistance strategies.
4. Immediately after the disaster, the **lead reconstruction agency** (in consultation with affected communities, local government, national and international nongovernmental agencies, and the private sector) must design and implement the mechanisms that will be used to coordinate the reconstruction activities of all participating entities.

5. **Local government** needs to decide on and cement the partnerships it will need to ensure reconstruction takes place at the local level in an efficient and equitable manner, beginning with its relationship with the affected community, and including cooperation with nongovernmental organizations (NGOs) and the private sector.

6. **Agencies involved in reconstruction** need to decide on the mechanisms they will use to guarantee a two-way flow of information with affected communities. They also need to design consistent messages that empower those communities.

Public Policies Related to Institutional Arrangements

If government has developed a proper institutional strategy for disaster management, its arrangements should be put into operation once a disaster strikes. The disaster strategy may need to be adjusted, depending on the scope and nature of the disaster, but to the extent possible, this strategy should govern. Ideally, the strategy addresses both reconstruction and response, since—as explained throughout this handbook—reconstruction begins the day of the disaster. Given the local nature of housing and community reconstruction, the institutional strategy for reconstruction should provide a central role for local government and the affected communities themselves.

DANIEL PITTET

More commonly, there is no predefined institutional strategy, or it covers only the initial response. In these cases, roles and responsibilities are likely to be assigned in an ad hoc manner. A risk in these situations is that responsibility is unnecessarily centralized. Where local capacity was weak pre-disaster or is greatly weakened by the disaster, centralized responsibility for reconstruction may be warranted. But in other cases, the existing framework for the assignment of responsibilities between levels of government, such as the country's policy frameworks related to decentralization and community development, should be given great weight in formulating the institutional arrangements for planning and implementing post-disaster housing and community reconstruction.

Technical Issues
Planning for Reconstruction before a Disaster

Pre-disaster planning for post-disaster response and reconstruction is becoming more common. The institutional arrangements for response and reconstruction can be defined as part of either an emergency management plan or a broader disaster risk management plan. In developing the plan, it may be necessary to overcome differences in perspective between officials responsible for emergency management and those who would manage reconstruction. The scope of a post-disaster reconstruction policy is described in 🖻 Chapter 2, Assessing Damage and Setting Reconstruction Policy.

The Need for Authority, Autonomy, and Political Support

To wield the necessary authority to get the job done right, a reconstruction agency needs both autonomy and support from the highest level of government and the political system, and broad-based support from the other agencies and communities affected by the reconstruction process. Support from the top helps minimize political interference: individual politicians or parties trying to capture resources for their own constituents who were not affected by the disaster, for instance, or contracts being intercepted for individual gain. Support from below strengthens cooperation, thus making the collective reconstruction effort more effective and equitable. Support builds when actors see that decisions are based on transparent policies and accurate data about the impact of the disaster on the surviving population.

The need for authority, autonomy, and political support applies at each level where responsibility is vested. This chapter focuses on the requirements to manage housing and community reconstruction in the context of the overall institutional requirements for reconstruction; however, these same issues apply to local governments and communities, to the extent they are delegated a critical role in reconstruction.

Rebuilding Governmental Capacity

A disaster may be confusing and overwhelming for government officials, especially local officials. Both national and local political leaders will want to show rapid results in response to public pressure for recovery. There may be pressure on local officials to ignore local policies (such as land use regulations) and procedures (such as building code enforcement) or to turn responsibilities over to international aid agency staff who may be "disaster veterans."

Rebuilding governmental capacity is an essential prerequisite for physical, social, environmental, and economic reconstruction. Having a governance structure in place during reconstruction helps ensure the sustainability of the investments that have been made and policies that have been established. For local government, this may mean such practical activities as rebuilding staffing levels and recovering records damaged by the disaster. National reconstruction agencies are unlikely to have a ready-made strategy for overcoming local institutional weaknesses. International financial institutions (IFIs) and other funding sources, however, can gain great leverage by providing technical, financial, and material support to reestablish local government capacity.

Organizational Options

Various models have been adopted by governments for reconstruction management. Each has its own distinctive merits as well as demonstrated weaknesses.

In large-scale disasters, or where government capacities are limited, creating a new dedicated agency may be the best alternative. The experience of earthquake reconstruction in Marmara, Turkey, is an example of how having a dedicated implementation unit can streamline the recovery process for the physical reconstruction, although it may not have contributed to institution building or improved longer-term mitigation. Managing international appeals for support; arranging large credits and grants from donors and IFIs; and managing procurement, disbursement, monitoring, and evaluation also present huge challenges in the aftermath of large-scale disasters. These functions alone may require a new institutional solution.

Some well-managed recovery operations have been undertaken using existing line ministries and departments. 🏠 Case studies on the National Institute for Disaster Management in Mozambique and the Housing Foundation in Iran, below, show how existing institutions can be effective managers of post-disaster reconstruction.

Not only do the structures that have been used to manage reconstruction vary greatly, they are also frequently seen to evolve over time. Models from one disaster are difficult to import directly into a new disaster context. The 🏠 case study below on FOREC, the institutional model adopted after the 1999 Armenia earthquake reconstruction, and FORSUR, created after the 2007 Ica/Pisco, Peru, earthquake, demonstrates how institutional solutions may be context-specific.

Following any disaster, governments need to decide fairly quickly on the institutional arrangements for recovery and reconstruction management. The following table indicates the principal organizational options with their inherent strengths and weaknesses.[1]

Having a governance structure in place during reconstruction helps ensure the sustainability of the investments that have been made and policies that have been established. For local government, this may mean such practical activities as rebuilding staffing levels and recovering records damaged by the disaster.

1. Wolfgang Fengler, Ahya Ihsan, and Kai Kaiser, 2007, *Managing Post-Disaster Reconstruction Finance, International Experience in Public Financial Management*, World Bank Policy Research Working Paper 4475 (Washington, DC: World Bank). http://go.worldbank.org/YJDLB1UVE0.

🏠 **For access to additional resources and information on this topic, please visit the handbook Web site at www.housingreconstruction.org.**

Organizational Models for Reconstruction

Option 1. Create new dedicated organization or task force
(applicable in centralized or decentralized context)

Strengths	Weaknesses	Recommendations
Highly independent, focused	Risks relegating line ministries to the sidelines and duplicating their efforts	Seriously consider for large-scale disasters
Provides mechanism for resource allocation, procurement, and staffing	Takes time to clarify roles and responsibilities	Employ if government is decimated by the disaster or involved in civil conflict
Handles complex financial arrangements with international donors	May lack local ownership	"Sunset" clauses critical to avoid agency surviving beyond its mission
Simplifies consultation with government	Expensive; requires premises, facilities, and staff	The 🖼 case studies on the 1985 Mexico City earthquake and the 2005 North Pakistan earthquake responses, below, show how a new entity is sometimes a very effective solution.
Effectively addresses tasks	Problematic exit strategy; will probably fight to survive	
	If re-created for each disaster, doesn't build on experience	
	(See 🖼 case study on Colombia, below)	

Option 2. Create dedicated organization or task force drawn from existing line ministries
(applicable in centralized or decentralized context)

Strengths	Weaknesses	Recommendations
Improves coordination with existing sector activities and policies	May lack political authority	When formed from existing line ministries, far more likely disaster recovery lessons will be applied to improve future disaster operations
Top executive drawn from outside bureaucratic ranks	Can weaken ministries and undermine ongoing non-disaster programs	The 🖼 case study on the Bam earthquake reconstruction, below, illustrates the success of a task force-type organizational model.
Exit strategy: staff returns to previous government positions	Proper expertise may not exist in line ministries	
	International agencies may not finance backfilling of normal ministry functions	

Option 3. Existing governmental agencies manage recovery under national disaster plan
(particularly applicable in decentralized context)

Strengths	Weaknesses	Recommendations
Places sector responsibility with sector expertise	Can overburden provincial and local governments with inadequate capacity to manage large reconstruction program	Effective, but needs detailed pre-disaster planning, staff training, and national disaster plan
Full local ownership	Can overload line ministries with double agenda (reconstruction and normal programs)	Requires existing line ministries to be strengthened with experienced staff
Exit strategy: staff returns to previous government positions	International agencies may not finance backfilling of normal ministry functions	Advisable option if reconstruction is manageable and local governments are strong and decentralized
If country is highly decentralized, sector rehabilitation corresponds to decentralized functions	Existing government system may be incapable of reconstruction duties	The 🖼 case studies on the National Institute for Disaster Management in Mozambique and the Housing Foundation in Iran, below, show how existing institutions can be effective managers of post-disaster reconstruction.
Disaster risk reduction lessons carried back to normal operations		
Increases probability government will apply disaster recovery lessons to future disaster operations		

The Need for Mandate, Policy, and Plan

Laying out the building blocks of a good institutional framework may be more important than suggesting institutional forms. To be effective, this framework needs several elements, specifically a mandate, a policy, and a plan that include the following elements.

Mandate. This is the official direction and support given to the reconstruction agency from a higher level of government, and should include the following elements:

- Support by appropriate legislation
- Sustained political support without interference
- Direct links to relevant line ministries
- Adequate financial, human, and material resources
- Knowledge of disaster recovery process dynamics
- Mechanisms for continual two-way consultation with affected communities
- Effective management information systems
- Administrative systems capable of managing international loans and grants
- Good governance, including mechanisms for interaction with civil society
- Credibility with surviving communities and other agencies involved in reconstruction

Policy. The reconstruction policy is the set of principles on which reconstruction planning and implementation is based (see 📖 Chapter 2, Assessing Damage and Setting Reconstruction Policy), including:

- Reconstruction aims and objectives
- Financial policies
- Rules for registration and monitoring of agencies involved in reconstruction
- Special provisions for highly vulnerable groups
- Housing assistance policies with eligibility criteria and allocation schemes
- Links among strategies to provide (or provide support for) immediate shelter, transitional housing, and full reconstruction
- Standards for agencies assisting communities regarding, among others, minimum and maximum financial and in-kind assistance and requirements for community participation, with the means for measuring and enforcing them
- Approach to address safety concerns, including, among other things, siting and land use controls, building design and technology, materials, training, enforcement procedures, and legal requirements
- Anticorruption policies

Plan. Practical arrangement for interinstitutional coordination and consistent implementation of policies, including:

- Participation plan and mechanisms for consultations with various stakeholders
- Communications strategy and plan
- Clear definition of roles and responsibilities for various entities
- Plan for allocation and delivery of financial resources
- Monitoring and evaluation plan and tools to maintain accountability to beneficiaries
- Strategy for transition between different stages of disaster recovery, especially the handoff from transitional to permanent housing programs
- Strategies for linking different sectors of recovery (in addition to housing, plans for infrastructure, livelihoods, health, and education)
- Mechanisms for ensuring coordination among agencies working in related sectors, particularly between housing and infrastructure reconstruction

Financial Planning

Implementing institutions cannot fulfill their purposes without adequate, predictable budgetary support for the entire reconstruction period. The extraordinary nature and scale of the requirements, and the likelihood that funding is coming partially from external sources, may mean that the reconstruction budgeting will be taken out of the regular budget cycle. This can provide flexibility and reduce administrative procedures, and therefore expedite recovery. But the alternate arrangements still should be transparent. These issues are discussed in 📖 Chapter 15, Mobilizing Financial Resources and Other Reconstruction Assistance.

Local Government as a Central Reconstruction Actor

Local problems and opportunities are best managed by local officials using local powers and applying local knowledge. At the same time, a balance is needed between policies that ensure quality and equity in housing reconstruction and implementation based on local reality, capacity, and culture.

The role of local government varies from one country to another with respect to the extent of the powers and resources delegated to them. In countries where extensive devolution has taken place, and local capability is strong, there is greater opportunity for management from the local level than there is in contexts where local government is little more than an arm of central government and does not have administrative and financial autonomy. Countries where there are multiple levels of government, with varying levels of authority (provinces, states, districts, communes, etc.), may face unique challenges. In many of these countries there are significant functional overlaps or gaps; this situation may be exacerbated by a disaster and make decision making cumbersome. Streamlined procedures and a very clear delineation of functions can help ensure an effective response to the demands that reconstruction places on all levels of government in these countries. It is critical to conduct an analysis of strengths and weaknesses of local governments, as well as of the risks and risk management strategies that may be employed, before defining the role of local government in reconstruction. The role of local government should be defined in the reconstruction policy.

Risks and Challenges

- Creating new institutions to manage recovery and reconstruction that duplicate or sideline the work of existing entities who are capable of doing the work.
- Creating new institutions that end up working without a clear mandate or sufficient autonomy.
- Planning reconstruction without paying attention to the need to rebuild national and local governmental capacity.
- Overcentralizing reconstruction planning and implementation.
- Government, in its weakened post-disaster state, allowing external entities to assume too much authority over management and coordination of recovery.
- Government failing to incorporate and institutionalize disaster risk reduction activities in reconstruction.
- Multiple entities, both inside and outside government, with different demands and priorities, at odds with one another, and operating independently of government's reconstruction policy.
- Political interference in reconstruction and corruption in large contracts, including those for housing and community reconstruction.

Recommendations

1. Define reconstruction policies and institutional mechanisms before disaster strikes.
2. If the demands of the disaster go beyond government capacity, establish a dedicated organization to manage disaster recovery.
3. Wherever possible, administer reconstruction using existing ministries and/or municipal departments and their existing staff, or at a minimum, provide them a central role.
4. Equip both the lead reconstruction agency and the local agencies charged with housing reconstruction with a structure, a mandate, a policy, and a plan.
5. Make certain that central or local governments weakened by a disaster are strengthened so that they can adequately manage reconstruction.
6. Have government regulate the work of all stakeholders, verifying their capacity, establishing standards for their work, and ensuring that their interventions are consistent with national policy.
7. Set up reliable monitoring and evaluation procedures to guarantee accountability and transparency.

Case Studies
1999 Armenia Earthquake, Colombia, and 2007 Ica/Pisco Earthquake, Peru

How the Context Affects the Success of a Reconstruction Institutional Model

Given the well-known success of the reconstruction program in Colombia after the 1999 Armenia earthquake, led by the Fund for the Reconstruction and Social Development of the Coffee-Growing Region (*Fondo para la Reconstrucción y Desarrollo Social del Eje Cafetero* [FOREC]), and the superficial similarities between that seismic event and the August 2007 earthquake in the Ica, Chincha, and Pisco provinces of Peru, it was understandable that the president of Peru would look to Colombia for advice on how to organize reconstruction. In fact, shortly after a meeting between officials from the two countries, Peruvian President Alan García announced the creation of a fund for the affected region similar to FOREC, the Fund for the Reconstruction of the South (*Fondo para la Reconstrucción del Sur* [FORSUR]).

PRISCILLA PHELPS

However, there were some crucial differences between the FOREC and FORSUR models that affected FORSUR's progress. FOREC was designed in ways that made it particularly effective, and it adopted an approach to reconstruction that served to increase its influence. These aspects included (1) administrative and normative autonomy; (2) a Board of Directors composed of local and central governments and the private sector; (3) an early, in-depth assessment of damage and needs followed by a planning process that was strategic, not just focused on rebuilding what was there before (within 10 months all plans were completed and all financing was programmed); (4) adequate funding from national and international sources; (5) a relatively advanced institutional framework for decentralization that included a system for local land use planning; (6) a decentralized approach to reconstruction that depended on both existing NGOs and, to a lesser extent, on local governments (the FOREC model has occasionally been criticized for not giving local governments sufficient scope); and (7) a transitional shelter strategy that provided adequate time for reconstruction to be done properly.

In contrast, Peru's decentralization process was not mature, and government preferred to minimize the role of NGOs and of its generally weak local governments. FORSUR reported to the president of the Council of Ministers and was considered part of government, subject to most governmental procedures for approval of projects and disbursement of resources. Its Board was large (19 members) and appointed from all sectors, although its Executive Director was from the private sector, and the public perceived that it was largely a private entity. In general, FORSUR had a limited mandate and was unprepared for the social aspects of reconstruction. Partly because a comprehensive damage and loss assessment was never carried out, the number of people affected and their needs were underestimated by FORSUR for some time after the disaster. FORSUR had decentralized offices with limited staff within seven months after the earthquake, but its principal office remained in the capital, Lima, for the first 18 months, and little effort was made to coordinate with local governments or to communicate with the public. And both the political will and the commitment of resources to carry out the reconstruction seemed to be missing. Some families given debit cards to procure construction materials found their accounts unfunded within a few months of the earthquake; some were still waiting for resources two years after the event.

The logical conclusion to draw from this comparison is that an institutional model taken from one context is difficult to import into another without a careful evaluation of its suitability and chances of success. Even a rapid institutional analysis would have predicted that the FOREC model was not a good fit for the social, economic, and political conditions of Peru in 2007.

Sources: Sandra Buitrago, 2009, personal communication; LivinginPeru.com, "Archive for Natural Disasters," http://www.livinginperu.com/news/natural-disasters; *and* Samir Elhawary and G erardo Castillo, 2008, "The Role of the Affected State: A Case Study on the Peruvian Earthquake Response" (HPG Working Paper, Overseas Development Institute), http://www.odi.org.uk/resources/download/1213.pdf.

1985 Mexico City Earthquake, Mexico

Creating a New Entity to Manage Urban Housing Reconstruction

The September 1985 Mexico City earthquake killed approximately 10,000 people and left some 250,000 homeless and another 900,000 with damaged homes. Reconstruction was declared by the president to be a project of national importance and a new agency, Renovación Habitacional Popular (RHP), was created to manage the challenge of housing reconstruction. This new agency was in charge of the clearance, reconstruction, and repair of more than 42,000 apartments, and oversaw the provision of temporary shelter to some 85,000 mostly low-income families. RHP started operations with staff borrowed from government ministries, many without previous planning or disaster experience. However, after the first chaotic weeks, its capacity grew and the project was successfully and rapidly completed. RHP's tasks included everything from rebuilding damaged housing units (many of them in multi-family apartment buildings) to retrofitting housing at risk of damage from future earthquakes. The results of the program are impressive: reconstruction of 78,000 housing units, 48,000 of them during RHP's original 2-year mandate; an expenditure of US$392 million; an improved tenure situation for those affected, accomplished by acquiring damaged substandard properties and reselling them to the residents; and a significant level of community participation in the project. The RHP experience shows that, in some situations, a new, dedicated agency is the best institutional solution. It also demonstrates that the success of an ambitious reconstruction program depends on the ready availability of adequate financial resources and a strategy that is well tailored to the requirements of the specific program.

Sources: Sosa Rodriguez and Fabiola Sagrario, n.d., "Mexico City Reconstruction after the 1985 Earthquake," Earthquakes and Megacities Initiative, http://emi.pdc.org/soundpractices/Mexico_City/SP2_Mx_1985_Reconstruction_Process.pdf.; The World Bank, 2001, "Bank Lending for Reconstruction: The Mexico City Earthquake," http://lnweb90.worldbank.org/oed/oeddoclib.nsf/DocUNIDViewForJavaSearch/9C4EA21B B9273C74852567F5005D8566; *and* Aseem Inam, 1999, "Institutions, Routines, and Crisis. Post-Earthquake Housing Recovery in Mexico City and Los Angeles," Cities 16(6):391-407.

CAMILLO BOANO

2005 North Pakistan Earthquake

Formation of Government Agency to Address Disaster Reconstruction

After the North Pakistan earthquake in 2005 in which 75,000 people were killed and another 100,000 injured, the government of Pakistan created a new dedicated body to manage disaster recovery: the Earthquake Reconstruction and Rehabilitation Authority (ERRA). ERRA was placed directly under the prime minister's office and given a broad-ranging agenda: "to plan, coordinate, monitor, and regulate reconstruction and rehabilitation activities in earthquake affected areas, encouraging self reliance via private public partnership and community participation, [and] ensuring financial [transparency]." ERRA coordinated all national and international assistance agencies and facilitated the work of implementing partners, presenting it with the challenge of managing relationships with all the entities involved in the reconstruction: provincial and district authorities, the military, donor agencies, and the NGO community. Initially, ERRA's rapid growth also presented transparency and accountability challenges, but, with time, an open system of reporting on resources and projects was developed and shared publicly from its own Web site. In general, ERRA is seen as a good example of effective management of all aspects of the reconstruction mandate it was given.

Sources: ERRA, 2006, *Rebuild, Revive with Dignity and Hope, Annual Review, 2005-2006* (Islamabad:ERRA), http://www.erra.pk/Reports/ERRA-Review-200506.pdf; *and* ERRA, "Welcome to ERRA," http://www.erra.pk/default.asp.

 For access to additional resources and information on this topic, please visit the handbook Web site at www.housingreconstruction.org.

2003 Bam Earthquake, Iran
A Dedicated Task Force Drawn from Line Ministries Oversees Reconstruction

Immediately after the 2003 earthquake in Bam, Iran, President Khatami, in consultation with his ministers and deputies, established a Steering Committee to plan and oversee reconstruction. A Provincial Reconstruction Focal Point Branch in Kerman Province (where Bam is located), under the supervision of the Steering Committee, was also established. The Steering Committee included the ministries of Housing and Urban Development, Finance, Interior, Judiciary, and Islamic Culture and Guidance; and heads of the Management and Programming Organization, the Cultural Heritage and Tourism Organization, the Iranian Red Crescent Society, the Kerman Housing Foundation (KHF), the Kerman Governorship, and Kerman provincial authorities. The Steering Committee was charged with establishing the basic reconstruction policies and was given significant autonomy: all decisions made by the Steering Committee had the same legal weight as those of the president and the cabinet. The Steering Committee established the Bam High Council of Architects and Planners, composed of five experts and skilled consultants. The council was asked to prepare guidelines for reconstruction. The Steering Committee also chose the then-president of the Housing Foundation of the Islamic Revolution (HF) as its secretary-general and designated the HF as the overall executing entity for construction, reconstruction, and retrofitting of the shelters and houses in Bam and in rural areas.

Source: Victoria Kainpour, UNDP Iran, 2009, personal communication, http://www.undp.org.ir/.

National Institute for Disaster Management, Mozambique
Integrating Disaster Management into Existing Institutional Structure

Floods and cyclones occur with some predictability in Mozambique. Since 2006, to manage the country's response to these disasters, Mozambique has used a predesigned disaster recovery facility, the National Emergency Operation Center (Centro Nacional de Operações de Emergência [CENOE]). During normal times, CENOE gathers information, monitors weather, and conducts disaster research. If a disaster strikes, CENOE and its branches act as physical and information nodes for coordination and decision making. The country's principal disaster agency is the National Disasters Management Institute (Instituto Nacional de Gestão de Calamidades [INGC]) (created under a different name during the civil war that ended in 1992), which is located in the Ministry of State Administration (Ministério da Administração Estatal) and under a Coordinating Council of Disasters Management. After any disaster, CENOE is activated to manage relief and recovery, with INGC acting as its secretariat. CENOE's operations are supported by a technical council composed of all the key sectors needed for disaster reconstruction. Thus, Mozambique's approach to dealing with disasters is not to establish a specific body to manage reconstruction on a disaster-by-disaster basis, but rather to activate a well-planned system that works with existing line ministries, departments, and local governments in a highly devolved manner, with district governments playing a key reconstruction role, and to have disaster recovery undertaken by the same body that organizes the national disaster risk management program.

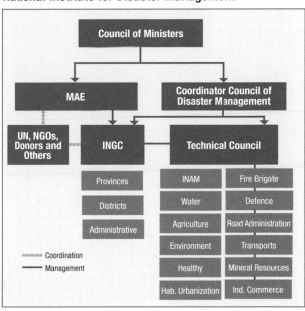

National Institute for Disaster Management

Sources: Samia Amin and Markus Goldstein, eds., 2008, *Data Against Natural Disasters: Establishing Effective Systems for Relief, Recovery, and Reconstruction* (Washington, DC: World Bank), http://siteresources.worldbank.org/INTPOVERTY/ Resources/335642-1130251872237/9780821374528.pdf; *and* F. Christie and J. Hanlon, 2001, *Mozambique and the Great Flood of 2000* (Oxford: The International African Institute James Currey and Indiana University Press), pp. 75-82.

The Housing Foundation of the Islamic Revolution, Iran
Iran's Housing Development Agency Also Responds to Disasters

The Housing Foundation of the Islamic Revolution (*Bonyad Maskan* or HF) was established in 1979 by Ayatollah Ruhollah Khomeini in the aftermath of the Islamic Revolution. Since then, HF has grown to more than 100 branches throughout the country, with its main office located in Tehran. HF is government's implementing arm in rural and urban housing for the poor and underprivileged members of the society. It also upgrades existing housing. In 2005, its annual goal was to construct or improve 200,000 rural housing units and 86,000 urban housing units. In addition to the direct implementation of social housing, the foundation is involved in the planning, evaluation, research, and provision of financial resources for housing development. Going beyond its housing focus, HF provides technical and research support to rural development through the renovation of historic villages, development of land use plans, and programs to expand land titling.

This permanent government agency also plays a critical role in disaster risk reduction and disaster response. HF is government's permanent disaster mitigation and post-disaster implementing agency, and the leading agency for planning, designing, and directing post-disaster housing reconstruction. In a post-disaster context, it works with agencies covering other sectors or issues, depending on the particular situation. In normal times, it investigates vulnerabilities in construction practices and promotes disaster-resistant construction. The HF's reconstruction approach reflects the accumulated technical knowledge of the agency and its staff, with more than a million housing units constructed or reconstructed in the past 30 years. Its work increasingly incorporates goals such as community participation, socio-cultural sensitivity, and emphasis on the environment.

Source: Housing Foundation of the Islamic Revolution, "Housing Foundation of the Islamic Revolution," Fars Province Housing Foundation, http://www.bonyadmaskanfars.ir/indexe.php.

Resources

Christoplas, I. 2006. *The Elusive "Window of Opportunity" for Risk Reduction in Post-Disaster Recovery.* Briefing paper for session 3 at the ProVention Consortium Forum 2006. "Strengthening Global Collaboration in Disaster Risk Reduction." Bangkok, February 2–3. http://www. proventionconsortium.org/themes/default/pdfs/Forum06/Forum06_Session3_Recovery.pdf.

Clinton, William Jefferson. 2005. "Lessons Learned from the Response to the Indian Ocean Tsunami." Transcript of remarks to UN Economic and Social Council (ECOSOC) Humanitarian Segment Panel, New York City. July 14. http://www.unisdr.org/eng/media-room/point-view/2005/WJC-ECOSOC-transcript.pdf.

Davis, I., ed. 2007. *Learning from Disaster Recovery, Guidance for Decision Makers.* Geneva and Kobe: International Recovery Platform (IRP). http://www.unisdr.org/eng/about_isdr/isdr-publications/irp/Learning-From-Disaster-Recovery.pdf. See pages 34-37 for a discussion of models for the management of recovery.

Fengler, W., A. Ihsan, and K. Kaiser. 2008. *Managing Post-Disaster Reconstruction Finance— International Experience in Public Financial Management.* Washington, DC: World Bank. http://go.worldbank.org/AE2YBURBA0.

Spangle, W. 1991. *Rebuilding after Earthquakes, Lessons from Planners.* Portola Valley, CA: William Spangle and Associates.

INTERNATIONAL, NATIONAL, AND LOCAL PARTNERSHIPS IN RECONSTRUCTION

14

Guiding Principles for Partnerships in Reconstruction

- Partnerships between government and international, national, and local organizations are essential to successful reconstruction.
- Partners that arrive later in the recovery period should respect the agreements that earlier-arriving partners made with government and affected communities before their arrival.
- Negotiated rules should govern the collaboration between government, nongovernmental organizations (NGOs), civil society organizations (CSOs), and affected communities in a reconstruction program. The terms of partnerships should be concretely defined and formalized in writing.
- The Global Humanitarian Platform (GHP) "Principles of Partnership" should always be adhered to.
- NGOs and CSOs are almost always more effective when working within their area of expertise and the limits of their capacity and resources.
- Governments have the right and responsibility to require that NGOs and CSOs follow ground rules, conform to the reconstruction policy, and report regularly on their activities.
- Regular reporting by partners, and monitoring and evaluation by government and affected communities, can improve the results of NGO and CSO partnerships.

This Chapter Is Especially Useful For:
- Policy makers
- Lead disaster agency
- Agencies involved in reconstruction
- Project managers

Introduction

No single organization or category of organization can provide the institutional, human, technical, and financial resources needed to carry out a successful post-disaster reconstruction program. Collaboration among these organizations is key to successful post-disaster housing and community reconstruction.

The roles of various agencies in reconstruction are discussed in 📖 Chapter 1, Early Recovery: The Context for Housing and Community Reconstruction, with a focus on the chronology of their activities. That chapter also provides an overview of how the reconstruction process is likely to play out. This chapter focuses on the mechanics of collaboration among organizations. It provides information to help distinguish the types of organizations and their motivations, and practical advice on ensuring that the interventions of these organizations are planned, coordinated, and carried out in a systematic way, consistent with the reconstruction policy.

Key Decisions

1. **Government** should decide on the lead agency or individual to work with the United Nations (UN) Humanitarian Coordinator (HC) to agree on the involvement of the UN in the disaster response.
2. The **lead disaster agency** or other designee should work with the UN HC to decide on the role of the UN agencies, including whether the cluster system will be activated, and, if so, in which sectors.
3. The **lead disaster agency**, in consultation with **affected communities**, should agree with partners on the parameters for NGO and CSO involvement in response and reconstruction.
4. The **lead disaster agency** should decide whether a registration process will be required for NGOs and CSOs involved in reconstruction and should agree with them on the coordination mechanisms and reporting procedures to be used.
5. **Partners** should decide, in consultation with **government**, on the coordination mechanisms they will use among themselves.
6. The **lead disaster agency, partners**, and the **affected communities** should jointly decide on the system and the benchmarks to be used for monitoring the participation of partners in reconstruction, at the national and community levels.

Public Policies Related to Partnerships

Through their participation in the UN System, governments have the services and support of the UN available to them in a post-disaster situation. Good coordination by government with the UN HC and the Humanitarian Country Team will help ensure the effectiveness of the UN response. The role of NGOs and CSOs in response and reconstruction will vary from country to country, as will the legal framework under which these organizations operate. In general, laws are in place that require the registration of local and international NGOs, and of some types of CSOs as well, and these laws should be adhered to in the post-disaster context. Legal requirements often increase in proportion to the size of the organization, with larger organizations being required to disclose their financial operations in a manner similar to private firms.

Registration of CSOs and NGOs helps ensure that government is aware of their presence and allows government to monitor their activities, although registration rules should not be so strict as to discourage needed interventions. The proliferation of partners is a risk in recovery and reconstruction, so governments may need to expand registration requirements to require participation in coordination mechanisms and additional forms of reporting by these organizations.

Technical Issues

Since the late 1980s, as international organizations have strengthened their commitment to lessening the impact of natural disasters worldwide through disaster risk reduction (DRR) and to improving their response to disasters, they have also worked to improve the quality and to expand the extent of their collaboration. These efforts are contributing to several important goals, including improved international disaster response, increased funding for DRR research and policy development, expanded DRR efforts at the national level, and strengthened interinstitutional cooperation. This section briefly reviews some of the most important interagency agreements and collaborations, most of which the World Bank participates in. It also highlights how the work of these entities relates to post-disaster housing and community reconstruction.

European Commission/World Bank/UN Joint Declaration

In September 2008, the European Commission, the UN, and the World Bank signed the Joint Declaration on Post-Crisis Assessments and Recovery Planning, which addresses coordination mechanisms for both post-conflict and post-disaster situations.[1] This declaration commits the signatories to:

- Communicate strategically at both headquarters and field levels as we monitor situations of fragility and conflict, and imminent or actual natural disasters, and identify opportunities for joint initiatives where our combined efforts may offer advantages.
- Participate in the relevant in-country planning processes and support the development and use of shared benchmarks/results frameworks and joint processes for monitoring and review.
- Support the development and use of the common methodologies for post-conflict needs assessments, and a common approach to post-disaster needs assessments and recovery planning.
- Invest in development of tool kits and staff training to deepen collective and institutional capacity for these processes.
- Monitor progress in the implementation of the common platform through a senior level meeting that would take place once a year.

The activities under this declaration related to the convergence of assessment methodologies are described in ⌨ Chapter 2, Assessing Damage and Setting Reconstruction Policy.

Global Facility for Disaster Reduction and Recovery

The Bank's activities in the Joint Declaration with respect to natural disasters are coordinated by the Global Facility for Disaster Reduction and Recovery (GFDRR).[2] Launched in 2006, the GFDRR is a partnership of the International Strategy for Disaster Reduction system to support the implementation of the Hyogo Framework for Action 2005–2015 (HFA).[3] The mission of the GFDRR, which is managed by the World Bank on behalf of the participating donor partners and other partner stakeholders, is to reduce vulnerabilities to natural hazards by mainstreaming disaster reduction and climate change adaptation in country development strategies.

In outlining the reconstruction strategy, it is necessary to identify the role NGOs and CSOs will play. Government should also manage competition among NGOs and CSOs by setting rules for their involvement in reconstruction; requiring information sharing; and establishing mechanisms for reporting, coordination, and monitoring of their activities.

1. European Commission, United Nations Development Group, and World Bank, 2008, "Joint Declaration on Post-Crisis Assessments and Recovery Planning," http://www.undg.org/docs/9419/trilateral-JD-on-post-crisis-assessments-final.pdf.
2. Global Facility for Disaster Reduction and Recovery, http://www.gfdrr.org. As of late 2009, GFDRR partners include the governments of Australia, Canada, Denmark, France, Italy, Japan, Luxembourg, Norway, Spain, Sweden, Switzerland, and the United Kingdom; the European Commission; and the World Bank.
3. In January 2005, 168 governments adopted the HFA at the World Conference on Disaster Reduction held in Kobe, Hyogo, Japan. The HFA is a 10-year plan to make the world safer from natural hazards. Endorsed by the UN General Assembly in Resolution 60/195, the HFA is now the primary international agreement guiding DRR efforts.

The Global Humanitarian Platform

The GHP is a forum that was created in 2006 to bring together the three main components of the humanitarian community—NGOs, the International Red Cross and Red Crescent Movement, and the UN and related international organizations.[4] The goal of the GHP is to enhance the effectiveness of humanitarian action. The founding premise of the GHP is that no single humanitarian agency can cover all humanitarian needs and that collaboration is, therefore, not an option, but a necessity.

Based on the principle of diversity, the GHP does not seek to convince humanitarian agencies to pursue a single mode of action or work within a unique framework. The GHP aims at maximizing complementarity based on the participating organizations' different mandates and missions, and emphasizes the importance of coordinating with and integrating local organizations, such as local NGOs and CSOs, in humanitarian response. The organizations participating in the GHP have agreed to the following "Principles of Partnership"[5] as the basis of their collective action.

Principles of Partnership

Principle	Explanation
Equality	Equality requires mutual respect between members of the partnership irrespective of size and power. The participants must respect each other's mandates, obligations, independence, and brand identity and recognize each other's constraints and commitments. Mutual respect must not preclude organizations from engaging in constructive dissent.
Transparency	Transparency is achieved through dialogue (on equal footing), with an emphasis on early consultations and early sharing of information. Communications and transparency, including financial transparency, increase the level of trust among organizations.
Result-oriented approach	Effective humanitarian action must be reality-based and action-oriented. This requires result-oriented coordination based on effective capabilities and concrete operational capacities.
Responsibility	Humanitarian organizations have an ethical obligation to each other to accomplish their tasks responsibly, with integrity and in a relevant and appropriate way. They must make sure they commit to activities only when they have the means, competencies, skills, and capacity to deliver on their commitments. Decisive and robust prevention of abuses committed by humanitarians must also be a constant effort.
Complementarity	The diversity of the humanitarian community is an asset if we build on our comparative advantages and complement each other's contributions. Local capacity is one of the main assets to enhance and on which to build. Whenever possible, humanitarian organizations should strive to make it an integral part in emergency response. Language and cultural barriers must be overcome.

Key International, National, and Local Partner Institutions and Their Roles

The Inter-Agency Standing Committee. The Inter-Agency Standing Committee (IASC) is an interagency forum for coordination, policy development, and decision making involving key UN and non-UN partners.[6] It develops humanitarian policies, agrees on clear divisions of responsibilities for the various aspects of humanitarian assistance, identifies and addresses gaps in response, and advocates for effective application of humanitarian principles. Together with the Executive Committee for Humanitarian Affairs (ECHA), the IASC forms the key strategic coordination mechanism among major humanitarian actors in a disaster situation. Its Full Members include, among others, the Food and Agricultural Organization (FAO), the UN Office for the Coordination of Humanitarian Affairs (UN OCHA), the United Nations Development Programme (UNDP), and the United Nations Human Settlements Programme (UN-HABITAT). Its Standing Invitees include, among others, the World Bank, the International Committee of the Red Cross (ICRC), the International Council of Voluntary Agencies (ICVA), the International Federation of the Red Cross and Red Crescent (IFRC), the American Council for Voluntary International Action (InterAction), and the United Nations Office of the High Commissioner for Human Rights (OHCHR).

The IASC contributes to the post-disaster shelter field largely through the Emergency Shelter Cluster (ESC). Recently, the IASC and UN-HABITAT produced the 2008 edition of "Shelter Projects," which monitors post-disaster emergency and transitional shelter projects,[7] and published the guidebook, "NFIs for Shelter,"[8] discussed in 📖 Chapter 15, Mobilizing Financial Resources and Other Reconstruction Assistance.

4. Global Humanitarian Platform, http://www.globalhumanitarianplatform.org.
5. GHP, http://www.globalhumanitarianplatform.org/doc00002172.doc.
6. IASC, http://www.humanitarianinfo.org/iasc/.
7. http://www.sheltercentre.org/library/Shelter+Projects+2008.
8. Non-food items. See http://www.sheltercentre.org/library/Selecting+NFIs+Shelter.

 For access to additional resources and information on this topic, please visit the handbook Web site at www.housingreconstruction.org.

IASC Global Cluster Leads. In December 2005, the IASC Principals designated global CLs for nine sectors or areas of activity where in the past either there was a lack of predictable leadership in situations of humanitarian emergency or there was considered to be a need to strengthen leadership and partnership with other humanitarian actors (see below). This nine sectors/areas complement those sectors and categories of population where leadership and accountability are already clear, e.g. agriculture (led by FAO), food (led by WFP), refugees (led by UNHCR) and education (led by UNICEF).

The Global Cluster Leads

Sector or Area of Activity		Global Cluster Lead
Camp Coordination/Management	IDPs (from conflict)	UNHCR
	Disaster situations	IOM
Early Recovery		UNDP
Emergency Shelter	IDPs (from conflict)	UNHCR
	Disaster situations	IFRC (Convener)
Emergency Telecommunications		OCHA/UNICEF/WFP
Health		WHO
Logistics		WFP
Nutrition		UNICEF
Protection	IDPs (from conflict)	UNHCR
	Disasters/civilians affected by conflict (other than IDPs)*	UNHCR/OHCHR/UNICEF
Water, Sanitation and Hygiene		UNICEF

* UNHCR is the CL of the global Protection Cluster. However, at the country level in disaster situations or in complex emergencies without significant displacement, the three core protection-mandated agencies (UNHCR, UNICEF, and OHCHR) consult closely and, under the overall leadership of the HC/RC, agree which of the three will assume the role of CL for Protection.

Emergency Shelter Cluster. The ESC is co-chaired by the UN High Commissioner for Refugees (UNHCR) and IFRC. UNHCR leads the ESC in the area of conflict-generated internally displaced persons (IDPs), while the IFRC is convener of the ESC in disaster situations.[9] The main partners in the ESC are UN-HABITAT, UN OCHA, the Norwegian Refugee Council, Oxfam International, Care International, CHF International, Shelter Centre, the International Organization for Migration, the United Nation Children's Fund (UNICEF), UNDP, the World Food Programme (WFP), the Danish Refugee Council, and any NGO involved in emergency shelter. Decisions made in the ESC set the stage for later housing and community reconstruction activities, as explained in 🏠 Chapter 1, Early Recovery: The Context for Housing and Community Reconstruction, and are therefore important to agencies involved in reconstruction, including the World Bank.

The IFRC acts as a "convener" rather than a CL for emergency shelter. In that capacity, it has made a commitment to provide leadership to the broader humanitarian community in emergency shelter in disaster situations, to consolidate best practices, to map capacity and gaps, and to lead a coordinated response. It does not act as the "provider of last resort," as do other CLs, nor is it accountable to any part of the UN System. The IFRC also does not participate in Consolidated Appeals launched by the UN, but instead appeals separately for support in providing leadership and strengthening capacity for the provision of emergency shelter in disasters resulting from natural hazards.

The UN System. The UN System supports disaster prevention, response, and reconstruction through a number of its component agencies and organizations.

The International Strategy for Disaster Reduction. The ISDR is a system of partnerships with the overall objective to generate and support a global disaster risk reduction movement to implement HFA. The ISDR was endorsed by the World Conference on Disaster Reduction in 2004 and by the UN General Assembly and serves as the overall framework for implementing disaster risk reduction at the local, national, regional, and international levels. The ISDR is partner to the World Bank in the GFDRR. The secretariat to ISDR is UNISDR, an entity within the UN Secretariat.

9. A variety of useful tools and guidelines on emergency shelter are available on the Emergency Shelter Cluster Web site, http://www.humanitarianreform.org/humanitarianreform/Default.aspx?tabid=77.

Two other agencies whose roles are especially important in housing and community reconstruction are the UN OCHA and the Cluster Working Group on Early Recovery (CWGER).

The Office for the Coordination of Humanitarian Affairs. UN OCHA supports and facilitates the work of UN agencies, NGOs, and the International Red Cross and Red Crescent Movement in delivering humanitarian services. It works closely with governments to support them in their lead role in humanitarian response. UN OCHA supports the UN HC in needs assessments, contingency planning, and formulation of humanitarian programs. The head of UN OCHA, as Emergency Relief Coordinator, chairs the IASC, which comprises all major humanitarian actors, including the Red Cross and Red Crescent Movement and three NGO consortiums.

UN OCHA deploys staff to disaster areas on short notice and supports several "surge capacity" mechanisms, including the United Nations Disaster Assessment and Coordination System (UNDAC), which can dispatch teams within 24 hours of a natural disaster to gather information, assess needs, and coordinate international assistance.

UN OCHA also solicits donor support through the Consolidated Appeals Process (CAP); issues emergency appeals on behalf of countries affected by disasters; and manages the Central Emergency Response Fund (CERF), which enables UN agencies to jump-start relief activities following natural disasters. UN OCHA also leads various activities to improve information flow in disasters, including managing humanitarian information centers in the field (as described in ⬛ Chapter 17, Information and Communications Technology in Reconstruction) and running ReliefWeb, the "global hub for time-critical humanitarian information on complex emergencies and natural disasters."[10]

DANIEL PITTET

While UN OCHA's work is not principally shelter-related, it contributes to the advancement of good practices on post-disaster emergency and transitional shelter, for instance, through its collaboration with the Shelter Centre on the development of the guidelines *Shelter After Disaster: Strategies for Transitional Settlement and Reconstruction.*[11]

The Cluster Working Group on Early Recovery. UNDP coordinates the CWGER, one element of the UN reform agenda. The CWGER is intended to strengthen humanitarian response capacity and effectiveness.[12] It operates at the global level to strengthen preparedness and technical capacity by designating "Cluster Leads" (CLs) that ensure leadership and accountability in sectors or areas of activity. At the country level, the CWGER ensures a more coherent and effective response by mobilizing international agencies, organizations, and NGOs in all key sectors under the UN HC and the Humanitarian Country Team. The CWGER has also worked to establish a clearer division of labor among organizations and to define roles and responsibilities within sectors. It also acts for the UN HC as the first point of call and the "provider of last resort" in all sectors or areas of activity.[13] The key actors in the cluster system are explained in the following table.

10. UN OCHA, "ReliefWeb," http://www.reliefweb.int.
11. United Nations Office for the Coordination of Humanitarian Affairs (UN OCHA) and Shelter Centre, 2010, *Shelter After Disaster: Strategies for Transitional Settlement and Reconstruction* (Geneva: UN OCHA), http://www.sheltercentre.org/library/Shelter+After+Disaster
12. UNDP, 2008, "UNDP Policy on Early Recovery," http://www.undp.org/cpr/documents/Early_Recovery/er_policy.pdf.
13. The "provider of last resort" concept within the cluster system means that it is the responsibility of the CL to call on all relevant humanitarian partners to address any critical gaps in the post-disaster response. If the partners cannot meet a critical need, the CL, as "provider of last resort," may need to fill the gap. If funds are not forthcoming for these activities, the CL is expected to work with the UN HC and donors to mobilize the necessary resources.
14. IASC, "Guidance Note on Using the Cluster Approach to Strengthen Humanitarian Response," 2006, http://www.humanitarianreform.org/humanitarianreform/Portals/1/Resources%20&%20tools/IASCGUIDANCENOTECLUSTERAPPROACH.pdf.

Emergency Relief Coordinator (ERC)	The ERC is the undersecretary general for humanitarian affairs, who also heads OCHA. The ERC ensures that an agreement is reached on country-level cluster/sector leads and that this decision is communicated to humanitarian partners, donors, and other stakeholders. (The UN HC informs the host government and country-level humanitarian partners of the agreed arrangements.)
Humanitarian Coordinator	The HC (or Resident Coordinator [RC] where an HC has not yet been appointed) is the most senior UN humanitarian official on the ground for an emergency. This person ensures the adequacy, coherence, and effectiveness of the overall humanitarian response and is accountable to the ERC. With the Humanitarian Country Team, the HC establishes coordination mechanisms and is responsible for adapting them to reflect government capability.
Cluster Leads	The CLs (sometimes referred to as Sector Leads)[14] support government coordination and response efforts, facilitate coordination between cluster partners within a given sector and between different sectors, encourage collaboration, ensure that responses adhere to existing guidelines and standards, collate and share information, identify response gaps and duplication, and act as provider of last resort. Sector/cluster lead agencies are accountable to the UN HC.

⬛ **For access to additional resources and information on this topic, please visit the handbook Web site at www.housingreconstruction.org.**

Nongovernmental Organizations

Defining "nongovernmental organization." An NGO is any nonprofit, voluntary citizens' group organized on a local, national, or international level. Generally outcome-oriented and driven by people with a common interest, NGOs perform a variety of service and humanitarian functions, bring citizen concerns to governments, advocate and monitor policies, and encourage political participation through organizing and providing information. Some are organized around specific issues, such as human rights, environment, or health. They provide analysis and expertise, serve as early warning mechanisms, and help monitor and implement international agreements. International, national, and local NGOs carry out work related to disaster response and recovery. The legal form of NGOs is diverse and depends on variations in countries' laws and practices. However, there are four main groups of NGOs generally recognized worldwide[15]:

- Trusts, charities, and foundations
- Not-for-profit companies
- Unincorporated and voluntary associations
- Entities formed or registered under special NGO or nonprofit laws

The Asian Development Bank characterizes NGOs according to whether their principal focus is operational or advocacy, believing that this distinction is key to determining the type of interaction the Bank can have with them: operational cooperation and collaboration versus policy dialogue.[16] Yet the Bank acknowledges that in many cases it is not possible to characterize an NGO entirely as purely operational or advocacy, since some are involved in both types of activities. The two groups have the following characteristics.

Operational NGOs

- Primary areas of activity are directed toward the contribution or delivery of development or welfare services, including emergency relief, and environmental protection and management.
- Display a range of programs, organizational structures, operational orientations, and areas of operation, both program-related and geographical.
- Exist at all levels: community, local, district, national, regional, and international.

Advocacy NGOs

- Primary orientation is advocacy of policies or actions that address specific concerns, points of view, or interests.
- Work to influence the policies and practices of governments, development institutions, other actors in the development arena, and the public.
- Exist more often at national and international levels, and increasingly are forming national and international networks and consortia that link groups with parallel or convergent interests.
- Exist to serve as a voice that they consider otherwise would not be heard in social, economic, and political processes.

15. Wikipedia, "Non-governmental organization," http://en.wikipedia.org/wiki/Non-governmental_organization.
16. Asian Development Bank, 2009, "Nongovernmental Organizations and Civil Society," http://www.adb.org/Documents/Policies/Cooperation_with_NGOs/ngo_sector.asp?p=coopngos.

FEMA Photo/Mark Wolfe

In Gujarat after the 2001 earthquake, Kutch Nav Nirman Abhiyan, a coalition of 14 NGOs, worked alongside the Gujarat State Disaster Management Authority (GSDMA) in 600 earthquake-affected villages and towns to encourage civic participation in reconstruction, as described in the 🖥 case study, below.

International NGOs. International NGOs include high-profile humanitarian actors who play an extremely important role in disaster response and recovery. As international entities, they may not be organized under or subject to national law; however, a number of international NGOs operate as networks of national organizations, each of which is subject to the national laws that govern the formation and obligations of nongovernmental corporations in the respective country. International NGOs act individually and collectively to mobilize financial, technical, and human resources after disasters, and work as peers with the UN in such initiatives as the GHP, discussed above. Major international NGOs include the International Red Cross and Red Crescent Movement, CARE International, Mercy Corps, Oxfam International, Plan International, World Vision International, Save the Children Alliance, Food for the Hungry, Amnesty International, Caritas International, Doctors Without Borders, International Rescue Committee, and Habitat for Humanity International.

International NGOs are often the first international agencies to mobilize after a disaster. Because of their experience at quickly establishing a presence in-country and in the disaster area (or expanding a presence they already have), they often assist government in conducting initial post-disaster assessments and in designing and putting in motion the early stages of the response, as described in 📖 Chapter 1, Early Recovery: The Context for Housing and Community Reconstruction. For these reasons, international NGOs will often have a grasp of the situation on the ground—second only to that of government.

The 📖 case study on Hurricanes Katrina and Rita, below, shows an example of where two international NGOs, Habitat for Humanity International and Church World Service, worked with 53 local community-based organizations (CBOs) to assist affected families with repair and reconstruction activities.

Civil Society Organizations

Defining "civil society organization." This handbook uses the phrase "civil society organization" as a generic term to refer to the wide array of national and local nongovernmental and not-for-profit organizations that express the interests and values of their members and/or others based on ethical, cultural, political, scientific, religious, or philanthropic considerations.[17] The application of the phrases "nongovernmental organization" and "civil society organization" varies from one country to another. Many of the points made here about CSOs apply equally to NGOs. See 📖 Chapter 12, Community Organizing and Participation, for a discussion of the ways in which affected individuals, communities, and CBOs participate in reconstruction.

The private sector directly carries out reconstruction in myriad ways, but may also do so through a CSO, such as an association of firms in a single industry or of professionals with a particular expertise, such as engineering or communications.[18] DHL is one private firm that directly contributes services worldwide after disasters, providing logistical services in airports to manage relief supplies, as shown in the 📖 case study, below.

CSOs carry out uncoerced collective actions around shared interests, purposes, and values. In theory, civil society is distinct from the state, the family, and the market. In practice, the boundaries between them can be indistinct. CSOs differ in their levels of formality, autonomy, power, and reach.[19] While most civil society activity remains local, over the decades CSOs have worked collectively to shape global policy through advocacy campaigns and the mobilization of people and resources.

In the World Bank, there has been a deliberate shift away from use of the term "NGO"—due to its more narrow application to professional, intermediary, and nonprofit organizations that advocate and/or provide services—toward the term "CSO." This reflects the Bank's effort to reach out to a broader group of organizations that includes not just NGOs, but also trade unions, community-based organizations, social movements, faith-based institutions, charitable organizations, universities, foundations, professional associations, and others. This broader group is also likely to be active in post-disaster reconstruction.

Types of CSOs. CSOs are categorized by their objectives, geography, and funding. The World Bank classifies a CSO according to whether its mission is charitable, service-oriented, participatory, or devoted to community empowerment. Of the four forms of NGOs mentioned above, the definition of CSO used here includes largely the last two categories: unincorporated and voluntary associations, and entities formed or registered under special NGO or nonprofit laws.

CSOs can also be classified geographically as local, national, or international, although—as mentioned above—this handbook refers to international NGOs and national organizations associated with international NGOs as NGOs, rather than CSOs. In post-disaster situations, it is important to know if a CSO is already working in a disaster area and therefore has local knowledge. Both local and international organizations can have local knowledge. CSOs can be grouped in many ways. A useful typology proposed by the World Bank for selecting CSOs for consultation processes is shown below.

17. London School of Economics Centre for Civil Society, 2004, "What is civil society?" http://www.lse.ac.uk/ collections/CCS/what_is_civil_society.htm.
18. A discussion of the essential role of the private sector in reconstruction is beyond the scope of this handbook. Some useful reference material on this topic is included in the Resources section, below.
19. London School of Economics Centre for Civil Society, 2004, "What is civil society?" http://www.lse.ac.uk/ collections/CCS/what_is_civil_society.htm.

Typology of Civil Society Organizations[20]

Functions	Category examples	Implications for selection criteria and process
Representation	■ Membership organizations, including labor unions, women's associations, peasant organizations ■ NGOs, federations, umbrella organizations, or networks ■ Faith-based organizations ■ Organizations of indigenous peoples	The selection of these organizations should be based on the size and type of the organization and the legitimacy of representation. Questions that can help classify the organization include: ■ Who belongs to the organization? ■ What are the criteria for membership? ■ In what activities does the organization engage? ■ Does it cater to members only, or does it take up action on behalf of a wider group? ■ What is the geographic and sectoral coverage of the organization?
Technical expertise	■ Professional and business associations ■ Think tanks and other research groups	The selection should be based on the expertise and knowledge of issues and the legitimacy of members' expertise.
Advocacy	■ Trade unions ■ NGOs ■ Human rights groups ■ News and media groups ■ Campaign organizations	The selection should be based on how actively a group is advocating issues, its capacity to mobilize and educate a constituency, its credibility, and its demonstrated interest in constructive engagement.
Capacity building	■ Foundations (local, international, and community) ■ CSO support organizations ■ Training organizations	The selection should be based on the issues associated with a proposed project or strategy under study. An organization may represent a key interlocutor that strengthens the capacity of civil society to participate in the consultation.
Service delivery	■ Local, national, and international NGOs ■ Credit and mutual aid societies ■ Informal, grassroots, and community-based associations	The selection should be based on the relation of these issues to a proposed project. Issues of representation may also come into play for some of these groups.

Role of CSOs in reconstruction. CSOs can play a central role in post-disaster reconstruction. They bring institutional, human, technical, social, and financial resources to reconstruction and—being local—can link reconstruction efforts to longer-term sustainable development activities in a disaster-affected region. However, the role and influence of CSOs in the disaster context varies considerably, depending on their scale, sponsoring organization, financial strength, purpose, and geographic reach. A list of common roles includes:

■ providing humanitarian, technical, manpower, material, advisory, scientific, and financial assistance to government or directly to the affected population in all phases of the reconstruction cycle;

■ influencing reconstruction policy, especially when acting collectively;

■ advocating for equity, human rights, transparency, accountability, and justice in the reconstruction process; and

■ coordinating and communicating between government, local people, and national and international organizations.

The 📖 case study on the Shelter Advisory Group in Tamil Nadu, below, describes a unique case where a multidisciplinary public-private partnership was formed to provide quality control over the work of NGOs involved in reconstruction. After Hurricane Katrina, a group of universities collaborated with citizens to develop a revitalization plan for the 9th Ward of New Orleans, as described in the 📖 case study, below.

Challenges in Collaborating with NGOs and CSOs

Need for systematic approach. Numerous post-disaster evaluations point out the risks of ignoring or not establishing ground rules for the work of NGOs and CSOs in reconstruction. Evaluations of the 2004 Indian Ocean tsunami response represent an extreme, although not unique, case.[21] The effective use of the resources that these organizations bring to reconstruction—institutional, human, technical, social, and financial—requires a planned approach. In outlining the reconstruction

20. World Bank, Civil Society Team, 2007, "Consultations with Civil Society: A Sourcebook (Working Document)," p. 64, http://siteresources. worldbank.org/CSO/Resources/ ConsultationsSourcebook_ Feb2007.pdf.

21. Elisabeth Scheper, Arjuna Parakrama, and Smruti Patel, 2006, *Impact of the Tsunami Response on Local and National Capacities,* London: Tsunami Evaluation Coalition (TEC), http://www.alnap.org/pool/files/ capacities-final-report.pdf.

strategy, it is necessary to identify the role these organizations will play. Government should also manage competition among NGOs and CSOs by setting rules for their involvement in reconstruction; requiring information sharing; and establishing mechanisms for reporting, coordination, and monitoring of their activities. Transparency and public participation are hallmarks of some, but not all NGOs; government may need to establish standards for community participation and information disclosure. The role foreseen for NGOs and CSOs in reconstruction should be laid out in government's reconstruction policy. A description of the scope of the reconstruction policy is included in 🏠 Chapter 2, Assessing Damage and Setting Reconstruction Policy.

MARVIN NAUMAN/FEMA

After the 2003 Bam earthquake, government turned to UN OCHA and UNDP to coordinate the work of international NGOs, which included an effort to more equitably allocate their support between rural and urban areas, as described in the 🏠 case study, below.

Organizational agendas versus government goals. NGOs and CSOs sometimes bring their own agendas to the reconstruction context, and their goals may not reflect government policy objectives. These agendas may include promoting a political or religious ideology. Agencies have been known to exclude beneficiaries who do not share specific religious or philosophical beliefs. Organizational agendas are influenced by the source of NGO or CSO funding in at least two respects: the need to show results to funders may sometimes cloud the judgment of organizations, and the funders may insist on specific approaches that reflect their values.[22] Government must be on guard for discriminatory practices and ensure that CSOs and NGOs are willing to align their activities with reconstruction policies. Government outreach, clear communication of objectives and conditions for NGO and CSO participation, and interagency coordination mechanisms can all help unify criteria, standards, and modalities of assistance.

Capacity limits and need for strengthening. One concern about the way CSOs operate is that they can take on too much responsibility at too large a scale with insufficient funding. Where government is funding NGO or CSO activities, funding should be calibrated to organizational capacity. Government review of CSO project initiatives can help detect in advance possible problems with their proposed activities. At the same time, government should not overlook the expertise of local CSOs or allow international NGOs to overshadow their local counterparts. Local academic institutions, professional organizations, and licensing authorities are examples of CSOs that can play a critical role in post-disaster reconstruction. Institutional strengthening or other support may be necessary if a CSO's capacity is taxed by its participation in the reconstruction effort.

Formalizing Government/NGO/CSO Collaboration

Assessing NGOs and CSOs. In working with NGOs and CSOs, government must establish their legitimacy and confirm their capacity. CSOs are generally most effective working within their established area of expertise in activities that will not overtax their capacity. Assessments may be needed to identify organizations' constituency, capacity, outreach capacity, and technical skills.

A formal registration system may also be necessary, especially for international NGOs entering the country for the first time after a disaster to provide services. At a minimum, a registration or tracking system should ensure that CSOs are duly registered in their own countries; have the experience to carry out legitimate, needed activities in a professional manner; are not imposing a particular philosophy or religion on the affected population as a condition for services; and are able and willing to report on their activities and the financial resources under their control. Listed below is information government should consider requesting of NGOs and CSOs involved in reconstruction.

22. For example, one NGO involved in the Hurricane Mitch reconstruction in Honduras was unwilling to address the housing needs of unwed mothers.

 For access to additional resources and information on this topic, please visit the handbook Web site at www.housingreconstruction.org.

Data for NGO/CSO Registration System

Data	Purpose
Name	Legal name and doing-business-as name, internationally and locally.
Legal status	National or international organization. Type of incorporation or other legal status. Legal basis for receiving funds from international and/or national sources. Permission to operate in the country.
Experience	Experience in the affected country and/or region and in similar post-disaster reconstruction activities. Supervisory structure, and experience of senior officials.
Expertise	Principal services: financial, technical assistance, human resources. Language skills of staff. Systems for project management.
Beneficiary screening criteria	Screening criteria for beneficiaries, especially philosophical or religious preconditions, if any.
Financial capacity	Financial management capacity. Availability and source of own funds. Experience with managing government and other outside funds. Ability to present accurate and timely financial information.
Institutional contacts	Headquarters and institutional information, including names of senior management and board members.
Contact information	Location of office, telephone, fax, Web site, e-mail address.

Formalizing the NGO/CSO role. It is advisable that governments collaborating formally with NGOs and CSOs in reconstruction activities define the terms of the partnership with written agreements or contracts. For CSOs in particular, the community contract model promoted by UN-HABITAT may be a useful resource for drafting these agreements.[23] In other cases, normal contracting frameworks used in government procurement can be applied. The agreement should define expected outcomes and benchmarks, establish financial disclosure and monitoring requirements, and define sanctions in the event there is nonperformance of obligations under the contract.

Oversight system. Delegating certain duties does not absolve government from its oversight responsibilities. Government monitoring is necessary to ensure that NGO and CSO partners adhere to established reconstruction guidelines and parameters. A government agency should be assigned by the lead disaster agency to coordinate the work of the NGOs and CSOs and to monitor their performance. The World Bank's good practice notes on involving NGOs in Bank-supported activities may be useful in this context.[24] Reporting by organizations should reference benchmarks, outputs, and outcomes, and cover both programmatic and administrative expenditures. Reporting should include an accounting of any counterpart contributions made by the CSO, government, or the affected community and families.

Start-up and periodic meetings are useful tools for raising issues, reviewing resource requirements, and negotiating adjustments in prior estimates of inputs or outcomes. Governments often request that NGOs establish and manage the coordination system in which government then participates. In those cases where NGOs and CSOs are directly supporting participatory reconstruction projects in communities, government should have procedures for consulting regularly with communities and families about their satisfaction with the services they are receiving. In cases where NGO presence in the country is only temporary, government may require a continuity plan to ensure the sustainability of the activities that NGOs initiated.

A successful south-south partnership (i.e., a partnership between developing country organizations) that took place following the 2004 Indian Ocean tsunami in Aceh, between a local coalition of NGOs and the Indian coalition Kutch Nav Nirman Abhiyan, is described in the 🖳 case study, below.

Risks and Challenges

- Government allows the UN or other partners to establish reconstruction policy.
- The support of the cluster system could be useful, but it is not activated or is activated too late to be effective.
- NGOs and CSOs proliferate after a disaster and government fails to set standards or financing limits or to coordinate their activities.

23. UN-HABITAT Regional Office for Asia and the Pacific, n.d., "Community Contracts," http://www.fukuoka.unhabitat.org/event/docs/EVN_081216172311.pdf.
24. World Bank, 2000, "Involving Nongovernmental Organizations in Bank Supported Activities," http://web.worldbank.org/WBSITE/EXTERNAL/PROJECTS/EXTPOLICIES/EXTOPMANUAL/0,,contentMDK:20064711~menuPK:4564189~pagePK:64709096~piPK:64709108~theSitePK:502184,00.html.

⚒ 🗑 **SAFER HOMES, STRONGER COMMUNITIES: A HANDBOOK FOR RECONSTRUCTING AFTER NATURAL DISASTERS**

- NGOs and CSOs overstate their capacity, receive more funding than they are capable of managing, or make commitments to affected communities that they cannot fulfill.
- NGOs and CSOs conduct their work in a nonconsultative, top-down manner, working on a turnkey basis to deliver finished products to "beneficiaries."
- NGOs and CSOs require that community members conform to an organizational agenda, including those of a religious nature, in order to qualify for the benefit being offered.
- Government does not require alignment of CSO or NGO activities with reconstruction policy or the disclosure of reconstruction outputs and financial results.
- NGOs and CSOs pretend they represent the community instead of supporting the articulation of community preferences.

Recommendations

1. Government should request whatever support it needs from the UN or other partners to define reconstruction policy and implement the reconstruction program, while maintaining overall coordination of the process.
2. In developing a reconstruction program, identify the roles best suited to the UN, other humanitarian agencies, NGOs, and CSOs, and deploy partner resources based on an assessment of their experience, ability to execute, local knowledge, and financial capacity.
3. Ensure that a CSO's role in the reconstruction process is consistent with the organization's established mandate.
4. Create a reporting mechanism to monitor CSO project design, development, and implementation activities.
5. When needed to ensure their involvement, provide technical, financial, and implementation support to local CSO initiatives.
6. When NGOs or CSOs are engaged by government to carry out specific activities, calibrate funding to organizational capacity, formalize programmatic relationships, and establish benchmarks for program activities.
7. Establish a monitoring and evaluation system for all NGO and CSO activity and mechanisms for keeping track of the satisfaction of the population being served.

Case Studies

2001 Gujarat Earthquake, India

Kutch Nav Nirman Abhiyan Empowers Villages through Mediation with Official Bodies

The Kutch Nav Nirman Abhiyan, a coalition of 14 grassroots NGOs formed in the aftermath of the June 1998 Kandla cyclone, was widely praised for its role in post-disaster relief and rehabilitation following the 2001 Gujarat earthquake that killed or injured more than 26,000 people in India's Kutch District. With the GSDMA retaining ultimate authority, Abhiyan worked in 600 earthquake-affected villages and towns in the Kutch District's wide geographical expanse to bring civic engagement to the reconstruction process, complementing GSDMA's knowledge with its in-depth knowledge of the district. Abhiyan linked technical experts with illiterate villagers and worked to ensure reconstruction efforts addressed the best interests of the people. It did this by encouraging the formation of village committees to select partnering agencies in reconstruction (as a result, a number of agencies not considered to be working in the best interests of the community were rejected), helping villagers conduct damage and loss assessments, and forming committees to disseminate information on reconstruction packages and policies between villagers and government agencies and among the villages to ensure equity in reconstruction policy and implementation. It also successfully lobbied for policy measures that greatly improved transparency, accountability, and community involvement, and it convinced banks and government agencies to route reconstruction funds directly to beneficiaries through bank accounts.

Source: Lena Dominelli, 2007, *Revitalising Communities in a Globalising World* (Farnham, UK: Ashgate Publishing, Ltd.), http://www.ashgate.com/default.aspx?page=637&calctitle=1&pageSubject=471&pagecount=2&title_id=7729&editjon_id=8762&lang=cy; *and* Abhiyan, http://www.kutchabhiyan.org.

TCGI

2003 Bam Earthquake, Iran

Experience of Coordinating International NGOs Involved in Reconstruction

At the time of the 2003 earthquake in Bam, Iran, the UN had not yet adopted the cluster approach to coordinate international aid after disasters. However, UN OCHA and UNDP set up a coordination mechanism to support the Iranian government's management of reconstruction, including the coordination of international NGOs. As part of this effort, UNDP and the Housing Foundation of the Islamic Revolution (HF) coordinated activities in the shelter sector. UNDP organized an initial meeting that included government agencies, as well as the Iranian Red Crescent, UN-HABITAT, and the international NGOs, to discuss government's shelter sector policies and its reconstruction approach.

As the recovery progressed, regular meetings continued in various sectors, including shelter. The international NGOs active in shelter provision in and around Bam before the earthquake were mostly working in rural areas, where reconstruction was faster and less complicated, than in urban areas, where detailed structure plans had to be respected and construction techniques were more complex. However, the damage from the Bam earthquake in urban areas was enormous (approximately 25,000 urban housing units were lost), so international NGOs were encouraged to diversify. The international NGOs built 3,200 replacement housing units after the Bam earthquake, 850 of which were in urban areas. The direct financial aid provided by international NGOs was between US$4,000 and US$7,000 per household. The population also had access to government grants and low-interest loans. (The World Bank also provided a US$220 million loan for reconstruction, a large part of which was used by the HF to purchase materials for housing reconstruction.) However, much of the added value of the international NGOs and UNDP during this time did not have to do with their financial assistance, but with their demonstration of participatory approaches in reconstruction and their support to the neediest groups within the affected population.

Sources: Victoria Kianpour, UNDP Iran, 2009, personal communication, http://www.undp.org.ir/; *and* World Bank, 2004, *Technical Annex for a Proposed Loan of US$220 Million to the Islamic Republic of Iran for a Bam Earthquake Emergency Reconstruction Project,* http://web.worldbank.org/external/projects/main?pagePK=64283627&piPK=73230&theSitePK=40941&menuPK=228424&Projectid=P088060.

2005 Hurricane Katrina, New Orleans, United States

Universities Unite to Help Rebuild New Orleans

The ACORN Housing-University Collaborative was formed to assist in rebuilding New Orleans in the aftermath of Hurricane Katrina. Cornell University, Columbia University, the Pratt Institute, New Jersey's Science and Technology University, and Louisiana State University participated with ACORN. Four months after the hurricane, the collaborative issued "The People's Plan for Overcoming the Hurricane Katrina Blues: A Comprehensive Strategy for Promoting a More Vibrant, Sustainable, and Equitable 9th Ward." The plan was well received by all stakeholders in New Orleans. Residents considered it truly representative of their needs. It featured 56 immediate, short-term, and long-term revitalization measures to address all aspects of community revival, including social, economical, environmental, and physical planning issues. In March 2007, both the New Orleans City Planning Commission and the New Orleans City Council passed resolutions to incorporate the plan's main elements into the comprehensive Unified New Orleans Plan. The plan can be viewed at http://www.rebuildingtheninth.org/resources/.

Source: Kenneth M. Reardon, Marcel Ionescu-Heroiu, and Andrew J. Rumbach, 2008, "Equity Planning in Post-Hurricane Katrina New Orleans: Lessons from the Ninth Ward," http://www.huduser.org/periodicals/cityscpe/vol10num3/ch4.pdf.

2004 Indian Ocean Tsunami, Tamil Nadu, India

Public-Private Partnerships for Safer Houses – The Shelter Advisory Group in Nagapattinam

Public-private partnerships were used with great success in Tamil Nadu during post-tsunami housing reconstruction, with government providing land, specifications, and infrastructure, and NGOs providing houses. About 20,000 houses had to be built in the district of Nagapattinam, and construction quality was a concern. Despite the recognized commitment of the NGOs, the pace of work necessitated regular monitoring and field-based support.

A unique entity, the NGO Coordination and Resource Center (NCRC) was created to coordinate the efforts of all the players in the worst-affected districts. A joint Construction Quality Audit was also launched, supported by the NCRC, the government of Tamil Nadu, and UNDP. The multidisciplinary Shelter Advisory Group (SAG) was established at the district level, chaired by the District Collector and headed by Prof. Shantha Kumar, Emeritus Professor at Indian Institute of Technology Chennai and the main author of the TN Technical Guidelines for disaster-resilient housing. The SAG was supported by the Shelter Support Group (SSG), a team of post-disaster construction specialists, who visited the sites every month and provided technical support. Creating a registry of construction laborers at the village level—and training them in disaster-resilient construction—created a trained workforce of about 200 masons and other laborers. The SSG also trained government architects, contractors, engineers, and the engineers' association on integrating disaster risk reduction techniques into building practices. Based on contact with the field, the SSG provided the SAG with data on reconstruction progress, raised issues of concern, and made general recommendations. This information was discussed at monthly district-level "Construction Clinics" attended by SAG, SSG, and the NGOs. SAG advised the NGOs, individually and collectively, based on the feedback and concerns of the SSG. This system visibly improved construction; improved the flow of information; and proved that technocrats, bureaucrats, and implementers can work together with a common agenda and approach.

Source: C. V. Sankar, India National Disaster Management Authority, 2009, personal communication.

2005-2007 Worldwide

Disaster Risk Teams Mobilized by Express and Logistics Giant DHL

In 2005, DHL and UN OCHA established a strategic partnership to deliver aid quickly to remote areas immediately following a catastrophe by overcoming transportation and logistics challenges. As part of its larger corporate responsibility program, DHL is working to establish a global network of Disaster Response Teams (DRTs) to reduce bottlenecks in airports close to natural disaster sites. DHL is the umbrella brand of Deutsche Post World Net, the world's largest express and logistics company. Headquartered in Bonn, with 520,000 employees in more than 220 countries and territories worldwide, DHL has set up three DRTs worldwide. The first one is located in Singapore; the South Florida team covers Latin America and the Caribbean; and the most recent DRT base is DHL Express UAE in Dubai, which will cover the Middle East/Africa region. In the event of a major catastrophe, teams composed of specially trained DHL employees will help manage crucial logistics operations in airports close to the affected region, ensuring that relief supplies are efficiently sorted, stored, and distributed. The DRTs helped deliver some 4.77 million pounds of relief materials for post-tsunami and post-Hurricane Katrina relief operations. DHL and its DRTs will also support UNDP in its leadership role in reducing disaster risk and building capacities to reduce risk in countries worldwide through disaster preparedness and awareness. DHL can be contacted through its "Disaster Management" Web site at http://www.dp-dhl.de/en/responsibility/helping_people_gohelp/disaster_management.html.

Sources: UN OCHA, "United Nations Helps Launch DHL Disaster Response Team," Press Release, May 31, 2006; UN OCHA, "Disaster Response Teams: An OCHA/DHL Partnership," http://ochaonline.un.org/tabid/4777/Default.aspx; *and* DHL, 2007, "DHL Launches Disaster Response Team for Middle East/Africa," *Middle East Events* (October 2007), http://www.middleeastevents.com/site/pres_dtls.asp?pid=2244.

2004 Indian Ocean Tsunami, Aceh, Indonesia

Successful South-South Partnership

Immediately after the news of the 2004 Indian Ocean tsunami reached the outside world, the overseas development agency of the Catholic Church in Germany, Misereor, realized that Urban Poor Linkage Indonesia (UPLINK), a national coalition of NGOs and CBOs that focuses on urban poor issues and Misereor's most important partner in the Indonesia, lacked experience in post-disaster reconstruction and would require support if it were to respond. As a result, Misereor arranged a partnership between UPLINK and Kutch Nav Nirman Abhiyan, a network of NGOs from Gujarat, India, that aims to enhance communities' resilience and disaster preparedness. Abhiyan had played a pivotal role in the 2001 post-earthquake reconstruction in Gujarat, India, through its community-driven approach to reconstruction and its advocacy efforts to avoid relocation. In Aceh, Abhiyan helped UPLINK with development of a project concept and design of an implementation strategy, and with overcoming challenges in implementation. Abhiyan was also charged with ensuring that the affected people would actively participate in the rehabilitation and reconstruction in Aceh in a way that reflected their needs and capacities. The partnership produced impressive results, including the reconstruction of more than 3,000 quality houses. With Abhiyan's help, UPLINK was able to navigate the complexities of the recovery and reconstruction process while strengthening its own capacities. (The project won a 2008 Dubai Best Practices award as "Integrated People-Driven Reconstruction in Post-Tsunami Aceh.") The relationship between these two like-minded organizations continued throughout most of the project implementation and extends beyond the collaboration in Aceh.

EARL KESSLER

Source: Jennifer Duyne Barenstein et al., 2007, "People-Driven Reconstruction and Rehabilitation in Aceh. A review of UPLINK's Concepts, Strategies and Achievement" (evaluation by World Habitat Research Center under contract to Misereor), World Habitat Research Center, http://www.worldhabitat.supsi.ch; *and* Profile Uplink Indonesia.

2005 Hurricanes Katrina and Rita, Gulf Coast, United States
CSOs Combine Forces to Carry Out Home Repair Program

Habitat for Humanity International (HFHI) and Church World Service (CWS) formed a partnership in April 2006 to assist low-income families with disaster recovery funds in areas heavily damaged by Hurricanes Katrina and Rita along the Gulf Coast. This 2-year grant program was a joint effort to provide local long-term recovery organizations (LTROs) with the funding to support affected families with the repair and reconstruction of their homes. These LTROs were composed of local community, faith-based, and voluntary agencies that were involved in disbursing available resources (gift-in-kind materials, volunteer labor, case management, and funding) to affected families with unmet needs remaining after receiving disaster assistance funds from primary sources, such as the federal government and the insurance market. One of the benefits of the project was that it allowed both HFHI and CWS to focus on their core areas of strength (new home construction and post-disaster community organizing, respectively), while increasing each other's capacity to respond to the high level of need across the region within their normal models of program delivery. Like many of the other rebuilding projects under way during the same time period in the Gulf, there were some minor delays in completing repairs due to fluctuations in available volunteer and contract labor, which led to the original project timeline being extended an additional two months. At the end of the project, almost US$4 million had been disbursed by HFHI and CWS to 53 different LTROs, which resulted in the repair of approximately 697 homes.

Source: Giovanni Taylor-Peace, Habitat for Humanity International and Bonnie Vollmering, Church World Service, 2009, personal communication; *and* "Habitat for Humanity International, Gulf Recovery Effort," https://www.habitat.org/gulfrecoveryeffort/default.aspx.

Resources

European Commission, United Nations Development Group, and World Bank. 2008. "Joint Declaration on Post-Crisis Assessments and Recovery Planning." http://www.undg.org/docs/9419/trilateral-JD-on-post-crisis-assessments-final.pdf.

IASC. 2006. "Guidance Note on Using the Cluster Approach to Strengthen Humanitarian Response." http://www.humanitarianreform.org/humanitarianreform/Portals/1/Resources%20&%20tools/IASCGUIDANCENOTECLUSTERAPPROACH.pdf.

UN OCHA. 2006. *Exploring Key Changes and Developments in Post-Disaster Settlement, Shelter and Housing, 1982-2006.* Scoping study to inform the revision of Shelter after Disaster: Guidelines for Assistance. New York: UN OCHA.

Scheper, Elisabeth, Arjuna Parakrama, and Smruti Patel. 2006. *Impact of the Tsunami Response on Local and National Capacities.* London: Tsunami Evaluation Coalition (TEC). http://www.alnap.org/pool/files/capacities-final-report.pdf.

UNISDR. 2007. "A Guide for Implementing the Hyogo Framework." Geneva: UN. http://www.unisdr.org/eng/hfa/docs/Words-into-action/Words-Into-Action.pdf.

UNISDR Platform for the Promotion of Early Warning. 2008. *Private Sector Activities in Disaster Risk Reduction: Good Practices and Lessons Learned.* Bonn: UN. http://www.preventionweb.net/english/professional/publications/v.php?id=7519.

UNISDR , World Bank, and World Economic Forum (WEF). 2008. *Building Resilience to Natural Disasters: A Framework for Private Sector Engagement.* Geneva: WEF. http://www.preventionweb.net/english/professional/publications/v.php?id=1392.

Wisner, Ben and Bruno Haghebaert. 2006. "State/Civil Society Relations in Disaster Risk Reduction." Discussion paper for "Strengthening Global Collaboration in Disaster Risk Reduction." Bangkok: ProVention Consortium Forum.

World Bank. 2000. "Involving Nongovernmental Organizations in Bank Supported Activities." http://web.worldbank.org/WBSITE/EXTERNAL/PROJECTS/EXTPOLICIES/EXTOPMANUAL/0,,contentMDK:20064711~menuPK:4564189~pagePK:64709096~piPK:64709108~theSitePK:502184,00.html.

World Bank, Civil Society Team. 2007. "Consultations with Civil Society: A Sourcebook (Working Document)." http://siteresources.worldbank.org/CSO/Resources/ConsultationsSourcebook_Feb2007.pdf.

Project Implementation

15

MOBILIZING FINANCIAL RESOURCES AND OTHER RECONSTRUCTION ASSISTANCE

Guiding Principles for Mobilizing Finance and Other Assistance

- Transparent, timely systems must be in place or be put in place for programming financial resources for reconstruction and monitoring reconstruction progress.
- Tracking systems for housing and community reconstruction projects should be compatible with the systems used to track the overall reconstruction program.
- The systems and procedures used in managing reconstruction funds may be on-budget or off-budget, special or normal, but in all cases government must apply good public financial management (PFM) practices.
- Even if government is not handling all reconstruction funds, it should consider tracking them and establishing rules for their use in reconstruction.
- Reconstruction financing decisions should be as consistent as possible with existing sector investment plans, both national and local.
- The financial strategy for reconstruction should ensure that the delivery of assistance is fair, efficient, and transparent, down to the household level.
- The form of housing assistance should be consistent with the reconstruction approach and should take into consideration the capacity of households to receive and manage the funds and the state of the construction materials market.

This Chapter Is Especially Useful For:

- Policy makers
- Lead disaster agency
- Reconstruction finance agency
- Financial specialists
- Agencies involved in reconstruction

Introduction

Without financing, post-disaster reconstruction cannot take place. A good reconstruction financing effort is one that is efficient, transparent, and firmly directed toward realizing the physical results envisioned in the reconstruction policy. Those in charge of reconstruction financial management should take a strategic approach, and take seriously their responsibility to the affected population and to the public.

A number of conditions contribute to the development and implementation of a successful financing strategy: clarity about objectives, good coordination among the sources of financing and executing agencies, and careful administration of the receipt and the distribution of funds. Careful tracking of assistance from all agencies involved in reconstruction increases accountability and the effectiveness of the reconstruction effort.

This chapter covers various dimensions of the two key aspects of reconstruction finance: (1) mobilization and tracking of financing sources and (2) targeting and delivery of financial and other reconstruction assistance to households. It also briefly mentions other financial sources that may support households during reconstruction, specifically migrant remittances and microfinance.

Public Policies Related to Mobilizing Financial Resources and Other Reconstruction Assistance

National and local governments increasingly have long-range plans to guide public expenditures in regions or specific sectors and systems for prioritizing public investments and for approving projects. These are useful when negotiating post-disaster financial commitments with donors and for benchmarking the housing and infrastructure investments of outside agencies.

Yet post-disaster funding from external sources is rarely incremental over the medium term. Therefore, poorer countries dependent on external funding for both post-disaster reconstruction and future public investment should analyze how donor contributions to reconstruction will affect future development goals and investment plans. In programming reconstruction funds, policy makers should keep national investment goals in mind, while at the same time being sensitive to long-term inequities that may be created by reconstruction between the disaster area and non-affected areas of the country.

Few governments have mechanisms already in place to distribute post-disaster housing assistance. However, systems used to make transfers to families under social safety net programs, such as income support or housing subsidies, may be adaptable to deliver post-disaster housing assistance. These programs may also have census data that can be used to target and qualify families for housing assistance.

The normal sources of funds households use for construction and livelihoods may be seriously constrained after a disaster. Countries in which microfinance institutions (MFIs) are prevalent may have a government agency charged with supporting MFIs that can step in to assist those whose clientele has been affected by the disaster. Similarly, academic institutions or nonprofit agencies knowledgeable about migration may be able to help government analyze the need for intervention to shore up the delivery systems for migrant worker remittances following a disaster.

Key Decisions

1. **Government** must designate the agency to manage and monitor reconstruction financing. If this is not the lead disaster agency itself, the relationship of this entity with the lead disaster agency and with the agency responsible for normal government financial management will need to be clearly established.

2. The **reconstruction finance agency** needs to help government develop a viable reconstruction finance strategy and support government in presenting it to donors.

3. The **reconstruction finance agency** needs to decide with **government** on the PFM approach to reconstruction finance, such as whether financial management will be on-budget or off-budget and the type of controls to be employed.

4. The **reconstruction finance agency** should decide on the system for tracking reconstruction finance at the national and project levels, and work with **agencies involved in reconstruction** to promote full use of the system.

5. The **reconstruction finance agency**, in consultation with the **lead disaster agency, local government**, and **agencies involved in reconstruction**, should decide how to use reconstruction funds to support existing sector-specific public investment plans in carrying out reconstruction.

6. **Agencies involved in reconstruction** should decide with **government** on guidelines for qualifying recipients of aid and the means for delivering assistance to them, so that results are equitable and secure.

7. **Government**, in consultation with **agencies involved in reconstruction**, should establish guidelines for communications with affected communities regarding assistance and for disclosure of financial information to affected communities and the general public, and monitor their implementation throughout reconstruction.

8. **Government**, in consultation with **agencies involved in reconstruction**, should establish conditions for the use of housing assistance, for instance, those related to the improvement of disaster resistance in reconstruction. (See 📖 Chapter 6, Reconstruction Approaches, and 📖 Chapter 10, Housing Design and Construction Technology.)

Technical Issues
Mobilizing Financial Resources

Mobilizing national and international resources. In the aftermath of a major disaster, a series of activities generally proceeds in rapid succession, leading ultimately to the identification and programming of financial resources for reconstruction at the national, community, and household levels. These activities generally start with the initial assessment, the definition of an outline strategy, and the issuance of a rapid appeal, followed by more detailed needs assessments; the scheduling of a donor conference; the development of a reconstruction policy, strategy, and financial plan by sector; and the establishment of implementation mechanisms for reconstruction.[1] This sequence of activities and guidelines for establishing reconstruction policy are discussed in 📖 Chapter 1, Early Recovery: The Context for Housing and Community Reconstruction, and 📖 Chapter 2, Assessing Damage and Setting Reconstruction Policy.

> *Policy makers must ensure that the entity responsible for operating and maintaining the infrastructure being rebuilt is involved in reconstruction decisions and discussions about budgetary support for future operations.*

1. Wolfgang Fengler, Ahya Ihsan, and Kai Kaiser, 2008, *Managing Post-Disaster Reconstruction Finance: International Experience in Public Financial Management*, Policy Research Working Paper 4475 (Washington, DC: World Bank). http://go.worldbank.org/YJDLB1UVE0.

Because the impact of a disaster may exceed a country's resources and capacity to respond, financial assistance from international donors often plays a significant role in recovery and reconstruction. Donor conferences provide the venue for mobilizing official international assistance. Outcomes from the donor conference, such as commitments to specific sectors, prioritization of needs by government, and sector policies, will affect the resources available for housing and community reconstruction. Because donors are more likely to make commitments to support sector strategies or project proposals that are clearly defined, having predefined strategies for reconstruction in key sectors such as housing is an aid to mobilizing donor resources.

Depending on the scale and visibility of the disaster, assistance will come from multiple sources, including international, national, and local nongovernmental organizations (NGOs); civil society organizations (CSOs); and the private sector. The roles of these entities are discussed in 🔖 Chapter 1, Early Recovery: The Context for Housing and Community Reconstruction, and 🔖 Chapter 14, International, National, and Local Partnerships in Reconstruction. Coordination of these entities falls to government.

One of the key issues in post-disaster situations is how to make financial resources available quickly. NGOs and bilateral agencies may have greater flexibility in the short run than government, although the amounts at their disposal are frequently limited. Mobilizing national resources will generally involve a budget reallocation process. An important issue will be whether the country's budget law is flexible enough to allow prompt reallocation of budgeted resources when a disaster occurs, while providing for necessary controls. In a country whose PFM system provides for emergencies, the arrangements should be defined in an emergency management policy.

With international financial institution (IFI) funds, new loans and grants can be provided. Alternatively, or in addition, resources in existing projects be reprogrammed. Representatives of the IFIs in the country assist with these requests. World Bank Community-Driven Development programs (which operate on the principles of local empowerment, participatory governance, demand-responsiveness, and enhanced local capacity) are good candidates for restructuring, since the execution arrangements are appropriate for housing and community reconstruction, as are loans and programs in infrastructure sectors where reconstruction is needed. See 🔖 Chapter 20, World Bank Response to Crises and Emergencies, for a description of how the Bank can respond to borrower requests to restructure existing loans and programs.

Public financial management. Government will be required to make a number of decisions regarding PFM in reconstruction. (These decisions will generally be made for the entire reconstruction program, within which housing and community reconstruction is only one sector.[2]) The goal should be an implementation scheme that provides flexibility without sacrificing control, although compromises are sometimes made at the beginning, while additional financial safeguards are put in place. These decisions include:

- The management and institutional set-up for reconstruction (See 🔖 Chapter 13, Institutional Options for Reconstruction Management, for a discussion of the options generally considered)
- How much time to spend in reconstruction planning versus proceeding rapidly to project implementation (this decision may be sector-specific)
- Whether reconstruction funds will be managed on-budget or off-budget
- Whether public funds will be spent early on or later in recovery
- If a regular or special procurement regime will be used for public funds
- The role given to institutions of accountability and control (e.g., supreme audit institutions, internal audit units, and inspection units) and their need for additional capacity
- Whether to use ex ante or ex post controls in overseeing the use of funds
- How to equalize funding during reconstruction among regions, sectors, and types of projects
- The extent to which reconstruction funds can be used to support existing sector-specific public investment strategies
- How to coordinate the funding of all agencies involved in reconstruction, including NGOs and the private sector

International financial institutions and donors often provide assistance to establish the procedures necessary for reconstruction PFM. This can include:

- assessing government's capacity for management of reconstruction finance[3];
- providing funds for technical assistance for improving PFM during the reconstruction period;
- establishing a multi-donor fund and execution arrangements[4]; and
- providing technical assistance to set up and help run a financial tracking system.[5]

PRISCILLA PHELPS

2. For that reason, only a brief discussion of post-disaster PFM is included in the handbook. Additional information is found in the Resources section.
3. Wolfgang Fengler, Ahya Ihsan, and Kai Kaiser, 2008, *Managing Post-Disaster Reconstruction Finance: International Experience in Public Financial Management,* Policy Research Working Paper 4475 (Washington, DC: World Bank). http://go.worldbank.org/YJDLB1UVE0.
4. See World Bank, 2005, Multi-Donor Trust Fund for Aceh And North Sumatra (MDTFANS),Operations Manual, http://www.multidonorfund.org/documents/operational_manual_final.pdf.
5. A discussion of tracking system characteristics is found in Cut Dian Agustina, 2008. *Tracking The Money: International Experience With Financial Information Systems And Databases For Reconstruction* (Washington, DC: Global Facility for Disaster Reduction and Recovery), http://www.preventionweb.net/english/professional/publications/v.php?id=2474.

Tracking funds at the project level. PFM is a concern not only within central government; financial tracking is also needed at the community and project levels during reconstruction.

Numerous project management systems are available as software and online systems; however, no proprietary system seems to be in common use for post-disaster housing and community reconstruction project management. Local governments with good financial management systems may have project financial tracking capacity, although in rural areas or countries with a weak decentralization scheme this is rare. The local tracking system should have similar capabilities to the national tracking system and should be able to communicate with it, so government can aggregate information on the progress of execution and expenditure at the local level. Since many tracking systems are Web-based, the local system could be a component of the national system, although local Internet speed and access may be a constraint. In the absence of a technology-based solution, a simpler spreadsheet-based or paper reporting system may be adequate, especially in the early days of reconstruction.

The system should be accessible by and understandable to households and communities overseeing their own projects. Indonesia developed a system with many of the features described above after the Yogyakarta earthquake as part of the World Bank Community-Based Settlement Reconstruction and Rehabilitation Project.[6]

The ideal system for tracking expenditures for housing and community reconstruction has the functional capabilities shown in the following table.

Capabilities of Housing and Community Reconstruction Project Management System

Project-level data

Government monitors project-level data suitable to produce output and outcome indicators by project and for the sector. These may be a product of the tracking system at the project level or produced by government by project and sector. Examples are:

- Project disbursements
- Families assisted
- Demographic information
- Project milestones
- Overhead costs

Housing reconstruction component

- Administrative component for keeping track of household registration, family registration, demographic data, names of responsible parties, proof of property ownership, registration of owner contribution, owner bank information, power of attorney, etc.
- Budget and budget execution at the individual household level
- Expenditures or progress payments tracking at the household
- Physical advance of individual projects, such as project schedule, architect and engineering reports, safety inspection reports, change orders, and photographs

Infrastructure reconstruction component

- Administrative component for registering connections, ratepayers, hours worked, etc.
- Budget entry and tracking at project level
- Expenditures by project
- Ability to monitor physical advance of projects, such as engineer's reports, safety inspection reports, change orders, and photographs

Project financial management

- Administrative component for tracking contracts, purchase orders, vendor and contractor information
- Project budget and budget execution
- Receipts and disbursements of funds
- Work plans and project milestones
- Project overhead costs and allocations

6. Java Reconstruction Fund, "Community-Based Settlement Reconstruction and Rehabilitation Project for NAD and Nias," http://www.rekompakjrf.org (in Bahasa) *and* "Progress Report 2008, Two Years after the Java Earthquake and Tsunami: Implementing Community Based Reconstruction, Increasing Transparency," http://www.javareconstructionfund.org/ducuments/pdf/2008-07-07_JRF-2nd%20Progress%20Report_ENG.pdf.

How "build back better" affects reconstruction costs. "Build back better" is a phrase widely used after the 2004 Indian Ocean tsunami; however, the phrase has many interpretations. Codified into 10 propositions in a report to the office of the UN Secretary-General by special envoy and former U.S. President, Bill Clinton, the "build back better" concept encourages reconstruction that reduces vulnerability and improves living conditions, while also promoting a more effective reconstruction process.[7] Not all the "build back better" propositions necessarily increase reconstruction cost ("Good recovery planning and effective coordination depend on good information"), but others may ("Good recovery must leave communities safer by reducing risks and building resilience"). One estimate of cost increases from implementing disaster risk reduction measures, such as using ringbeams, plinth beams, and stronger roof connections, is around 10 percent of reconstruction cost, including additional materials, training, and supervision.[8] These improvements should be implemented whenever possible, and may be sound conditions for the receipt of housing assistance, even if they result in marginally higher costs.

CHRIS JENNINGS

Budgeting for operation and maintenance. Policy makers must ensure that the assets created in rebuilding and the responsibility for operating and maintaining them are properly transferred to the relevant owner and/or operating entity (e.g., subnational government, public corporation). The respective entity should be involved in the reconstruction decisions and discussions about operations and budgetary support, including tariff adjustments. If the entity is not involved, new public investment can deteriorate from lack of maintenance in the years following reconstruction. This issue is discussed in more detail in 🏠 Chapter 8, Infrastructure and Services Delivery.

Targeting and Delivering Assistance to Households

The criteria for eligibility and levels of assistance are discussed in 🏠 Chapter 4, Who Gets a House? The Social Dimension of Housing Reconstruction. The following sections provide information on the administrative aspects of providing assistance, such as the form of assistance, the mechanisms for delivery, and the procedures for qualification. A communications strategy, including such mechanisms as telephone hotlines, will be needed to promote accountability and detect problems in the targeting and delivery of assistance.

A communications strategy, including such mechanisms as telephone hotlines, will be needed to promote accountability and detect problems in the targeting and delivery of assistance. The suggestions on communicating with the affected population included in 🏠 Chapter 3, Communication in Post-Disaster Reconstruction, should be kept in mind when developing and announcing the targeting and delivery mechanisms for housing assistance.

Cash transfers and vouchers. Cash transfers and vouchers are the most common ways to provide assistance to affected populations to carry out housing reconstruction. They can also be used to pay for work on infrastructure projects and to provide a broader "social safety net."[9] A detailed discussion of the factors to consider in designing a social protection system for natural disasters is found in 🏠 Chapter 4, Who Gets a House? The Social Dimension of Housing Reconstruction, Annex 1, Considerations in Designing a Social Protection System for Natural Disasters. The types of transfers commonly used are the following.

7. William J. Clinton, 2006, *Key Propositions for Building Back Better*, New York: UN, http://www.reliefweb.int/rw/RWFiles2006.nsf/FilesByRWDocUnidFilename/TKAE-6WW9H3-Full%20Report.pdf/$File/Full%20Report.pdf.
8. Estimate provided by Practical Action, 2009.
9. International Committee of the Red Cross (ICRC) and the International Federation of Red Cross and Red Crescent Societies (IFRC), 2007, "Guidelines for Cash Transfer Programming" (Geneva: ICRC/IFRC). See especially "Section B: Guidance Sheets," http://www.ifrc.org/docs/pubs/disasters/cash-guidelines-en.pdf.

 For access to additional resources and information on this topic, please visit the handbook Web site at www.housingreconstruction.org.

Type of cash transfer	Description
Unconditional cash transfers	Given with no conditions as to how the money should be used. Often used immediately after an emergency.
Conditional cash transfers	Given on the condition that recipients do something (for example, rebuild their house, plant seeds, provide labor, or establish or reestablish a livelihood). See Designing the Conditional Cash Transfer System for Reconstruction, below.
In-kind transfers or vouchers	Stipulate the items or services for which the recipient can exchange his or her voucher, including construction materials; have a specific value; can either define a service or good that the voucher can be exchanged for or allow the recipient freedom as to purchases; exchanged with specified vendors or at organized fairs.
Public works	Payment for work on public works programs. Wages should be slightly below market levels to avoid competing with labor market.
Social safety net or other social transfers	Repeated, unconditional cash transfers provided to vulnerable households or individuals (for example, the elderly or pregnant women). Often focus on reactivating and/or replacing livelihood activities. Best implemented in partnership with government agencies.

Special arrangements are needed to ensure adequate reconstruction assistance for vulnerable households. This may entail higher payments so that the household can hire all the labor or someone to oversee the reconstruction project.

The delivery mechanism for cash and vouchers requires careful planning and execution, especially when using direct deliveries. Two important aspects of the delivery system for funds are the institutional intermediary who will manage the delivery and the form in which the funds will be delivered. The following are needed to administer a transfer program:

- A clear targeting rationale and a reliable recipient identification system
- Institutional capacity sufficient to carry out the program in a timely manner
- Good coordination between governmental and nongovernmental actors, if both are involved
- A system for monitoring, reporting on, and evaluating the program, and making adjustments, if necessary

The source of the funds and the agency responsible for delivery may be distinct. The World Bank has analyzed the potential for social funds to participate in the delivery of financial assistance and points out the importance of using an agency experienced in community-based development that has an operational presence in the disaster zone.[10] Having an effective housing assistance strategy to which key institutions are committed may be more important than having vast sums of money, as shown in the 🖳 case study, below, on Hurricane Katrina assistance.

The forms in which assistance may be delivered to recipients could include the following:

- Cash transfer into bank/post office accounts
- Cash transfer to local remittance and money transfer companies and burial societies
- Direct cash, check, or voucher distribution to recipient
- Mobile ATMs (for cash withdrawals), smart cards, or money orders
- Direct credit to mobile phones (including distribution of phones, if necessary)
- Delivery through local businesses or community-based organizations

The delivery of funds may not necessarily be to individual households. Management of funds by groups of households has been used successfully in Indonesia and elsewhere. Groups can assist in delivery of, approval of, and social control over disbursements, thereby improving transparency and lowering security and transaction costs.

In designing the delivery system, program sponsors should factor in travel costs, gender mix, security risks (especially if women are delivering or picking up the cash or vouchers), cultural familiarity with the mechanism chosen (ATMs, for example, may require instruction), and direct delivery methods for those who cannot travel.

10. World Bank, 2009, *Building Resilient Communities: Risk Management and Response to Natural Disasters through Social Funds and Community-Driven Development Operations*, (Washington, DC: World Bank), http://siteresources.worldbank.org/INTSF/Resources/Building_Resilient_Communities_Complete.pdf.

At the same time, for the transfer program to provide households with the materials they need for reconstruction, the following conditions must be in place in the local market:

- Availability and/or production chain capacity for goods
- A functioning market for the goods and/or services people need and geographical access
- Willing traders with financial and logistical capacity to get goods into the region and assurance that traders will accept vouchers (in the case of voucher systems)
- A reliable recipient identification system
- A reliable and secure delivery system for paying traders who accept vouchers
- No excessive taxation of goods
- An ability to monitor price levels to control price gouging
- An ability to monitor and if necessary offset inflation in costs of materials and labor

DANIEL PITTET

Compensating households for price increases may be done by government, or by NGOs or other agencies involved in reconstruction as was done in Sri Lanka after the 2004 Indian Ocean tsunami, as discussed in the 🖳 case study, below.

Providing construction materials in-kind. When local markets are not functioning properly, or do not have the capacity to provide the quantity or quality of materials required for reconstruction, providing in-kind assistance may be advisable. This can be done simply by buying construction materials in local market and or by sourcing the materials outside the local market. Depending on the quantities procured, in-kind assistance may require the agency involved to arrange warehousing and other forms of logistics. In-kind assistance should be provided to households according to an allocation scheme with criteria and qualification procedures similar to those for cash and vouchers. (In fact, stricter controls may be required for the delivery of materials, which may be more subject to fraud than are cash and voucher programs, due to the demand for and marketability of the goods.) From these "materials banks," materials may be provided without a financial exchange or traded for cash or vouchers. Goods may be physically delivered directly to households, or the banks may act like markets where homeowners come and choose materials.

While the provision and distribution of in-kind assistance can be challenging for the agencies involved, the benefits may outweigh the costs. The frequently cited benefits of in-kind provision at the household level include the following.

- For poor people, it makes it more likely they will obtain what they need.
- For residents of remote areas, it reduces the time spent travelling to markets and the cost of transporting materials.
- For all homeowners, it helps ensure that they have the proper quantity and quality of materials required to improve building safety.

For an agency experienced with the procurement of construction materials, there are important benefits that can be realized from the scale of procurement, relative to purchases in small lots. Specifically, an agency can:

- order according to specifications (e.g., pre-bent iron bars) that will increase safety, reduce labor inputs on the construction site, and save training time;
- demand quality standards, including materials testing;
- ensure that the desired materials are procured, even if they have to be brought from another region of the country or imported;
- negotiate lower prices or organize procurement before post-disaster prices increases kick in; and
- by receiving materials on a schedule, deliver materials to homeowners in standardized packages (for instance, a package corresponding to a core house).

Designing the Conditional Cash Transfer System for Reconstruction

House Design and Cost

In many programs, the amount of housing assistance is set at a level to allow construction of a core house. Using the core house as a benchmark, government or agencies involved in reconstruction can work with architects, chartered surveyors, and engineers to develop typical floor plans, house designs, labor requirements, materials specifications and quantities, and construction cost estimates. Those developing the house designs and cost estimates will need guidance regarding the reconstruction approach to be employed, role of contractors, availability and cost of materials, etc., so that the specifications accurately reflect conditions in the disaster location.

Homeowners should ideally be given a choice of core house floor plans. The core house designs should generally be improved versions of local traditional designs. Using these designs, the sufficiency of the proposed housing assistance amount should be verified. Where families have financing capacity beyond the core house or other financing sources are available, higher-cost housing designs and cost estimates may also be developed. The house designs should be used to develop construction guidelines and to build model houses. (See 🏠 Chapter 16, Training Requirements in Reconstruction.)

Progress Payments

Payments should be linked to the progress of construction, in order to control the quality and disaster-resistance of the construction. Progress payments also reduce the potential that beneficiaries use the housing assistance for other purposes and do not rebuild or repair their house and reduce the amount of assistance spent on houses that are never completed (although there will always be some). At the same time, there may be reasons to allow flexibility in the use of the housing assistance. See 🏠 Chapter 4, Who Gets a House? The Social Dimension of Housing Reconstruction. The amount of the construction progress payments can be derived using the house designs and cost estimates.

Households often have no funds with which to pre-finance the construction, so the payment schedule generally has to advance funds. While this part of the payment is at risk, effective supervision should keep this to a minimum. Payments also have to be timely, so that households do not run out of funds.

Providing four or five installments, each linked to an identifiable stage of construction, keeps transaction costs under control, while providing sufficient leverage on the homeowner. The release of the installment takes place only after an inspection that certifies the consistency of construction with established safety standards. The amount of the installments depends on local conditions (house size, type of construction, and price of materials) and whether part of the assistance is being provided in-kind as materials.

The following is an example of a payment schedule:

- Installment 1: 10 percent advance for excavation
- Installment 2: 25 percent after the completion of the excavation of the foundations
- Installment 3: 25 percent after the construction of the plinth level
- Installment 4: 25 percent after the top of the wall (wall band) is cast
- Installment 5: 15 percent after roof completion (including gables, if applicable)

Special Payments

Special arrangements are needed to ensure adequate housing reconstruction assistance for vulnerable households (e.g., widows, elderly, female-headed households). This may entail, for example, higher payments so that the household can hire all the labor and/or someone to oversee the reconstruction project.

Care should be taken that the payment scheme encourages homeowners to save on reconstruction costs, e.g., by contributing more labor or recycling disaster debris. Incentives may also be provided to those who finish sooner, since this saves costs for government or other agency overseeing the construction process.

Source: Norbert Wilhelm, 2008, "Post-Disaster Housing Reconstruction in Developing Countries: Proposal for a Systematic and more Efficient Approach" (unpublished paper).

The humanitarian sector has extensive post-disaster experience with the delivery of goods, such as construction materials, which are referred to as "non-food items" or NFIs. An overriding concern is to avoid disrupting the local market chains that people rely on for their livelihood. A timely and reasonably priced supply of materials is dependent on a number of factors, including (1) availability of raw materials; (2) availability of labor; (3) access to utilities, such as power and water, at the production facility; (4) availability and affordability of transportation services; and (5) functioning logistics and distribution systems. It is recommended that agencies involved in reconstruction try to identify precisely where the supply chain for the reconstruction materials may be constrained, and to focus first on intervening to resolve those particular bottlenecks before creating a provisioning system that substitutes for the local market. 🏠 Annex 1, How to Do It: Deciding Whether to Procure and Distribute Reconstruction Materials, summarizes a range of issues to consider in developing a delivery program for construction materials, based on the experience of the humanitarian sector.

Providing fully built houses. The ultimate form of in-kind assistance is the provision of a fully built house. Programs that provide fully built houses with no contribution from the family are sometimes sponsored after large-scale disasters, often by international organizations concerned about the speed of the response but without existing programs in the disaster area.

Awarding fully built houses as a way to rehouse an affected community has had mixed results. Households sometimes sell a house soon after it is awarded. This is common if the recipient family does not consider the location suitable, could build itself a house for less than the value of the fully built house it has been given, or has pressing needs for cash. Some agencies attempt to force the family to remain in the house, for instance, by restricting the ability to sell for a period of time; however, these requirements can be bypassed by renting or arranging a transaction that is not formally a sale. Fraud can also be a serious problem in programs that award fully built houses, so good control over eligibility and qualification procedures is critical. If the assistance strategy is to provide a fully built house, households should be involved in the construction process to increase the chance that it meets their needs or given the liberty to use this asset as they prefer once it is transferred to them, as they would with cash. See 📖 case study in Chapter 8, Infrastructure and Services Delivery, on relocation following Hurricane Mitch, for an example of where families did not remain in fully built houses.

Qualifying Recipients of Assistance

Once the criteria for assistance are established, a system is needed to qualify recipients. This may be especially important in a social protection-type transfer program, where eligibility may seem subjective but recipients of housing assistance should also be subject to a qualification process. Even when the criteria are relatively straightforward (a fixed amount of assistance for every homeowner living in a specific location, for instance), the qualification process may be complex. For the criteria just mentioned, for example, both residency and prior homeownership need to be verified. Outreach or targeting may also be necessary, to ensure that the qualification process reaches all those who are eligible. Local communities, government, and external agencies can help with targeting and qualifying potential recipients.[11] The procedures can be community-based or administrative, or a combination of the two. Some options for qualifying recipients and their associated risks are shown below.

11. For numerous summaries of cash transfer programs, see Swiss Agency for Development and Cooperation, "SDC Cash Transfer Projects," http://www.sdc-cashprojects.ch/en/Home/Experiences/SDC_Cash_Transfer_Projects.

Methods of Qualifying Recipients in Transfer Programs

Type	Method for establishing qualification	Possible risks
Community-based	Through community leaders	Community leaders may include ineligible friends and family.
		Women and the vulnerable may be excluded.
		Community structures may have broken down.
	Through a committee elected by the community	This is a time-consuming and resource-heavy process.
		Interpretation of criteria may vary from location to location.
		Communities may not feel ownership of the criteria.
	Triangulation of lists compiled by different groups of, for example, men, women, elders	Socially marginalized may be excluded.
	Self-targeting: individuals or households opt in to the program	May only benefit those who register first.
		The housebound, elderly, etc. may not be able to come and register, and therefore risk being excluded.
Administrative	Selection based on existing government data	Can lack flexibility for people whose eligibility is marginal.
		There is less room for participatory processes.
		Data used may be out of date.
	Selection based on real-time data collection (survey or census)	Verification can be challenging in an emergency due to population mobility, displacement.
		There may be a lack of local knowledge and data.
		Surveys and monitoring incur costs.

Qualifying in urban environments. Qualifying recipients in urban environments is frequently more complex than in rural settings, where people tend to know one another. The following are options for improving qualification in urban areas.

- Form local selection committees composed of religious leaders, respected families, women, and representatives of respected professions (for example, teachers).
- Present eligibility/targeting criteria to numerous community groups.
- Divide the urban area into smaller units.
- Target locations within the city where the eligible may congregate (for example, aid centers or refugee camps).

Complaint and Grievance Redressal

Processes that allow stakeholders to file complaints[12] are an important part of an assistance program. They help to ensure transparency and fairness and reduce the risk of errors or manipulation. Complaints are common in programs that provide direct financial assistance.[13] Those being assisted are in a vulnerable situation, and program criteria related to assistance schemes—no matter how well defined—are subject to misinterpretation. Complaints increase when amounts of assistance are large, such as those related to housing reconstruction.

A post-disaster complaints system can have different levels of mediation and grievance redressal, which should be operating both at the project level and at the national policy level. Components of such a system include:

- ongoing outreach, consultation, and review procedures to resolve complaints;
- the use of an independent advisory panel or committee;
- access to the country's ombudsman or similar function, where it exists; and
- access to legal redress through the courts.

Detailed advice on setting up grievance redressal mechanisms at the project level is included in ⌨ Annex 2, How to Do It: Establishing a Grievance Redressal System. Information on complaint handling related to public procurement, such as channels for whistleblowers, is provided in ⌨ Chapter 19, Mitigating the Risk of Corruption.

Lending and Bank Servicing in Reconstruction

Providing credit for post-disaster reconstruction can be done through the banking system or administered by government. Use of credit is more common in countries with good property insurance systems, where insurance proceeds provide the bulk of the reconstruction funds. Demand for credit is likely to be greater in urban reconstruction, where incomes are higher and because multifamily housing is difficult to rebuild without it.

Banks or government may provide reconstruction credit. Governments with experience lending to a population similar to the one affected by a disaster are in the best position to provide credit for reconstruction. Unless potential borrowers' income is unaffected by the disaster, banks should not be pressured to provide credit for reconstruction under conditions that expose them to unacceptable risks, without government guarantees or other risk reduction strategies.

Banks may play other roles in reconstruction finance, such as in safely delivering housing assistance. They have experience handling large quantities of cash and have financial controls in place. Care should be taken to ensure that banks are experienced or properly prepared to administer reconstruction finance, whether they are providing credit, or simply acting as an intermediary for the delivery of assistance. Bank administrative systems should be capable of handing the volume of transactions; checkbooks, debit cards, and other necessary materials must be available in sufficient quantities; and potential customers should be advised in advance of documentation and other bank requirements. Experts in retail banking may be needed to set up post-disaster programs that entail a large number of banking transactions.

12. "Complaint" and "grievance redressal" are generally considered interchangeable names for the same process, although some agencies have a sequence of procedures, the first part of which is called the "complaint process" and the second part of which is called the "grievance process."

13. Egon Rauch and Helmut Scheuer, 2007, "SDC Cash Workbook: A Practical User's Guide for the Preparation and Implementation of Cash Projects," Swiss Agency for Development & Cooperation (SDC), http://www.sdc-cashprojects.ch/en/Home/SDC_Cash_Workbook/media/Documents/Cash_Workbook/061207_SDC%20Cash%20Workbook%20FINALVERSION%20SCREEN%20VarLayoutLHP.pdf.

Migrant Worker Remittances in Reconstruction

Migrant worker remittances, by diversifying the recipient family's income, play a significant role in helping manage risk within the family's finances.[14] They smooth consumption patterns when work is seasonal and help households absorb unexpected shocks and demands, including those created by natural disasters. Remittances therefore act as a type of insurance that improves a family's ability to respond to a crisis.

Labor migration may be an important coping strategy for families affected by a disaster. Migration varies by season, degree of permanence, and location (such as urban/rural, domestic/regional/international), and it has different demographic characteristics (migration by families, men, or women), all depending on the country and culture. Analyzing the components of household income reveals the level of dependence of the affected population on remittances or income from labor migration and the extent to which a disaster has affected them.

Aid agencies should avoid discouraging post-disaster migration or penalizing families for whom migration is an economic coping strategy. This may mean permitting cash grants to be used to support and sustain remittances, including returning to jobs abroad or communicating with relatives. Reestablishing the local communications and banking systems through which remittances flow should be a priority in recovery.

Assistance schemes and reconstruction approaches should reflect the fact that people will want to blend the assistance resources with remittances and other household resources in reconstruction and productive activities. Cash-based support generally provides greater flexibility and choice.

Role of the diaspora community in reconstruction. Assistance for entire communities is sometimes financed by collective remittance programs, organized by diverse diaspora groups, sometimes known as hometown associations.[15] Government may take action to facilitate the entry of remittances following a disaster, as Indonesia did in 2006.[16] See the 🏠 case study on Hela Sarana, below, for an example of how the diaspora supported post-tsunami reconstruction in Sri Lanka.

14. Abid Qaiyum Suleri and Kevin Savage, 2006, *Remittances in Crises: A Case Study from Pakistan,* Humanitarian Policy Group (HPG) Background Paper (London: Overseas Development Institute), http://www.odi.org.uk/hpg/papers/BGPaper_RemittancesPakistan.pdf; *and* Kevin Savage and Paul Harvey, eds., 2007, *Remittances during Crises: Implications for Humanitarian Response,* HPG Report 25 (London: Overseas Development Institute), http://www.odi.org.uk/hpg/papers/hpgreport25.pdf.

15. Also called township associations and *organisations de solidarité international issues de l'immigration,* among other names.

16. Kevin Savage and Paul Harvey, eds., 2007, *Remittances during Crises: Implications for Humanitarian Response,* HPG Report 25 (London: Overseas Development Institute), http://www.odi.org.uk/hpg/papers/hpgreport25.pdf.

🏠 **For access to additional resources and information on this topic, please visit the handbook Web site at www.housingreconstruction.org.**

Microfinance in Reconstruction

The principle role of MFIs in reconstruction is reactivating the local economy. This is because these institutions are often the principal source of credit for the livelihood activities of low-income, disaster-affected households.[17] Few MFIs have the capacity to finance housing reconstruction; however, they commonly finance microenterprises that are based in the home and provide income that will make reconstruction possible.

There are significant risks for MFIs operating in post-disaster situations. Funders should not pressure MFIs with reconstruction lending targets, for example. Some recommended guidelines for MFIs under these conditions, which agencies supporting MFI activities should also understand, include the following.

- Avoid activities beyond the normal capacity and mission, such as giving medium-term loans to rebuild assets if they have not been provided before.
- Limit relief activities to locating clients, linking clients to relief, and transporting people to where they can receive services.
- Wait until an emergency is over to assess clients' property damage and credit standing before making reconstruction loans, limiting loans for purposes that don't generate cash income.
- For disaster-affected clients with loans outstanding, MFIs may adjust savings requirements or reschedule loans, but should avoid subsidizing interest rates or providing other forms of economic relief, to avoid sending mixed messages to clients and damage the credit culture.
- Adjust services to a client's circumstances, since some clients will be more severely affected by a disaster than others.
- Process insurance claims quickly to give clients access to emergency cash, while screening out false claims (for MFIs with insurance programs).
- Enter new areas to provide emergency financial assistance with caution, and explain the MFI's purposes clearly, so the MFI is not viewed as a relief agency or donor program.

Restoration of Livelihoods

One of the primary preoccupations of a household after a disaster is restoring its livelihood. It is logical, therefore, that government, humanitarian, and other agencies involved in reconstruction are also concerned about livelihood restoration. Reconstruction provides many livelihood opportunities (as is discussed in several other chapters of this handbook); however, a full treatment of this issue is, regrettably, beyond the scope of the handbook. The authors encourage government and agencies involved in reconstruction to include experts in post-disaster livelihood restoration in assessment teams and in teams developing the assistance strategy for reconstruction, so that livelihood opportunities—especially for poorer households—are maximized in reconstruction. The ⌨ case study on Habitat for Humanity Nicaragua, below, shows how purchasing of reconstruction materials after Hurricane Felix provided income to the members of a Miskito lumber cooperative.

Risks and Challenges

- Duplication of efforts among government and agencies involved in reconstruction due to lack of coordination in financial programming and in monitoring execution.
- Lack of transparency and accountability in the use of reconstruction funds.
- Abandonment of good PFM practices to speed up disbursements, or overtaxing the PFM system in a way that slows down reconstruction.
- Poor financial decision making due to out-of-date tracking information.
- Confusion on the part of affected households about how to access assistance programs and the rules for using the funds, due to poor communications with the families.
- Delays in distributing reconstruction assistance. Distribution of funds in a way that does not reflect actual needs, due to insufficient outreach, poor qualification procedures, or fraud.
- Misunderstanding the composition of household income, including the importance of remittances.
- Ignoring security risks when delivering cash and voucher assistance.
- Failing to understand the nature of bottlenecks in the market for reconstruction goods and services and therefore trying to solve the wrong problem.
- Long-term negative economic impact of the disaster in poor households.

17. Consultative Group to Assist the Poor (CGAP), 2005, *Sustaining Microfinance in Post-Disaster Asia: Guidance for MFIs and Donors* (Washington, DC: CGAP), http://www.cgap.org/p/site/c/template.rc/1.26.1882#key. Also, a range of research on the topic of microfinance and disasters by the Banking With the Poor Network can be found at http://www.bwtp.org/.

Recommendations

1. Be prepared to define the housing reconstruction program in as much detail as possible during the donor conference, as this will encourage the commitment of donor resources.
2. Define the basic parameters of reconstruction, such as minimal housing and public services standards, and establish maximum assistance to reduce competition among agencies involved in reconstruction.
3. Establish an expenditure tracking system at the national level, integrated with tracking at the project level, that ensures timely, accurate, and transparent exchange of information between levels of government, and between government and communities or agencies involved in reconstruction.
4. Define a communications strategy as well as the procedures for regularly and publicly disclosing financial information related to the reconstruction program to affected communities and the general public.
5. Design the assistance scheme so that it provides resources in a form consistent with the reconstruction approach or approaches people will employ. Grievance redressal is a necessary element of an assistance scheme.
6. Realize that members of the community may know best who needs post-disaster assistance and can help in identifying and qualifying households.
7. If production chains and materials markets are functioning, provide assistance in cash. Consider the distribution of construction materials based on a careful analysis of bottlenecks in materials markets.
8. Create a delivery system for cash that is accessible for recipients and secure for those delivering and receiving it. Make sure the system includes special arrangements for those unable to travel.
9. Understand that certain households will rely on remittance flows and microfinance to complement housing assistance, so measures may be needed to ensure access.
10. View institutions as mechanisms that can aid reconstruction, but don't press them into activities that undermine their long-term survival.
11. Seek the advice of experts in designing the reconstruction program in such a way that it provides livelihood opportunities to poor households.

Case Studies

2007 Hurricane Felix, Nicaragua

Building Back Better while Supporting Livelihoods

Hurricane Felix hit Nicaragua's Autonomous Region of the North Atlantic (RAAN) on September 4, 2007, leaving behind a path of destruction and despair among indigenous Miskito communities already suffering from significant levels of poverty. One of the worst-affected communities in the RAAN was the Auhya Pihni settlement. Of 150 houses in the settlement, only 3 were left standing, and they were heavily damaged.

During the assessment of recovery options, Habitat for Humanity Nicaragua (HFHN) agreed with the community that local materials and hurricane-resistant design would be used in the new homes. Timber and processed wood are the preferred local materials for house construction in the area, and the Miskito communities make their livelihood from the controlled exploitation of the surrounding forest.

HFHN agreed to purchase wood milled from the trees knocked down by the storm and additional timber processed by a community-based Miskito cooperative. Through this action, the project benefited from a plentiful supply of materials purchased at a highly discounted price, while the cooperative generated much-needed income to support local families. The community was also able to participate with their skills and labor in the implementation of the project. The house design incorporated flood- and wind-resistant design elements, such as reinforced stilts, hurricane strapping, and use of a specially designed nails to attach the roofing to the structure. The final outcome was 150 houses built within one year of the hurricane and economic support to the families in Auhya Pihni, taking advantage of the natural linkage between housing recovery and livelihood.

Sources: Mario C. Flores, Habitat for Humanity International, 2009, personal communication, http://www.hfhi.org; *and* "Habitat for Humanity Nicaragua Hurricane Felix," http://www.habitat.org/disaster/active_programs/nicaragua_hurricane_felix.aspx.

2004 Indian Ocean Tsunami, Sri Lanka

Role of Diaspora Financing in Reconstruction

After the 2004 Indian Ocean tsunami hit Sri Lanka, several donors and other agencies developed programs to provide cash for permanent housing reconstruction. The Sri Lankan government cash grant for self-builders was US$2,500 for a new house and US$1,000 for repairs to a damaged house. For full rebuilding, grants were released in four installments over six months. For damaged houses, the money was released in two installments over six months as repairs were completed. The value set by government proved to be inadequate in some cases as demand in the market as a result of the disaster more than tripled the pre-disaster price for construction materials, skilled labor, and land . Government was reluctant to adjust the amount of its housing cash grants because it perceived that the influx of NGO funds for overall tsunami reconstruction was contributing to the cost increases. Another problem was the lack of an effective mechanism to coordinate NGO activities to ensure equitable distribution of the NGO funds, including in the conflict area in northeast Sri Lanka. As a result, NGOs were called on in many cases to provide additional support, either through top-up payments or in-kind assistance, so that people could complete construction.

Sources: Will Somerville, Jamie Durana, and Aaron Matteo Terrazas, 2008, "Hometown Associations: An Untapped Resource for Immigrant Integration?" *MPI Insight* (Washington, DC: Migration Policy Institute), http://www.migrationpolicy.org/pubs/Insight-HTAs-July08.pdf; Leslie Dep, Royal Institution of Chartered Surveyors, 2009, personal communication; *and* Hela Sarana, http://www.helasarana.org.uk.

INFROGMATION

New Orleans Katrina Housing Facts

- New Orleans had approximately 87,500 owner-occupied housing units in 2000.

- 61% of owner-occupied homes suffered major or severe damage from Katrina.

- 51% of renter-occupied housing suffered major or severe damage from Katrina.

- 80% of subsidized affordable housing suffered major or severe damage from Katrina.

- By late 2008, there were 60,016 applications for reconstruction funds recorded for the city, of which 72% had been approved.

- 80% of homeowners received insufficient funds from various sources to rebuild.

- The average gap between resources and rebuilding costs is $55,000.

Source: New Orleans Master Plan and Comprehensive Zoning Ordinance, "Fact Sheet: Housing," 2009, http://www.nolamasterplan.org.

2005 Hurricane Katrina, Gulf Coast, United States

More than Money Is Needed to Resettle Disaster-Affected Households

In the United States, federal aid ordinarily lasts only 18 months after a disaster. But temporary housing programs set up in the aftermath of Hurricane Katrina only expired in June 2009, some 45 months after the worst hurricane in U.S. history hit New Orleans and other parts of the U.S. Gulf Coast. While 139,000 households had successfully cycled out of temporary trailers by the end of the program, more than 3,000 households were threatened with eviction from trailers, and another 15,000 faced a cut-off from federal rent subsidies at the deadline.

Expensive missteps in the temporary housing strategy included sheltering the affected population in cruise ships, hotel rooms, military facilities, and unhealthy trailers, until many families were finally moved into apartments. The states of Louisiana, Mississippi, and Texas received reconstruction funds and low-income housing vouchers, but lagged in developing long-term solutions.

When the problems with the trailer program were revealed, US$400 million in federal funds was offered to mass produce prefabricated cottages that could be used either temporarily or permanently. These cottages cost less than US$34,000 to construct, but could not be located where flood risk was the greatest. Mississippi built 3,075 of the cottages, but local jurisdictions refused to grant permits or alter zoning codes, fearing the cottages would lower property values. (Temporary use was later allowed.) Louisiana received funds to build the cottages, but after two years had not delivered a single one. In early 2009, the federal government offered to sell federally owned mobile homes to trailer occupants for US$5 each, and offered an additional US$50 million in vouchers and US$40 million in economic stimulus funds to Louisiana and Mississippi, even though the states had not used the money that was already available.

An important lesson from Hurricane Katrina is that money alone is not sufficient to provide permanent housing solutions to all those affected by a disaster. A reconstruction strategy that provides the proper incentives to the actors involved and to which all institutions are committed may be a more critical element.

Source: "Administration to Reveal Plans for Katrina Housing Transition," *The Washington Post*, June 3, 2009; *and* "Permanence Eludes Some Katrina Victims," *The Washington Post*, June 13, 2009.

2004 Indian Ocean Tsunami, Sri Lanka
Effect of Post-Disaster Price Increases

After the 2004 Indian Ocean tsunami hit Sri Lanka, several donors and other agencies developed programs to provide cash for permanent housing reconstruction. The Sri Lankan government cash grant for self-builders was US$2,500 for a new house and US$1,000 for repairs to a damaged house. For full rebuilding, grants were released in four installments over six months. For damaged houses, the money was released in two installments over six months as repairs were completed. The value set by government proved to be inadequate; demand in the market as a result of the disaster had increased the price for construction materials, skilled labor, and land by at least three times their pre-disaster costs. As a result, NGOs were called on to provide additional support, either through top-up payments or in-kind assistance, so that people could complete construction.

Source: Lesley Adams and Paul Harvey, 2006, "Cash and Shelter, Learning from Cash Responses to the Tsunami, Issue Paper 4" (London: Humanitarian Policy Group), http://www.odi.org.uk/hpg/Cash_vouchers_tsunani.html.

Resources

Agustina, Cut Dian. 2008. *Tracking the money: International experience with financial information systems and databases for reconstruction.* Washington, DC: Global Facility for Disaster Reduction and Recovery. http://www.preventionweb.net/english/professional/publications/v.php?id=2474.

Amin, Samia and Markus Goldstein, eds. 2008. *Data Against Natural Disasters Establishing Effective Systems for Relief, Recovery, and Reconstruction.* Washington, DC: World Bank. http://siteresources.worldbank.org/INTPOVERTY/Resources/335642-1130251872237/9780821374528.pdf.

Banking With the Poor. *Microfinance and Disaster Management.* Banking With the Poor Network. http://www.bwtp.org/arcm/mfdm/index.html.

Benson, Charlotte and Edward J. Clay. 2004. *Understanding the Economic and Financial Impacts of Natural Disasters.* Disaster Risk Management Series, #4. Washington, DC: World Bank. http://www.preventionweb.net/english/professional/publications/v.php?id=1848.

Emergency Shelter Cluster of the Inter-Agency Standing Committee. 2008. *Selecting NFIs for Shelter.* Geneva: IASC. http://www.sheltercentre.org/library/Selecting+NFIs+Shelter.

Fengler, Wolfgang, Ahya Ihsan, and Kai Kaiser. 2008. *Managing Post-Disaster Reconstruction Finance: International Experience in Public Financial Management.* Policy Research Working Paper 4475. Washington, DC: World Bank. http://go.worldbank.org/YJDLB1UVE0.

Foundation for Development Cooperation (FDC). 2007. *Capacity Building for Microfinance in Post-Tsunami Reconstruction, Summary Report.* Brisbane: FDC. http://www.bwtp.org/arcm/mfdm/Capacity_Building_Brief_Report.pdf.

International Committee of the Red Cross (ICRC) and International Federation of Red Cross and Red Crescent Societies (IFRC). 2007. *Guidelines for Cash Transfer Programming.* Geneva: ICRC. See especially Section B: Guidance Sheets. http://www.ifrc.org/docs/pubs/disasters/cash-guidelines-en.pdf.

Mathison, Stuart. n.d. *Microfinance and Disaster Management.* Brisbane: The Foundation for Development Cooperation. http://www.fdc.org.au/Files/Microfinance/ Microfinance%20and%20 Disaster%20Management.pdf.

Naik, Asmita, Elca Stigter, and Frank Laczko. 2007. *Migration, Development and Natural Disasters: Insights from the Indian Ocean Tsunami.* Geneva: International Organization for Migration. http://www.iom.int/jahia/Jahia/cache/offonce/pid/1674?entryId=14556.

Rauch, Egon, and Helmut Scheuer. 2007. *SDC Cash Workbook: A Practical User's Guide for the Preparation and Implementation of Cash Projects.* Berne: Swiss Agency for Development and Cooperation. http://www.sdc-cashprojects.ch/en/Home/SDC_Cash_Workbook.

World Bank. 2006. *Global Economic Prospects 2006: Economic Implications of Remittances and Migration.* Washington, DC: World Bank. http://go.worldbank.org/0ZRERMGA00.

World Bank. 2006. *Hazards of Nature, Risks to Development: An IEG Evaluation of World Bank Assistance for Natural Disasters.* Washington, DC: World Bank. Appendix G on cash support. http://www.worldbank.org/ieg/naturaldisasters/docs/natural_disasters_evaluation.pdf.

Yang, Dean, and Hwa-Jung Choi. 2005. *Are Remittances Insurance? Evidence from Rainfall Shocks in the Philippines.* Ann Arbor, MI: Ford School of Public Policy. http://ssrn.com/abstract=703782.

Reconstruction depends on the disaster-affected population having access to construction materials. When the usual system that supplies construction materials is either inadequate or interrupted, interventions by government or agencies involved in reconstruction may be required. The decision about how and where to intervene in the market is relatively complex, and the best intervention may be something other than materials procurement, warehousing (also called stockpiling in the humanitarian community), and distribution. This annex provides guidance on making these decisions.

Understanding Non-Food Item Distribution

Construction materials are referred to by the humanitarian community as "non-food items" (NFIs).[1] They are distributed in large numbers every year as part of humanitarian assistance programs. Humanitarian agencies know that NFI distributions help to save lives and restore a sense of home following a natural disaster. The categories of NFIs generally include (1) general household support items, such as blankets and cook sets; (2) household shelter construction support Items, such as tool kits and construction materials (see adjacent box); and (3) household water, sanitation, and hygiene support items, such as mosquito nets and household water treatment.

As the Inter-Agency Standing Committee (IASC) guide on NFIs notes, the demand for and source of NFIs changes over the course of a disaster response. Initially, a higher proportion NFIs are likely to have been stockpiled or brought in from outside the region, because there may be constraints on local availability and logistics. Later, after the emergency phase, the NFIs are more likely to be procured and distributed locally and regionally.

Analyzing the Need for Materials Distribution

In addition to creating logistical and economic challenges for the agencies involved, the importation and distribution of construction materials can create distortions in the local materials market that can have long-term unintended consequences. In making an assessment of available resources, and deciding whether materials need to be brought in from outside, it is important to evaluate the impact of the disaster on materials availability, demand, and distribution. Agencies involved in reconstruction should also confirm whether household shelter construction support has taken place or is planned as part of the humanitarian response. There are several factors to consider.

Increases in demand. Demand for materials increases dramatically after a disaster. However, the pattern of the increase changes over the reconstruction period, which is likely to last a number of years. Both the population affected by the disaster and implementing agencies will be looking for materials for reconstruction. In addition to what is needed for reconstruction, the population will need materials for

---

Household Shelter Construction Support Package

- Common practice for delivery of construction materials by humanitarian agencies includes:
  - Delivery in phases
  - Separate distributions of materials for foundation, frame, and roof
  - Monitoring of construction progress as each phase is delivered
- Delivery in phases reduces logistics delays.
- Packages for shelter construction are of three types:
  - Packages that make an entire shelter
  - Packages of materials to be used with existing structures
  - Materials that contribute to or repair a shelter

Source: IASC Emergency Shelter Cluster, 2008, *Selecting NFIs for Shelter* (Geneva: IASC), http://www.sheltercentre.org/library/Selecting+NFIs+Shelter.

---

constructing shelter. Displaced families will need materials for transitional settlement, such as a temporary shelter on the property of a host family. Returned or nondisplaced families will need materials for transitional reconstruction, for example, for weatherproofing their damaged house, while they undertake full repairs. One of the justifications for the transitional shelter approach, whereby materials are used for shelter and reused in reconstruction, is precisely the reduction in the demand for reconstruction resources.

Decreases in supply. Demand for materials can often outstrip sustainable supply, as the baseline supply is usually just a fraction of what is required after a disaster. Constraints on supply may produce interruptions in production, including (1) the workforce is not available immediately after the disaster because their priorities change or they are displaced and unable to get to work; (2) there are shortages of raw materials because of the disaster; (3) there is insufficient or damaged capacity of the manufacturing plant, due to cut-off of power or water; and (4) there is a demand for materials that are not normally harvested or manufactured or used in a particular season, for example, thatch. The transportation of significant quantities of materials can also disrupt markets and damage infrastructure, for example, absorbing all available transport capacity or damaging the roads themselves.

Logistics breakdowns and transportation interruptions. Post-disaster logistics may also constrain supply. This may affect both normal movement of materials and shipments that agencies attempt to bring into the region. The capacity to move materials is affected by (1) damage to access routes; (2) reductions in transport capacity (damage to trains or competition for trucks, for example); (3) security problems in transportation; and (4) barriers to transportation, such as border crossings and hazardous areas.

The distance materials are transported, such as those caused by diversions of routes, and the number of stages in the distribution chain also affect supply. Losses in transit are caused by damage and theft and by inefficiencies in the on- and off-loading of materials. When combined, loss in transit and double handling may account for well over 10 percent of materials sourced not reaching their destination, a significant inefficiency.

Warehousing and distribution issues. Finally, in considering the outside procurement and distribution of materials, the distribution chain may require warehousing at international, regional, national, local, and satellite community levels. Damage to warehouses or higher demands for warehousing or distribution can also cause interruptions in supply. Local distributions may be to communities or households, with each option having advantages and disadvantages. Temporary warehouses may need to be constructed, and entities who can handle the physical distribution to recipients, possibly private shipping and distribution companies, nongovernmental organizations (NGOs), or civil society organizations (CSOs) with distribution experience or under the direction of agencies with this experience, must be identified.

Addressing Bottlenecks in the Construction Materials Chain

As this analysis demonstrates, it is not enough to simply identify that materials are in short supply or that prices have risen. It is necessary to understand which factors in the supply chain are creating bottlenecks, and to identify the appropriate interventions. Interventions in response to inadequacies and interruptions in the market supply of construction materials may be general or localized, short term or medium term. For example, the most effective way to increase supply may be to temporarily provide transportation for a workforce or to rebuild the power station that would allows a sawmill to get back online to produce planks and door and window frames

locally. The Emergency Market Mapping and Analysis (EMMA)[2] can be used to identify these bottlenecks.

An analysis tool like EMMA is useful in conducting this analysis. EMMA is used in shelter recovery to inform early decisions about the possibility of using cash, to help identify opportunities and actions needed to restore or rehabilitate critical market systems, and to track the impact of a crisis and humanitarian interventions on critical markets.

Similar to a value chain analysis, EMMA provides insights on the impact of a disaster on markets and supply chains for critical reconstruction inputs and other products. Still in development with the support of Oxfam Great Britain and the United States Agency for International Development (USAID) Office of U.S. Foreign Disaster Assistance (OFDA), among others, the analysis produced can inform early decisions about whether cash or in-kind is the most appropriate form of assistance, can help identify opportunities and actions needed to restore or rehabilitate critical market systems, and can be used to monitor the impact of humanitarian interventions on markets.

The EMMA methodology analyzes three critical dimensions of a market system:
1. Institutions and rules (policies, institutions, regulations, and norms that affect a trading environment)
2. Value chain actors (businesses and individuals who own the goods as they move along a chain from producers to consumers)
3. Services and infrastructure (business services and infrastructure that support or enable the chain's operations to work more effectively)

The figures below were produced as part of an EMMA and show the affect of Cyclone Nargis on the supply of thatch in Myanmar.[3]

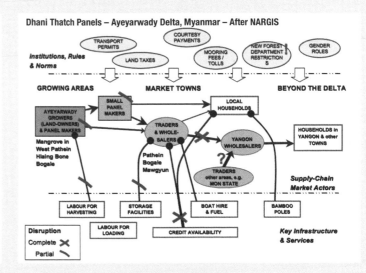

Annex 1 Endnotes
1. Emergency Shelter Cluster of the Inter-Agency Standing Committee, 2008, *Selecting NFIs for Shelter*, Geneva: IASC, http://www.sheltercentre.org/library/Selecting+NFIs+Shelter.
2. Rick Bauer, "Understanding the Role of Local Markets in Shelter Recovery: The Emergency Market Mapping and Analysis Tool" (presentation, Oxfam Great Britain, Shelter Meeting 08b), http://www.sheltercentre.org/meeting/material/The+Emergency+Market+Mapping+and+Analysis+tool+EMMA.
3. Practical Action Consulting, 2008, "EMMA Pilot Test 2, Myanmar, Key Findings and Recommendations," http://www.sheltercentre.org/library/Emergency+Market+Mapping+and+Analysis+EMMA+Pilot+Test+2+Myanmar.

Complaint and grievance redressal mechanisms help to ensure transparency and fairness and reduce the risk of errors or manipulation in an assistance program. All assistance programs produce complaints. Beneficiaries are in a precarious situation, and program criteria—no matter how well defined—are subject to misinterpretation, creating fertile ground for complaints.[1] Complaints increase when amounts are higher, as those for housing reconstruction generally are. Lessons from experience include the following.[2]

Nature of Complaints

The matters people will complain about are fairly predictable.
- Exclusion: People neglected or forgotten who should actually be registered.
- False claims: People who know that they are not eligible trying to get assistance anyway.
- Reinstatement: Eligible beneficiaries deleted or omitted from registration list due to clerical errors.
- Hardship: People who don't strictly meet the criteria asking for inclusion for reasons of hardship. (Names may be registered and procedures for exceptions should be transparent and well documented.)
- Inclusion: People objecting to presence of certain other people on the beneficiary list.
- Program staff, rules, or procedures: People lodging complaints against the program administration, employees, or rules, or about the amount of assistance.

Design Requirements

Complaint mechanisms cannot be added when complaints arise; they must be designed into any program that provides cash, vouchers, or other assistance from the beginning. The characteristics of good complaint mechanisms are as follows.
- Staff and beneficiaries understand the mechanism.
- Mechanism is set up early, has well-documented procedures, and good records of intake and outcomes.
- It is timely in responding, so grievances don't build.
- Processing is confidential, impartial, and transparent, so people feel fairly treated.
- Decisions are based on good information and validated locally wherever possible.
- Agency is able to provide redress for issues it is taking complaints for and to guarantee safety of staff involved.

Communicating Procedures

Complaint mechanisms cannot be added when complaints arise; they must be designed into any program that provides cash, vouchers, or other assistance from the beginning. The characteristics of good complaint mechanisms are as follows.

Elements of a Grievance Redressal System

Announcement	The right to complain and procedures for complaints must be explained in detail to beneficiaries.
Complaint intake	The procedure must be clear. Appointments may be advisable. All complaints must be registered in a database regardless of source. A telephone hotline, if used, should include a reliable system to register complaints. Complainants should receive a receipt, ideally a copy of the written record.
Location	A safe place is provided to present complaints and to be interviewed, ideally away from where cash or vouchers are distributed. Complainants should not be allowed to congregate at this location.
Enquiry and verification	Each complaint should be verified within a given period, using local information.
Communicating decisions	The complainant should be notified in writing as to whether the complaint has been accepted or denied.

Second Hearing

It is often advisable to have a second level of appeal mechanism that provides arbitration if the complaint cannot be satisfactorily resolved within the program's complaint system. In the case of an arbitration process:
- The steps and controls in the process should be similar if not identical to the complaint procedure.
- The appeal committee may include a wider range of actors, including community representatives or local officials.
- More extensive investigation may be carried out.
- The decision of the grievance committee should be final.

An alternative is to have the second hearing be advisory and to have an additional arbitration process. The number of levels of recourse will depend on the particular circumstances.

Other Feedback Mechanisms

Feedback mechanisms should be available at each level where accountability is required, at a minimum, agency, staff, and program. There should be a mechanism to report corruption in the system, either a separate system, or reports may be accepted through the complaint mechanism and forwarded to an appropriate body. See Chapter 19, Mitigating the Risk of Corruption, for information on anticorruption measures. Complaint procedures should be evaluated once they have been in operation for a period of time to ensure that outcomes are fair and acceptable to those being served. The database of complaints is an important input for this evaluation.

Annex 2 Endnotes
1. Egon Rauch and Helmut Scheuer, 2007, "SDC Cash Workbook: A practical user's guide for the preparation and implementation of Cash Projects," Swiss Agency for Development & Cooperation (SDC), http://www.sdc-cashprojects.ch/en/Home/SDC_Cash_Workbook/media/Documents/Cash_Workbook/061207_SDC%20Cash%20Workbook%20FINALVERSION%20SCREEN%20VarLayoutLHP.pdf.
2. Christian J. Hansen, 2006, *Concept Paper: Complaint Mechanisms* (Copenhagen: Danish Refugee Council).

Guiding Principles for Training in Reconstruction

- Training is the intervention that most determines whether the housing reconstructed after a disaster is an improvement over what people had before, especially with respect to disaster resilience.
- A training program should be developed based on an assessment of housing damage, the reconstruction approach, and the recommended technical guidelines.
- An appropriate organizational set-up is a necessity for a training program, as are sufficient resources to mobilize trainers and a system for producing and distributing written material.
- Training should be simple and based on people's everyday experience—the simpler the instructions, the better. Good results come as much from instilling a "safety mindset" in builders as from the sophistication of the knowledge imparted.
- Training on the job is an essential component of a training program. Follow-up is needed throughout the reconstruction process.
- Development of a training program should start early on, and be finalized only when the reconstruction approaches have been agreed to and detailed damage assessments have been conducted.
- The training design must be adapted to specific country conditions and reconstruction requirements.

This Chapter Is Especially Useful For:
- Lead disaster agency
- Agencies involved in reconstruction
- Project managers
- Civil society organizations

Introduction

The training of those who will be directly involved in housing reconstruction programs can play a decisive role in ensuring the quality and disaster resilience of the reconstructed housing. Key considerations in developing a training program concern its organization, the technical content, and the form in which damage data is gathered and information is provided to those involved in construction.

To ensure that a training program reaches wide coverage, scaling up is a major concern. The system usually needs to have a multiplier effect, whereby it starts with the training of trainers, who then go on to train others. The content of the training generally does not need to be technically difficult; a worker with modest skills should be able to learn the requirements in a few days. However, learning requires an open mind, since innovations in construction methods often need to be introduced. Small but specific adjustments can have an enormous effect on building resilience. The training curriculum should be designed to help builders acquire a commitment to safety improvement.

Key Decisions

1. The **lead disaster agency** should decide how reconstruction training will be managed within the context of the housing and community reconstruction strategy and ensure that adequate staff and resources are available for the lead training agency.
2. The **lead training agency** should help decide on the reconstruction approach after conducting a thorough review of housing damage in the disaster region and, based on that review, design the training program.
3. The **lead training agency** should decide on the requirements and begin recruitment of the core team, the trainers, and the field teams as early as possible, even while the training program is being designed.
4. **Agencies involved in reconstruction** should decide on the human and financial resources they can provide to assist with the development of the training program and with the complete dispersion of training activities and materials throughout the disaster region.

5. **Agencies involved in reconstruction** should coordinate with the **lead training agency** to incorporate training into their projects and to agree on standards and procedures for monitoring and evaluation of training activities.
6. **Local governments** should coordinate closely with the **lead training agency** and should ensure that relevant local building codes, guidelines, and approval procedures are incorporated in the training program.
7. **Civil society organizations associated** with the building trades and **academic institutions** should decide how they can support the training process and contribute their expertise during development and implementation of the training program.

Public Policy Related to Reconstruction Training

Post-disaster reconstruction training is unlikely to be contemplated in any public policy document, except when a country has post-disaster institutional arrangements in detail before the disaster. Regulations such as building codes and technical guidelines for builders should be analyzed and incorporated in training if they are relevant.

A Core Team can train 10 (2-person) assessment training teams.

Those developing a post-disaster reconstruction training program should clarify early on whether there is training capacity in the construction trades, in either the public or private sector. If there are licensing requirements for building trades, for example, there may be public or private training institutes, training materials, and experienced trainers who can be called on to help develop and implement the post-disaster builder training. Associations of building trades may also be a resource, especially if they have programs to train or credential their members.

Each assessment training team can then train 10 (2-person) field assessment teams, to produce 100 field assessment teams.

To ensure the quality of training programs, a number of countries are developing accreditation programs for training. The accreditation ensures that training programs are evaluated by an external body to determine if applicable training standards are met. Some countries have independent organizations that oversee the accreditation process, while others accredit through a government agency. Professionals carrying out these efforts should be involved in the development of post-disaster construction training. Faculty from institutions of higher learning, such as schools of engineering and architecture, and built environment professionals such as chartered surveyors may also be able to contribute expertise.

In the absence of a sufficient local institutional framework, international experts can also be called on, but their expertise should not be allowed to completely displace that of local technical experts and builders. Additional resources that may be useful in developing the training program are listed in the Resource section, below, and in Chapter 10, Housing Design and Construction Technology.

Technical Issues

The advice in the chapter is based on extensive experience with training. It describes in detail a particular model that may not be appropriate after a particular disaster. However, the principles on which this model is built can be applied in nearly every situation.

Scope of Training Requirements

There are training needs in almost every stage of a reconstruction project. This chapter focuses specifically on those training activities that help ensure (1) the accurate identification of damage (that is, the detailed damage assessment of houses, not the initial post-disaster needs assessment [PDNA]), and (2) the quality of the physical housing reconstruction. Such a training program should include the following:

- Training of technical advisors and supervisors
- Training for detailed damage assessment (at this stage, the degree of damage of each house is identified and the houses are assigned a category for housing assistance)[1]
- Training of owners and other builders in the field
- Training for data collection and information management[2]

In a large disaster, the first three elements ideally entail the training of trainers, in order to reach adequate coverage.

The training program must be tailored to the reconstruction approach and is especially critical if the reconstruction will be carried out by owners (owner-driven reconstruction) or by small-scale,

1. See also Chapter 2, Assessing Damage and Setting Reconstruction Policy.
2. For an extensive discussion of use of information and communications technology in reconstruction, see Chapter 17, Information and Communications Technology in Reconstruction.

semi-skilled contractors. It includes two components: a component for training in assessment and a component for training in reconstruction. Only training in assessment is necessary if the reconstruction is to be fully implemented by professional construction firms with knowledge of resilient building practices, an approach likely to be employed, for example, in an urban context for the reconstruction of multi-story buildings. See 📖 Chapter 6, Reconstruction Approaches, for additional information.

Typical Training Program Staffing

The staffing requirements should be carefully planned and adjusted according to what is learned during the scoping study, the detailed damage assessment, and/or the reconstruction process.

A "four eyes" principle is strongly recommended, which implies that no field team should have fewer than two people. The number of field staff on each team may be as many as five, depending on such factors as the accessibility of the sites.

The team staffing includes the following elements.

Chief Training Officers. The core team includes two chief training officers (CTOs)—one for assessment and one for reconstruction. Each CTO should have an organizational assistant and assistant for preparing training material (demonstration models and handouts). The CTO for reconstruction is also supported by a senior mason and a carpenter to do practical demonstrations during the demonstration phase of the training.

The CTOs should have experience in training and in the execution of several reconstruction projects, and a sound technical background and demonstrated skills to develop training material and to translate technical problems into demonstrations that can be understood by unskilled people. It is preferable that the person be as independent as possible, so s/he is usually an expatriate.

After the initial training of trainers (first level), the CTOs should oversee subsequent training and pilot activities, and ensure that a commitment to training standards is maintained.

Assessors. As well as carrying out detailed damage assessments, a frequent (and recommended) practice is to employ members of assessor teams to later carry out the field supervision once their assessment assignment is fulfilled. If this is done, members of assessment teams will need to receive training in both assessment and construction, and the training sessions will need to be scheduled in a way that this is possible.

Trainers. The first-level trainers are responsible for training field supervisors and advisors (second level). Once the second-level training sessions are completed, an ideal task for the first-level trainers is supervision and cross-checking the field teams. The staffing of the training teams is two persons at a minimum: one with a background in building construction (an engineer) and one with experience in working with villagers (social worker or community facilitator).

Some trainers should remain available for additional training during the project, as the field teams are subject to turnover. One trainer can later oversee 10–15 field teams, depending on the geography and distribution of field sites.

While it is advantageous for the team members to be experienced engineers or to have a construction background, the basic technical knowledge can in fact be learned from the

JOCELYN AUGUSTINO/FEMA

training. In special cases when more in-depth technical know-how is required, experts from the core team can be brought in.

Crafts persons. One crafts person is assigned to each reconstruction training team to oversee hands-on activities during training. The crafts person's assistance is required mostly during the model-building phase (first six months).

Information experts. An information management expert should either be added to each assessment team or be available for daily data input. Administrative oversight and management of databases should take place at the agency responsible for overseeing reconstruction and the training system.

Timing of Training

The more time that passes after the disaster, the greater the probability that people will begin to rebuild without any guidance or supervision. Therefore, training for assessors and reconstruction technical advisors should start as soon as possible after the disaster, once the critical decisions are made about reconstruction and training material can be prepared. Similarly, reconstruction field teams should be deployed as soon as possible after the damage assessment. The first reconstruction field teams can be ready about two weeks after the completion of the detailed damage assessment, which provides sufficient time to construct pilot houses in the field, as reconstruction scales up over several months.

Structure of Training Program

The following figure shows the relationship of the different elements of the training program.

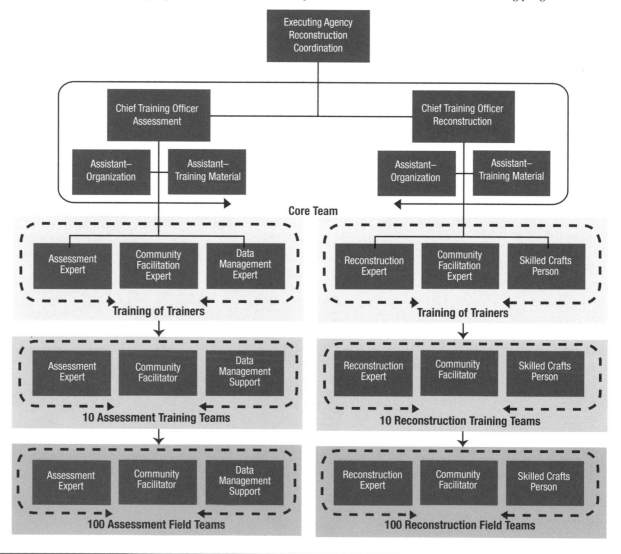

 For access to additional resources and information on this topic, please visit the handbook Web site at www.housingreconstruction.org.

242   SAFER HOMES, STRONGER COMMUNITIES: A HANDBOOK FOR RECONSTRUCTING AFTER NATURAL DISASTERS

Calculation of staffing requirements

In a project in which 50,000 houses are to be reconstructed, one team of two persons (one engineer and one social worker serving as a field team) can manage the reconstruction of 100–120 houses, under average conditions. Assuming the lower number allows for staff level fluctuations during the project. Therefore, 500 teams of 2 experts need to be trained.

The core team conducts the first-level training of trainers, which results in 10 teams of trainers after two weeks. The second-level training (conducted by these 10 teams) can train 10 field teams in every training session of two weeks. In 10 weeks, the trainers can conduct five such trainings, resulting in 500 field teams (5 x 10 x 10). Three months (2 weeks for one training of trainers + 10 weeks for field team training) is approximately the same as the time needed to conduct the detailed damage assessment. If the project is larger, the core team should train a second group of 10 trainer teams.

The figure of 100 houses per team is conservative.[3] Average time of construction per house, access to housing sites, available time for construction (weather conditions and harvesting time are examples of factors to consider), and frequency of visits or interactions of the supervisors with the beneficiaries will all affect this ratio. It is assumed that supervisors will visit each site or village twice a week, spending four days a week in the field and one day in administrative duties. These calculations must be adjusted based on the experience gained during execution, so flexibility is a must.

Principles for Organizing the Training Process

The core team should organize the training process based on the following guidelines.

The core team. The core team is the starting point of training. The core team designs the content of the training and organizes and conducts the training of trainers. Later, the core team is a resource pool for handling queries, updating recommended practices, and designing remedial measures for mistakes that are likely to occur.

Training of trainers. The model proposed depends on the multiplier effect of training trainers who themselves go on to train others using the same methodology. It is extremely important during training to ensure that not only technical knowledge is transmitted. Most important is that the trainees get skills to explain the key issues to the next level, so that the proper implementation takes place in the field. The emphasis should be on establishing a commitment to proper construction and safety on the part of the trainers and builders, on understanding the typical working conditions, and on overcoming predictable challenges.

Who trains whom. Trainers receive training from the core team. The trainer teams then train the field teams, which should have a similar staffing structure. The field teams then provide supervision, advice, and on-the-job training to the builders and owners in the field.

Supervision and feedback. The usual 1-day training of crafts persons is considered to be almost useless if not followed up in the field. Similarly, the provision of superficial information to homeowners (a half-day presentation or use of posters) does not provide the necessary understanding of how to construct or reconstruct safely. Instead, there should be continuous on-the-job training and efficient supervision, adequate feedback mechanisms, and constant oversight of reconstruction in the field to ensure that the desired results are being produced.

Technical basis for training. Preparation of training material needs to start at a very early stage and should be well coordinated with the overall social communications strategy, which may employ radio, television, or other media. See 📖 Chapter 3, Communication in Post-Disaster Reconstruction.

The content of the training is based on the reconstruction approach and the technical guidelines, the roles and responsibilities defined as part of the reconstruction strategy, and the institutional arrangements for the reconstruction program. See 📖 Chapter 6, Reconstruction Approaches; 📖 Chapter 10, Housing Design and Construction Technology; and 📖 Chapter 2, Assessing Damage and Setting Reconstruction Policy.

Training must cover a range of challenging issues that trainees must be prepared to cope with:
- Technical knowledge of program rules and the reasoning behind them (the latter empowers trainers when they are explaining the requirements)
- Training skills and methods
- Administrative procedures, exceptions, and resolution mechanisms (e.g., who can decide if and how to mitigate in cases of noncompliance, how the grievance procedures work, etc.)

3. Note that the ratio of 1:100 refers only to construction supervision and assumes that there are also facilitators working directly with households to organize the overall reconstruction process. See Annex 1, How to Do It: Establishing a Community Facilitation System for Post-Disaster Housing and Community Reconstruction, in 📖 Chapter 12, Community Organizing and Participation, for information on the community facilitation approach.

- Typical mistakes and problems, including both technical and cultural issues, and procedures for dealing with them, including relevant mitigation measures

Traditional buildings normally need some structural improvement, depending on typical modes of failure, caused by vulnerabilities due to shortcomings in the original design and/or by the force of the disaster itself. Traditional construction methods are normally:
- mud masonry or mud structures;
- wooden structures with mud-type infill; or
- lightweight structures (normally wood or bamboo, sometimes with plaster cladding).

Recommended Elements and Content of Training

Preparation phase for the core team. During preparation, the training program for both assessment and reconstruction is designed so that it properly reflects the applicable guidelines and conditions in the field, and the training material is developed. A scoping study in the field is needed to understand the types and extent of damage that assessors and field teams will encounter. This scoping work may be conducted separately or as part of the global PDNA.

Information gleaned from the scoping study should be used to produce a document that identifies the vulnerabilities of local housing and the improvements needed in structure, design, materials, and construction practices. Many houses will be completely rebuilt, so architects should develop model house designs and construction specifications, at a minimum, for the core house. See Chapter 15, Mobilizing Financial Resources and Other Reconstruction Assistance, for a description of the process for developing model house specifications.

The critical inputs needed to develop the training program are the document that shows improvements needed to retrofit or repair existing houses and the plans and specifications for rebuilt houses, information on the reconstruction approach, and the terms of the housing assistance. Photos collected during the scoping study and analysis of the specific conditions to be addressed in housing reconstruction are used to ensure that training materials reflect actual conditions in the field. Simplicity is key; demonstrations and methods must be understood by the unskilled.

Training in detailed housing damage assessment. The first stage of the assessment component of the training program is the training of trainers, using a training approach that the newly educated trainers will follow when they train the assessment field teams.

Training in damage assessment should take place in three steps. Step 1 is classroom instruction for approximately three days, including lectures in the morning and practical demonstrations as well as tests using photos of damage in the afternoon. Step 2 is assessment field training, which should take approximately one week, during which the group is divided into teams that evaluate the same group of houses on different days followed by comparative analysis and discussion of the results. The goal is for different assessment teams to reach uniform results. Step 3 should be lectures to consolidate acquired experience and cover assessment problem solving.

The group of assessors trained by the core group will go on to train the field assessment teams. Trainers from the core group should oversee the next level training process by visiting training activities and taking corrective measures, where required.

Specific content of training. The content of the training program should include learning to assign damage categories, estimate the housing assistance (if based on damage), and collect required data. (The same team may also be involved in organizing the qualification process. See Chapter 15, Mobilizing Financial Resources and Other Reconstruction Assistance.) Topics include:
- Basic information about disaster effects and typical damages (earthquake, storm, flooding)
- Understanding what can be repaired and strengthened and what cannot
- Criteria for the specific categories of damaged buildings (normally three categories) and understanding of the related structural design issues (guidelines are required)
- Procedures regarding what should be documented and how (photography, global positioning system [GPS] data, forms and procedures, information from owners)
- Information on grievance procedures (should be defined by the time of the assessment)
- How to deal with social issues in the field (complaints, assaults, bribery attempts, etc.)

It is also necessary to provide for data collection and information management. Assessors should receive basic training on data collection, overcoming missing information, plausibility checks, and data handling, but they should be assisted in the field by experts in this area.

Training of reconstruction supervisors and technical advisors. The first stage of the training of supervisors and technical advisors is the training of trainers, using a training approach that the newly educated trainers will then follow with their training of the reconstruction field teams. The training should take one to two weeks and should include lectures and the simultaneous building of the model buildings of the type(s) that have been explained in the training.

Early in the training period, training teams are formed, consisting of two experts—one whose principal focus in the training will be technical and a second whose focus in the training will be on facilitating the relationship with the members of the community—and a trained crafts person. In the field, each member of the team needs to be able to both provide advice and training on the guidelines for reconstruction and supervise reconstruction, with the goal of ensuring that the homeowners are able to receive

If it is necessary to carry out any testing of or research into the safety of local building materials or housing technologies, and to identify measures to improve housing resilience during reconstruction, these activities should begin as early as possible—as soon as a study is carried out to identify the ways in which local building materials and technologies failed in the disaster.

Local universities and international research organizations can play a key role in the research and testing of improved building technologies. See the 🖥 annex to Chapter 10, Housing Design and Construction Technology, for a list of international organizations that work on improving vernacular building technologies.

4. A distinction is made here between two types of model buildings: "pilot houses," which have scale and design very similar to houses that will be built in the field, and "demonstration buildings," which may be of a different type or scale (health clinic, for example), but are built to teach the construction methods. Model-building exercises are also important because they allow crafts persons who will later be assigned to field teams to be trained in the program approach.

5. IAEE's Web site is http://www.iaee.or.jp. The "Guidelines for Earthquake Resistant Non-Engineered Construction" can be found at http://www.nicee.org.

payment at defined points in the process. The trained crafts person participates in the training of trainers and later supports the training team in carrying out the training of the field teams.

Trainees should practice the training by giving each other lectures, to prepare for providing builders with the information necessary to carry out safe reconstruction. The whole group should practice model building,[4] either in nearby villages or at a training camp. Field reconstruction teams have a similar composition to the training team. Each field team should include a crafts person, who will later support the team in carrying out the training of builders in the field. The practical aspects of the training are disseminated at the village level by constructing a pilot building and by site visits by the field team during the construction period. The field teams are monitored by the training team who earlier trained them.

Specific content of training. Training should cover basic construction skills and how to teach them, explain how to organize reconstruction in the field, and provide the capacity to diagnose and address the specific damage from the disaster, and should include:

- Understanding basics of disaster impacts (earthquake, storm, flooding) on buildings
- Principles of reconstruction/mitigation (building structure, location, etc.)
- Learning and practicing basic technical skills (assisted by crafts persons) (if trainees are mainly engineers, their structural skills may need to be updated)
- Model demonstrations so that unskilled builders can understand the rules that are being enforced
- How to hold training sessions in the field
- Social pressures they will be subject to and how to handle them
- Information on grievance procedures
- Typical mistakes, mitigation measures, and where flexibility is, and is not, permitted
- Any updates to procedures that are issued during the project
- Material quality and testing, mainly simple field tests
- Procedures for carrying out model-building projects in field training

If, in addition to the normal construction method with reinforced concrete and cement mortar masonry (for earthquake-prone areas, normally using International Association for Earthquake Engineering [IAEE] guidelines[5]), improved traditional methods are acceptable, then related technical information and building methods must be included in the formal and demonstration components of training.

Scaling up. For both assessment and construction, about 10 teams (20 persons) can be trained in each training session, both by the core team and by the training teams, so that the initial training by the core team over a period of four weeks (two weeks for training of trainers and two weeks for training of field teams) results in about 100 (10 x 10) field teams.

On-the-job training of crafts persons and orientation of home owners. The field teams trained during the second stage of training then carry out the training of builders in the villages.

A lecture-type demonstration should take place in several successive evenings with builders, community leaders, and interested homeowners, showing disaster effects and key rebuilding requirements. If homeowners are to be directly involved in reconstruction (i.e., acting as builders), they receive the same training as other builders. Otherwise, orientation of homeowners results mainly from watching the construction of the pilot houses, from written materials (with illustrations), and from community leaders and similar persons. The following principles apply.

- Practical training of builders is done on the job and in the field, by constructing model buildings. The buildings built in the model-building activity can be houses of vulnerable families, in which case the model building accomplishes two results simultaneously—training and critical reconstruction. See the 🖥 case study, below, on the use of model buildings in the Pakistan Poverty Alleviation Fund/Sarhad Rural Support Program after the 2005 North Pakistan earthquake.
- Focus is on improvement of normal building skills, such as filling of vertical joints in masonry, installation of headers in stone masonry, roof anchoring, and good practices for mixing concrete.
- Good practices related to specific types of disasters should be taught and incorporated in the guidelines (e.g., bands, opening rules, vertical bars, hooks at stirrups, tension lap length). See the 🖥 case study, below, on how shake-table demonstrations and guidelines were used to educate builders, homeowners, and others in the 2003 Bam earthquake reconstruction.

Depending on the construction/reconstruction technology and on the availability and skill level of local masons, additional masons may need to be brought in or trained in earthquake-proof reconstruction using IAEE-recommended practices. In the case of mitigation for other types of (or multiple) disasters, reconstruction will entail measures for wind resistance, such as roof anchoring, and for flood mitigation, such as raising building level. The 🖥 case studies, below, show how Habitat for Humanity has provided training after two emergencies so that women and internally displaced persons could participate as trades people in the reconstruction effort.

Typical Training Problems and How to Address Them

Typical problem	Potential solution
Supervisors belong to a social group that considers physical work not consistent with their status, and as a result practical demonstrations in the field don't take place.	Ensure that a willingness to do hands-on work is used as a selection criteria and is included in the position description used to recruit supervisors. In some societies, there are easily identifiable experts or expert groups, who may be targeted during the recruiting process.
Contradictory or unclear messages are given during training, which leads to confusion. For instance, training on framed structures (which are not recommended for simple housing due to difficult detailing) can't easily be adapted to supervise reinforced masonry construction (shear wall system).	Begin training only once the rules for reconstruction and detailed damage assessment are clear. Ensure that the core team is highly qualified and that adequate time is spent on the preparation stage.
Good students of training during aren't necessarily good trainers or supervisors in the field.	Follow-up and thorough supervision of the second-level supervisors is required (task of first-level supervisors).
There is social pressure on the supervisors to certify noncompliant construction work for payment. This ranges from bribery to physical attacks.	This problem can be mitigated if beneficiaries know that certifications will be cross-checked at the next level and, if not genuine, will only delay payments.
Explanations are too theoretical and not easily understood by ordinary people.	Demonstrate all key effects with simple models that are related to everyday practice.
Pilot or model buildings are different from what people will actually be building.	Avoid mixing training for common reconstruction with experimental building practices, which might make sense in a research setting but should be tested before being applied in the field.

Recommendations

1. Conduct a rigorous, thorough review of housing damage in the disaster region, and use the knowledge gained as the basis for designing the detailed assessment reconstruction techniques and for developing builder training.
2. Start training only after establishing the rules for assessing the damage to individual houses and developing a consensus on the reconstruction approach and appropriate reconstruction technologies.
3. Hire the strongest and best-prepared core team possible, as it is decisive for the success of the program.
4. The reconstruction approach, the program rules, and the training material and public information are all interrelated. Procure key training staff as early as possible, so that they can be involved in establishing the rules and in preparing training materials and public information.
5. Draw up and implement a comprehensive staffing plan that is coordinated with the overall program. Don't let recruitment become a bottleneck, and be prepared to keep training trainers during the reconstruction program, as there will likely be turnover and dismissals.
6. Adapt the content and organization of the training program to the specific situation and keep both flexible, able to be improved as the reconstruction program progresses.
7. Take the time to research how to improve local building technologies and materials or investigate whether research has been done inside or outside the country, so that the training program is based on scientific knowledge, not intuition.
8. Use demonstrations and simple messages in training, so that the concepts and instructions will be understood by builders without formal training (crafts persons and homeowners).

 For access to additional resources and information on this topic, please visit the handbook Web site at www.housingreconstruction.org.

Case Studies

2005 North Pakistan Earthquake, Pakistan

Use of Demonstration Buildings

Model building is the best way to demonstrate building technologies and provide on-the-job training. A demonstration project can also provide a house for a vulnerable family who might otherwise have difficulty rebuilding. Public buildings, such as a training center, meeting hall, or storage building, are also good demonstration projects, and can serve as a location for information sharing, material banks, or accommodations for the field staff. Demonstration buildings must use techniques and materials directly related to the approved construction methods and be affordable using the funding available. It is unwise to build demonstration buildings that raise expectations about the quality or quantity of housing that cannot be fulfilled with the available funds, as has happened in numerous programs. If there is more than one approved construction method (a reinforced masonry method and an improved traditional construction method, for instance), each should be used in demonstration projects. The adjacent photo shows a wooden frame building in the Pakistan North-West Frontier Province, Siran Valley reconstruction project later used as an information center. The project, executed by the Pakistan Poverty Alleviation Fund/Sarhad Rural Support Program, was financed through KfW/Germany.

© GRONTMIJ-BGS AND DR.-ENG. N. WILHELM.

Source: Dr.-Eng. Norbert Wilhelm, 2009, personal communication.

2003 Earthquake, Bam, Iran

Raising Public Awareness on Earthquake-Resilient Construction through Shake-Table Tests and Technical Guidelines

Empowering communities that had been affected by disaster through increased access to information resources on disaster recovery programs and projects, thereby raising public awareness on earthquake-resistant construction techniques, was one of the United Nations Development Programme's (UNDP) top priorities in its support to the reconstruction process in Bam, Iran, after the 2003 earthquake. In addition to its community-based information and communications strategy, UNDP organized shake-table demonstrations. The main objective of these demonstrations was to show the effect of earthquake-resilient technologies to the affected population and builders. A wide range of stakeholders participated, including local laborers and masons; homeowners from communities affected by the earthquake; university students, instructors, and researchers; representative of local authorities; representatives of the media; and staff from international and local nongovernmental organizations (NGOs) involved in the Bam recovery. The demonstrations were considered crucial to making participants aware of their vulnerabilities and of measures that could be taken to reduce risk in reconstruction. Several guidelines were also produced for engineers, architects, and recovery decision makers, including Guidelines for Urban Planners on Child-Friendly City Concept, Guidelines for Training of Building Workers on Masonry Earthquake-Resilient Construction, Earthquake and Conventional Building in Iran: A Guideline for Architects and Engineers, and Typology and Design Guide for Housing in Bam.

Source: Victoria Kianpour, UNDP Iran, 2009, personal communication, http://www.undp.org.ir/.

How technical requirements can be demonstrated with models and photos

Bending stiff corners of typical earthquake bands are important for maximum resistance. It is a common problem that is often not understood. The adjacent photo shows how the load-bearing capacity increases if the beam continues at the corners. A bent, stiff corner provides this continuation but requires special reinforcement. At the same time, the demonstration should show how the reinforcement should be placed and why.

© DR.-ING: N. WILHELM

DANIEL PITTET

2004 Indian Ocean Tsunami, India
Women Trained as Masons in Post-Tsunami Livelihood Project

Habitat for Humanity India partnered with the Centre for Action, Development, Research and Education in India (CADRE), a social services nongovernmental organization (NGO), to provide training as masons for a group of women whose family incomes had been adversely affected by the 2004 Indian Ocean tsunami. Fifty women who were members of self-help groups started the program in Colachel, Tamil Nadu, and 35 completed the month-long course and nearly six months of on-the-job training, from July through December 2005. The women were each paid Rs 140–Rs 190 (US$3.16–US$4.30) a day.

Six months later, a few of the women were hired as full-time masons; others formed crews to do repairs and renovations in their villages. Mary, a middle-aged, unmarried woman, was building houses for an NGO in a tsunami recovery project. She also took on small construction projects that she could share with her fellow women masons. Together, they had built concrete block walls and a kitchen; six of them plastered a house. "After the training, I was confident," she said. "We can do this work. We can figure the cost and materials and we can build a house." By April 2008, Mary had taken on her nephew as an apprentice and could afford to build her own house. The women maintain that because of their skills and knowledge, whether or not they are in charge of a village building site, shoddy construction is a thing of the past in their communities.

Sources: Kathryn Reid, Habitat for Humanity International, 2009, personal communication; *and* Habitat for Humanity India, "Welcome to Habitat for Humanity," http://www.habitatindia.in.

2006 War, Lebanon
Vocational Training for Home Repair and Reconstruction

The 2006 war in Lebanon forced approximately 25 percent of the Lebanese population from their homes and damaged or destroyed more than 97,000 homes. As part of Habitat for Humanity's (HFH) strategy of providing housing while assisting in the recovery of the local economy, HFH partnered with YMCA Lebanon to implement a livelihoods development program intended to train unemployed returning internally displaced persons in construction techniques in order to create livelihood opportunities. The 10-week vocational training program focused on sanitary and electrical installation, supplemented the existing construction workforce, and provided an alternate livelihood strategy while agricultural lands were being demined. YMCA developed the course curriculum and utilized practicing trades persons as course instructors. Trades persons were given a 4-day training of trainers orientation to the training curriculum and instruction on conducting effective vocational education. Following five weeks of theoretical learning, the 42 students spent five weeks in the field, gaining hands-on experience in installing sanitary and electrical networks in homes in their communities under the supervision of their instructors. All trainees were given a set of tools, which they were allowed to keep after completion of the training. Students benefitted not only from the training curriculum but also from the wealth of experience brought to the training by practicing professionals. Some 37 students completed the practical training, and 28 had found employment by course completion.

Source: Judy Blanchette, Habitat for Humanity International, 2009, personal communication; *and* "Habitat for Humanity International Lebanon," http://www.habitat.org/intl/ame/113.aspx.

Resources

Extensive information on training and training programs on a range of topics is available within the humanitarian community. A small sample of these resources is listed below.

Inter-Agency Standing Committee (IASC) Humanitarian Reform

Emergency Shelter Cluster. Training is organized for coordinators, information managers, and technical specialists in the context of the IASC Humanitarian Reform process. The Emergency Shelter Cluster periodically conducts training for Information Managers. http://www.humanitarianreform.org/humanitarianreform/Default.aspx?tabid=77.

IASC Clusters. The IASC Clusters have rosters in order to deploy rapidly qualified and trained professionals in emergency situations on global level. http://www.humanitarianreform.org/humanitarianreform/Default.aspx?tabid=53.

The Emergency Shelter Cluster Field Coordination Toolkit. This website contains information that may be useful in designing training programs. http://www.humanitarianreform.org/Default.aspx?tabid=301.

Training sponsored by humanitarian agencies

Many humanitarian agencies provide training to their staff and to external counterparts. The International Federation of the Red Cross and Red Crescent Societies, for example, provides Field Assessment and Coordination Teams training that is aimed at experienced relief core staff. http://www.ifrc.org/what/disasters/responding/drs/tools/fact.asp.

People in Aid

A global network of development and humanitarian assistance agencies, People in Aid promotes good practice in the management and support of humanitarian aid workers and provides a wide range of practical resources for agencies seeking to improve the quality of their human resources management. They act through collaboration with various partners, such as local or regional training organizations.

Additional information is also available through the handbook Web site, http://www.housingreconstruction.org.

FEMA NEWS PHOTO

UN-HABITAT

"People's Process in Aceh and Nias (Indonesia). 2007. Manuals and Training Guidelines." (In English and Bahasa). http://www.unhabitat-indonesia.org/publication/index.htm#film.

Volume 1	Orientation and Information:	http://www.unhabitat-indonesia.org/files/book-153.pdf
Volume 2	Community Action Planning and Village Mapping	http://www.unhabitat-indonesia.org/files/book-1407.pdf
Volume 3	Detailed Technical Planning for Housing and Infrastructure	http://www.unhabitat-indonesia.org/files/book-1417.pdf
Volume 4	Housing and Infrastructure Implementation	http://www.unhabitat-indonesia.org/files/book-1420.zip
Volume 5	Completion of Reconstruction Works	http://www.unhabitat-indonesia.org/files/book-1421.pdf
Volume 6	Monitoring, Evaluation and Controls	http://www.unhabitat-indonesia.org/files/book-225.pdf
Volume 7	Socialization and Public Awareness Campaign	http://www.unhabitat-indonesia.org/files/book-226.pdf
Volume 8	Training and Capacity Building	http://www.unhabitat-indonesia.org/files/book-229.pdf
Volume 9	Complaints Handling	http://www.unhabitat-indonesia.org/files/book-231.pdf

PART 2
MONITORING AND INFORMATION MANAGEMENT

Guiding Principles

1   A good reconstruction policy helps reactivate communities and empowers people to rebuild their housing, their lives, and their livelihoods.

2   Reconstruction begins the day of the disaster.

3   Community members should be partners in policy making and leaders of local implementation.

4   Reconstruction policy and plans should be financially realistic but ambitious with respect to disaster risk reduction.

5   Institutions matter and coordination among them improves outcomes.

6   Reconstruction is an opportunity to plan for the future and to conserve the past.

7   Relocation disrupts lives and should be kept to a minimum.

8   Civil society and the private sector are important parts of the solution.

9   Assessment and monitoring can improve reconstruction outcomes.

10 To contribute to long-term development, reconstruction must be sustainable.

The last word: Every reconstruction project is unique.

17 INFORMATION AND COMMUNICATIONS TECHNOLOGY IN RECONSTRUCTION

Guiding Principles for Information and Communications Technology

- Information is the basis of effective decision making. Low- or no-cost access to reliable and accurate information is crucial after a disaster, as is the use of open, shared, and coordinated systems.
- Use of technology in post-disaster assessment, planning, monitoring, and implementation is rapidly advancing and improving both the speed and quality of agency interventions.
- The principles of information management and exchange in emergencies—accessibility, inclusiveness, interoperability, accountability, verifiability, relevance, objectivity, neutrality, humanity, timeliness, sustainability, and confidentiality—should always be adhered to.[1]
- Interoperability ensures that data from various sources can be integrated and used by multiple stakeholders to enhance outcomes, monitoring, and evaluation.
- The quality of information and communications technology (ICT) systems in place before a disaster will affect response and reconstruction, yet awareness of the importance of strengthening ICT systems gained during the disaster response often fades once the disaster is over.
- Policies and systems for post-disaster information sharing should incorporate as wide a range of stakeholders as possible.
- ICT systems deployed in the field after a disaster must function in an environment of weak communications infrastructure and low bandwidth.

Introduction

Access to reliable, accurate, and timely information at all levels of society is crucial immediately before, during, and after a disaster. Information needs to be readily collected, processed, analyzed, and shared in order for stakeholders to effectively respond. Without information, individuals and institutions are often forced to make crucial decisions based on sketchy, conflicting, reports.

Patterns of critical information exchange during crisis situations are different than in normal business. Identifying and deploying appropriate public, private, and volunteer resources in a coordinated, timely manner depends on a commitment to addressing—in advance of a disaster—such concerns as interoperability and the use of common standards. Also, ICTs are only as good as their weakest link. So preparedness for disaster communications needs to anticipate scenarios in which any individual ICT element, including the "backbones"—broadcast radio, television, mobile telephony, electric power, database management, and Internet communications—are compromised.

Institutional knowledge and the supporting ICT infrastructure are as vital to the economy today as capital, energy, raw materials, labor, and transportation, and are ever-changing. The use of information technology in post-disaster situations, and the corresponding institutional arrangements for using it, are expanding possibly even more rapidly and continuously evolving. Consequently, this chapter is not meant to provide a definitive explanation of this topic, but rather is intended to point out broad directions and examples and to provide references to key organizations and initiatives, each of which will undoubtedly already have evolved since this handbook was published.

This Chapter Is Especially Useful For:
- Lead disaster agency
- ICT specialists
- Agencies involved in reconstruction
- Project managers

1. United Nations Office for the Coordination of Humanitarian Affairs (UN OCHA), "Information Management Overview," http://www.humanitarianinfo.org/IMToolBox/index.html.

 For access to additional resources and information on this topic, please visit the handbook Web site at www.housingreconstruction.org.

253

Key Definitions[2]

Framework data	The seven themes of geospatial data that are used by most geographic information system (GIS) applications (geodetic control, orthoimagery, elevation and bathymetry, transportation, hydrography, cadastral, and governmental units). These data include an encoding of the geographic extent of the features and a minimal number of attributes needed to identify and describe the features.
Geographic information	Coordinate and attribute data for location-based features, usually in the categories of point (e.g., a well), line (e.g., a road), polygon (e.g., a forest), cell (e.g., a raster-based "rectangle"), or coordinates (e.g., the latitude-longitude of a point on the ground).
Geographic information system	A computer system for the input, editing, storage, retrieval, analysis, synthesis, and output of location-based (also called geographic or geo-referenced) information. GIS may refer to hardware and software or include data.
Information and communications technology	The range of tools, applications, systems, and resources used to create, acquire, store, exchange, analyze, and process information and share data in all its forms.[3] ICTs as used in this chapter include the Internet, geospatial data, GIS, satellite- and land-based communications, Web 2.0, data tracking systems, and data warehousing systems. The phrase "information technology" is sometimes used interchangeably with ICT.
Metadata	Information about data, such as content, source, vintage, accuracy, condition, projection, responsible party, contact phone number, method of collection, and other characteristics or descriptions.
Open standards	Standards made available to the general public and are developed (or approved) and maintained via a collaborative and consensus-driven process. Open standards facilitate interoperability and data exchange among different products or services and are intended for widespread adoption.
Spatial data	Information that identifies the geographic location and characteristics of natural or constructed features and boundaries on the Earth. This information may be derived from, among other sources, remote sensing, mapping, charting, surveying technologies, global positioning system (GPS), or statistical data.
Spatial data infrastructure	The technology, policies, standards, human resources, and related activities necessary to acquire, process, distribute, use, maintain, and preserve spatial data.

The variety of data required after a disaster is exceeded only by the speed with which it is needed and the number of stakeholders collecting it. If agencies involved early make their information widely available, other stakeholders will save time and money and will be working with the same baseline data.

Key Decisions

1. **Government** must decide on the lead agency to coordinate information management (IM). Although the agency will not control the use of ICTs in recovery and reconstruction, it may need to address constraints on their use. **Local government** may also designate an information management coordinator.

2. The **information management agency, agencies involved in relief and reconstruction**, and **local communities** should establish interagency relationships, and should sign agreements and protocols (if necessary) to share information and technology.

3. The **information management agency** and **agencies involved in relief and reconstruction** must define information needs for assessment and monitoring, including GIS data and maps, and must decide how to locate or procure it.

4. **International organizations**, in consultation with the **information management agency**, must decide on the types of ICT technology, equipment, and data to be deployed in the field, and on procurement procedures, if additional equipment is required.

5. The **information management agency** should work with **agencies and communities involved in relief and reconstruction** to establish the IM strategy: rules, protocols, procedures, and locations for the sharing, storing, and disclosure of information.

6. **Agencies involved in relief and reconstruction** should encourage the collaboration of **local communities** through the use of ICTs and work with communities to establish a system for monitoring joint ICT activities and to provide feedback during implementation.

2. U.S. Office of Management and Budget, 2002, "Coordination of Geographic Information and Related Spatial Data Activities, Circular No. A-13, Revised," http://www.whitehouse.gov/omb/rewrite/circulars/a016/print/a016_rev.html.

3. Berna Baradan, 2006, "The Role of Information and Communication Technologies in the Process of Post-Disaster Housing Reconstruction" (paper at the First International CIB-Endorsed METU Postgraduate Conference, "Built Environment & Information Technologies," Ankara, Turkey, March 17–18), http://www.irbdirekt.de/daten/iconda/06059007139.pdf.

Public Policies Related to ICT
International Law and Agreements

Tampere Convention. The transborder use of telecommunications equipment by humanitarian organizations can be impeded by regulatory barriers that make it difficult to import and deploy the equipment for emergencies without prior consent of local authorities. The Tampere Convention is a globally binding treaty governing the provision and availability of telecommunications equipment during disaster relief operations.[4] The Tampere Convention was ratified in 1998 and went into effect on January 8, 2005. The convention calls on states to facilitate the prompt provision of telecommunications assistance to mitigate the impact of a disaster, and covers both the installation and operation of reliable, flexible telecommunications services. It is a means to influence states to pursue a set of common expectations regarding freedom and access of individuals providing emergency services in disaster situations. As of January 2009, 40 countries had signed the convention. The International Telecommunications Union (ITU) assists in fulfilling the objectives of the Tampere Convention.

ITU Framework for Cooperation in Emergencies. The ITU Framework for Cooperation in Emergencies (IFCE) seeks to deploy on-demand ICT applications and services anywhere, anytime in the immediate aftermath of a disaster and is organized around its technology, finance, and logistics clusters.[5] The ITU is the leading United Nations (UN) agency for ICT issues and the global focal point for governments and the private sector in developing networks and services. The ITU addresses a range of issues related to the integration of telecommunications and ICT in disaster prediction, detection, and alerting.

International Charter on Space and Major Disasters. The International Charter on Space and Major Disasters aims to provide a unified system of space data acquisition and delivery to those affected by natural or man-made disasters.[6] Users request the mobilization of space and ground resources of member agencies to obtain data and information on a disaster. The result is high-quality satellite imagery of the disaster location available to emergency responders, often within hours. A 24-hour operator verifies the request, an emergency officer prepares an archive and acquisition plan, and a project manager assists the user throughout the process. The services of the charter can be requested by a civil protection, rescue, defense, or security body of a charter member country. Authorities of other countries submit requests through charter members or international agencies. See Annex 1, How to Do It: A Primer on Acquiring Satellite Images, for guidance on using this and other sources of satellite data.

National Laws

Legal frameworks for telecommunications at the national level are rapidly being modernized in many countries. Even so, they can become a constraint in a post-disaster situation. This is precisely why the conventions discussed above have been developed. The following are potential concerns that may have to be addressed.

Telecommunications. Government policies on telecommunications can affect IM during response and reconstruction. If access to telecommunications services is at odds with national defense policy, for instance, then crucial post-disaster telecommunications deployment may be hampered. For example, following the 2005 North Pakistan earthquake, some NGOs reported that government restricted the use of cell phones in certain areas, hampering coordination and exchange of information. In these cases an appeal for application of international standards or lower-tech communications approaches may be required.

Equipment and software. Delays may be encountered if ICT equipment and software must be imported. Sources of delays can include customs regulations, license approval, frequency allocation, and restrictions on use of specific technologies. If equipment and software must be imported, agencies should make a best effort to comply with national requirements. To avoid delays, it may be necessary to seek assistance in navigating paperwork and procedures to comply with national requirements.

The United Nations Geographical Information Working Group

UNGIWG is a network of professionals working in the fields of cartography and geographic information science to build the UN Spatial Data Infrastructure needed to achieve sustainable development.

UNGIWG addresses common geospatial issues (maps, boundaries, data exchange, and standards) that affect the work of UN Organizations and Member States. UNGIWG also works directly with non-governmental organizations, research institutions and industry to develop and maintain common geographic databases and geospatial technologies to enhance normative and operational capabilities.

UNGIWG aims to:

- improve the use of geographic information for better decision-making;
- promote standards and norms for maps and other geospatial information;
- develop core maps to avoid duplication;
- build mechanisms for sharing, maintaining and assuring the quality of geographic information;
- provide a forum for discussing common issues and emerging technological changes.

Source: "About UNGIWG," http://www.ungiwg.org/about.htm

4. International Telecommunications Union (ITU), "Emergency Telecommunications," http://www.itu.int/ITU-D/emergencytelecoms/response/index.html.
5. ITU, "Emergency Telecommunications," http://www.itu.int/ITU-D/emergencytelecoms/response/index.html.
6. International Charter on Space and Major Disasters, "About the Charter," http://www.disasterscharter.org.

Data standards. Census, cadastres, and other government housing-related data may be maintained—even within a single country—using different media, standards, and definitions. Data are also held by private, commercial, and nongovernmental organizations that have their own protocols. Local laws may complicate the sharing and transmission of certain types of data as well. While technology is generally no longer a hurdle to information sharing, as Web-based applications, bandwidth capacity, and standardization of technology make data movement more efficient, differences in rules, media, and standards may still present barriers. Numerous initiatives are under way worldwide, both formal (commercial) and informal (user groups), to develop information standards.

Data security. Information security rules or cultural concerns may also have to be overcome. Users and borrowers of information should always respect reasonable privacy and security concerns, even in an emergency situation. A practical approach whereby the policies of each jurisdiction are quickly analyzed and common procedures transparently agreed to for the emergency should promote the free flow of the information needed for reconstruction purposes.

Technical Issues
Coordinating Information Management

IM is the process of collecting, processing, analyzing, and sharing information, from and among various stakeholders. It has a vital role in response and reconstruction. For this reason, a strategic approach to IM in a post-disaster environment is highly recommended.

An effective humanitarian IM system can aggregate and disseminate information that will inform the recovery reconstruction process. IM can ensure that national information systems and standards are employed, can help build local capacities, and can assist government in establishing systems to manage long-term coordination of recovery and reconstruction. Proactive measures must be taken to ensure that key stakeholders can readily access and use the information and resources offered, for instance, by ensuring that the most sought-after information is available in local languages.

Examples of data that may be needed for response and planning in different sectors include:

- Local disaster plans, policies, and procedures
- Social, demographic, and economic data
- Contacts at community, local, regional, and national levels
- Land use plans and critical infrastructure inventories
- Property ownership records (cadastres)
- Building inventories, data on structures (e.g., housing, commercial, public, infrastructure)
- Cultural asset information, including age, construction technique/material, cultural significance, and condition
- Hazard maps and vulnerability data
- Safety and environmental standards and building codes

Communicating with Stakeholders

ICTs create a plethora of options for maintaining two-way communication among stakeholders, including the affected community, and inviting their participation in post-disaster decision making. The use of ICTs should be incorporated in the overall post-disaster communication plan. See ⌨ Chapter 3, Communication in Post-Disaster Reconstruction, for guidance on the development of the communication plan. Some examples of relevant current technologies are discussed below.

Web 2.0. Web 2.0 is an approach to Web site design that facilitates information sharing, interoperability, user-centered design, and collaboration. Examples of Web 2.0 include Web-based communities, hosted services, Web applications, social networking sites, video-sharing sites, wikis, and blogs. Assuming there is a functioning Internet service, a Web 2.0 site allows users to interact with other users or to change Web site content. For response and reconstruction, users could, for instance, report on local conditions, validate maps and other assessment data, report on materials delivery and availability, or coordinate assistance interventions.

The U.S. Geological Survey (USGS) has established a Web site called "Did You Feel It?" to collect data on earthquakes.[7] Using questionnaire responses, USGS develops the "Community Internet Intensity Map" (CIIM) by U.S. postal zip code. For events from which they receive a large enough response, USGS can geo-code the surveys using house addresses to create a more accurate map. The CIIM system was recently expanded worldwide. After the 2008 Wenchuan, China, earthquake, USGS received more than 700 responses.

Humanitarian Information Centers (HICs), organized by UN OCHA on request, can establish a collaboration Web site for agencies involved in reconstruction. HICs can also promulgate standards for information collection, establish data warehouses, and disseminate information (see box). The terms of reference for a HIC are available from UN OCHA.[8]

> ## Humanitarian Information Center
>
> The mission of the HIC is to support the humanitarian community in the systematic and standardized collection, processing and dissemination of information with the aim of improving coordination, situational understanding and decision making. In undertaking this mission, the HIC will complement the information management capabilities of the national authorities, as well as in-country development and humanitarian actors, in order to optimize the response and meet the needs of the affected population. The HIC will only be deployed in new complex emergencies or disasters where IM demands exceed the capacity of the Member State(s) and the IASC. In fulfilling its mission, the HIC will be guided by the principles of humanitarian information management and exchange in emergencies: accessibility, inclusiveness, interoperability, accountability, verifiability, relevance objectivity, humanity, timeliness and sustainability.
>
> *Source:* UN Office for the Coordination of Humanitarian Affairs (UN OCHA), Field Information Services Unit and Information Technology Section, "Humanitarian Information Centres and Partners," http://www.humanitarianinfo.org/.

Mobile telephones. Mobile telephones and smart phones are fast replacing radio and television as the best medium to communicate and coordinate with large populations and to manage digital transactions. Mobile technology includes everything from smart phone applications to simple text messaging and is becoming widely used in the developing world in health, banking, education, and community organizing. This technology can be used for warning populations about risks using common alerting protocol (CAP); communicating using short message service (SMS), Really Simple Syndication (RSS) feeds, or Twitter; transferring funds using mobile banking services; and many other purposes.

The Industrial and Commercial Bank of China (ICBC) used mobile banking to encourage donations to post-earthquake relief and reconstruction. ICBC opened free e-banking channels, such as Internet banking and mobile banking, exclusively for transferring donations and helped clients use the channels for the contributions, while ensuring the donors that all donations would be delivered promptly and safely. In the first two months after the 2008 earthquake, ICBC had transferred more than 200,000 donations to 175 charitable organizations across China, with the volume of donations reaching RMB72 million (US$10.5 million).[9]

Ham radio. The "ham" radio community is an often-overlooked resource that has a long tradition of providing communications in times of disaster. In addition to voice, amateur radio can provide a robust, though very low bandwidth (1200 baud) digital medium with the addition of a legacy modem. For remote areas, this is often the normal mechanism for email and electronic bulletin board access. Including this group in disaster preparation and post-disaster communications is strongly advised.

All of these ICTs are useful in organizing communities and in strengthening community participation in reconstruction. For a discussion of a range of participation approaches in reconstruction, see 🖳 Chapter 12, Community Organizing and Participation.

Assessing Post-Disaster Damage

Accurate, comparable, and appropriately scaled information provides the basis for damage and loss assessments (DaLAs), and related decision making concerning recovery and reconstruction.. Assessments are time- and labor-intensive, must be conducted rapidly, and must meet quality standards. For these reasons, numerous initiatives have been launched to expand the use of technology to improve the timeliness, quantity, and quality of assessment results.

7. USGS, "Did You Feel It: The Science Behind the Maps," http://earthquake.usgs.gov/eqcenter/dyfi/background.php.
8. UN OCHA, 2008, "Terms of Reference: Humanitarian Information Centres," http://www.humanitarianreform.org/humanitarianreform/Portals/1/cluster%20approach%20page/Res&Tools/IM/IASC%20-%20Humanitarian%20Information%20Centre%20Terms%20of%20Reference%20-%20May%202008.pdf.
9. ICBC, "ICBC devotes itself to the reconstruction after May 12 Wenchuan Earthquake," http://www.icbc.com.cn/icbc/icbc%20news/icbc%20devotes%20itself%20to%20the%20reconstruction%20after%20may%2012%20wenchuan%20earthquake.htm.

Integrating spatial data in assessments. The DaLA methodology, developed by the UN Economic Commission for Latin America and the Caribbean (ECLAC) and discussed in its Handbook for Estimating the Socio-economic and Environmental Effects of Disasters,[10] provides a standardized (statistical) assessment of the direct and indirect effects of a disaster event and their consequences on the social well-being and economic performance of the affected country or area. The assessment is currently based on a variety of "analog" (paper) and digital information sources (documents and data from public authorities and agencies, press articles, personal communication with officials, etc.). Survey data is also collected and verified by expert teams sent to the field once the emergency phase is winding down.

DaLas are used to determine the value of damaged and lost assets and to define reconstruction requirements. One DaLA objective is to identify the affect of the disaster on a geographical basis and by sector, together with corresponding reconstruction priorities. While the handbook already emphasizes the potential of new tools available for use in a post-disaster DaLA, up to now the utilization of geospatial information and spatial analysis techniques is not included within the DaLA framework. Therefore, the Global Facility for Disaster Reduction and Recovery (GFDRR) and the World Bank Spatial Team have been working to establish standards and to develop technical and training manuals for mission teams and GIS and IM operators for integrating spatial analysis into assessments.[11] The 📷 case study, below, on the assessment process following the 2004 Indian Ocean tsunami describes how Earth Observation technology was used to estimate the extent of collapsed structures.

The following sections describe the key tools that are expected to be utilized in this new approach to damage assessment.

Geographical information systems. GIS are instruments for storing, retrieving, mapping, and spatially analyzing geographic data by associating spatial features, referenced on a particular place on the earth, with descriptive attributes in tabular form. The power of GIS comes from its ability to integrate spatial information with statistical and analytical processes to derive spatial patterns not apparent from statistics; in other words, to make data more "visual." Maps inform the viewer about spatial patterns, trends, and correlations with other features, and are an important step toward more focused analyses. Hence, GIS is more than a map-making tool. Its power lies in its ability to query geographic information and its associated statistics databases. Information from various sources can be superimposed using GIS to identify risks and investment priorities and to establish baselines for reconstruction. See the 📷 case study on the Central American Probabilistic Risk Assessment (CAPRA), below, to see how a regional agency is working to provide risk information to local users using GIS technology.

The information on maps produced by GIS is displayed in thematic layers and spatially referenced (or "geo-referenced") to the earth. Geospatial information enables analysts and decision makers to get information on the real-world situation instead of using only statistical information. Adding the spatial dimension to databases supports finding answers relevant for policy support, decision making, and risk management. It is important to have baseline data with which to analyze disaster impacts. GIS databases and remote sensing can be used to locate baseline data, as can other local, national, and (increasingly) international data sources. GIS databases may be proprietary, but are increasingly becoming publicly available. Options for accessing GIS data are described in 📷 Annex 2, A Primer on GIS and GIS Data Sources.

Not all GIS data are fully compatible. Using international standards for data collection (e.g., International Standards Organization [ISO] 19115 or the standard being developed by users such as those of the Open Geospatial Consortium [OGC][12]) helps ensure the compatibility of data from various sources.[13] In a recent pilot study, the GFDRR and the World Bank Spatial Team used GIS to map natural hazards and climate change risks in Dakar, Senegal. The study will be replicated in other regions.[14]

Spatial analysis comprises analytical techniques and tools ranging from simple mapping to spatial econometrics and spatial process modeling. Spatial analysis applies techniques to interpret and analyze geographically-referenced data and their topological, geometric, or geographic properties to extract or generate new geographical information.

10. UN ECLAC, *Handbook for Estimating the Socio-economic and Environmental Effects of Disasters* (Mexico: ECLAC), http://www.eclac.cl/cgi-bin/getProd.asp?xml=/publicaciones/xml/4/12774/P12774.xml&xsl=/mexico/tpl-i/p9f.xsl&base=/mexico/tpl/top-bottom.xsl.

11. This chapter borrows from the documents produced under this initiative. When available, the technical and training manuals will be available at http://www.housingreconstruction.org.

12. OGC is a nonprofit, international, voluntary consensus standards organization that is leading the development of standards for geospatial and location-based services.

13. International Standards Organization (ISO), "Geographic Information/Geomatics,TC211," http://www.iso.org/iso/iso_catalogue/catalogue_tc/catalogue_tc_browse.htm?commid=54904&published=on&includedesc=true.

14. The World Bank is preparing a handbook for GIS use in assessments that will detail standardized GIS procedures, including use of GPS cameras, Google Earth for visualization and publication, and satellite-based rapid mapping of areas before and after an event.

Within increasingly data-rich environments and information captured from different environmental monitoring systems and terrestrial networks, mobile devices in the field, or remotely sensed imagery, a GIS provides a state-of-the-art digital platform for storing and managing these data, analyzing the data employing spatial analytical techniques, and finally presenting and visualizing the analytical results using statistics or cartography. Thematic maps and other cartographic products created with GIS are valuable tools to show, at a glance, spatial patterns, relationships, and trends that would not become as obvious from alphanumeric tables.

Remote sensing is an information-gathering method of delivering geospatial information on real-world phenomena. This technique allows analysts to quickly determine the areas affected by a disaster, even in remote regions that are difficult to reach. With its newest sensor generation, remote sensing also provides high-resolution

KUTCH NAV NIRMAN ABHIYAN

information on the impact on physical infrastructure and environmental assets. When tied to GIS, Earth Observation (EO) imagery can be used to identify damaged or destroyed structures and to correlate this information with socioeconomic, hazard, and other geo-referenced data, thereby providing a rapid estimate of damage for eventual verification on the ground.[15] The International Charter on Space and Major Disasters is an example of a remote sensing system in use worldwide, as discussed in the 📖 case study, below, on the 2009 Namibia floods.

Satellite-based EO (an example of remote sensing) is an important, independent source of information proven in disasters around the world to assist in disaster prevention efforts, mitigation, and relief, as well as in damage assessments and reconstruction. Satellite mapping can support on-the-ground DaLA by providing pre- and post-disaster information. Satellite data can deliver relevant geospatial information that can be used to map the disaster or to provide input to DaLAs. See 📖 Annex 1, How to Do It: A Primer on Acquiring Satellite Images.

Remote sensing supports and complements, but is still no substitute for, a field-based assessment. Outputs should be validated with a second data source, preferably on the ground. Portable GPS devices and mobile phones can be deployed to collect, transmit, and upload information to central databases that supply information to GIS

Planning and Monitoring Reconstruction

Monitoring and evaluation in housing reconstruction must be done to measure both financial and physical progress. ICT systems can be used to aggregate information from multiple platforms to facilitate reporting. They can also promote transparency among agencies involved in reconstruction and the recipient community, thereby helping to ensure wise use of scarce resources.

Information systems. Systems such as the Development Assistance Database (DAD) and project-level monitoring systems are used to monitor physical progress and expenditures.[16] 📖 See Chapter 15, Mobilizing Financial Resources and Other Reconstruction Assistance, for more information on these systems. Databases developed by the British Red Cross Society after the 2004 Indian Ocean tsunami in Indonesia were used to manage beneficiary registrations and cash transfers, as described in the 📖 case study, below.

ICTs can be used to coordinate the business processes by which construction projects are approved, inspected, and maintained. Efficient and transparent permit processing and enforcement can not only increase taxation revenues, but can also address institutional corruption and reduce disaster risks. Particularly during the post-disaster construction "boom," ICTs can play an important control function.

15. For an example of how these technologies were used to photograph the effects of the 2009 earthquake in Aquila, Italy, see Digital Globe, "L'Aquila, Italy Earthquake," 2009, http://www.digitalglobe.com/downloads/DG_Italy_Earthquake_Apr_2009.pdf.

16. "The DAD enables stakeholders in the development process to capture the most critical international assistance data on a donor and project-specific basis, including pledges, committed and disbursed amounts, sector and region of implementation, project description, relevant Key Performance Indicators, implementing agency, and other contacts," Development Assistance Database – Fact Sheet, Synergy International Systems, n.d., http://www.synisys.com/resources/DAD%20Factsheet.pdf.

Remote sensing and GIS. The same GIS data and spatial information described above as an input to assessments can also be used to plan reconstruction and to monitor the progress of reconstruction, the impact of reconstruction on ecosystems, and other effects of recovery. These data are critical to the success of the land use and site planning processes described in 🏛 Chapter 7, Land Use and Physical Planning.

Data can be gathered through remote sensing over time and calibrated with data from the ground to establish visual indicators of reconstruction and recovery. Data gathered on the ground, for instance from Web 2.0 applications, can be geo-referenced and used to monitor reconstruction in GIS applications.[17]

Advances in large-scale, automated 3D extraction and creation of virtual urban environments, mean that these technologies can be used economically for large areas to streamline the verification of insurance claims and to detect illegal construction. These tools can also be used to monitor illegal settlements and to identify potential relocation sites. These are the same remote sensing technologies used for rapid disaster assessments, as discussed above.

Other Considerations When Using ICTs

Innovation vs. standardization. While creating customized ICT solutions for post-disaster housing and community reconstruction may seem expedient, there is a risk that such solutions will be duplicative or will function suboptimally. Although initially time-consuming, developing standardized systems in a collaborative fashion can produce both immediate and longer-term benefits. Any deployed system should also be open, to facilitate sharing of data.

Tradeoffs in acquiring data. The principal tradeoffs in data acquisition are quality, cost, and time. Consider the following when deciding on data requirements.
- Identify the most critical data needed for planning and implementation.
- Confirm the ready availability of data (including baseline data for maps).
- Determine whether available data adhere to a standard (if not, it may be an indicator that it is of poor quality, and not worth acquiring).
- Identify measures to improve the quality and accuracy of data (e.g., identity control, internal controls, and data analysis and verification).
- Determine the level of disaggregation needed (the more disaggregated, the higher the cost to collect and analyze).
- Consider alternative methods for data collection, map making, and monitoring, including hand-drawn maps and community data collection.

Risks and Challenges

- Reliance on a single communication system, such as mobile phones, that may become overloaded or inoperable after a disaster.
- Failing to keep databases and systems current, backed up, and protected against hazards, which can result in the need to recreate them after a disaster.
- Delays in the delivery and operationalization of ICT post-disaster systems, and the failure to have a back-up plan in case the high-tech approach is not feasible.
- Failing to utilize up-to-date technologies that are available, or—conversely—insisting on using technologies beyond the capability of their intended users.
- Downplaying the importance of incorporating ICT specialists in post-disaster assessment, planning, and monitoring activities.
- Managing post-disaster information in an uncoordinated, ad hoc, and closely held manner, rather than consciously planning for and encouraging collaboration.
- Duplicating data collection because data are not shared and treated as a "public good" by agencies involved in reconstruction.
- Low levels of ICT interoperability due to lack of standards in system design and data collection.

17. The Recovery Project at Cambridge University Centre for Risk in the Built Environment is an effort to identify indicators of recovery that exploit ICTs, and can be used to measure, monitor, and evaluate recovery after a major disaster, http://www.arct.cam.ac.uk/curbe/recovery.html. See also Daniel Brown, Keiko Saito, Robin Spence and Torwong Chenvidyakarn, 2008, "Indicators for Measuring, Monitoring and Evaluating Post-Disaster Recovery," (presentation at Sixth International Workshop on Remote Sensing for Disaster Response, Pavia). http://tlc.unipv.it/6_RSDMA/Finals/4.3%20-%20Brown.pdf.

🏛 **For access to additional resources and information on this topic, please visit the handbook Web site at www.housingreconstruction.org.**

Recommendations

1. Integrate ICT widely into disaster response while avoiding ad hoc systems or systems that require a high level of technical capacity.
2. Incorporate specialists with ICT experience in assessment and project teams to promote full use of emerging ICTs in recovery and reconstruction.
3. Ensure that ICT systems are compatible with existing government systems, particularly if they will continue to be used after the disaster.
4. Involve stakeholders in assessment, validation, monitoring, and other reconstruction-related activities by using accessible, collaborative technologies, such as Web 2.0.
5. Support the use of open systems and standards to ensure interoperability. Require developers to standardize and geo-reference information through specifications in contracts and terms of reference.
6. Collaborate with the UN system to evaluate the advisability of establishing a HIC. If established, fund it adequately and ensure all key agencies commit to standardizing and sharing information.
7. To quickly acquire satellite images and maps of a disaster area, activate the International Charter on Space and Major Disasters directly or through the United Nations Office for Outer Space Affairs (UNOOSA).
8. Promote the use of field-level ICT systems that assist reconstruction project management, provide transparency to affected communities, and permit the unification of data.
9. Encourage government to develop resilient information systems that can be easily restored after a disaster, and to establish agreements with local and international ICT-related stakeholders that specify mechanisms for post-disaster cooperation.
10. Encourage governments to establish policies and laws that provide the right to information on hazards and risks, after a disaster and at other times, to support the incorporation of disaster risk reduction (DRR) measures in planning and construction.

Case Studies

2009 Floods, Namibia

Space and Major Disasters Charter Activation

In early 2009, the north-central and northeastern regions of Namibia experienced torrential rains that caused flooding along most of Namibia's northern borders. The water levels of the Cunene, Chobe, Zambezi, and Kavango rivers increased dramatically due to the combined effects of rain and water from tributaries originating in Angola and Zambia. The floods affected 350,000 people (nearly 17 percent of the country's population), caused the death of 102 people, and displaced more than 13,500 people.

On March 20, 2009, the International Charter on Space and Major Disasters was activated by UNOOSA on behalf of the United Nations Development Programme (UNDP) Namibia. This map illustrated satellite-detected increases in flood water extent along the Chobe River in the period between March 17 and 25, 2009 in the Caprivi Region, Namibia. Flood analysis was made using Radarsat & ENVISAT-ASAR data. Because of the difference in satellite sensors, there was some uncertainty about the flood extent change over time. This flood detection was a preliminary analysis that was later validated in the field.

In May 2009, government, the UN, and the World Bank conducted a PDNA. While almost all families in the areas of Oshana, Oshikoto, Ohangwena, and Omusati had returned home, inundations in Caprivi and Kavango delayed the return of those relocated to camps. By the end of June, government reported 28,103 people displaced in the Caprivi and Kavango regions and relocation camps remained open. Families that had returned home still required humanitarian assistance due to the loss of property, livestock, and crops, and limited access to basic services.

Sources: Relief Web, "Consolidated Appeals Process (CAP): Mid-Year Review of the Namibia Flash Appeal 2009," http://reliefweb.int/rw/rwb.nsf/db900sid/LSGZ-7UEDEN?OpenDocument&rc=1&emid=FL-2009-000007-ZWE; *and* International Charter on Space and Major Disasters, "Flood in Namibia," http://www.disasterscharter.org/web/charter/activation_details?p_r_p_1415474252_assetId=ACT-249.

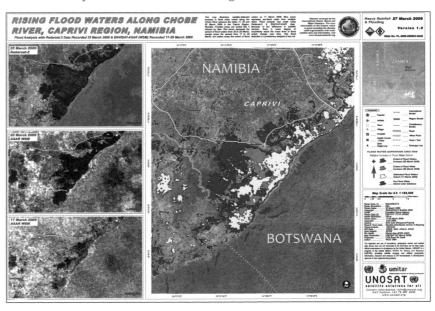

Disaster-Related Data Sharing and Coordination

Central American Probabilistic Risk Assessment Platform

Central America is highly vulnerable to a wide variety of natural hazards that present a serious challenge to the region's sustainable social and economic development. In response, the region has taken a proactive stance on risk prevention and mitigation. The Central American Probabilistic Risk Assessment (CAPRA) platform represents a strategic opportunity to strengthen and consolidate methodologies for hazard risk evaluations supporting this stance and existing initiatives.

Led by the Center for Coordination for the Prevention of Natural Disasters in Central America (*Centro de Coordinación para la Prevención de los Desastres Naturales en América Central* [CEPREDENAC]), in collaboration with Central American governments, the International Strategy for Disaster Reduction (ISDR), the Inter-American Development Bank, and the World Bank, CAPRA provides a set of tools to communicate and support decisions related to disaster risk at local, national, and regional levels in Central America. It provides a GIS platform and a methodology based on probabilistic risk assessment to support decisions in such sectors as emergency management, land use planning, public investment, and financial markets. Current CAPRA applications use data for (1) the creation and visualization of hazard and risk maps, (2) cost-benefit analysis tools for risk mitigation investments, and (3) the development of financial risk transfer strategies. Future applications by CAPRA partners may include real-time damage estimates, land use planning scenarios, and climate change studies.

Sources: CAPRA, http://www.ecapra.org/en/; *and* CEPREDENAC, http://www.sica.int/cepredenac/.

2004 Indian Ocean Tsunami, Aceh, Indonesia

Using Databases to Track Beneficiary Cash Transfers

Databases were developed by a number of organizations to track the flow of assistance funds after the 2004 Indian Ocean tsunami. The British Red Cross Society (BRCS) database in Aceh, Indonesia, involved a significant investment in design (three consultant-months) and was developed principally to track program resources. But the BRCS Aceh team found that the database was also extremely useful for tracking and managing beneficiary cash transfers for shelter. The database linked all stages of the post-disaster assistance process, from registration of beneficiaries to instructing banks to disburse progress payments. The BRCS database could also link the various sectoral elements of the BRCS program: shelter, livelihoods recovery grants, registration for land title, and so on. The lack of an adequate database for food relief programs was a significant weakness, particularly because it was the initial contact with most beneficiaries and could have been the foundation for the registration of all sectoral programs. The capacity to cross-reference data between different agency databases proved vital.

Source: Lesley Adams, 2007, "Learning from Cash Responses to the Tsunami: An HPG Background Paper, Final Report," Humanitarian Policy Group, http://www.odi.org.uk/hpg/Cash_vouchers_tsunami.html.

2004 Indian Ocean Tsunami, Chennai, India

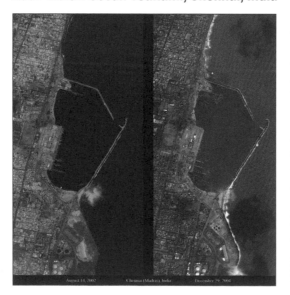

Before and After Imagery from IKONOS Satellite

After the deadly tsunamis generated by the December 26, 2004, earthquake near Sumatra devastated the island of Sri Lanka off the southeastern tip of India, the waves continued westward and slammed into southeastern India, along a stretch of coastline called the Coromandel Coast. Cities, towns, and fishing villages up and down the coast of the state of Tamil Nadu were victims of the waves. These images taken before and after the tsunami from the IKONOS satellite show the city of Chennai, a harbor city on the southeastern Indian coast, located about 350 kilometers north of the Palk Strait, which separates Sri Lanka and India.

Source: NASA Earth Observatory, "Earthquake Spawns Tsunami," http://earthobservatory.nasa.gov/NaturalHazards/view. php?id=14412.

2004 Indian Ocean Tsunami, Banda Aceh, Indonesia
Using EO Technology to Estimate Collapsed Structures

EO technology was deployed to estimate the collapsed structures after the 2004 Indian Ocean tsunami struck Indonesia. By defining a Primary Impact Zone (PIZ), and using observations of available before and after satellite imagery (QuickBird, LandSat7, ETM+, and Shuttle Radar Topography Mission [SRTM] data), an estimate of collapsed structures was obtained. The first step was to estimate the range of heavily damaged structures in the PIZ. Pre- and post-event QuickBird imagery was available for a limited area. All observable existing structures pre-event were counted in specific areas, and an estimate was reached of 5.6 structures per hectare. For areas outside of the available QuickBird coverage, estimates were based on interpretation of available LandSat imagery and low elevation areas defined by SRTM. Where pre-event images were not available, an estimated density of 4 structures per hectare was applied. The resulting analysis led to an estimate that 82 percent of structures had collapsed—a total of 29,545 collapsed structures in the PIZ.

Source: BAPPENAS, 2005, "Preliminary Damage and Loss Assessment, December 26, 2004 Natural Disaster," http://siteresources.worldbank.org/INTINDONESIA/Resources/Publication/280016-1106130305439/damage_assessment.pdf.

Resources

Agustina, Cut Dian. 2008. *Tracking the Money: International Experience with Financial Information Systems and Databases for Reconstruction.* Washington, DC: World Bank, ISDR. http://www.preventionweb.net/english/professional/publications/v.php?id=2474.

Ahmed, K. Iftekhar. 2007. *Emergency Telecommunications and Early Warning Systems for Disaster Preparedness in Chittagong, Bangladesh.* Geneva: International Telecommunication Union. http://www.housingreconstruction.org/housing/EWSChittagongReport01.

Amin, Samia, and Markus Goldstein, eds. 2008. *Data Against Natural Disasters: Establishing Effective Systems for Relief, Recovery, and Reconstruction.* Washington, DC: World Bank. http://siteresources.worldbank.org/INTPOVERTY/Resources/335642-1130251872237/9780821374528.pdf.

Baradan, Berna. 2006. "The Role of Information and Communication Technologies in the Process of Post-Disaster Housing Reconstruction." Paper at the First International CIB-Endorsed METU Postgraduate Conference, "Built Environment & Information Technologies." Ankara, Turkey, March 17–18. http://www.irbdirekt.de/daten/iconda/06059007139.pdf.

Currion, Paul. 2005. "Assessment Report: Pakistan Earthquake Response." Inter-Agency Workgroup on Emergency Capacity, Information and Technology Requirements Initiative. http://www.ecbproject.org/pool/ecb4-itr-assessment-pakistan-mb-28aug06.pdf.

GIS standards and guidelines. http://www.ungiwg.org/activities.htm *and* http://geonetwork.unocha.org/geonetwork/srv/en/main.home.

Guha-Sapir, D. 2006. "Collecting Data on Disasters: Easier Said Than Done." *Asian Disaster Management News* 12, no. 2 (April–June).

International Charter on Space and Major Disasters. 2009. http://www.disasterscharter.org/.

Mohanty, Sujit, Hemang Karelia, and Rajeev Issar. 2005. *ICT for Disaster Risk Reduction—The Indian Experience.* New Delhi: Government of India. Ministry of Home Affairs. National Disaster Management Division. http://www.ndmindia.nic.in/WCDRDOCS/ICT%20for%20Disaster%20Risk%20Reduction.pdf.

Mohanty, Sujit et al. 2005. *Knowledge Management in Disaster Risk Reduction: The Indian Approach.* New Delhi: Ministry of Home Affairs. http://www.ndmindia.nic.in/WCDRDOCS/knowledge-manageme.pdf.

UN OCHA. 2002. "Symposium on Best Practices in Humanitarian Information Exchange." http://www.reliefweb.int/symposium/.

United Nations Asian and Pacific Training Centre for Information and Communication Technology for Development (UN-APCICT). "Disaster Risk Reduction." http://www.unapcict.org/ecohub/communities/disaster-risk-reduction/disaster-risk-reduction/?searchterm=disaster risk. This site contains resources on the use of ICTs in the different phases of DRR.

United Nations Office for the Coordination of Humanitarian Affairs (UN OCHA). "Emergency Telecommunications Cluster Overview." http://www.humanitarianinfo.org/IMToolBox/ *and* http://oneresponse.info/GlobalClusters/Emergency%20Telecommunications/Pages/default.aspx. These sites include extensive information on OCHA's approach to field information management.

Earth Observation (EO) imagery is valuable for post-disaster assessments and for monitoring and evaluation (M&E). Aerial imagery is generally lacking in developing countries; therefore, satellite imagery is more commonly used. Imagery can be used to derive estimates of pre-disaster buildings and to begin to estimate structures and infrastructure damaged or destroyed. For M&E, imagery can show physical changes and reconstruction progress over time through project completion.[1]

Satellites images cover wide areas that are hard to access, allow for frequent updates, and provide an objective source of information in both time and spatial dimensions.[2] There are numerous commercial and government satellite operators. The technology is advancing rapidly, and products and services are varied. So, while there is little market consensus regarding standardized products and pricing, this annex provides parameters that will be useful in making choices when acquiring images.

Where to Acquire Imagery

Imagery can be acquired from both government and commercial sources. Many government sources are considered public domain and are available at no cost. These often come with tradeoffs in terms of precision (location and timing) and resolution. Imagery from commercial satellites can be acquired on a one-time basis (e.g., for assessments) or on a longer-term contract (e.g., for baseline and M&E).

National sources for aerial and satellite imagery include relevant ministries, technical institutions, and potentially regional or larger local governments. Capability and responsiveness will vary depending on the country and the nature and extent of the disaster. These entities may have access to imagery that can be used to create a baseline. Local service providers and technical experts working with disaster management authorities must know how to manage images from various sources, and to extract pertinent housing data from these images. They must also be capable of integrating this information into their existing systems and operational practices.[3]

International sources include satellite operators and resellers of imagery. These operators are numerous and provide a wide range of services (see **Note 2, Partial List of Organizations Providing Post-Disaster Imagery Services**, for a basic list).

What You Need to Know

Requestors of data should coordinate with local counterparts and other agencies involved in response and reconstruction to:

- locate and coordinate with the institution designated to manage post-disaster imagery;
- determine whether the required post-disaster imagery has already been requested or is in the possession of the institution; and
- determine whether pre-disaster baseline imagery is available.

Additional information. Providers have access to imagery from multiple satellites and this basic information will permit them to recommend the optimal solution. Consider the following.

Specifications/Questions	Considerations
Type of disaster and sector	Knowing the disaster type will help image providers and GIS vendors better assist you with identifying the appropriate solution.
Purpose What is the objective and the urgency, e.g., baseline images, damage assessment, or M&E?	This will tell image providers if this is a priority tasking assignment (additional fees) or if it will be a programmed task for a satellite at regularly time intervals. Satellites are programmed with tasks that dictate their course. Images required at a time or location outside of the program will require a priority re-tasking. A non-refundable fee is charged for re-tasking a satellite. If there is cloud cover, the satellite will have to return and the fee will be charged again.
Baseline image Is a processed baseline image available? Has a satellite been programmed in anticipation of a disaster or to regularly survey a high-risk area?	Without a processed baseline (pre-disaster) image, it is impossible to assess the extent of damage and losses. For slow-onset disasters (e.g., tropical storms), a satellite can be tasked to capture imagery before the disaster hits. For rapid onset disasters (e.g. earthquakes, volcanic eruptions), consider procuring regular interval image capture in hotspot and high-risk areas. There are numerous GIS data sources on the Internet, but these data often lack uniform quality standards.
Location/geographic coordinates What are the geographic coordinates of the area for which imagery is required?	The location will need to be specified with geographical coordinates. Latitude and longitude data can be located on Google Earth or elsewhere on the Internet.
Area of Interest (AOI) What is the extent of the housing-affected disaster area in square kilometers or miles?	Using the coordinates, the size of the area to be imaged can be defined. The purpose, type of disaster, and extent of the impact will be factors in defining the area. Satellite services may have a minimum area.

Temporal aspects	If the imagery is not available, a satellite must be programmed to capture it. This will be dependent on the specific need and the return period of the satellite (the amount of time between passes). The return period will also affect the interval for which M&E imagery can be collected.
What is the specific date and/or time for which imagery is required?	

Resolution	Satellite imagery uses geometric resolution expressed in meters to indicate the area of the earth's surface represented in a single pixel. For example, with a satellite resolution of 30m, a single pixel represents is 30m x 30m. 30m is the minimum resolution to observe disaster impact on housing; a higher resolution will improve image quality.[4] At 10m, it is possible to discern the presence and location of individual buildings. Smaller pixels—one meter or less—will show damage of individual buildings (e.g., wind damage to roofs). Widespread flooding can be detected and monitored using moderate-resolution.[5] Higher resolution costs more. There are a range of resolutions and spectral options available (see Note 3, Effect of Resolution on Photography Quality, for resolution guide).
For housing-related imagery, the minimal resolution is 30m, while the optimal resolution is 10m or less, depending on the nature of the analysis.	

Other aspects	Imagery can be requested with spectral and topographical data, which may be useful for planning reconstruction in relation to the surrounding geographical features or natural resources. Materials have different reflectance values, so information such as construction material, water, and vegetation can be identified. The use of multispectral imagery is considered critical for the correct interpretation of images for damage assessment.[6] GIS can overlay demographic or cartographic information onto satellite imagery, providing information on access, for example.
Enquire about what other data can be added to images when captured or when processed.	

Processing	Raw imagery requires processing, including merging images; enhancement and texture analysis; and overlaying geo-referenced data, such as the location of buildings. Imagery providers can often bundle processing and GIS services to be more cost effective than purchasing separately.
Is there local technical (hardware/software) and/or professional capacity for processing of raw imagery?	

Lead time and delivery	Some basic images and processing can be delivered via download within 24 hours of its capture. Larger images and multiple scenes take longer. Depending on where the images are needed, and the connectivity available, the files may need to be transferred to a local partner, downloaded to digital media (DVD), or printed for final delivery. Files need to be processed. The time required will depend on the resources available and resolution of the imagery.
Allow for at least 7 days for capture and processing. If faster service is required, that should be specified.	

Copyright	Commercial satellite companies do not place their imagery in the public domain or sell it; instead, the user must be licensed.[7]

Cost	Cost is largely determined by resolution, coverage area, color vs. black and white, and the return period of the satellite, with tradeoffs for each. Note 1 shows illustrative costs for a 10km x 10km black and white image from a variety of satellites, which does not include special priority tasking fees or processing.[8] Between $3,000 and $4,500 should be budgeted for priority tasking fees.

Procurement	Satellite imagery data can be procured as part of a consulting contract for conducting an assessment or independently, following general procurement guidelines and protocols. In either case, the terms of reference should specify the spatial data required, including mapping resolutions. A technical background is needed to use and analyze satellite imagery, so it is recommended that a consultant or firm be contracted that can be supervised by a team member proficient in the technology. The consultants would acquire the data and submit them to the procuring agency, along with other the deliverables. In this case, the imagery will be a component of the larger procurement.

Note 1: Illustrative Costs for a 100km² Black and White Image

Satellite	Resolution	Raw Image Cost
Landsat 7	15m	Free
Landsat 5	30m	Free
IRS-P6 (pan)	5m	$5,100–$6,700
RapidEye	5m	$2,000
SPOT 5	2.5m	$2,550
Euros A	2m	$1,500
IKONOS	1m	$1,000–$2,000
QuickBird	.61m	$2,250–$9,350
GeoEye-1	.41m	$1,250–$2,500
Radarsat-2	3m	$5,400

Note 2: Partial List of Organizations Providing Post-Disaster Imagery Services

Organization	Service Type (Commercial/Public)	Cost	URL
MDA Geospatial Services	Commercial	Yes	http://gs.mdacorporation.com/
Digital Globe	Commercial	Yes	http://www.digitalglobe.com
Spot Image	Commercial	Yes	http://www.spot.com
Image Sat	Commercial	Yes	http://www.imagesatintl.com
GeoEye	Commercial	Yes	http://www.geoeye.com/
Disaster Charter	Public	No	http://www.disasterscharter.org
World Agency of Planetary Monitoring and Earthquake Risk Reduction	Public	No	http://www.wapmerr.org/
Free Global Orthorectified Landsat Data via FTP	Public	No	http://landsat.org/
Earth Resources Observation and Science	Public	Yes	http://edc.usgs.gov/index.html

Note 3. Effect of Resolution on Image Quality

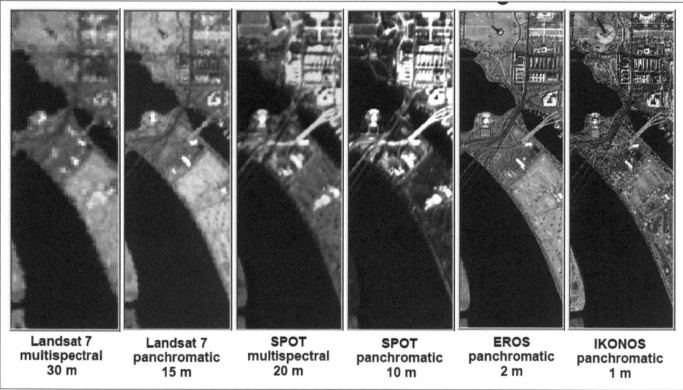

Medium Resolution

High Resolution

| Landsat 7 multispectral 30 m | Landsat 7 panchromatic 15 m | SPOT multispectral 20 m | SPOT panchromatic 10 m | EROS panchromatic 2 m | IKONOS panchromatic 1 m |

Source: Jim Cooper, 2009, "Overview of Change Detection Using Remote Sensing," Presentation to the World Bank, March 26, http://www.mdafederal.com.

Annex 1 Endnotes
1. For an example of time-lapse images at an industrial construction site, see http://www. satimagingcorp.com/gallery/quickbird-timelapse-china.html.
2. Avjeet Singh, 2009, Presentation at the World Bank, March 26.
3. W. U. Guoxiang, 2005, Presentation at the Asian Workshop on Satellite Technology Data Utilization for Disaster Monitoring, Kobe, Japan, January 20.
4. World Bank Spatial Team, 2009, "Report on Use of Satellite Imagery in World Bank Assessments" (draft).
5. Dr. Beverley Adams and Dr. J. Arn Womble, 2006, "Challenging the Odds of Hurricane Damage Data Collection: A Detailed Account from a First Responder," *Imaging Notes,* Volume 21, Number 2, http://www.imagingnotes.com/go/article_free.php?mp_id=66.
6. Dr. Beverley Adams and Dr. J. Arn Womble, 2006, "Challenging the Odds of Hurricane Damage Data Collection: A Detailed Account from a First Responder," in *Imaging Notes,* Volume 21, Number 2, http://www.imagingnotes.com/go/article_free.php?mp_id=66.
7. Wikipedia, n.d., "Satellite Imagery," http://en.wikipedia.org/wiki/Satellite_imagery.
8. Jim Cooper, 2009, "Overview of Change Detection Using Remote Sensing," Presentation to the World Bank, March 26.

SAFER HOMES, STRONGER COMMUNITIES: A HANDBOOK FOR RECONSTRUCTING AFTER NATURAL DISASTERS

In simplest terms, GIS is the merging of graphic map entities and databases. GIS consists of software and hardware; data and information; and a conceptual framework that allows a wide range of information to be presented in a spatial format.

The Uses of GIS Software[1]

Before any geographic analysis can take place, the data need to be derived (created) from field work, maps or satellite imagery, or acquired from data providers. Afterward, data need to be edited, and then stored. Data obtained from outside sources need to be viewed and eventually integrated (conflation) with existing data. To answer particular questions, data are queried and analyzed. Some specific analysis tasks may require a data transformation and manipulation before any analysis can take place. The query and analysis results are then displayed on a map.

GIS software can be proprietary or open-source, that is, software whose source code is accessible, allowing the software to be customized and new applications to be developed.[2] A number of free GIS software programs are available.[3] GIS data is available for a cost, or may also be offered for free.

GIS Data Representation[4]

GIS allow real world objects (roads, land use, elevation) to be represent with digital data. Real world objects can be divided into two abstractions: discrete objects (a house) and continuous fields (rain fall amount or elevation). There are two broad methods used to store data in a GIS for both abstractions: Raster and Vector.

Raster representation. Raster data is any type of digital image represented in grids. In digital photography, the pixel is the smallest individual unit of an image. A combination of pixels creates an image. While a digital image uses the output as representation of reality, the raster data type will reflect an abstraction of reality. Aerial photos are one commonly used form of raster data, used to display a detailed image on a map or for the purposes of digitization. Other raster data sets contain information regarding elevation, a DEM, or reflectance of a particular wavelength of light.

Raster data consists of rows and columns of cells, with each cell storing a single value. Raster data can be images (raster images) with each pixel (or cell) containing a color value. Or it can be a discrete value, such as land use, a continuous value, such as temperature, or a null value if no data is available. Raster data can be stored in raster cell (a single value), or raster bands where an extended attribute table with one row for each unique cell value. Raster data is stored in various formats. Database storage, when properly indexed, typically allows for quicker retrieval of the raster data but can require storage of millions of significantly-sized records.

Vector representation. Geographical features in GIS are often expressed as vectors, by considering those features as geometrical shapes. Different geographical features are expressed by different types of geometry; specifically, points (a simple location), lines or polylines (rivers, roads, topographic lines), and polygons (lakes, buildings, land uses). Polygons convey the most information and can measure perimeter and area.

Each of these geometries is linked to a row in a database that describes their attributes. For example, a database that describes lakes may contain a lake's depth, water quality, pollution level. This information can be used to make a map to describe a particular attribute of the dataset. For example, roads could be colored depending on whether they are paved or not. Geometries can also be compared. For example, the GIS could identify all wells (point geometry) within 1 mile of a lake (polygon geometry) that has a high level of pollution.

Advantages and disadvantages. There are advantages and disadvantages to using a raster or vector data model to represent reality. Raster datasets require more storage space than vector data, allow easy implementation of overlay operations, and produce images that may have a blocky object boundaries. Vector data can be displayed as vector graphics used on traditional maps; is easier to register, scale, and re-project, which can simplify combining vector layers from different sources; is more compatible with relational database environments; file sizes are usually smaller than raster data, by a factor of 10 to 100 (depending on resolution); is simpler to update and maintain, whereas a raster image will have to be completely reproduced; allows much more analysis, especially for "networks" such as roads, power, rail, and telecommunications, while raster data will not have all the characteristics of the features it displays.

Data Capture

Capturing data and entering it into a GIS system is a principal activity of GIS practitioners. GIS data stored in a digital format is entered into the system by a variety of methods, including digitalizing, which creates vector data; survey data which is given coordinates and entered; entering global positioning system data. Remote sensing data, usually digitalized aerial photographs or satellite images or waves (raster data), is an important source for GIS. Attribute data is entered into a GIS along with spatial data. For vector data, this includes additional information about the objects represented in the system. After entering, the data usually requires editing, to remove errors, or further processing.

Metadata is information that describes the data in the GIS. Seven entries are usually associated with geospatial metadata: (1) identification, (2) data quality, (3) spatial data organization, (4) spatial reference, (5) entity and attribute information, (6) distribution, and (7) metadata reference.[5]

Sources of Data

The internet offers numerous digital GIS data sources on global level for mapping and more specific databases representing a broad variety of social, economic and environmental parameters, such as administrative areas, land surface features including vegetation, land cover, hydrography and technical infrastructure elements, topography, global hazard risks, population distribution, or economic performance parameters.

These data bases are widely dispersed over the internet in different scales, precision, timeliness or data formats, and are in most cases lacking uniform data quality standards. Data bases must fulfill a range of requirements in order to be relevant at all for baseline mapping. Data need to be provided

- for free from public web-portals with high-quality reference documentation
- in GIS-ready format following international standards on data formats (ESRI ArcGIS compatible formats) and metadata (e.g. ISO 19115)
- up-to-date and/or regularly updated in sufficient time intervals,
- with appropriate resolution, precision and accuracy conforming to the requirements of the area and topic of investigation.

Free, global data sources are useful in GIS systems for general information and orientation. Baseline mapping is usually done by simply overlaying various spatial data sets and does not involve any specialized mapping or analysis procedures. Baseline maps employing global data sets can be reproduced for the entire globe. Local assessments baseline mapping relies on information about local facilities, premises and infrastructure, such as water supply systems, power plants, cultural heritage assets, educational premises and more.

The World Bank has compiled an extensive list of data sources that can be employed for baseline mapping in damage and loss assessments from open sources. Lists of GIS data sources can also be found on-line. As an example, a short list of sources for hazard risk are shown in the table below.

GIS Data Sources for Hazard Risk

Database	Producer	Description	Distribution
Global risk data platform	UNEP/GRID-Europe	Covers tropical cyclones and related storm surges, drought, earthquakes, biomass fires, floods, landslides, tsunamis and volcanic eruptions	http://preview.grid.unep.ch/index.php?preview=data&events=earthquakes&evcat=8&lang=eng
Natural Disaster Hotspots – Core Data Sets	Center for hazards & risk research - Columbia University	Database focused on earthquakes, landslides, volcanic eruptions, climatic and hydrologic hazards, and man-made environmental hazards	http://www.ldeo.columbia.edu/chrr/research/hotspots/
Low elevation coastal zone (LECZ)	Socioeconomic data and applications center (SEDAC)	Country-level estimates of urban, rural and total population and land area in a low elevation coastal zone (LECZ) were generated globally using Global Rural-Urban Mapping Project (GRUMP) alpha population and land area data products and a Digital Elevation Model (DEM) derived from Shuttle Radar Topographic Mission (SRTM) remote sensing data.	http://sedac.ciesin.columbia.edu/gpw/lecz.jsp

Virtual Globes

A virtual globe (or virtual earth) is a virtual representation of the earth that allows the user with to move around in the virtual environment by changing the viewing angle and position. Virtual globes are capable of showing different views of the earth's surface and layers of information. These may be geographical features, man-made features, or abstract representations of demographic quantities such as population.

Virtual globes vary in the quantity, quality, and timeliness of the information they display, and whether their code is open source. Nearly all have a freeware version, and for-purchase versions with additional functionality. They can be used for observing and mapping typography and development patterns (including time-lapse maps), and some allow the importation of data. A few popular virtual globes are listed below.

Google Earth: http://earth.google.com/
NASA World Wind: http://worldwind.arc.nasa.gov/java/
Bing Maps: http://www.bing.com/maps/.

Annex 2 Endnotes
1. Stefan Steiniger and Robert Weibel, 2009, "GIS Software: A Description in 1000 Words," http://www.geo.unizh.ch/publications/sstein/gissoftware_steiniger2008.pdf.
2. See "The Open Source Geospatial Foundation," http://www.osgeo.org/.
3. See "FreeGIS Project," http://www.freegis.org/about/project.
4. Gary E. Sherman, 2008, *Desktop GIS* (Raleigh: The Pragmatic Bookshelf).
5. Federal Geographic Data Committee, "Geospatial Metadata," http://www.fgdc.gov/metadata.

Guiding Principles for Monitoring and Evaluation

This Chapter Is Especially Useful For:
- Lead disaster agency
- Agencies involved in reconstruction
- Project managers
- Affected communities

- As with assessments, monitoring and evaluation can take place at whatever level is relevant to the organization seeking the information, which will be similar levels to those at which assessments are conducted.
- Define and agree with stakeholders what will be monitored and evaluated early in project development.
- A mix of qualitative and quantitative approaches is likely to be the most useful for monitoring and evaluation in a post-disaster situation. Participatory performance monitoring and households surveys are two especially useful qualitative tools.
- Assessment data are a critical source of baseline information for evaluation, another reason to promote the sharing of this information among agencies.
- Government can simplify the task of tracking reconstruction if it provides agencies with guidance on the indicators it wishes to be monitored at the project level. The indicators to be monitored should be based on the reconstruction policy.
- Good monitoring and evaluation principles are not different in a post-disaster situation, but to apply them may require more flexibility and imagination.
- If government is not prepared to aggregate data collection from multiple agencies to monitor reconstruction, agencies in one sector or region should consider coordinating the monitoring among themselves.

Introduction

Monitoring and evaluation shouldn't be confused with each other. Monitoring is the routine, daily assessment of ongoing activities and progress, while evaluation is the periodic assessment of overall achievements. Monitoring looks at what is being done, whereas evaluation examines what has been achieved or what impact has been made.

There are countless audiences for the information that comes from the monitoring and evaluation of post-disaster projects, including funders, government, executing agencies, the general public, and—of course—the affected community.

In Chapter 2, Assessing Damage and Setting Reconstruction Policy, it is suggested that information gathered and produced in a post-disaster assessment might be looked at as a "public good." A similar case can be made for monitoring and evaluation results, given the large number of stakeholders for most reconstruction projects.

Yet monitoring and evaluation of humanitarian and development activities, while often attempted, are not always that effective as tools to communicate results. Monitoring and evaluation can be even more difficult for disaster-related projects: project assessments and designs may have been hastily prepared, baselines are often not established, and the necessary data might be hard to collect. But good monitoring and evaluation not only improve project outcomes for stakeholders, they have the potential to contribute to international understanding of what "works" in reconstruction—knowledge that is still in somewhat short supply.

Many good tools and resources are available for monitoring and evaluation under "normal conditions." However, few methodologies have been adapted specifically to the disaster environment. Even so, this chapter argues for a rigorous, yet participatory and flexible approach to monitoring and evaluation in all aspects of housing and community reconstruction.

For access to additional resources and information on this topic, please visit the handbook Web site at www.housingreconstruction.org.

269

Key Decisions

1. The **lead disaster agency** should decide how monitoring and evaluation will be carried out within the reconstruction program.
2. The **lead disaster agency**, in consultation with **agencies involved in reconstruction**, should decide how information on expenditures and progress at the project level will be tracked and reported, in order to facilitate consolidation.
3. **Agencies involved in reconstruction** should jointly define protocols for collecting and consolidating sector information, in the absence of government guidance.
4. **Agencies involved in reconstruction** should decide how they can involve affected communities in monitoring and evaluation activities.
5. **Agencies involved in reconstruction** should decide how the results of monitoring and evaluation activities will be shared with the affected community and the general public.
6. **Affected communities** should demand that monitoring and evaluation provide objective project results, which may imply contracting third-party evaluators to conduct them.

Public Policies Related to Monitoring and Evaluation

Government may have policies that require monitoring and evaluation (M&E),[1] and some even require the disclosure of the results of projects built with public funds. Most donors, international financial institutions (IFIs), and nongovernmental agencies have internal M&E policies as well. However, there may not be any policy that governs the reporting of project results by agencies to government or to project beneficiaries.

Monitoring, like assessments, may be an area where there are efficiencies in collaboration, but not necessarily the right incentives. Government should consider establishing protocols for the collection and reporting of post-disaster data, in order to facilitate collection, consolidation, and analysis at the national level. Rules may also be needed to establish minimum parameters for the M&E of projects and to require the disclosure of results. With these rules in place, government can track the progress and the effectiveness of all expenditures related to the disaster and of all the projects being carried out.

Technical Issues
A Comprehensive Project Evaluation

A comprehensive project evaluation includes several distinct elements,[2] all of which are covered in this chapter. The elements are the following.

- Monitoring: to assess whether a program is being implemented as was planned. A program monitoring system enables continuous feedback on the status of program implementation, identifying specific problems as they arise.
- Process evaluation: to analyze how the program operates. Focuses on problems in service delivery.
- Cost-benefit or cost-effectiveness evaluation: to assess program costs (monetary or non-monetary), in particular their relation to alternative uses of the same resources and to the benefits being produced by the program.
- Impact evaluation: to determine whether the program had the desired effects on individuals, households, and institutions, and whether those effects are attributable to the program intervention. (A detailed discussion of impact evaluation is found in 📖 Annex 1, How to Do It: Conducting an Impact Evaluation of a Reconstruction Project.)

What to Monitor and Evaluate in Reconstruction

While evaluation is more strategic than monitoring, which has an operational focus, both monitoring and evaluation are about two things: learning and accountability. This no different in a post-disaster reconstruction project. While it may sound obvious, it still bears mentioning: agencies must carefully define what should be monitored and evaluated, and why, before designing the M&E program.

1. Monitoring and evaluation are separate but related activities, often discussed together. The handbook uses the convention of referring to the two activities together as "M&E."
2. Judy L. Baker, 2000, *Evaluating the Impact of Development Projects on Poverty: A Handbook for Practitioners,* (Washington, DC: World Bank), http://go.worldbank.org/8E2ZTGBOI0.

In an evaluation of the 1998 Armenia, Colombia, earthquake reconstruction,[3] Gonzalo Lizarralde proposes the following as the aspects of post-disaster housing project to evaluate:

1.	Efficiency	Were the local and external resources optimized?
2.	Results	Were the targeted outputs attained?
3.	Timing	Were the outputs available at the right time?
4.	Quality	Was this a good project in the environment where it was used?
5.	Pertinence	Were the outputs made available to the right people?
6.	Acceptability	Did the local community use the outputs/services offered?
7.	Strategy	Did the outputs offered correspond to the needs of the population?
8.	Scope	How much of the real need was covered? Is that percentage satisfactory?
9.	Impacts/objectives	Did the project reduce the vulnerabilities of the population?
10.	External aspects	How did the environment affect the results of the project?

Monitor and Evaluate Programs, Projects, or Households?

As with assessments, M&E takes place at whatever level is relevant to the organization seeking the information. With one exception, these levels are similar to those at which assessments are conducted, although unlike with assessments there is little movement toward common M&E standards. These levels are:

- National reconstruction program (multi-sectoral) M&E
- Housing and community sector-level M&E
- Program or project-level M&E for a specific reconstruction project (not an assessment level)
- Household-level M&E (generally collected using household surveys) (see the 🔑 case study, below, on the unexpected results of a household survey following the 2004 Indian Ocean tsunami reconstruction in Indonesia)

The following table compares the separate characteristics of M&E at each level and shows the responsible party.

Level	Monitoring	Evaluation	Responsible party
National reconstruction (multi-sectoral) M&E	Equivalent to tracking system discussed in 📖 Chapter 15, Mobilizing Financial Resources and Other Reconstruction Assistance.	Reconstruction program evaluation is conducted once reconstruction is substantially complete.	■ Government ■ United Nations (UN) agencies ■ Donors as a group
Housing and community sector M&E	Tracking system should provide monitoring at the sector level to ensure equitable distribution of resources among sectors. Process monitoring may be useful at the sector level if a set of programs is using standard processes.	Joint evaluation of all programs in the housing and community reconstruction sector in a locality might be considered. Conducted once reconstruction is complete, or midway through if problems arise.[4]	■ Government ■ United Nations agencies or Clusters ■ Donors as a group ■ Academic institution
Program or project M&E[5]	Monitoring system should be established for each project or program as part of project design. Monitoring should include the effectiveness of project processes. If government defines monitoring indicators, information will be compatible with national tracking system. Donor and IFI programs may cover more than one sector and be monitored at both program and project level. Project monitoring should be accessible by the affected community, and the monitoring system may be Web-based, as was done in the Community-Based Settlement Reconstruction and Rehabilitation for NAD and Nias Program in Indonesia.[6]	The feasibility and need for evaluation of a project or program should be defined during project design. Donor and IFI programs may cover more than one sector, and need to be evaluated at both program and project level. An "Implementation Completion Report and Results Report" is prepared for all World Bank projects. The report provides detailed information about project outcomes. Many are publicly available.[7] Project sponsors should consider conducting an addition ex post evaluation several years after project completion.	■ Program or project sponsor ■ Affected community or its representatives can organize local M&E using participatory approach ■ Sponsor should be required to report results to government ■ Evaluation should be carried out by third party(ies)

3. Gonzalo Lizarralde, 2002, "Organizational Design, Performance and Evaluation of Post-Disaster Reconstruction Projects," http://www.grif.umontreal.ca/pages/i-rec%20papers/gonzalo.PDF.

4. 📖 Chapter 2, Assessing Damage and Setting Reconstruction Policy, Annex 2, How to Do It: Assessing Post-Disaster Housing Damage, provides an assessment methodology based on Land Ownership and Housing, Final Report (Informe Final, Tenencia de la Tierra y la Vivienda), conducted in Peru to analyze the effect of the 2008 Ica/Pisco earthquake by Centro de Estudios y Promoción del Desarrollo; the UN Human Settlements Programme (UN-HABITAT); the Department for International Development; and the Ministry of Housing, Construction and Sanitation. This assessment was carried out one year after the earthquake to evaluate the problems with the reconstruction program.

5. A concurrent construction audit can be used to monitor a construction project. A construction audit scope of work is included in 📖 Chapter 19, Mitigating the Risk of Corruption, Annex 3, How to Do It: Conducting a Construction Audit.

6. World Bank, "Community-based Settlement Reconstruction and Rehabilitation Project for NAD and NIAS," http://web.worldbank.org/external/projects/main?Projectid=P096248&Type=Overview&theSitePK=40941&pagePK=64283627&menuPK=64282134&piPK=64290415.

7. World Bank, "Documents and Reports," http://go.worldbank.org/LRCBQPWF40.

Level	Monitoring	Evaluation	Responsible party
Household M&E	Monitoring the needs and perceptions of the affected community in real time can be carried out using feedback mechanisms, two-way communications, surveys,[8] community scorecards, and other tools. 📖 Annex 2, How to Do It: Conducting a Social Audit of a Reconstruction Project, includes information on three participatory performance monitoring methodologies. An addition tool is participatory impact assessment.[9]	Outcomes at the community level and perceptions of the affected community should be central topics of the project evaluation. Household satisfaction surveys or beneficiary monitoring studies should be conducted as part of the evaluation. Public sources of survey data and World Bank formats can be used to standardize the household surveys used to collect evaluation data.	■ Agencies involved in reconstruction ■ Government (housing ministry, for example) may conduct household-level monitoring to see effects of its own or agency programs ■ Affected community or its representatives can organize local M&E using participatory approach

How Agencies Organize Project Information

IFIs, nongovernmental organizations (NGOs), and development and humanitarian agencies use two principal frameworks for defining and organizing project goals, objectives, and monitoring indicators. They are the "results framework" and the "logical framework." They are both explained here to promote a shared understanding of organizations' approaches to M&E.

The results framework. Monitoring and evaluation take place in the context of a strategic dialogue among development agencies and their governmental clients about "aid effectiveness." Many development agencies, including the World Bank, have in the past few years oriented their development interventions to conform and contribute to the "Managing for Development Results" agenda. This approach combines a coherent framework for development effectiveness with practical tools for strategic planning, risk management, progress monitoring, and outcome evaluation. For maximum effect, it requires:

- objectives that are clearly stated in terms of expected outcomes and beneficiaries;
- intermediate and higher-order outcome indicators and targets;
- systematic monitoring and reporting;
- demand for results by partner countries and development agencies alike;
- an effective and continuous dialogue on results; and
- strengthening of country capacity to manage for results.

These principles were endorsed in the Rome Declaration on Harmonization in February 2003 and further developed by the Organisation for Economic Co-Operation and Development (OECD) in various reference materials.[10]

8. Extensive information on designing and conducting household surveys is available at International Household Survey Network, http://www.internationalsurveynetwork.org/home/; and World Bank "Poverty Net, Accessing Surveys," http://go.worldbank.org/B50PMCIUV0.

9. Andrew Catley, John Burns, Davit Abebe, Omeno Suji, 2008, *Participatory Impact Assessment: A Guide for Practitioners* (Boston: Feinstein International Center at Tufts University), https://wikis.uit.tufts.edu/confluence/display/FIC/Participatory+Impact+Assessment.

10. Organisation for Economic Co-Operation and Development, "Aid Effectiveness," http://www.oecd.org/department/0,2688,en_2649_3236398_1_1_1_1,00.html and "Managing for Development Results," http://www.mfdr.org/.

DANIEL PITTET

As a result of these agreements, a number of agencies, including the World Bank and the U.S. Agency for International Development, now use the "results framework" to organize and report on project processes and outcomes. Results-based management and results frameworks are similar to logical frameworks (discussed below), but they take a broader look at the context of the project in an organization. While often used for strategic planning, results frameworks are useful for project-level design as well.

A results-based approach aims to improve management effectiveness and accountability by defining realistic expected results, monitoring progress toward the achievement of expected results, integrating lessons learned into management decisions, and reporting on performance. Inputs and the activities that transform them into outputs reflect the process of implementing projects and program rather than desirable end results in themselves. The results framework presents project objectives and indicators in the following format.

Monitoring and Results Framework Matrix

Project development objective	Outcome indicators	Use of outcome information
Overall objective for project	List of indicators that will be used to monitor outcomes	Assess whether expected results are being achieved
Intermediate results	**Results indicators for each component**	**Use of results monitoring**
Component 1		
Results 1-1 to 1-n	Indicators to monitor each result	How monitoring will occur for each result
Component 2		
Results 2-1 to 2-n	Indicators to monitor each result	How monitoring will occur for each result
Component 3 (project management may be counted as a component)		
Results 3-1 to 3-n	Indicators to monitor each result	How monitoring will occur for each result

This matrix is accompanied by a second matrix that describes in detail the baseline data for key indicators, the target values, and the data collection and reporting arrangements.

The logical framework matrix. The logical framework matrix (LFM) is a project "snapshot" that is still used by a number of international agencies. It is an instrument for arranging the 10 questions listed above in a logical, succinct way, to define project, program, or policy objectives, and to identify expected causal links (the "program logic"), outcomes, and impact. It also helps identify indicators for M&E at each stage, as well as potential risks.

Logical Framework Matrix[11]

	Activity description	Indicators – answer the question	Sources of verification	Assumptions and risks
Goal	The broad pro-poor development "impact"/higher-level objective to which the activity will contribute	"Is progress being made towards the goal?"	How the information will be collected, when and by whom, and how it will be reported.	
Development objectives or purpose	The more specific development outcome(s) to be achieved by the activity.	"Have the activity outcomes been achieved?" measured in terms of quality, quantity, and time.	Sources of information and how it will be reported.	Factors outside the activity management's control that may affect the activity objectives to goal link.
Results or outputs	The products and/or services delivered by the activity that are under the implementation management's control.	"Have the outputs been delivered?" measured in terms of quality, quantity, and time.	How the information will be collected, when and by whom, and how it will be reported.	Factors outside the project management's control that may affect the output to activity objective link.
Tasks/activities	The tasks that have to be completed to deliver the planned outputs.	**Inputs:** Summary of the program/project budget.	(Sometimes a summary of costs/budget is given in this box).	Factors outside the activity management's control that may affect the tasks/activities to output link.

11. New Zealand's International Aid & Development Agency (NZAID), 2006 [updated 2007], "Logical Framework Approach, Annex 5: Developing a Logical Framework Matrix," http://nzaidtools.nzaid. govt.nz/logical-framework-approach/annex-5-developing-logical-framework-matrix.

The 🔖 case study, below, on the reconstruction in Colombia following the 1999 Armenia earthquake, shows how the results of the project were reported using an LFM.

Data Management Issues in Monitoring and Evaluation

Availability of baseline data. Good M&E depends on establishing a valid baseline, to make it possible to know whether the project being monitored or evaluated has really had an effect. Baseline data for housing and community reconstruction will generally consist of social and economic indicators for households and physical development indicators for the community. Baseline data can be collected specifically for the project or come from post-disaster assessments, census bureaus, studies carried out during project preparation, the Humanitarian Information Center, or other donors. Information and communications technology, including photographic and geographic information systems, can be used in monitoring and to collect baseline data. (See 📖 Chapter 17, Information and Communications Technology in Reconstruction.) If there is a commitment to good monitoring, government and donors should be able to combine forces to develop adequate baselines for the disaster area.

Availability of monitoring and evaluation data. M&E should cover processes, costs and benefits, and impacts. The project's design and results framework or logical framework will help define what specifically should be monitored and evaluated. "Output" and "activity" data will be generated by the project's own monitoring and financial systems; the project should be set up to facilitate the collection of these data. Other data may come from the national-level tracking system and/or surveys and data-gathering exercises that government and donors may conduct jointly. The following table shows some potential sources of baseline and M&E data.

Potential Sources of Data for Monitoring and Evaluation[12]

	Sources of baseline data	Sources of M&E data
Goal Recovery of nation from disaster and contribution to larger development goals	National assessment data National census Household surveys National accounts	National monitoring data National census Household surveys (existing or new) National accounts Reconstruction programevaluation
Objectives or purpose Normalize economic and social activities through the restoration of essential housing and basic infrastructure	National assessment data National census Household surveys National accounts Regional accounts Local and sector assessments State/municipal social indicators	Reliable sources of social and economic indicators Data collection by third parties may be advisable
Outputs Build or repair houses and public and social infrastructure		Output data from project monitoring system Community surveys Joint assessments
Tasks/activities The tasks that have to be completed to deliver the planned outputs		Data from project financial system Project indicators from monitoring/tracking system

The disaster environment. Above all, the disaster environment itself may be what makes M&E so difficult. The World Bank states that an impact evaluation is intended to determine whether a program had the desired effects on individuals, households, communities, and institutions, and whether those effects are attributable to the program intervention. But when there are multiple agencies implementing multiple interventions in the same locality, it may be difficult to attribute impact to any one project. In addition, some of the results sought from post-disaster projects are qualitative or difficult to measure ("commitment to building back better" or even "greater community participation"). Disaster projects are sometimes designed rapidly with insufficient information, necessitating adjustments during implementation, and making agencies reticent to have their work "judged." And, there is apt to be turnover and inexperience in the executing agency and higher priorities in government than providing census or other data to donors. Therefore, project designers should be practical when identifying indicators and means of verification for post-disaster projects. Third parties may be needed to collect the data or run the monitoring program

12. NZAID, New Zealand's International Aid & Development Agency (NZAID), 2006 [updated 2007], "Logical Framework Approach, Annex 5: Developing a Logical Framework Matrix," http://nzaidtools.nzaid.govt.nz/logical-framework-approach/annex-5-developing-logical-framework-matrix.

altogether. But government should not hesitate to establish rules for monitoring and to send the message that agencies must be accountable for the resources they are spending. For a detailed discussion of post-disaster impact evaluation, see 📖 Annex 1, How to Do It: Conducting an Impact Evaluation of a Reconstruction Project.

Audits versus Monitoring and Evaluation

At times, the word "audit" is used interchangeably with "monitoring." Audits can serve a monitoring function, especially if they are carried out in a concurrent manner. However, audits generally measure results in a more structured way against predefined rules and practices. Formally, an audit analyzes:

- the legality and regularity of project expenditures and income, in accordance with laws, regulations, and contracts, such as loan contracts and accounting rules;
- the efficiency of the use of project funds measured against accepted financial practices; and
- the effectiveness of the use of project funds, that is, whether they were used for the intended purposes.

EARL KESSLER

See 📖 Chapter 19, Mitigating the Risk of Corruption, for a discussion of the purposes of audits and the types that may be useful in post-disaster reconstruction projects.

Social audits. Social audits are a special form of audit that is used for "participatory performance monitoring" purposes. With social audits, the public and the affected community oversee and report on an organization's activities or a reconstruction project. With proper supervision, participatory performance monitoring can be used to collect either qualitative or quantitative information for M&E. For details on conducting a social audit and a summary of other participatory performance monitoring mechanisms, see 📖 Annex 2, How to Do It: Conducting a Social Audit of a Reconstruction Project.

Risks and Challenges

- Poor coordination within the housing sector or with government prevents data from being aggregated across the sector.
- Assessment data are not shared among agencies, government, and others stakeholders, causing inconsistencies and excess data collection costs.
- Lack of baseline data for projects or insufficient time to develop baselines.
- Project staff lacks commitment to monitoring, leading to delays in the implementation and limited availability of M&E information by project managers.
- Participatory M&E methods are not employed because of limited capacity or lack of commitment.
- Multiple baselines are developed and multiple monitoring indicators are used.
- Evaluations are conducted "in-house" and don't convey actual project results.
- Information from M&E systems is not shared so learning about which reconstruction interventions are effective does not take place.

Recommendations

1. Take M&E seriously in housing and community reconstruction, in spite of the complexities of the post-disaster environment.
2. At the same time, be realistic about the challenges and design a monitoring system that is easily manageable while producing reliable results.
3. Consider conducting an impact evaluation using qualitative methods, alone or in combination with quantitative methods.
4. Take advantage of the knowledge gained during assessment when designing the M&E processes and establishing the project baseline.
5. Work with government and other donors to harmonize M&E indicators, so that information on project results can be compared and aggregated.
6. To avoid bias when conducting program evaluations, use objective, experienced evaluators. It is most likely that this will entail hiring outside evaluation experts.
7. Don't just monitor the affected community. Involve them in project M&E, using social audits or other participatory monitoring methods, and make sure the community receives the results.

DANIEL PITTET

Case Studies
1999 Earthquake, Armenia, Colombia
Monitoring and Evaluation of the Colombia Earthquake Recovery Project

Damages from the 1999 Armenia, Colombia, earthquake were estimated at US$1.6 billion dollars. Some 560,000 people suffered direct earthquake losses, and 1.5 million more residing in 5 departments (provinces) and 28 municipalities in the region were affected indirectly.

The development objective for the World Bank's Earthquake Recovery Project for reconstruction after the earthquake was "to assist project beneficiaries to normalize economic and social activities through the restoration of essential housing and basic infrastructure built according to adequate seismic standards." Components of the project included (1) grants of up to US$6,000 for shelter assistance to homeowners who met established criteria and for new houses for renters in vulnerable groups, (2) rehabilitation and retrofitting of social infrastructure, (3) rehabilitation of public infrastructure, (4) capacity building for natural disaster management, (5) social capital restoration, and (6) project management. The project was approved on March 21, 2000, and closed on August 20, 2002.

The reconstruction program was under permanent control, monitoring, and auditing by public entities, such as the General Controller's Office, as well as private entities and citizen oversight groups, which permitted a guarantee from all stakeholders that the projects were being properly executed both physically and financially. Official project monitoring responsibility was contracted to a consortium of universities. Bank staff conducted nine monitoring missions to Colombia during execution of the loan.

The project was given a highly satisfactory rating in the Implementation Completion and Results Report dated January 10, 2003.[13] The following were the results of the project.

13. World Bank, 2003, "Implementation Completion Report on a Loan in the Amount of US$225 Million to the Government of Colombia for the Earthquake Recovery Project," http://go.worldbank.org/KTH3BR97W0.

Project Indicators	Projected	Actual	Results
Outcome/impact indicators			
Increase in the amount of new and repaired housing meeting seismic codes	43,480		Subsidies for units partially damaged
	17,550		Subsidies for units destroyed
	18,420		Subsidies for units structurally damaged
		100,000	Subsidies to repair housing
		13,000	Subsidies to rebuild housing of owners
		17,000	Subsidies to rebuild housing of tenants
Total housing	**79,450**	**130,000**	
Number of families relocated from temporary shelters		600	Families in temporary shelters reduced from 14,000 in 1999 to 600 in 2002.
Lower unemployment in the project area		19%	Rate of unemployment fell from 52% in 2/99 to 19% by 2001.
Number of reconstruction and micro-zoning plans implemented in the project area		All	Land use plans developed for all municipalities in the region and used in relocation/reconstruction effort.
Total impact	**79,450**	**130,000**	
Output indicators	**Units**	**Units**	
Schools	650	604	
Churches	161	60	
Other public buildings	417	355	
Total outputs	**1,228**	**1,019**	
Activities/expenditures	**Million US$**	**Million US$**	
Housing	243.00	243.05	
Social infrastructure	75.00	82.40	
Public infrastructure	115.00	107.60	
Disaster management	7.00	7.00	
Social capital	8.00	8.00	
Management	19.75	19.75	
Total baseline cost	467.75	467.80	
Total project costs	467.75	467.80	
Front-end fee	2.25	2.25	
Total expenditures	**470.00**	**470.05**	

2004 Indian Ocean Tsunami, Aceh, Indonesia

Who Cares about Quality?

The Indonesian organization Urban Poor Linkage Indonesia (UPLINK) is a national coalition of NGOs and community-based organizations that focuses on urban poor issues. UPLINK provided emergency help and housing reconstruction assistance to 25 villages in one of the coastal areas that was most affected by the 2004 tsunami in Aceh, Indonesia. The project completed more than 3,300 houses using a participatory approach. The reconstruction work of UPLINK won an international award and was recognized by various national and international technical evaluations for the outstanding quality of the new houses.

While expressing general satisfaction with UPLINK's work, people voiced some reservations about its high quality. In fact, a survey carried out in 2007 revealed that a significant number of people considered such factors as size, number of rooms, provision of a kitchen or a porch and furniture, and an overall "modern" house more important than quality or protection from future disasters. Houses with inferior quality but with free furniture were more appreciated than the high-quality, unfurnished houses provided by UPLINK. In addition, people expressed a willingness to forgo the participatory process used by UPLINK if a contractor-built house was bigger and looked more modern! This indicates that in evaluating housing assistance options, people evaluate numerous features, including size, design, and amenities of the housing package, and that important considerations for funding agencies, such as the quality and safety of construction, are not necessarily a priority for the homeowners and accordingly are unlikely to be considered when families begin expanding their homes. In the case of UPLINK, people surveyed considered the quality as "excessive" and would have preferred a more standard quality in exchange for a little extra space, a kitchen, or a porch.

Source: Jennifer Duyne Barenstein et al., 2008, *People-Driven Reconstruction and Rehabilitation in Aceh: A Review of Uplink's Concepts, Strategies and Achievements* (Aachen: Misereor).

Resources

Baker, Judy L. 2000. "Evaluating the Impact of Development Projects on Poverty: A Handbook for Practitioners." http://siteresources.worldbank.org/INTISPMA/Resources/handbook.pdf.

Catley, Andrew et al. 2008. *Participatory Impact Assessment—A Guide for Practitioners.* Boston: Feinstein International Center at Tufts University. https://wikis.uit.tufts.edu/confluence/display/FIC/Participatory+Impact+Assessment.

Field, Erica and Michael Kremer. 2006. *Impact Evaluation for Slum Upgrading Interventions.* Washington, DC: World Bank. http://siteresources.worldbank.org/INTISPMA/Resources/383704-1146752240884/Doing_ie_series_03.pdf.

Goicoechea, Ana. 2008. "Preparing Surveys for Urban Upgrading Interventions: Prototype Survey Instrument and User Guide." http://siteresources.worldbank.org/INTURBANDEVELOPMENT/Resources/336387-1169585750379/UP-6_Surveys.pdf.

Molund, Stefan, and Göran Schill. 2007. *Looking Back, Moving Forward: SIDA Evaluation Manual.* 2nd ed. Stockholm: Swedish International Development Agency. http://www.sida.se/sida/jsp/sida.jsp?d=118&a=3148&language=en_US.

Organisation for Economic Co-operation and Development. *Principles for Evaluation of Development Assistance.* Paris: OECD. http://www.oecd.org/dataoecd/31/12/2755284.pdf.

ProVention Consortium. n.d. *M&E Sourcebook.* Geneva: ProVention Consortium Secretariat IFRC. http://www.proventionconsortium.org/?pageid=62.

UN-HABITAT. 2001. *Guidelines for the Evaluation of Post Disaster Programmes: A Resource Guide.* Nairobi: UNCHS (UN-HABITAT). http://www.unhabitat.org/content.asp?cid=1264&catid=286&typeid=16&subMenuId=0.

World Bank. 1996. *Performance Monitoring Indicators: A Handbook for Task Managers.* Washington, DC: World Bank. http://siteresources.worldbank.org/BRAZILINPOREXTN/Resources/3817166-1185895645304/4044168-1186409169154/24pub_br217.pdf.

World Bank. 2000. *Key Performance Indicator Handbook.* Washington, DC: World Bank. http://info.worldbank.org/etools/library/view_p.asp?lprogram=3&objectid=38956.

World Bank. 2004. *Monitoring and Evaluation: Some Tools, Methods and Approaches.* Washington, DC: World Bank. http://lnweb90.worldbank.org/oed/oeddoclib.nsf/InterLandingPagesByUNID/A5EFBB5D776B67D285256B1E0079C9A3.

World Bank. 2004. *Ten Steps to a Results-Based Monitoring and Evaluation System.* Washington, DC: World Bank. http://go.worldbank.org/C5TSRIQPR0.

World Bank. "Impact Evaluation," http://go.worldbank.org/2DHMCRFFT2. Contains extensive tools and resources on impact evaluation.

Selected World Bank Projects with Housing Reconstruction, Transitional Shelter, Relocation, and/or Slum Upgrading Components

India, Emergency Tsunami Reconstruction Project, 2005, P094513.
http://web.worldbank.org/external/projects/main?projid=P094513&theSitePK=40941&piPK=51351143&pagePK=51351001&menuPK=51351213&Type=Overview

India, Gujarat Emergency Earthquake Reconstruction Project, 2002, P074018.
http://web.worldbank.org/external/projects/main?pagePK=64283627&piPK=73230&theSitePK=40941&menuPK=228424&Projectid=P074018

Iran, Bam Emergency Earthquake Reconstruction Project, 2004, P088060.
http://web.worldbank.org/external/projects/main?pagePK=51351038&piPK=51351152&theSitePK=40941&projid=P088060

Jamaica, Inner City Basic Services for the Poor Project, 2006, P091299.
http://web.worldbank.org/external/projects/main?projid=P091299&theSitePK=40941&piPK=51351143&pagePK=51351001&menuPK=51351213&Type=Overview

Turkey, Marmara Earthquake Emergency Reconstruction Project, 1999, P068368.
http://web.worldbank.org/external/projects/main?pagePK=64283627&piPK=73230&theSitePK=40941&menuPK=228424&Projectid=P068368

As discussed in this chapter, a comprehensive project evaluation includes several distinct elements.[1] This chapter provides guidance on whether to carry out an impact evaluation of a post-disaster housing and community reconstruction project, and recommends resources that are available to support the process, should it be decided to conduct one.

The elements of a comprehensive project evaluation are the following.

- Monitoring: to assess whether a program is being implemented as was planned. A program monitoring system enables continuous feedback on the status of program implementation, identifying specific problems as they arise.
- Process evaluation: to analyze how the program operates. Focuses on problems in service delivery.
- Cost-benefit or cost-effectiveness evaluation: to assess program costs (monetary or non- monetary), in particular their relation to alternative uses of the same resources and to the benefits being produced by the program.
- Impact evaluation: to determine more broadly whether the program had the desired effects on individuals, households, and institutions, and whether those effects are attributable to the program intervention.

Is an Impact Evaluation Required?

A true impact evaluation is one designed to answer the so-called "counterfactual questions": How would individuals who participated in the program have fared in the absence of the program? And if those who were not involved in the program had been incorporated, what would have been the outcome? Impact evaluation therefore requires establishing a valid comparison group of individuals who were not in the program, but on whom the program would have had a similar impact had they participated. The identification of this comparison group is critical to any impact evaluation.

The additional effort and resources required for conducting impact evaluations are best mobilized when the project is innovative and replicable, involves substantial resource allocations, and has well-defined interventions. Impact evaluations can also explore unintended consequences, whether positive or negative, on beneficiaries.

Before carrying out an impact evaluation, it must be determined whether one is warranted. The costs and benefits should be assessed, and consideration should be given as to whether another approach, such as monitoring of key performance indicators or a process evaluation, would be adequate or more appropriate. An impact evaluation requires:

- a need and desire to assess the causality associated with the project;
- strong political and financial support; and
- an ability to "net out" the effect of the interventions from other factors through the use of comparison or control groups.

Qualitative techniques are also used for carrying out impact evaluation, without attempting to make a causal connection. The focus is on understanding processes, behaviors, and conditions as they are perceived by the individuals or groups being studied. There are risks associated with using qualitative methods alone to evaluate impact, including subjectivity in data collection, the lack of statistical robustness, and generally small sample sizes, which make it difficult to generalize to a larger, representative population. The validity and reliability of qualitative data are very dependent on the skill of the evaluator and field staff in interpreting the information they collect.

However, there are benefits from using qualitative information that might be especially relevant in the post-disaster context, where a quantitative impact evaluation may be impossible. Qualitative assessments are flexible, can be carried out using rapid techniques, and can employ novel data collection approaches. They can also enhance other elements of the impact evaluation by providing an understanding of stakeholder perceptions and priorities that may in turn have affected program impact. The affected population can even play a role in qualitative evaluation, using such tools as participatory monitoring. Three participatory monitoring techniques are described in this chapter in ☖ Annex 2, How to Do It: Conducting a Social Audit of a Reconstruction Project.

Clarifying Evaluation Objectives

Once it has been determined that an impact evaluation is appropriate and justified, it is necessary to establish clear objectives and agree on the issues that will be the focus of the evaluation. Clear objectives are essential to identifying information needs, setting output and impact indicators, and constructing a solid evaluation strategy to provide answers to the questions posed. Statements of objectives that are too broad do not lend themselves to evaluation.

A logical framework or results framework can provide the basis for identifying the goals of the project and the information needs for the evaluation. If either of these has been prepared during project preparation, it should serve as the starting point for defining objectives and issues for the evaluation. If not, it can be developed in preparation for the evaluation. Reviewing other evaluation components, such as cost-effectiveness or process evaluations, may also be important objectives of a study and can complement the impact evaluation. A process evaluation can assess the procedures, dynamics, norms, and constraints under which a particular program is carried out. Qualitative and participatory methods can also be used to assess impact.

No evaluation technique or set of techniques is perfect. The evaluator must make decisions about the tradeoffs for each method chosen during the planning of the evaluation.

Impact Evaluation Best Practices

Although each impact evaluation will have unique characteristics requiring different approaches, a best practice impact evaluation should include:

- an estimate of the counterfactual by (1) using random assignment to create a control group (experimental design), and (2) appropriately and carefully using other methods, such as matching to create a comparison group (quasi-experimental design);
- to control for pre- and post-program differences in participants, and to establish program impacts, relevant data collected at baseline and follow-up (including sufficient time frame to allow for program impacts);
- sufficiently sized treatment and comparison groups to establish statistical inferences with minimal attrition;
- cost-benefit or cost-effectiveness analysis to measure project efficiency; and
- qualitative techniques to allow for the triangulation of findings.

Identifying a control group is challenging under ordinary circumstances; for a post-disaster project, it may be considerably harder. This may be due to the fact, for instance, that the project takes place only in a specific location (where the disaster occurred), that those who participate have special characteristics (all members of the affected population whose houses were destroyed), or that it may not be ethical to withhold assistance from some who were affected. Selecting the control group can be accomplished using methodologies that fall into two broad categories: experimental designs (randomized) and quasi-experimental designs (nonrandomized).

Main Steps in Designing and Implementing Impact Evaluations

During Project Identification and Preparation
1. Clarify objectives of the evaluation
2. Explore data availability
3. Design the evaluation
4. Form the evaluation team
5. If data will be collected:
   (a) Select sample design
   (b) Develop data collection instrument
   (c) Staff and train fieldwork personnel
   (d) Pilot test
   (e) Data collection
   (f) Data management and access

During Project Implementation
6. Collect data on an ongoing basis
7. Analyze the data
8. Write up the findings and discuss them with policy makers and other stakeholders
9. Incorporate the findings in project design

Slum Upgrading as a Model for Post-Disaster Reconstruction Projects

Limited material on impact evaluations following disasters is available to draw on. However, significant work has been done by the World Bank and others on impact evaluations for specific types of infrastructure projects and for slum upgrading projects, which provide a framework for evaluating housing and community reconstruction projects. Because slum upgrading projects have many similarities with multi-sectoral community reconstruction projects, this annex recommends drawing on this experience.

According to the World Bank,[2] slum upgrading consists of physical, social, economic, organizational, and environmental improvements within neighborhoods. These projects may be undertaken by citizens, community groups, businesses, and local and national authorities. Typical actions include:

- regularizing security of tenure through property mapping, titling and registration;
- installing or improving basic infrastructure, including water, waste collection, storm drainage, electricity, security lighting, and public telephones;
- removal or mitigation of environmental hazards;
- providing incentives for community management and maintenance;
- constructing or rehabilitating community facilities, such as nurseries, health posts, and community centers;
- home improvement, including material upgrading, new construction, and expansion of existing structures;
- improving access to health care and education and programs to address community issues, such as crime and substance abuse;
- enhancement of income-earning opportunities through training and micro-credit; and
- crime control.

Some of the challenges faced in slum upgrading evaluations that are relevant to evaluations of housing and community projects are discussed below.

Slum Upgrading Evaluation Challenges Relevant to Reconstruction Project Evaluation

1. Mobility	High turnover in the project site will create distortions in the findings, but may also be a project outcome, so should be evaluated carefully.
2. Rural-urban ties	Mobility of residents to and from rural areas and the transfer of funds through worker remittances are potential impacts that should be evaluated.
3. Informal sector	Residents may be participating in both the formal and informal commercial and credit sectors, and evaluations should capture both and the changes in both from the project. Residents may be more forthcoming about formal than informal economic activity, income, etc., creating distortions in data.
4. Population heterogeneity	In urban settings with more diverse populations, it may be important to disaggregate findings by gender, race, ethnicity, and/or class. Certain interventions may be more effective with some subgroups than others. Ethic group relationships may also be affected by the project.
5. Spillovers	Project benefits may extend outside the project boundaries and make it difficult to measure impact and to select the control group, particularly in more dense, urbanized areas.
6. Contamination	The behavior of the control group may change if its members know about the project and anticipate it being delivered to them in the future.
7. Crime	Residents may be reticent to disclose certain information about economic or criminal activity in neighborhoods where crime is a problem and data sources other than direct reporting may be required.
8. Multiple simultaneous interventions	Projects covering several sectors, such as community reconstruction projects, are difficult to evaluate because the impacts of specific elements are difficult to separate out. Potential solutions include comparing to projects with different sets of interventions, or projects with elements sequenced differently, but these may be difficult to find. Where there are multiple actors in charge of different types of interventions, close cooperation is required among sponsors to conduct an impact evaluation.

Outcome Indicators for Housing and Community Reconstruction

The following indicators (some taken from slum upgrading projects) are relevant for post-disaster reconstruction projects. Specific outcomes within these categories should be chosen based on the details of the intervention and anticipated effects.

Housing indicators
- Housing reconstructed/rehabilitated by number and type of housing
- Housing reconstructed/rehabilitated by tenancy type
- Housing safety improvements by number and type
- Displacement and return
- Completion rates of housing
- Household occupancy
- Household size
- Household satisfaction (process, project, housing, services, amenities)
- Real estate market effects

Social indicators
- Time use in household
- Time to work
- Childhood occupation by gender
- Intrahousehold bargaining and gender issues
- Fertility
- Mental health, including stress and depression

Community-level indicators
- Residential segregation
- Social services access/quality
- Public services access/quality
- Community environmental management and risk reduction
- Distance to work and social services
- Indicators of social capital (participation, bartering)
- Political enfranchisement

Economic indicators
- Household income and distribution
- Employment and income generation activities
- Formal sector integration
- Credit market demand and access
- Cost recovery
- Composition of assets
- Formal and informal taxes

Program indicators
- Distribution and use of subsidies
- Household contribution to reconstruction
- Population displacement time/cost
- Cost per unit of construction/rehabilitation (housing and community facilities)
- Overhead cost per unit of construction/rehabilitation (housing and community facilities)
- Impact on local markets

Household Surveys and Survey Data

Household and community surveys are the most common instruments for collecting data for impact evaluations. They can be used to collect both quantitative data to evaluate project impacts and qualitative information to evaluate household satisfaction and perceptions of reconstruction projects.[3] The World Bank Urban Sector Board has designed a prototype survey instrument, organized by sector, that includes comprehensive printable household and community questionnaires for evaluating slum upgrading projects and that is useful for data collection for evaluating post-disaster projects.[4] Guidance on questionnaire design and sampling is available from the International Household Survey Network (IHSN).[5]

There are multiple data sources that may help reduce data collection costs for an impact evaluation, such as administrative data, household survey data, census data, facility survey data, industry data, specialized survey data, participatory assessments, and geographic information system and global positioning system data.

Household surveys are essential analysis tools for collecting information on satisfaction and other project results at the household level. A census covers the whole population in the country. A survey covers only a subset—generally a small fraction—of all households. Common survey types include single-topic surveys, multi-topic surveys, demographic and health surveys, employment surveys, rapid monitoring surveys, service satisfaction surveys, and specialized.

The use of household surveys has become increasingly widespread around the world, as has the effort to standardize survey design and survey indicators. Government may be able to supply survey data for an impact evaluation. There are also a number of online sources of survey data, as shown below.

Public Sources of Survey Data

IHSN	The IHSN is a partnership of international organizations seeking to improve the availability, quality, and use of survey data in developing countries. It provides a central survey catalogue that lists existing and planned surveys by country, as well as other technical resources on household surveys. http://www.internationalsurveynetwork.org/home
Poverty Net Web site of the World Bank	This site provides an extensive list of household survey data sources around the world. http://go.worldbank.org/PCRSXRI320
World Bank Development Data Platform (DDP)	DDP lists existing household surveys along with questionnaires, other documentation, and datasets by country. http://go.worldbank.org/AM8Z12FUL0
World Bank Survey-Based Harmonized Indicators Program (SHIP) for Africa	SHIP facilitates the monitoring of social and economic outcomes of national development programs using standardized household survey data. http://go.worldbank.org/4FSNHCFAA0

Expertise Required for an Impact Evaluation

An impact evaluation should be carried out only by evaluation experts. The team may include a combination of international and national consultants. A proposed evaluation design should be provided to the consultants or should be developed by them and approved by the agency contracting the consultancy.

A basic impact evaluation team includes an impact evaluation manager, a lead researcher and a research assistant, field supervisors, and interviewers.

- The manager clarifies the objectives of the evaluation, taking into account the client's needs, drafting the terms of reference for the team members, reaching agreement among the team members and the client about implications of the implementation of the evaluation, and coordinating the field work.
- The lead researcher, usually an economist, is responsible for selecting the evaluation methodology; defining and supervising the data collection strategy, including survey and sampling design; supervising the field work; conducting the quantitative analysis; and writing the evaluation report.
- The research assistant is responsible for giving support to the lead researcher, especially when it comes to designing the data entry programs and processing data to produce basic results for the qualitative analysis.

- Field supervisors oversee the interviewees and other data collection in the field. Use of national consultants to carry out the field work is a common practice to overcome language and cultural barriers.

Some evaluation teams include a sociologist and/or an anthropologist to ensure community participation and to perform the qualitative analysis. A fieldwork manager may be needed to supervise data collection, including scheduling the field supervisors and interviewers. The team should coordinate its work with government officials in relevant sectors. Examples of terms of reference for an impact evaluation are available from the World Bank.[6]

Annex 1 Endnotes

1. Judy L. Baker, 2000, *Evaluating the Impact of Development Projects on Poverty: A Handbook for Practitioners*, (Washington, DC: World Bank), http://go.worldbank.org/8E2ZTGBOI0.
2. Erica Field and Michael Kremer, 2006, *Impact Evaluation for Slum Upgrading Interventions* (Washington, DC: World Bank), http://siteresources.worldbank.org/INTISPMA/Resources/383704-1146752240884/Doing_ie_series_03.pdf.
3. Examples of post-disaster household survey instruments and results in the public domain are limited. See Sarah Zaidi, 2006, "Results of the RISEPAK-LUMS January Household Survey in the Earthquake Affected Areas of Mansehra and Muzaffarabad" (RISEPAK: Lahore), for one example from the 2005 North Pakistan earthquake.
4. Ana Goicoechea, 2008, "Preparing Surveys for Urban Upgrading Interventions: Prototype Survey Instrument and User Guide" (Washington, DC: World Bank), http://siteresources.worldbank.org/INTURBANDEVELOPMENT/Resources/336387-1169585750379/UP-6_Surveys.pdf.
5. IHSN, http://www.internationalsurveynetwork.org/home.
6. Judy L. Baker, 2000, *Evaluating the Impact of Development Projects on Poverty: A Handbook for Practitioners*, (Washington, DC: World Bank), http://go.worldbank.org/8E2ZTGBOI0.

Background

Participatory performance monitoring refers to the involvement of citizens, users of services, or civil society organizations (CSOs) in the monitoring and evaluation of service delivery and public works. Participatory performance monitoring can make an important contribution to reducing corruption and improving the quality of post-disaster reconstruction. This guidance covers the use of social audits in detail, and briefly mentions two other methods of participatory performance monitoring: citizen report cards (CRCs) and community score cards (CSCs).

A social audit (sometimes also referred to as social accounting) is a process that collects information on the resources of an organization or on a particular project, such as a housing or infrastructure reconstruction project.[1] The information is analyzed and shared publicly in a participatory fashion. Although the term "audit" is used, social auditing does not merely consist in examining costs and finance—the central concern of a social audit is how resources are used for social objectives.

Purpose

The scope of social audits may differ. They may be used for investigating the work of all government departments over a number of years in several districts. They may also be used to manage a particular project in one village at a given time. Most social audits will usually consist of the following activities and outcomes:

- Produce information that is perceived to be evidence-based, accurate, and impartial
- Create awareness among beneficiaries and providers of local services
- Improve citizens' access to information concerning government documents
- Be a valuable tool for exposing corruption and mismanagement
- Permit stakeholders to influence the behavior of government
- Monitor progress and help to prevent fraud by deterrence

How to Implement a Social Audit

Social audits methodologies vary considerably and are influenced by the country context, the availability of information, and the legal and political framework. In general, implementation would include the following steps.

Activity	Considerations
Definition of objectives	The objectives of the social audit exercise should be clearly delineated. As a first step, one should identify the relevant agencies/projects that will be subjected to audit, the time frame for the audit, and the factors/indicators that will be audited.
Identifying stakeholders	The stakeholders should be identified and included in the whole process. The stakeholders should be a mix of government actors from different levels, service providers and/or contractors, representatives of CSOs, beneficiaries, and workers of the service providers or contractors. Special consideration should be given to marginalized social groups.
Data collection	Social audits use a combination of different methods for obtaining relevant data, including interviews, surveys, quality tests, compilation of statistics, case studies, participant observation, evaluation panels, and relevant official records. This is a crucial stage in the process but is often difficult and frustrating, since the agencies under investigation may not have kept records properly or may be unwilling to provide such records. It is important to include the officials from the agencies that are being evaluated, since officials may be more willing to provide information if they are included and gain and understanding of the potential benefits of the process.
	Quality tests may be expensive to conduct and not feasible given budget constraints. In cases where quality tests were conducted (e.g., testing the quality of the cement used in a construction or the bitumen premix for a road), they often produce hard evidence of resource misuse.
	The process of collecting data is extensive and takes a lot of time. Audit committees in each community can be made responsible for interviewing representatives, such as the municipal mayor and the head of the procurement and contracts unit, for visiting the sites (roads, buildings, etc.), and for collecting information on the project outputs.
Data analysis/collation	Deciphering official records can be challenging and complex. The information gathered through different methods and from different sources should be summarized in one comprehensive document that is easy to understand for everyone who is involved in the process. For the data to be user-friendly, they may have to be converted.
Distributing and getting feedback on the information	The findings from the audit are provided to the stakeholders for feedback. Citizens who worked on project sites play an important role in this step, since they can verify the figures relating to material and non-material resources stated in the project documents. This information exchange provides an opportunity for building civic momentum and publicizing the public hearing. Some social audit initiatives have used creative media, such as songs, street plays, and banners, to explain the process and advertise for the public hearing.

Activity	Considerations
Public hearing	If the area under consideration is large, several public hearings should be held, since it is important that the location is convenient and accessible to encourage attendance by as many constituents as possible. At the beginning of the hearing, the rules of conduct are explained to the participants to avoid conflict. After workers or residents have described social audit findings, which can include evidence of corruption, inefficiencies in utilization of funds, or poor planning, public officials are given adequate opportunity to justify their performance in projects. Marginalized groups should be actively encouraged to contribute their points of view.
Follow-up	Following the public hearing, the final social audit report will be written up. This will include recommendations for government regarding actions to address specific instances of corruption and mismanagement. Copies of the report should be widely disseminated to government officials, the media, participants involved in the process, and other organizations deemed relevant to the issues at stake. Key findings and recommended actions should be disseminated in written and oral formats.

Who Implements the Social Audit?

The steps described above may differ depending on the agency and the available resources.[2] In some countries, government periodically carries out social audits for self-evaluation. In many developing countries, however, CSOs have initiated the social audit process to hold government accountable for its actions. Depending on the scope of the audit, different CSOs, research institutes, or government agencies may work together under the direction of one lead institution. The choice of the implementing agency is crucial for the success of the auditing process. The organization should be perceived as impartial and above party politics by all groups involved in the process.

Where Have Social Audits Been Used?

Social audits have been applied in many countries. While social audits have sometimes been used to investigate the quality of services, such as the police, customs, or schools, the majority of social audits have focused on public works. The time frame under investigation typically ranges from two to five years. Social audits can be employed after a project is finished and during the planning and implementation phase. Auditing during the planning or implementation phase is often not feasible, since it requires close cooperation with the government agency that will be audited. However, when possible, auditing during the planning phase is valuable, as it has the advantage of preventing inappropriate acts by monitoring decision making, bidding, contracting, and execution. Social audits that are conducted after the project is finished can be carried out independently of a strong willingness of the agency under scrutiny, although a minimum level of cooperation is often required for obtaining the necessary documents, especially if there is no access to information legislation.

Public works social audits have often produced the following findings:

- Works are paid for but have not taken place, e.g., roads or wells exist only on paper.
- Work is done only in part (only a fraction of the amount stated in the records is delivered or only a part of the tasks agreed on is completed).
- Work is done in worse quality than the quality specified in the contract.
- Work that is done is billed twice and payments are made twice.
- Payrolls include "ghost workers" (people who are dead, have long left the village, have never worked on the project, etc. appear on payrolls).
- The wages actually paid are considerably below what is stated in the records.

Strengths and Challenges of Social Auditing

Strengths	Challenges
■ Improves transparency of public works/services	■ Preventing elite capture
■ Exposes and reduces corruption and mismanagement	■ Lack of legal obligation for government to act on the findings
■ Improves the quality of public works/services	■ The process requires time, costs, and significant organization efforts
■ Strengthens the capacities of communities in participatory local planning	■ Possibility of manipulating stakeholder views

Other Participatory Performance Monitoring Mechanisms to be Considered

Citizen Report Cards. CRCs are participatory surveys that solicit user feedback on the performance of public services. CRCs can significantly enhance public accountability through the extensive media coverage and civil society advocacy that accompanies the process.

CRCs are used in situations where demand-side data, such as user perceptions on quality and satisfaction with public services, are absent. By systematically gathering and disseminating public feedback, CRCs serve as a "surrogate for competition" for state-owned monopolies that lack the incentive to be as responsive as private enterprises to their client's needs. They are a useful medium through which citizens can credibly and collectively "signal" to agencies about their performance and advocate for change. A prerequisite is the availability of local technical capacity to develop the questionnaires, conduct the surveys, and analyze results.

Strengths and Challenges of CRCs

Strengths	Challenges
■ CRCs can be used to assess either one public service or several services simultaneously.	■ CRCs require a well thought out dissemination strategy so that public agencies take note of citizen feedback and take the required action to correct weaknesses.
■ The feedback can be collected from a large population through careful sampling.	■ In locations where there is not much technical capacity, CRCs may be difficult to design and implement.
■ CRCs are quite technical and thus there may not be a need for a major citizen mobilization effort to get the process started.	■ If there is an error in sampling, the quality of service may not be reflected in the survey results.
■ Perceived improvements in service quality can be compared over time or across various public agencies involved in service provision.	

Community Score Cards. The CSC process is a community-based monitoring tool that is a hybrid of the techniques of social audits and CRCs. Like the CRC, the CSC process is used to demand social and public accountability and responsiveness from service providers. By linking service providers to the community, citizens are empowered to provide immediate feedback to service providers.

The CSC solicits user perceptions on quality, efficiency, and transparency, and uses the "community" as its unit of analysis.

It is focused on monitoring at the local/facility levels. It facilitates community monitoring and performance evaluation of services, projects, and even government administrative units (like district assemblies). Since it is a grassroots process, it is also more likely to be of use in a rural setting.

The Operational Manual for Implementing the Community Scorecard Process in the Maharashtra Rural Water Supply & Sanitation Project is a useful resource for organizing a CSC process.[3]

Strengths and Challenges of CSCs

Strengths	Challenges
■ Can be conducted for one public service or several services simultaneously.	■ CSCs rely on good-quality facilitators who may not always be available.
■ This is a community-level process that brings together service providers and users to discuss possible ways of improving service quality.	■ Reaching out to stakeholders before beginning the score card process is critical, but may not always be feasible.
■ Perceived improvements in service quality can be compared over time or across various public agencies involved in service provision.	■ In locations where there is not much local technical capacity, CSCs could be difficult to design and implement.
	■ CSCs cannot be easily applied to large geographical areas.

Annex 2 Endnote

1. This annex is adapted from World Bank, n.d., "Chapter 3, Methods and Tools" in "Social Accountability Sourcebook," http://www.worldbank.org/socialaccountability_sourcebook/.
2. See Centre for Good Governance, 2005, *Social Audit: A Toolkit. A Guide for Performance Improvement and Outcome Measurement* (Hyderabad: Centre for Good Governance) http://unpan1.un.org/intradoc/groups/public/documents/cgg/unpan023752.pdf.
3. World Bank, 2004, "Operational Manual for Implementing the Community Scorecard Process in the Maharashtra Rural Water Supply & Sanitation 'Jalswarajya' Project," http://www.sasanet.org/documents/Case%20Studies/Accountability%20in%20 Maharashtra%20RWSS.pdf *and* World Bank, 2007, "Case Study 4: Maharashtra, India: Improving Panchayat Service Delivery through Community Score Cards," http://www. sasanet.org/documents/Newreport/Maharashtra/Case4_Maharashtra_SAc_CSC_ August%202007.pdf.

19 MITIGATING THE RISK OF CORRUPTION

Guiding Principles for Mitigating the Risk of Corruption

- Good governance is more than preventing corruption; accountability for the effectiveness of the reconstruction effort should be the overriding goal.
- A clear reconstruction policy and strategy with which all agencies involved are familiar, accompanied by agreements to combat corruption and to implement transparent reporting systems, and rigorous monitoring are tools for ensuring accountability in reconstruction.
- The larger the cost and the faster the pace of reconstruction, the more vigilant all agencies need to be against corruption.
- Tracking financial inflows and outflows, while important, is not sufficient for providing transparency and accountability throughout the reconstruction process.
- Disaster-affected communities, corruption's ultimate victims, can play a key role in combating corruption if systems for social audit or other kinds of participatory performance monitoring are established.
- Measures to reduce corruption in post-disaster reconstruction can be successfully introduced even if the country's overall integrity system is weak.

This Chapter Is Especially Useful For:
- Policy makers
- Lead disaster agency
- Governance specialists
- Local government officials
- Project managers

Introduction

Governance refers to the manner in which public officials and institutions acquire and exercise the authority to shape public policy and provide goods and services. The mobilization and utilization of financial resources for the public good is an essential part of governance. In countries with "good governance," citizens respect the government because, among other reasons, those in authority manage public resources effectively. Where governance systems are not working effectively and transparency and accountability mechanisms are weak or lacking, corruption in the use of public resources often increases. One of the predictable outcomes under these circumstances is that poor people's needs are marginalized and development outcomes suffer.

During disaster recovery, citizens often perceive that public resources are not being managed well and that corruption is rampant. Corruption is the misuse of an entrusted position for private gain, by employing bribery, extortion, fraud, deception, collusion, or money laundering. Transparency International states that private gain should be interpreted broadly to include gains accruing to a person's family members, political party, or institutions in which the person has an interest.[1] The World Bank defines corruption in terms of corrupt, fraudulent, collusive, coercive, and obstructive practices.[2] These activities are criminal offenses in most countries, although the institutional capacity to prevent and sanction corruption may be insufficient, or may be overwhelmed by the disaster.

This chapter examines where corruption is found in recovery and housing and community reconstruction, particularly in public procurement, and discusses approaches to mitigate it.

Key Decisions

1. **Government** must decide on an approach to managing the risk of corruption in reconstruction, which may entail the designation of an agency to oversee the governance of the reconstruction program. The lead anticorruption agency may be an independent entity other than the lead disaster agency; however, the capacity to manage corruption should be a factor in the choice of institutional option to manage reconstruction. (See 🔲 Chapter 13, Institutional Options for Reconstruction Management, for a review of these options.)

2. **Government** (with the **lead anticorruption agency** or whichever institution or institutions will be providing leadership to manage corruption risk) should confirm that existing government anticorruption systems are adequate to be applied to the reconstruction program or establish the systems and sanctions that will be applied. The system should incorporate

1. Kenneth Kostyo, ed., 2006, *Handbook for Curbing Corruption in Public Procurement: Experiences from Indonesia, Malaysia and Pakistan* (Berlin: Transparency International), http://www.transparency.org/publications/publications/other/procurement_handbook.
2. 2. World Bank. 2006. "Guidelines: Procurement under IBRD Loans and IDA Credits" (Washington, DC: World Bank). http://go.worldbank.org/RPHUY0RFI0.

corruption risk analysis, promote preventive measures, include detection controls, and be adaptable to various kinds of organizations.

3. The **lead anticorruption agency** should collaborate with **donors** and **international financial institutions** (IFIs), as well as **local governments** and **agencies involved in reconstruction**, to decide on the legal framework and operational rules for procurement and for combating corruption throughout the reconstruction program and to identify any requirements for institutional strengthening.

4. The **lead anticorruption agency** should decide how to equip government agencies involved in reconstruction with systems to fight corruption in reconstruction and work to ensure they are implemented.

5. The **lead anticorruption agency** should work to ensure that all nongovernmental agencies involved in reconstruction are equipped with and are using systems to fight corruption in their activities.

6. **Agencies involved in reconstruction** should establish and publicize whistleblower or other mechanisms to ensure perceptions or instances of corruption in their projects are readily reported.

7. The **lead communications agency** should decide with **government** how to communicate to the public the measures in place to fight corruption in reconstruction and should encourage the public to report perceptions or instances of corruption.

Public Policies Related to the Mitigation of Corruption

The majority of the concepts and tools discussed in this chapter are best implemented well in advance of a disaster, through laws and policies, such as anticorruption laws, governance and anticorruption strategies, or integrity systems. National and international anticorruption organizations, including Transparency International, work with governments worldwide to promote legislative reform in this area, and can assist after a disaster. There are also a number of international conventions and agreements that address governance and anticorruption measures in public procurement that form the basis for international cooperation to improve transparency. To the extent international agencies are involved in reconstruction, these agreements serve as a framework for addressing corruption issues.[3]

3. Transparency International, "TI Anti-Corruption Handbook: National Integrity System in Practice: Introduction," http://www.transparency.org/policy_research/ach/introduction.
4. World Bank, "Financial Management," http://go.worldbank.org/0HI4LODL60

The legal and policy framework at the national level for public financial management (PFM) is an important instrument for mitigating corruption and providing transparency and accountability in reconstruction. Although core fiduciary principles apply to reconstruction financial management, planning, budgeting, and project implementation often use special arrangements in the early years of reconstruction. Even under special modalities, however, PFM that conforms as closely as possible to the legal framework is critical. The Public Expenditure and Financial Accountability (PEFA) process is an international framework that is used to assess whether PFM arrangements are adequate. A PEFA analysis conducted in advance of a disaster can be used to identify weaknesses in PFM and areas for improvement and to monitor the effectiveness of reforms.[4] In a post-disaster situation, measures to assess corruption risk and to prevent and detect corruption are likely to be less systematic and more situational. Assessments and measures that can be implemented immediately are discussed in this chapter.

Social accountability or participatory performance monitoring mechanisms such as social audits and complaint mechanisms are also useful for strengthening post-disaster governance and transparency. Countries with established integrity systems may have arrangements in place; others can establish them as part of the reconstruction strategy. See 📖 Chapter 18, Monitoring and Evaluation, for guidance on the use of participatory performance monitoring.

EARL KESSLER

Technical Issues
Who Is Responsible for Preventing Corruption in Reconstruction?

The number of places where corruption can take place is as long as the list of potential corrupt practices, which follows in the next section. This implies that "everyone and no one" is responsible for preventing corruption. This dispersed responsibility creates an enormous challenge for all agencies involved, including government. Nevertheless, the leadership of government to establish an anticorruption culture throughout the entire reconstruction process is essential. A number of efforts can be made to establish common anticorruption standards among all agencies, but a monitoring system is needed to ensure that they remain effective over time. Beside the measures described in this chapter that government itself can take, such as requiring integrity pacts and financial disclosure by government officials or use of audits, other collaborative measures could include the following.

- Government requires that any agency involved in reconstruction submit an anticorruption plan and report regularly on its implementation.
- All agencies involved in reconstruction require their private contractors to sign codes of conduct and their staffs to sign integrity pacts, with both subject to spot checks by government or outside auditors.
- An online system is established to share corruption schemes that are discovered or warning signs among agencies on a real-time basis.
- A common database of affected households is created among agencies, with unique identifiers to monitor the distribution of aid, including housing assistance.
- Common monitoring indicators are developed and reported for all projects, and data analysis is used to identify divergence from averages for costs of materials, administrative expenses, etc.
- Registration of aid workers and contractors is required, and information is shared among agencies to avoid rehiring staff or private firms involved in questionable practices.
- A common anticorruption monitoring board composed of agency and community representatives is established and/or a common pool of outside specialists is hired to standardize disbursement procedures and analyze samples of transactions.

Government may want to require that, along with implementing public sector codes of conduct and private sector integrity pacts, local and international agencies involved in reconstruction be asked to subscribe to a common set of standards regarding transparency and accountability.

Where Corruption Can Occur in Reconstruction

A disaster creates fertile conditions for corruption, waste, and mismanagement including (1) the large quantities of aid inflows and of goods being procured; (2) the pressure to spend quickly; (3) institutions that have different administrative procedures; (4) agencies that are unfamiliar with contracting large projects; (5) competition among aid agencies; (6) poor staff communication, screening, and/or training; (7) weak administration and oversight systems; and (8) the economic desperation of the affected population. A wide range of actors can perpetrate corruption, including government officials; aid agency staff and officials; citizens, including the affected population and their representatives; contractors; and vendors.

Post-disaster construction projects are especially prone to corruption because of their scale and complexity. There are also difficulties in specifying the work ex ante and nontransparent practices in the construction industry, and there may be limited government capacity to oversee numerous large-scale projects. Not all corruption, however, is related to procurement. For instance, deceptively attempting to qualify for post-disaster assistance is fraud. At the same time, not all appearances of corruption are, in fact, corruption.

Some examples of questionable activities that may—or may not—entail corruption are listed in the following table. The ways in which people will try to use the reconstruction process for their own benefit will be specific to the situation and even the culture. Government and agencies involved in reconstruction can keep this list of activities in mind in developing the reconstruction program so that reconstruction policy and assistance mechanisms are designed to discourage and detect them.

For access to additional resources and information on this topic, please visit the handbook Web site at www.housingreconstruction.org.

Questionable or Corrupt Practices in Reconstruction

Activity	Questionable or corrupt practices
Assessment	Overstating the extent of damage and needs by providing falsified data to assessors
	Damaging property to give the false impression that it is disaster-affected
	Homeowners or local officials influencing those conducting the assessment
	Assessor recommending projects in which he or she has a personal interest
Planning and pre-bidding	Unaffected population claiming eligibility for assistance
	Affected people claiming additional assistance (extra house) using false information.
	Reconstruction projects that are unnecessary, overdimensioned, or not based on the reconstruction procurement plan
	Inflated cost estimates, including for land purchases
	Information that is leaked to a private owner or buyer about land needed for a public project
	Projects that are approved without proper permits or designs Projects that are prepared for bidding without comment by the public or responsible local officials
	Projects specifications that are defined to limit the number of bidders
	Deviation from standard bidding documents
	Direct contracting of bids without proper justification
	Restricted advertising, insufficient notice, inadequate time for preparing bids
	Advance release of bid information to one bidder
	Bids being accepted after the submission deadline
Awarding and project implementation	Bid evaluation committee with conflicts of interest with bidders
	Amending evaluation criteria after receipt of bids
	Company presenting competing bids
	Government allowing bid evaluation report to be revised or reissued
	Government imposing subcontracting requirements on prime contractor
	Staff members involved in contract award becoming involved in contract supervision
	Contract variations and change orders being approved without proper verification
	Contractor's claim for costs beyond the common labor cost raise and inflation rates
	Materials and equipment used and workmanship not as specified; paperwork not consistent with items delivered
	Contractors providing false information to project inspectors on progress of work or inspectors being coerced to approve progress payments or certify conformance with building permits
	Inaccurate as-built drawings being presented or accepted
Monitoring	Staff responsible for oversight having conflicts of interest
	Control systems that are inadequate, unreliable, or inconsistently applied
	No follow-up to indications, suspicion, or accusations of corruption
	Lack of confidentiality on accusations of corruption
	Delayed or superficial audit; delayed publication of the audit report
	Failure to disqualify companies impugned in audit reports

5. Kenneth Kostyo, ed., 2006 *Handbook for Curbing Corruption in Public Procurement: Experiences from Indonesia, Malaysia and Pakistan* (Berlin: Transparency International), http://www.transparency.org/publications/publications/other/procurement_handbook.

6. United Nations (UN) Office on Drugs and Crime (ODC), 2004, "Tool #8, UN Model Code of Conduct for Public Servants" and "Tool #13, Disclosure of Assets and Liabilities by Public Officials," *The Global Programme Against Corruption: UN Anti-Corruption Toolkit*, 3rd ed. (Vienna: UNODC), http://www.unodc.org/documents/corruption/publications_toolkit_sep04.pdf.

Characteristics of Transparent Procurement Processes

The principal hallmarks of proficient public procurement are the economy, efficiency, fairness, transparency, accountability, and application of ethical standards.[5] Controls and sound, standardized procedures are the first line of defense against corruption in procurement. Transparency International promotes minimum standards for public contracting, including the following.

- A code of conduct is in force that commits the contracting authority and its employees to a strict anticorruption policy.[6]
- Only companies that enforce a strict anticorruption policy are allowed to tender proposals.
- A blacklist that bars companies from tendering proposals for a specified period of time is maintained by government.
- Public contracts above a low threshold are open to competitive bidding, with limited, clearly justified exceptions.

SAFER HOMES, STRONGER COMMUNITIES: A HANDBOOK FOR RECONSTRUCTING AFTER NATURAL DISASTERS

- All procurement information, including direct contracting or limited bidding processes, is made public; only legally protected information is kept confidential.
- No bidder is given access to privileged information related to the contracting or selection process.
- Bidders are allowed sufficient time to prepare their bids and to prequalify.
- Sufficient time is allowed to give an aggrieved competitor the opportunity to challenge the award.
- Contract change orders beyond a cumulative threshold (for example, 15 percent of contract value) are monitored at a high level, preferably by the body that awarded the contract.
- Control and auditing bodies are independent and functioning effectively; their reports are publicly accessible.
- Key functions of a project—demand assessment, preparation, selection, contracting, supervision, and control—are managed separately within government.
- Safeguards, such as the use of committees and staff rotation, are applied, and staff members responsible for procurement are adequately trained and remunerated.
- Civil society is allowed to participate as independent monitors of both the tender and execution of projects.

Assessing the Risk of Corruption

It may be necessary to conduct an assessment to evaluate whether the controls and procedures in place are adequate to prevent corruption in reconstruction procurement and which, if any, additional anticorruption measures need to be implemented. Two sources of information are the PEFA framework and corruption risk assessments.

Public expenditure and financial accountability framework. The PEFA framework identifies weaknesses in PFM, including procurement, and uses performance indicators to identify areas for reform and to monitor improvements.[7] (See 📖 Chapter 15, Mobilizing Financial Resources and Other Reconstruction Assistance.) The World Bank or other members of the PEFA partnership may have conducted a PEFA or similar analysis. If not, a rapid assessment of a country's systems may be necessary, with special emphasis on procurement capacity. When weaknesses are detected, international agencies can play an important role by providing funds for technical assistance, along with their reconstruction funds, for improving PFM during the reconstruction period.[8]

Even developed countries can have trouble controlling corruption in a post-disaster environment. The 📖 case study on Hurricanes Katrina and Rita below presents an example of where the systems used for post-disaster disbursements to households failed to prevent fraud.

Corruption risk assessments. Corruption risk assessment tools tend to be oriented toward evaluating systematic risks within the public sector. To date, there is no definitive post-disaster governance or corruption risk assessment methodology for individual development projects or development institutions. However, some useful resources are listed below.
- The World Bank's Governance and Anticorruption (GAC) Strategy and Implementation Plan. See 📖 Annex 1, How to Do It: Developing a Project Governance and Accountability Action Plan.
- The Committee of Sponsoring Organizations of the Treadway Commission (COSO) framework. See 📖 Annex 2, How to Do It: Conducting a Corruption Risk Assessment.
- The United Nations UN Anti-Corruption Toolkit, especially "Tool #2: Assessment of Institutional Capabilities and Responses to Corruption"[9]
- The Corruption Risk Assessment Questions table developed by Management Accounting for NGOs (MANGO) for Transparency International and the U4 Anti-Corruption Resource Centre[10]
- A list of risk assessment tools compiled by the U4 Anti-Corruption Resource Centre on its Web site[11]

Tools to Promote the Integrity of Public Officials

Two anticorruption tools that may be implementable as part of the reconstruction program, even in the absence of a broader public sector integrity system, are the code of conduct and the disclosure of assets. These measures should be accompanied by anticorruption training for all public officials involved in the reconstruction program.

Codes of conduct. Codes of conduct for public officials[12] can be used to establish general standards of behavior consistent with principles of integrity, transparency, accountability, and responsible use of organizational resources. They may also address standards applicable to specific groups of employees, such as those involved in the reconstruction program. The code should define

7. World Bank, "Financial Management," http://go.worldbank.org/0HI4LODL60.
8. The key benchmarks that PFM systems can significantly influence are (1) credibility of information, (2) timeliness and equitability of implementation, and (3) control of corruption. For information on performance measurement indicators for post-disaster PFM, see PEFA Web site, http://www.pefa.org/pfm_performance_frameworkmn.php.
9. UNODC, 2004, "Tool #2: Assessment of Institutional Capabilities and Responses to Corruption," The Global Programme Against Corruption: UN Anti-Corruption Toolkit, 3rd ed. (Vienna: UNODC), http://www.unodc.org/documents/corruption/publications_toolkit_sep04.pdf.
10. Pete Ewins et al., 2006, Mapping the Risks of Corruption in Humanitarian Action (London: Overseas Development Institute and MANGO).
11. U4 Anti-Corruption Resource Centre, http://www.u4.no/helpdesk/helpdesk/queries/query85.cfm#1.
12. UNODC, 2004, "Tool #8, UN Model Code of Conduct for Public Servants," The Global Programme Against Corruption: UN Anti-Corruption Toolkit, 3rd ed. (Vienna: UNODC), http://www.unodc.org/documents/corruption/publications_toolkit_sep04.pdf.

procedures and sanctions to be applied in cases of noncompliance. Administration of the code should be done by an independent individual or body and should be readily accessible so that a public employee can enquire whether an activity would be in breach of the rules before engaging in it. Standards may include positive obligations, such as the requirement to disclose conflicts of interest, and prohibitions, such as those against disclosure of certain information or acceptance of gifts. Public sector codes of conduct usually apply not only to conduct inconsistent with the office but also to conduct that might give the perception of impropriety or damage the credibility of that office.

Declaration of assets and income. The declaration of assets and income by public officials is a tool to deter illicit enrichment from bribery, kickbacks, etc.[13] It helps ensure that unlawful behavior is monitored, quickly identified, and dealt with. The disclosure of financial information by public servants raises privacy concerns, so it may not entail full public disclosure, except in cases where improper conduct is discovered or proven, but rather disclosure to specially established bodies that are trusted and empowered to take action if wrongdoing is suspected, such as inspectors or auditors general. It is generally not necessary or practicable to subject every public employee to a disclosure process, but instead to apply the policy to officials at or above a certain seniority or those in positions with a high risk of corruption, such as reconstruction procurement officials. While it is common for public officials subject to declarations of assets to report annually, the accelerated nature of reconstruction procurement may require more frequent reporting.

Integrity Pacts Promote Transparency with the Private Sector

The Integrity Pact (IP) is promoted by Transparency International as a useful tool for fighting corruption in public contracting.[14] It consists of an agreement between government and bidders involved in public procurement and contracting that neither side will pay, offer, demand, or accept bribes. Nor will they collude with competitors in obtaining or carrying out the contract. It requires bidders to disclose all expenses paid in connection with the contract and to agree to be sanctioned if there are violations. Sanctions can include loss of the contract, forfeiture of the company's performance bond, damage liability, blacklisting, and criminal or disciplinary action against government employees.

IPs cover all phases of a project, from planning to operation, and can be used for any kind of reconstruction contract. IPs enable companies to abstain from bribing by assuring them that their competitors will do the same, and that government and its officials will take the necessary precautions to prevent corruption. IPs reduce the costs of corruption in public procurement, strengthen trust in the public sector and its procurement activities, and improve the overall investment climate.

In addition, IPs are flexible and adaptable to many legal settings, with conflict resolution and sanction imposition generally handled through arbitration mechanisms rather than the judicial system. Independent monitoring of the pacts is required and can be carried out by a civil society organization (CSO) or other independent and accountable entity. Although IPs should be mandatory in reconstruction, not all governments require them.

Integrity Pacts With and Among Nongovernmental and International Organizations

The humanitarian sector has been concerned with the integrity of its activities for many years. Government may want to require that, along with public sector codes of conduct and private sector IPs, local and international agencies involved in reconstruction be asked to subscribe to a common set of standards regarding transparency and accountability. These rules could cover, among other topics, agency procurement, codes of conduct and disclosure of assets by agency staff, and communications with the affected community and the public regarding their activities. Government agreements with international and bilateral agencies may provide the basis for these agreements. One example of integrity guidelines for the nongovernmental sector is the "Code of Ethics & Conduct for NGOs"[15] promoted by the World Association of Non-Governmental Organizations (WANGO). The Web site of One World Trust provides extensive resources on nongovernmental organization (NGO) accountability initiatives.[16] See also 📖 Chapter 14, International, National, and Local Partnerships in Reconstruction, for a discussion of registration procedures for NGOs.

Audits Improve Project Transparency

Audits work primarily through transparency. They make corruption riskier and more difficult by determining and exposing whether project funds were handled in accordance with laws, regulations, contracts (such as loan contracts), and accounting rules. They also examine the efficiency (measured against accepted financial procedures and practices) and the effectiveness (compared to the agreed-

13. UNODC, 2004, "Tool #13, Disclosure of Assets and Liabilities by Public Officials," *The Global Programme Against Corruption: UN Anti-Corruption Toolkit*, 3rd ed. (Vienna: UNODC), http://www.unodc.org/documents/corruption/publications_toolkit_sep04.pdf.
14. Transparency International, 2009, *Global Priorities: Public Contracting*, http://www.transparency.org/global_priorities/public_contracting/integrity_pacts.
15. WANGO, "Code of Ethics & Conduct for NGOs," http://www.wango.org/codeofethics/ComplianceManual.pdf.
16. One World Trust, "NGO Initiatives," http://www.oneworldtrust.org/index.php?option=com_content&view=article&id=87&Itemid=84#f.

upon purposes) of the use of project funds. See 📖 Chapter 18, Monitoring and Evaluation, for a comparison of monitoring, evaluation, and auditing.

Some auditors can act on their own findings, but they are usually restricted to investigation, reporting, making recommendations, and referring findings to another body for action. Auditors generally report to a body inside the organization, but outside of management, such as a board of directors or the legislature. Auditing should generally be carried out by an entity independent from the organization under audit, based on standards that are defined before the audit begins.[17] A large measure of an auditor's power resides in the fact that audit reports are generally made public, especially in the public sector. Even entities in possession of confidential information, such as national security matters or sensitive commercial information, should not be exempt from being audited.

The generic categories for audits are "financial audits" and "performance audits." Audits may have a combination of financial and performance audit objectives or may have objectives limited to only some specific aspect of one audit type.

Financial audits

Financial audits are conducted in the private sector and the public sector for similar purposes, but generally using somewhat different standards. Financial audits focus on the use of funds and the resulting financial performance.

Financial statement audits	Provide reasonable assurance about whether the financial statements of an audited entity present fairly the financial position, results of operations, and cash flows in conformity with generally accepted accounting principles. They include audits of financial statements prepared in conformity with an established basis of accounting.
Financial-related audits	Determine whether (1) financial information is presented in accordance with established or stated criteria, (2) the entity has adhered to specific financial compliance requirements, or (3) the entity's internal control structure over financial reporting and/or safeguarding assets is suitably designed and implemented to achieve the control objectives.

Performance audits

Performance audits (also called operational audits) provide an independent assessment of the performance of a government organization, program, activity, or function in order to provide information to improve public accountability, facilitate decision making, or initiate corrective action.

Economy and efficiency audits	Used to determine (1) whether an entity is acquiring, protecting, and using its resources (such as personnel, property, and space) economically and efficiently; (2) the causes of inefficiencies or uneconomical practices; and (3) whether the entity has complied with laws and regulations on matters of economy and efficiency.
Program audits	Used to determine (1) the extent to which the desired results or benefits established by the legislature or other authorizing body are being achieved; (2) the effectiveness of organizations, programs, activities, or functions; and (3) whether the entity has complied with significant laws and regulations applicable to the program.

Options for Conducting Audits

Audits may be conducted at different times in the project or budget cycle, or carried out by different agencies, including the public or the affected community. Below are some of the options.

Pre-audit/ post-audits	Audits can be carried out before and/or after the activity itself takes place. A forensic audit is a form of post-audit in which evidence is gathered specifically for investigation and prosecution of criminal acts.
Concurrent or simultaneous audits	Concurrent or simultaneous audits are a type of ex post audit that avoids the delays inherent in pre-audits, while drastically reducing the time between the activity and the post-audit. See 📖 Annex 3, How to Do It: Conducting a Construction Audit, a methodology that can be used for a post-audit or a simultaneous audit of a construction project.
Internal/ external audits	Audits may be carried out by specialized internal units of government, an independent government institution, or private accounting or auditing professionals. Depending on the country, and the type of audit, these professional may be called accountants, auditors, internal auditors, management accountants, certified fraud examiners, or certified public accountants.
Social audits	Social audits are arrangements whereby the public and the affected community oversee and report on an organization's activities or a reconstruction project. For details on conducting a social audit and a summary of other "participatory performance monitoring" mechanisms, see 📖 Chapter 18, Monitoring and Evaluation, Annex 1, How to Do It: Conducting a Social Audit of a Reconstruction Project.

17. UNODC, 2004, *The Global Programme Against Corruption: UN Anti-Corruption Toolkit*, 3rd ed. (Vienna: UNODC), http://www.unodc.org/documents/corruption/publications_toolkit_sep04.pdf.

Special audit entities. The volume and speed of disaster procurement or questions about the auditing capacity of government may make a special audit entity necessary. This may be an operational unit auditing concurrently or a higher-level body that oversees the budgeting, procurement, and auditing processes. If government procedures already contemplate such a mechanism, it should be mobilized. If not, procedures should be established so it can be created and staffed with either private auditors or trained auditors from within the public sector. The UN suggests that such an entity be composed of a combination of national and international experts.[18] The design and staffing process must ensure the entity's independence, the avoidance of conflicts of interest by those working in the entity, and the transparency of the entity's operations. A national public accountants association may be able to advise on design and start-up. The 🖳 case study on Malaysia. below, shows how government deployed the national Practices, Systems and Procedure Examination Unit after the 2004 Indian Ocean tsunami to analyze PFM procedures in the agencies that would be handling reconstruction funds to determine whether the measures already in place were adequate.

World Bank audits. The World Bank regularly conducts audits to review the procurement, contracting, and implementation processes in Bank-financed projects. These audits verify whether procurement and contracting were carried out according to the loan agreement and whether the expected economy and efficiency were achieved, evaluate the Bank's oversight of the project, and identify ways to improve procurement and contracting.[19]

Complaint Mechanisms

Complaint mechanisms allow corruption to be reported by social actors, including public employees, in a confidential manner. (Grievance processes related to housing assistance are a special use of complaint mechanisms. Grievance redressal is discussed in 🖳 Chapter 15, Mobilizing Financial Resources and Other Reconstruction Assistance.) Ideally, complaint mechanisms are formalized in a larger "integrity system," but they can also be employed on a situational basis during post-disaster recovery. Just a few of the potential instruments available are described below.

Whistleblower laws. In establishing laws or other legal instruments to protect whistleblowers, a balance should be sought between protection of the whistleblower and accountability of the whistleblower, to minimize fraudulent complaints.

Telephone hotlines. A hotline should be introduced as part of the larger anticorruption strategy and be well publicized. This publicity can be incorporated into the communication strategy of the reconstruction program. See 🖳 Chapter 3, Communication in Post-Disaster Reconstruction. A hotline must be staffed by trained operators and have a secure phone line. The information gathered in these conversations must be collected systematically and treated with confidentiality.

Civil society monitoring. CSOs can provide advice and counsel to whistleblowers or can conduct social audits. The same rules of confidentially and accountability apply. See 🖳 Chapter 14, International, National, and Local Partnerships in Reconstruction, for information on how government can work effectively with these institutions.

Ombudsmen. Ombudsmen receive and consider a wide range of complaints that fall outside the jurisdiction of courts or administrative bodies. Their specific roles depend on whether other similar official bodies exist and how effective they are. Ombudsmen require a clear and relatively broad mandate, independence, public accessibility, transparency, integrity, and sufficient resources to carry out their duties. They can provide the following services:

- Investigate relatively minor complaints while avoiding expensive legal proceedings
- Provide remedies in certain cases
- Serve as a clearinghouse, referring complaints to a more appropriate forum for further action
- Educate government staff about standards of conduct, raise questions about the appropriateness of established codes or service standards, and recommend adjustments
- Raise awareness about the public's rights to information and the level of efficient and honest public services they should expect
- Conduct proactive research regarding complaints and complaint patterns

Redundancy of complaint mechanisms. Whistleblowers should always have at least two complaint mechanisms available to them: the first, an entity within the "offending" organization,

> *In establishing laws or other legal instruments to protect whistleblowers, a balance should be sought between protection of the whistleblower and accountability of the whistleblower, to minimize fraudulent complaints.*

18. World Bank, 2001, Procurement Policy and Services Group, Operations Policy and Country Services VPU, "Bank-Financed Procurement Manual," http://siteresources.worldbank.org/PROCUREMENT/Resources/pm7-3-01.pdf.

19. UNODC, 2004, "Tool #14: Authority to Monitor Public Sector Contracts," *The Global Programme Against Corruption: UN Anti-Corruption Toolkit,* 3rd ed. (Vienna: UNODC), http://www.unodc.org/documents/corruption/publications_toolkit_sep04.pdf.

such as a supervisor or internal oversight body, and the second, to provide backup if the first body fails to investigate, complete the investigation, take appropriate action, or report back in a timely fashion. Within the public sector, the first may be the general auditor, for example. The second mechanism also provides the whistleblower protection against retribution or a cover-up.

Risks and Challenges

- Bypassing sound procurement practices to speed up reconstruction.
- Slowing procurement processes to eliminate all possibility of corruption.
- Executing agencies with no knowledge of prices and other local market characteristics.
- Making integrity pacts optional for private firms bidding for and participating in post-disaster reconstruction.
- Not following through on the threat to sanction violators of anticorruption measures.
- Hiring staff members whom other organizations fired for corrupt practices.
- Failing to implement control measures and train government and other executing agency employees in preventing and reporting corrupt practices.
- Not protecting the confidentiality of whistleblowers reporting corruption.

Recommendations

1. Take a proactive approach to minimize corruption in the reconstruction program by assessing corruption risks early in reconstruction planning.
2. Recognize that corruption can be perpetrated by any stakeholder in the reconstruction program if given the opportunity. Be creative and proactive in identifying opportunities for corruption.
3. Use the communications strategy for the reconstruction program to alert the public regarding their role in mitigating corruption and mechanisms available to report wrongdoing.
4. Some anticorruption mechanisms can be implemented on an ad hoc basis, even if a comprehensive integrity system is not in place. However, advocate for a systematic approach to establish credibility with affected communities and the public.
5. Don't assume public officials know what corruption is, or that they shouldn't do it. Consider implementing—at a minimum—codes of conduct and asset disclosure procedures for staff involved in reconstruction procurement.
6. Use auditing as an anticorruption mechanism that can be tailored to the requirements of specific situations.
7. Look for substantive ways to involve social actors in the anticorruption effort.
8. Establish systems that ensure confidentiality for whistleblowers.
9. Funding sources should work to establish common transparency standards. They should require, whether individually or collectively, that the use of their funds be widely disclosed to the public.

Case Studies
2005 Hurricanes Katrina and Rita, Gulf Coast, United States
Extensive Fraud in Post-Katrina Audit

After Hurricanes Katrina and Rita devastated the U.S. Gulf Coast in 2005, the Federal Emergency Management Agency (FEMA) began a process for registering the people affected by the storms and providing them with "expedited assistance" (EA) payments. Using both Internet and telephone registration systems, FEMA registered 2.5 million households in the three months following the disaster. By December 2005, FEMA had disbursed US$2.3 billion (officially, US$2,000 per household). Those registered for EA were also potentially eligible for further assistance of up to US$26,200.

In December 2005, the General Accountability Office (GAO), the investigative arm of the United States Congress, began an audit of the process. GAO identified significant flaws in procedures for preventing, detecting, and deterring fraud, including limited controls to verify the identity and residence of those registering. Registrants using bogus social security numbers and property addresses were able to register, some multiple times, and were not screened out of the registration lists. FEMA's lack of controls also meant that many legitimately registered recipients erroneously received multiple payments. FEMA later estimated that as many as 900,000 of the 2.5 million people registered were duplicates. Using data-mining techniques, GAO estimated in 2006 that as much as US$1.5 billion of FEMA's EA payments were fraudulent.

Source: U.S. GAO, 2006, Hurricanes Katrina and Rita Disaster Relief: Improper and Potentially Fraudulent Individual Assistance Payments Estimated to be between $600 Million and $1.4 Billion (Washington, DC: GAO), http://www.gao.gov/new.items/d06844t.pdf.

DANIEL PITTET

2004 Indian Ocean Tsunami, Malaysia

Preventing Corruption through Existing Systems

On December 26, 2004, when the Indian Ocean tsunami struck the states of Penang, Perlis, Kedah, and Perak in Malaysia, a solid framework for preventing corruption was already in place. In 1961, the Malaysian government had established an independent Anti-Corruption Agency (ACA) to enforce the Prevention of Corruption Act. The ACA now has branches in each of Malaysia's 14 states and sub-branches throughout the country. In 1998, Integrity Management Committees (IMCs) were established in all federal and state agencies.

When the National Disaster Aid Fund was set up in the aftermath of the tsunami to manage the RM 90 million (US$24 million) for disaster relief, ACA Penang took action to head off the corruption threat. The national Practices, Systems and Procedure Examination Unit, deployed to analyze procedures in the disbursing and executing agencies, determined that the measures already in place were adequate. The assistance process for the population affected by the tsunami began with a police report detailing each affected person's loss and property damage. Three separate state committees, each with elected and local community representatives, then reviewed these reports, as did other government entities, before they were sent to the National Disaster Aid Fund Management Committee for approval. Other anticorruption measures included announcing assistance amounts for affected populations in the media, publicly displaying information on the assistance at the time of disbursement, and requiring that the government official and the recipient sign a form that warned of consequences of false claims and false information. Fewer than 15 complaints were received from the four affected states.

Source: Abu Kassim Bin Mohamad, 2005, "Effective Anti-Corruption Enforcement and Complaint-Handling Mechanisms: The Malaysian Experience," in "Curbing Corruption in Tsunami Relief Operations" (proceedings of the Jakarta Expert Meeting, Jakarta, April 7–8), http://www.u4.no/document/literature/adb-ti-2005-curbing-corruption-tsunami-relief-operations.pdf.

Resources

Asian Development Bank. 2005. "Expert Meeting on Corruption Prevention in Tsunami Relief." http://www.adb.org/documents/events/2005/Tsunami-Relief/default.asp#purpose.

Centre for Good Governance. 2005. *Social Audit: A Toolkit. A Guide for Performance Improvement and Outcome Measurement.* Hyderabad: Centre for Good Governance. http://unpan1.un.org/intradoc/groups/public/documents/cgg/unpan023752.pdf.

Kaufmann, Daniel, Aart Kraay, and Massimo Mastruzzi. 2009. "Governance Matters VIII: Aggregate and Individual Governance Indicators, 1996–2008." Washington, DC: World Bank.

Kostyo, Kenneth, ed. 2006. *Handbook for Curbing Corruption in Public Procurement.* Berlin: Transparency International. http://www.transparency.org/publications/publications/other/procurement_handbook.

Stansbury, Catherine, and Neill Stansbury. 2008. *Anti-Corruption Training Manual (Infrastructure, Construction and Engineering Sectors)*, international version. Transparency International. http://www.transparency.org/global_priorities/public_contracting/projects_public_contracting/preventing_corruption_in_construction.

Transparency International. 2009. "Contracting: Preventing Corruption on Construction Projects." *Tools.* http://www.transparency.org/tools/contracting/construction_projects.

Transparency International. 2009. "Public Contracting: The Integrity Pacts." *Global Priorities.* http://www.transparency.org/global_priorities/public_contracting/integrity_pacts.

UNODC. 2004. *The Global Programme Against Corruption: UN Anti-Corruption Toolkit,* 3rd ed. (Vienna: UNODC). http://www.unodc.org/documents/corruption/publications_toolkit_sep04.pdf.

World Bank. Governance and Anticorruption Web site and related resources. http://www.worldbank.org/wbi/governance.

World Bank. 2006. "Guidelines: Procurement under IBRD Loans and IDA Credits." Washington, DC: World Bank. http://go.worldbank.org/RPHUY0RFI0.

World Bank. 2009. "Governance Matters, 2009: Worldwide Governance Indicators, 1996–2008." http://info.worldbank.org/governance/wgi/index.asp.

World Bank. n.d. "Chapter 3, Methods and Tools" in "Social Accountability Sourcebook." http://www.worldbank.org/socialaccountability_sourcebook/.

How to Do It: Developing a Project Governance and Accountability Action Plan

Strengthening World Bank Group Engagement on Governance and Anticorruption[1]

In 2007, the Board of Directors of the World Bank Group (WBG) approved the Governance and Anticorruption (GAC) strategy and Implementation Plan, whose purpose is to help countries improve their systems of governance in a manner that is effective and sustainable over the long term. The strategy involves working at the country, operational, and global levels.

The behavior of government and other key stakeholders, such as the private and financial sectors, shapes the quality of governance and affects development outcomes. Therefore, the WBG's GAC work aims to help government develop the capability to devise and implement sound policies, provide public services, set the rules governing markets, and combat corruption, thereby helping reduce poverty.

At the project and sector levels, the Bank can help government:
- strengthen country systems, such as financial management;
- incorporate good governance and anticorruption objectives into sectoral programs;
- identify high-risk operations and prepare a Project Governance and Accountability Action Plan (GAAP);
- improve the quality of project design, supervision, and evaluation, and enhance third-party monitoring of projects;
- use fiduciary quality reviews of the Bank's financing portfolio;
- target resources to improve the supervision of high-risk projects; and
- establish teams to review project designs, risk ratings, and anticorruption action plans.

Developing a Project Governance and Accountability Action Plan[2]

An effective anticorruption program can be developed by selecting the elements appropriate for a specific project situation, and integrating them in a manner designed to have the maximum impact. Below are the steps to develop a GAAP.[3]

Step 1: Understand and prioritize corruption risks by corruption mapping and analyzing incentives and disincentives.
Outputs
- Summary prioritizing high-risk areas requiring careful review of incentives, disincentives, and remedies
- Action plan elements to mitigate the chances of corruption identified above
- Corruption mapping matrix, including analysis of incentives and disincentives

Step 2: Empower recipients and communities through smart project designs, involvement of recipients in the procurement process, and construction of simple works through communities.
Outputs
- List of project design features and mechanisms that will empower recipients and communities
- Mechanisms for recipient involvement in the procurement process
- Plan for citizen involvement
- Plan for construction of simple works by communities

Step 3: Build partnership for civil society oversight and feedback by initiating consultation with representatives of civil society.
Outputs
- Agreement on the watchdog role and the mechanism of oversight by civil society
- Agreement on the disclosure provisions to be included in the legal documents of the project
- Media strategy, including independent monitoring by the media
- Credible system to handle complaints under the project
- Plan for corruption surveys to get independent feedback
- Plan for periodic feedback from the private sector, including firms that are participating and those that are not

Step 4: Establish proven procurement policies to mitigate collusion.
Outputs
- Formal incorporation of policies against collusion in the project by their inclusion in legal documents, operational manuals, minutes of negotiations, materials of project launch, and plans of dissemination to all stakeholders

Step 5: Build strong task teams with the means of paying increased attention to fiduciary risks.[4]
Outputs
- Assurance that the task team has adequate capacity to deal with fiduciary issues by including procurement and financial management specialists in the task team
- Assurance that the borrower has made acceptable arrangements to put in place adequate capacity for project management, including capacity and systems for procurement, financial management, maintenance of records, and contract management
- Supervision plan to ensure compliance with agreed procurement and financial management procedures, progress on the ground, and handling of complaints
- Plan for assessing "value for money" by examination of the results of the procurement process through asset verification and comparison of prices obtained

Step 6: Clearly define remedies to ensure compliance with corruption prevention measures and remedies to deal with cases of fraud or corruption.

Outputs

- Definition of the use of remedies for suspension and cancellation, including actions required to avoid application of these remedies
- Definition of actions required from government, including credible sanctions against firms and individuals against whom evidence of fraud or corruption has been found, including government officials
- Inclusion of provisions for national competitive bidding in the loan agreement to facilitate declaration of misprocurement in project
- Definition of remedies to ensure implementation of agreed-to disclosure provisions
- Definition of remedies to reduce procurement delays
- Definition of remedies for deviations found in the fiduciary audit

Step 7: Consult stakeholders to finalize the GAAP.

Outputs

- Draft and final GAAP

Availability of Technical Assistance

World Bank project teams can assist governments in accessing the technical resources available under the GAC implementation effort. Extensive information on this initiative is available at the Bank Web site, "Governance and Anti-Corruption," http://go.worldbank.org/CI3TOJK4I0.

Annex 1 Endnotes

1. World Bank, Operations Policy and Country Services, 2007, "Implementation Plan for Strengthening World Bank Group Engagement on Governance and Anticorruption," http://siteresources.worldbank.org/PUBLICSECTORANDGOVERNANCE/Resources/GACIP.pdf.
2. World Bank, n.d., "Preparing Your Project Governance and Accountability Action Plan, Reducing Fiduciary Risk through Increased Transparency and Accountability: A Guidance Note for New Projects in India."
3. For alternative approaches for organizing and presenting GAC project strategies, see World Bank, 2007, "Project Appraisal Document, La Guajira Water and Sanitation Infrastructure and Service Management Project," pp. 136–142, http://www-wds.worldbank.org/external/default/WDSContentServer/WDSP/IB/2007/02/27/0000 20439_20070227093741/Rendered/PDF/38508.pdf and World Bank, 2005,"Project Appraisal Document, Multi Donor Trust Fund For Aceh And North Sumatra For A Community Recovery through the Kecamatan Development Project," http://www-wds.worldbank.org/external/default/WDSContentServer/WDSP/IB/2005/09/27/000012 009_20050927123517/Rendered/PDF/332030rev.pdf.
4. See World Bank, n.d., "Preparing Your Project Governance and Accountability Action Plan, Reducing Fiduciary Risk through Increased Transparency and Accountability: A Guidance Note for New Projects in India, Appendix 6: Fiduciary Checklist."

Objectives of the Corruption Risk Assessment

Agencies involved in reconstruction or government itself may need to (1) assess whether an organization being considered as an executor for a post-disaster reconstruction project has the capacity to execute the project without an unacceptable risk of fraud or corruption or (2) identify specific weaknesses in the organization's project management that would have to be addressed before or during project execution.

Methodology for Preparing a Corruption Risk Assessment

Much of the international technical advisory work done in recent years to analyze and propose interventions for reducing corruption in the public sector has taken a system-wide (or occasionally sector-wide) look at corruption and at the institutional reforms needed to mitigate it. The Public Expenditure and Financial Accountability (PEFA) framework[1] and much of the work on integrity systems by Transparency International[2] are good examples of work that takes this type of approach.

At a more micro level, a number of very targeted anticorruption tools have been developed for use in organizations (e.g., integrity pacts and policies to require disclosure of assets and liabilities by public officials), as discussed in this chapter. Indicators called "corruption warning signs" are also very useful for making an initial assessment of organizational corruption risk.[3] However, a corruption risk assessment should analyze and attempt to predict whether an individual organization will be prone to corruption and fraud in a systematic manner and to identify the specific weak points in its management that should be strengthened. Some possible options for carrying out a corruption risk assessment of an individual organization are shown in the following table.

Options for Assessing Corruption Risk

Option	Explanation	Resources
Modify a system-wide diagnostic tool to analyze corruption risks in an individual institution.	No example was found of this being done, but some useful starting points are listed at right.	The Public Financial Management Performance Measurement Framework[4] is used to analyze the entire public financial management system, but is relevant to a single organization Transparency International, Corruption Fighters Tool Kit[5] covers many topics relevant to assessing and/or monitoring organizational corruption
Assume there is a corruption risk and take a proactive preventive stance.	A concurrent audit provides the full-time presence of an auditing team in the organization to monitor procedures and identify evidence of corruption at the institutional or project level.	Annex 3, How to Do It: Conducting a Construction Audit, can be used as a tool to conduct a concurrent audit of an individual construction project. Transparency International's Project Anti-Corruption System (PACS) is an integrated system to assist in the prevention of corruption on construction projects.[6]
Hire a consultant or auditor to conduct an ex ante audit or to evaluate internal control systems.	Methodologies such as that developed by the Committee of Sponsoring Organizations of the Treadway Commission (COSO) apply a structured framework for evaluating the state of internal control.	See the framework proposed by the Committee of Sponsoring Organizations of the Treadway Commission (COSO).[7] See guidelines and standards offered by the the International Federation of Accountants (IFAC), the International Public Sector Accounting Standards Board (IPSASB), and the International Auditing and Assurance Standards Board (IAASB).[8]

Scope of the Corruption Risk Assessment

The focus of the corruption risk assessment of an organization should not be simply on financial management practices, but on the broader framework called "internal control." In promoting the improvement of internal control, the COSO framework defines internal control broadly as activities that create, protect, and enhance "stakeholder value" by managing the uncertainties that could influence the achievement of an organization's objectives. Internal controls are effected across the organization by its board of directors (or legislature or city council), management, and other personnel.

The internal control process should provide reasonable assurance to stakeholders that the organization is meeting or is capable of meeting the three objectives shown below (reinterpreted for the public sector environment). The corruption risk assessment should evaluate whether the organization under consideration has measures in place to ensure accomplishment of these objectives.

Objectives of an Internal Control System

Objective	Indicators	Basis for evaluation
1. Effectiveness and efficiency of operations	Accomplishment of an organization's basic business objectives, including performance and financial goals and safeguarding of resources	■ Annual plan for organization ■ Defined role of organization in the reconstruction plan as defined in reconstruction policy ■ Performance indicators of organization ■ Indicators defined in logical framework matrix for specific reconstruction project (see 📖 Chapter 18, Monitoring and Evaluation)
2. Reliability of financial reporting	Preparation of reliable published financial statements and other financial information, and public disclosure of selected financial indicators	■ Quality of budget execution reports for organization ■ Policies and capability of producing project financial statements for a specific reconstruction project ■ Policies and procedures for public disclosures
3. Compliance with applicable procedures, laws, and regulations	Compliance with laws and regulations to which the entity is subject	■ Legal framework of organization ■ Presence of and compliance with documented procedures ■ Capacity to manage compliance with project procedures (e.g., eligibility rules, construction standards) ■ Practice and quality of internal/external audit of organization

Evaluating the internal control framework. A comprehensive control framework requires five components according to COSO. Assessment of these characteristics of an organization will allow a judgment to be made about the presence of corruption risks and the identification of specific weaknesses:

1. Establishment and maintenance of a sound control environment (corporate culture)
2. Regular, ongoing assessment of risk
3. Design, implementation, and maintenance of control-related policies and procedures to compensate for identified risks
4. Adequate communication
5. Regular, ongoing monitoring of control-related policies and procedures to ensure that they continue to function as designed and to ensure that identified problems are handled appropriately

Identifying evidence of deficiencies in internal control and of fraud.[9] The scope of work for the corruption risk assessment should require that the following control deficiencies and indicators of fraud risk be looked for and analyzed. The depth of this analysis will vary depending on the size and complexity of the organization being assessed. Unless the assessment is intended to produce only a "go/no-go" decision on the use of a particular organization, the consultant's scope of work should require that the consultant describe the specific measures that need to be taken to correct the deficiencies identified.

Indicators of Deficiencies in Internal Control

■ Insufficient control consciousness within the organization, for example, the tone "at the top" and the control environment. Control deficiencies in other components of internal control could lead the auditor to conclude that weaknesses exist in the control environment.

■ Ineffective oversight by those charged with governance of the organization's financial reporting, performance reporting, or internal control, or an ineffective overall governance structure.

■ Control systems that did not prevent or detect material misstatements of financial results identified by prior audits.

■ An ineffective or nonexistent internal audit or risk assessment function.

■ Identification of fraud of any magnitude on the part of senior management.

■ Inadequate controls for the safeguarding of assets.

■ Evidence of intentional override of internal control by those in authority to the detriment of the overall objectives of the system.

■ Inadequate design of information systems in general and application controls that prevent the information system from providing complete and accurate information.

■ Employees or management who lack the qualifications and training to fulfill their assigned functions.

Indicators of Fraud Risk

- Threats to an organization's financial stability, viability, or budget due to economic, programmatic, or entity operating conditions.

- Opportunities to engage in fraud due to an organization's cultural or operational environment.

- Inadequate monitoring by management for compliance with policies, laws, and regulations.

- The stability and complexity of an organization's structure.

- Lack of communication and/or support for ethical standards by management.

- A history of impropriety, such as previous issues with fraud, waste, abuse, or questionable practices, or past audits or investigations with findings of questionable or criminal activity.

- Operating policies and procedures that have not been developed or are outdated.

- Lacking or nonexistent key documentation.

- Improper payments.

- False or misleading information.

- Unusual patterns and trends in contracting, procurement, acquisition, and other activities of an organization or program under audit.

Qualifications of Consultants

Assessments of internal control are generally conducted by auditors with experience in the subject matter. Depending on the country, these specialists may be called one or more of the following: accountants, auditors, internal auditors, management accountants, certified fraud examiners, or certified public accountants. It is important to establish that the consultant has experience in the type of assessment being requested. The accounting profession is increasingly oriented toward the evaluation of internal controls, so this expertise should exist in most countries. However, there may be less experience in applying this expertise to nongovernmental organizations or the public sector. While COSO was developed in the United States, there is international experience with the framework. Another auditing framework that provides similar guidance with which auditors may be familiar is that of the IAASB.[10]

The type of consultancy described in this annex is referred to in the accounting profession as an "attestation engagement" rather than an audit. According to the United States General Accounting Office, an attestation engagement can cover a broad range of financial or nonfinancial objectives and may provide different levels of assurance about the subject matter or assertion depending on the users' needs. The three types of attestation engagements are (1) examinations, (2) reviews, and (3) agreed-upon procedures.[11] Attestation engagements result in a report on a subject matter or on an assertion about a subject matter that is the responsibility of another party. Well-defined standards for attestation engagements are generally established in both national and international accounting rules.[12]

A less formal assessment may be conducted by a consultant with sufficient knowledge in internal control procedures. The experience and qualifications of the consultant should be carefully evaluated. Accountants are governed by professional principles and standards[13] that will apply to the handling of the assignment and the way findings are reported (e.g., objectivity, independence, professional judgment, quality control). The terms of reference of a non-accountant should require his or her compliance with substantially similar standards.

Outputs of the Corruption Risk Assessment

The consultant should provide at least the following outputs:

1. Plan for the assessment
2. Draft presentation of findings (the organization being analyzed should have the opportunity to comment on these)
3. Draft presentation of recommended measures to address findings (the organization being analyzed should have the opportunity to comment on these)
4. Final presentation of findings and recommendations

Annex 2 Endnotes
1. PEFA, http://www.pefa.org/index.php.
2. Transparency International, "Corruption Fighters Tool Kit," http://www.transparency.org/tools/e_toolkit.
3. World Bank, Latin America and Caribbean Region, 2007, *Corruption Warning Signs: Is Your Project at Risk?* (Washington, DC: World Bank), http://siteresources.worldbank.org/INTFIDFOR/Resources/4659186-1204641017785/2ProjectAtRisk.pdf.
4. PEFA, 2005, "Public Financial Management Performance Measurement Framework," http://www.pefa.org/pfm_performance_file/the_framework_English_1193152901.pdf.
5. Transparency International, http://www.transparency.org/tools/e_toolkit.
6. Transparency International, n.d., "Project Anti-Corruption System (PACS) (Construction Projects)," http://www.transparency.org/tools/contracting/construction_projects/section_a_pacs. The PACS uses a variety of measures that affect all project phases and all major participants at a number of contractual levels.
7. COSO, "Internal Control–Integrated Framework," http://www.coso.org/IC-IntegratedFramework-summary.htm. COSO was established in 1985 by five financial professional associations to improve the quality of financial reporting by focusing on corporate governance, ethical practices, and internal control. COSO has since broadened its scope to a framework for "Enterprise Risk Management"; however, the internal control framework is still widely used and is sufficient for anticorruption assessment purposes.
8. IFAC, http://www.ifac.org/About/; IPSASB, http://www.ifac.org/PublicSector/; *and* IAASB, http://www.ifac.org/IAASB/.
9. Comptroller General of the United States, 2007, "Government Auditing Standards, Appendix I: Supplemental Guidance," http://www.gao.gov/govaud/govaudhtml/d07731g-11.html#pgfId-1035039.
10. IAASB, http://www.ifac.org/IAASB/.
11. Comptroller General of the United States, 2007, "Government Auditing Standards, Chapter 6: General, Field Work, and Reporting Standards for Attestation Engagements," http://www.gao.gov/govaud/govaudhtml/d07731g-8.html#pgfId-1034320.
12. For example, see The American Institute of Certified Public Accountants, "Attestation Standards," http://www.aicpa.org/Professional+Resources/Accounting+and+Auditing/Authoritative+Standards/attestation_standards.htm.
13. Comptroller General of the United States, 2007, "Government Auditing Standards, Chapter 2: Ethical Principles in Government Auditing," http://www.gao.gov/govaud/govaudhtml/d07731g-4.html#pgfId-1034318, and "Government Auditing Standards, Chapter 3: General Standards," http://www.gao.gov/govaud/govaudhtml/d07731g-5.html#pgfId-1034319.

A construction audit is used to verify the procedures for procuring post-disaster construction services, the consistency of the construction with contract terms, and the use of the funds budgeted for the project.[1] Construction audits may be conducted on an ex post basis or carried out concurrently with the construction in situations where there is judged to be a high risk for corruption in the management of the project.

The nature of the construction project will determine the details of the auditor's scope of work. The focus of this guidance is on audit procedures for a single capital improvement project, although a construction audit may be ordered for an entire program consisting of multiple sites or projects, with minor adjustments in scope.

Construction Audit Objectives

1. To determine that construction contracts were awarded in compliance with applicable rules and regulations.
2. To determine that all revisions to the original contract were justified and properly approved.
3. To ensure that payments to the contractor were made in accordance with the contract, properly approved, transferred to the contractor using approved procedures, and received by the contractor in full.
4. To determine that the project was accounted for in accordance with proper accounting procedures and that the completed project was properly transferred to the responsible entity.

Procedures for Construction Audit

Activity	Considerations
1. Agree on scope and basis for audit	a) Agree on the objectives and scope of the audit with the agency being audited. The scope and procedures for the audit are based on an assessment by the auditor of the risks associated with the agency and with the activities to be audited. b) Identify the accounting and control standards for the audit. These may be the international generally accepted standards on auditing (ISA 800), auditing standards of the country, or auditing standards of the funding agency. The standards for the audit must be identified before the audit begins. c) Prepare the audit plan.
2. Audit preparation[2]	a) Analyze prior audit reports of the organization, both internal and external, for previously identified deficiencies in construction procedures. b) Identify all laws, policies, regulations, and accounting rules applicable to the area under review. The audit should include evaluation of compliance with all applicable regulations. c) Obtain the construction contract (and related addenda) and highlight significant and specific terms. The contract will be an important reference throughout the audit. d) Obtain the architect and/or engineer contracts for the project. Highlight significant and specific terms and ensure that they are adhered to throughout the project. e) Determine if fee and reimbursement guidelines appear reasonable. f) Prepare a schedule of accounts affected and indicate balances as of the audit date. Verify all accounts by reconciliation, analytical review, or testing. g) Obtain an understanding of the procedures and the flow of documents for construction project operations. h) Obtain copies of all reports generated for project; analyze use and distribution. May be generated from centralized computer systems, personal computers, or manually, including procurement, cash flow, schedule, and cost reports. i) Determine that records received from the contractor are adequate to monitor progress (progress and inspection reports, meeting minutes, photographs, updated schedules, material and equipment delivery schedules, drawing revisions, etc.). j) Review minutes of construction meetings to identify situations that may require additional testing or follow-up during audit.
3. Bid receipt and award	a) Obtain a copy of the bid package for the project, including the following items: i) Proposal; ii) Agreement; iii) General conditions; iv) Supplementary (Special) Conditions; v) Technical Specifications; vi) Drawings b) Review the following with respect to the receipt of bids: i) Format to submit their proposals enhances comparison of the bids; ii) Requirement for bid bonds and certificates of insurance; iii) Advertising copy meets government requirements for notification; iv) Process for qualifying bidders (if used); v) Process for replying to questions from bidders prior to bid opening; vi) Conduct of pre-bid conferences and site visits with bidders; vii) Process for receiving and controlling bids received prior to bid opening c) Review the bid tabulation, calculation, and award process.

Activity	Considerations
4. Contract administration	a) Contractor submittals. Review and analyze required submittals, including: i) Insurance certificates; ii) Signed bond forms; iii) Quality assurance or control program documentation; iv) Shop drawings; v) Subcontractor and material supplier bid and change order information; vi) Material certificates; vii) Fabrication, shipping, and construction schedules; viii) Equipment operating and maintenance manuals and instructions; ix) System for logging and tracking to ensure that required items have been received b) Insurance and bonds. Verify the adequacy and authenticity of: i) Required contractor and subcontractor insurance coverage, including general liability, automobile liability, workers compensation, and umbrella coverage; ensure that insurance was actually purchased; ii) Required contractor bonds, including the bid bond, payment bond, performance bond and maintenance bond; ensure that bonds were actually purchased c) Payment procedures. Review and analyze the following: i) Method used to calculate progress payments, to ensure a clear correlation between the payment processed and the percentage of work completed; ii) Original schedule of values and all requests for payments; iii) Most recent schedule of payment requests or progress billings; ensure that necessary evidence of inspection and approval is adequately documented; iv) Lien waivers submitted with progress payment invoices; v) All allowances identified in the contract to ensure that they have been properly adjusted; vi) The method used to calculate retention; ensure that the appropriate retention amounts were withheld from requests for payment; vii) Determine if the contract includes any reimbursable charges and ensure reimbursements agree with the contract provisions and prices; viii) Identify any claims on the project and ensure that hey have been settled d) Change orders. Improprieties are extremely common in the management of change orders. Carefully review and analyze: i) The change order log, including all requests and all change orders issued, documenting the reason for the change and the amount of the change; ii) Ensure that change orders represent changed or added work, and not work covered under the scope of the base contract or previously issued change orders; iii) Ascertain if change order work descriptions suggest problems that should be back charged to another party; iv) Ensure that change orders were not split to avoid approval requirements e) Subcontractor performance. Subcontractors frequently handle specialty areas, such as site work, foundation, framing, roofing, interiors, mechanical work, electrical work, and plumbing. The auditor should review the subcontractor documents at the contractor's office, including all subcontractor bids. i) Look for subcontractor bids that may indicate unauthorized reengineering of the construction; ii) Affidavits may be mailed to each subcontractor identified by the general contractor to request that they independently verify their cost of materials used in the project; iii) Ensure that controls are in place to avoid any nonconforming work or substituting lower cost materials; iv) Ensure subcontractor submittals include product data on all materials, shop drawings, etc., so that we can confirm that it meets the specifications of the job; v) Verify the inspection procedures by architect and the project manager of subcontractor performance. f) Reconciliation of contractor payments for project. Reconcile total payments/advances to date with contractor cost ledger accounts to ensure that contractor has used all owner funds only for project-related work. g) Final review. i) Determine how changes identified during the final review phase are handled, especially changes found too late to correct; ii) Review procedures if unauthorized changes are detected for withholding funds or negotiating a credit for the project.
5. Contract close-out	a) Document procedures for formally closing out construction contracts, to ensure that it is a structured process involving acceptance of the contractor's work, receipt of required documentation, and evaluation of the contractor's performance. b) Ensure the following documentation is received before final contract payment and release of retention: i) Releases of liens from the contractor and its suppliers and subcontractors; ii) Titles to major equipment incorporated in the facility; iii) Warranty documentation; iv) As-built drawings[3]; v) Inspection and acceptance records; vi) Operating and maintenance manuals; vii) (Possibly) spare parts, special tools, and consumable supplies c) Ensure that a post-performance evaluation is conducted. Examine the completed evaluation for adequacy.
6. Budget and accounting	Obtain most recent month-end project financial reports from accounting department of organization. a) Compare to the payment schedules of department managing project and review for differences, overdrafts, or fluctuations from amounts initially budgeted and unusual entries. b) Review procedures followed by the accounting department to ensure the propriety and validity of invoices submitted for payment on construction projects. c) Ensure controls are adequate to confirm approval and adequacy of funds available. d) Test a sample of transactions for accuracy, reasonableness, and adequacy of supporting documentation. e) Ensure procedures for transfer of asset from construction accounts to proper asset category. f) Validate transfer of asset to relevant operating entity, in accordance with government internal accounting procedures, including valuation and accounting category.

Activity	Considerations
7. Termination of audit	a) Complete any remaining tests or review any pending procedures or activities.
	b) Prepare draft audit report. To the extent possible, conclusions should be based on testing or observation. Only as a last resort, obtain answers through direct inquiry and interview.
	c) Forward draft to audited agency and request responses within agreed period of time.
	d) Discuss all exceptions and concerns with the appropriate personnel.
	e) Incorporate responses of audited agency into the final audit report, as appropriate.
	f) Prepare and distribute a final audit report.

Selection of Auditor and Audit Cost

The audit should be carried out by qualified accountants or auditors experienced in management and accounting procedures related to public sector capital construction and should be based on procurement rules and accounting standards of the public sector and/or of the agency funding the capital project. It may be advisable to use a competitive process for the selection of the auditor, with the selection based principally on qualifications and experience with similar types of audits. The selection process should ensure the avoidance of conflicts of interest, including any familial relationships, between staff of the audited agency and that of the auditing firm.

Payment of the auditor, especially for the concurrent audit, should never be based on the value of the construction contract or create an incentive for delays, because of the inherent conflicts of interest such compensation schemes create.

Annex 2 Endnotes

1. For the components of a sample construction audit, see City of Tampa, Internal Audit Department, "Audit Programs, Capital Construction Projects," http://www.tampagov.net/dept_Internal_Audit/information_resources/audit_programs.asp.
2. AuditNet, "Construction Auditing," http://www.auditnet.org/construction_auditing.htm.
3. "As-built" drawings are important because they represent the best record of the constructed facility and are needed for operations, maintenance, and repair throughout the facility's life. The cost for preparing as-built drawings should be included in the construction contract. The auditor should confirm that final as-built drawings have been secured, filed, and protected.

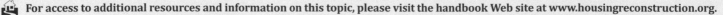

For access to additional resources and information on this topic, please visit the handbook Web site at www.housingreconstruction.org.

SAFER HOMES, STRONGER COMMUNITIES: A HANDBOOK FOR RECONSTRUCTING AFTER NATURAL DISASTERS

PART 3

INFORMATION ON WORLD BANK PROJECTS AND POLICIES

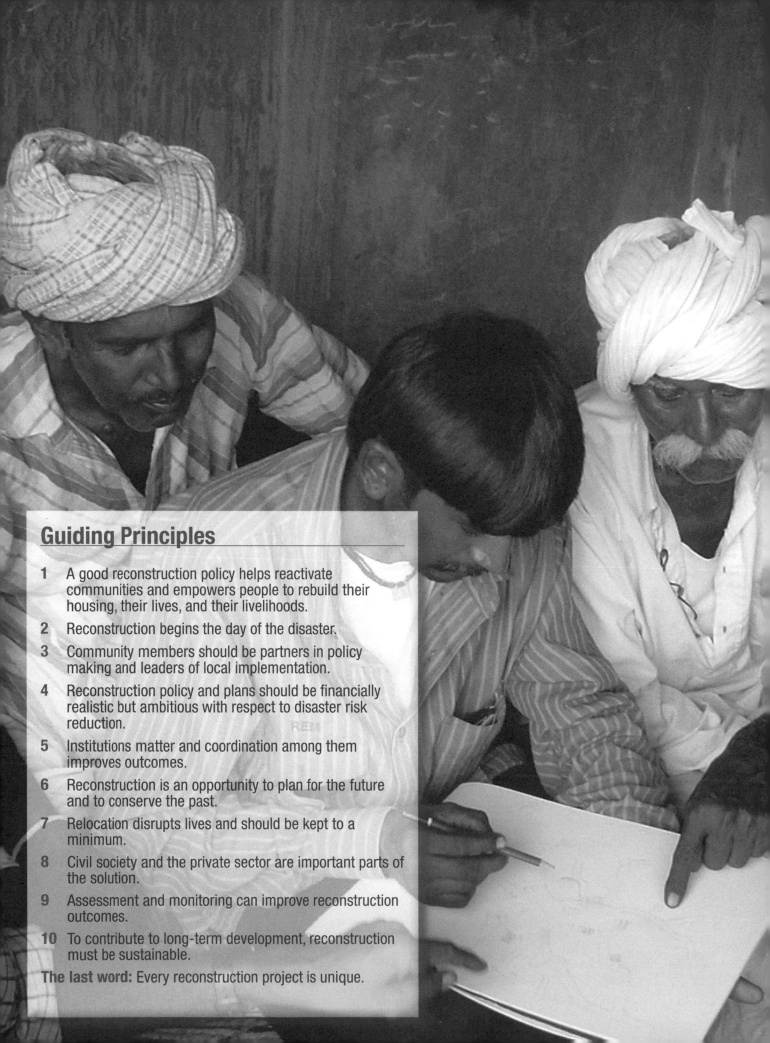

Guiding Principles

1   A good reconstruction policy helps reactivate communities and empowers people to rebuild their housing, their lives, and their livelihoods.

2   Reconstruction begins the day of the disaster.

3   Community members should be partners in policy making and leaders of local implementation.

4   Reconstruction policy and plans should be financially realistic but ambitious with respect to disaster risk reduction.

5   Institutions matter and coordination among them improves outcomes.

6   Reconstruction is an opportunity to plan for the future and to conserve the past.

7   Relocation disrupts lives and should be kept to a minimum.

8   Civil society and the private sector are important parts of the solution.

9   Assessment and monitoring can improve reconstruction outcomes.

10 To contribute to long-term development, reconstruction must be sustainable.

The last word: Every reconstruction project is unique.

20 WORLD BANK RESPONSE TO CRISES AND EMERGENCIES

Background

In 2007, the World Bank Executive Directors approved a new policy and set of procedures on emergency lending set out in Operational Policies and Bank Procedures (OP/BP) 8.00, Rapid Response to Crises and Emergencies (replacing the previous policy OP/BP 8.50, Emergency Recovery Assistance). OP/BP 8.00 attempts to align the Bank's outdated prior emergency policy with its evolving role in responding to crises and emergencies, and to improve the speed, effectiveness, and impact of the Bank's support to emergency recovery efforts.

The key policy features introduced in OP/BP 8.00 include:
- a broader definition of an "emergency" that allows the Bank to address the economic and social impacts resulting from an actual, or imminent, natural or man-made crises or disaster;
- application to a broader set of objectives, including support to the preservation of human, institutional, and social capital, and facilitation of peace building;
- emphasis on coordination with development partners in the delivery of integrated response efforts; and
- a call for a more strategic approach to disaster management and crisis prevention.

Under the policy, the Bank may provide a rapid response to a borrower's request for urgent assistance to address an event that has caused, or is likely to imminently cause, a major adverse economic and/or social impact associated with natural or man-made crises or disasters.

Forms of Bank Rapid Response

OP/BP 8.00 is premised on principles that include (1) the need to focus Bank assistance on its core development and economic competencies while remaining within its mandate; (2) establishment of appropriate partnership arrangements with other development partners, including the United Nations (UN); and (3) adoption of adequate oversight arrangements.

Objectives. The Bank may provide rapid response in support of one or more of the following objectives:
- Rebuilding and restoring physical assets
- Restoring the means of production and economic activities
- Preserving or restoring essential services
- Establishing and/or preserving human, institutional, and/or social capital, including economic reintegration of vulnerable groups
- Facilitating peace building
- Assisting with the crucial initial stages of building capacity for longer-term reconstruction, disaster management, and risk reduction
- Supporting measures to mitigate or avert the potential effects of imminent emergencies or future emergencies or crises in countries at high risk

Selection of instruments. The assistance strategy is developed by the Country Director (CD) in consultation with borrower and may include one or more of the following:
- Non-lending support, such as:
  - Provision of assistance for damage/needs assessment, and other technical assistance
  - Mobilization of donor assistance including establishment of multi-donor trust funds
  - Accessing grants from the Bank's programmatic post-conflict, low-income countries under stress (LICUS), and other trust funds

- Lending/financial support through a combination of:
  - New lending via an Emergency Recovery Loan (ERL) or credit (ERC)
  - Restructuring or reallocation within existing projects with or without additional financing, including provision of additional financing for such activities under OP/BP13.20, Additional Financing for Investment Lending
  - Redesigning investment projects not yet approved to include recovery activities
  - Supplemental development policy loans or credits
  - Contingent emergency loan to countries at high risk of natural disasters (see paragraph 13 of the policy)
  - Transfers from the surplus, in exceptional cases

The form and scope of the response can be adapted to the emergency's particular circumstances, taking into account the Bank's assistance strategy for the country. The country lending program may be adjusted to accommodate emergency operations, normally within the country's general lending allocation, taking credit risk and International Development Association (IDA) lending policies into account.

Recipient. The Bank's assistance should be focused in its core development and economic competencies, and may include
- assistance to borrower agencies and institutions involved in the emergency recovery effort and/or
- support, in partnership with other donors, of an integrated emergency recovery program that includes activities outside the Bank's traditional areas, such as relief, security, and specialized peace-building.

The Bank recognizes the lead of other international institutions, in particular the UN, in such activities, and forms partnership arrangements with other donors for the preparation, appraisal, and supervision of activities outside its core competencies.

Features of Bank Response

The Bank recognizes the risks involved in emergency situations, including the risks and lost opportunities associated with a delayed response, and the critical importance of speed, flexibility, and simplicity to an effective rapid response. As a result, emergency operations may have the following features.

Simplified, streamlined procedures are processed under accelerated and consolidated procedures and are subject to streamlined ex ante requirements (including in fiduciary and safeguards areas).

Risk management involves a different balance between ex ante and ex post controls and risk mitigation measures compared to regular operations, including on issues of fraud and corruption, requiring intensified supervision support to address such risks.

Financing percentage. Unless the country director determines otherwise, include Bank financing of up to 100 percent of the expenditures needed to meet the development objectives of such operations, including recurrent expenditures, local costs, and taxes.

Retroactive financing may include up to 40 percent of the loan amount for payments made by the borrower not more than 12 months prior to the expected date of signing the legal documents.

Larger Project Preparation Advance (PPA) limit may benefit from a PPA of up to US$5 million to cover start-up emergency response activities.

Quick disbursement and streamlined procedures may include a quick-disbursing component designed to finance a positive list of goods, (1) required for the borrower's emergency recovery program and (2) procured following procedures, that satisfy the requirements of economy and efficiency (normally the national emergency procurement procedures of the borrower). Streamlined financial management, procurement, and disbursement procedures may include:

 For access to additional resources and information on this topic, please visit the handbook Web site at www.housingreconstruction.org.

306   SAFER HOMES, STRONGER COMMUNITIES: A HANDBOOK FOR RECONSTRUCTING AFTER NATURAL DISASTERS

- rapid processing of withdrawal applications and additional flexibility on financing eligibility, direct payments, and use of letters of credit;
- higher prior review thresholds;
- simplified procurement methods;
- pre-qualified procurement and project management agents through sole-source or qualification-based selection; and
- expedited procedures for establishing and activating trust funds.

Initiation and Management of Bank Response

OP/BP 8.00 identifies the steps involved once the policy has been triggered. These include the following.

Internal Communications. The Regional Vice President (RVP) communicates with the Managing Director (MD) of the affected region, the Chief Financial Officer (CFO) and VP (OPCS); and, depending on the nature of the emergency, the Conflict Prevention and Reconstruction Unit (CPR), the Fragile States Unit (OPCFS) and the Hazard Management Unit.

Establishment of a Rapid Response Committee (RRC). An RRC is immediately convened by the responsible MD in the case of corporate emergencies.[1] Otherwise, an RRC is convened under the chairmanship of the RVP or the CD, depending on the nature of the emergency and/or the extent to which interdepartmental resource transfers are necessary. The RRC may assist in identifying and supplementing staff to prepare and implement the Bank's response, including from the callable roster.

Processing Timelines. For emergency projects of a simple design, task teams should aim to seek Board approval within 10 weeks of initiation of project discussion with government. For simple project restructuring, task teams should aim to seek approval within 4 weeks. (Detailed processing steps and turnaround times are found in the annex to this chapter.)

Staff and Consultant Rosters

In an effort to improve the availability and readiness of experienced staff and consultants to deploy on short notice, a Staff Roster has been established with staff registered from 20 sectors or units of the Bank, as of late 2009. This can be accessed through a Bank staff member to identify expertise in response to requests from country units or governments.

A Consultant Roster is also being developed, to improve the sharing of expertise with bilateral and multilateral agencies.

Emerging Implementation Issues under OP/BP 8.00

An analysis of the experience to date with OP/BP 8.00 shows the limitations of policy and procedural reforms in overcoming the challenges of working and delivering assistance in high-risk, insecure, low-capacity environments.[2] These findings point out risk factors for project teams.

1. Rapid response operations are processed within shorter time frames, but the actual delivery of assistance was still slow. Time saved from flexible procedures is lost in weak-capacity environments, including in delays negotiating UN-World Bank fiduciary agreements and contracting out fiduciary arrangements, or in setting up project implementation units.
2. Exogenous factors in the operating environment of rapid-response operations are difficult for the Bank to control. These include insecurity, rapid turnover in government, constrained capacity of the private sector, restricted access to project sites, and a limited market for goods and technical staff. In addition, the Bank's client governments must commit to overcoming their own internal constraints.
3. Bypassing government may set back state-building, creating long-term aid dependency and decreasing the legitimacy of the state in the eyes of its citizens. The Bank needs to engage with governments and development partners to design early interventions that support the legitimacy of the state, rather than undermine it. Simplifying project design is useful. It is also critical that country-based staff have the right skills and experience.
4. Significant resources—both technical and financial—and management attention are needed to support implementation and monitoring.
5. Collaboration with the UN has been an operational challenge, although the agreements are now in place to facilitate these arrangements.[3]

1. A "corporate emergency" is one that requires the mobilization of technical, financial, or institutional resources that are beyond what the RVP can provide and/or involve a degree of reputational risk that argues for a corporate review or response.
2. World Bank, 2009, "Rapid Response to Crises and Emergencies (OP/BP 8.00): Progress Report (Draft)," World Bank internal report.
3. See Chapter 14, International, National, and Local Partnerships in Reconstruction.

Resources

Brook, Penelope J. and Suzanne M. Smith, eds. 2001. *Contracting for Public Services: Output-Based Aid and Its Applications.* Washington, DC: World Bank. http://rru.worldbank.org/Documents/OBAbook/04foreword.pdf.

World Bank. "Guidelines: Financial Management Aspects of Emergency Operations Processed under OP/BP 8.00."

World Bank. 2007. Operational Manual, "OP 8.00 – Rapid Response to Crises and Emergencies." http://go.worldbank.org/54R3G3UES0.

World Bank. 2007. Operational Manual, "BP 8.00 – Rapid Response to Crises and Emergencies." http://go.worldbank.org/ILPIIVUFN0.

World Bank. 2007. "Processing Projects under OP/BP 8.00. Additional Guidance Note #1."

World Bank. 2009. "Processing Projects under OP/BP 8.00: A Review of Early Experience and Lessons Learned." Note of Discussion.

World Bank. 2008. "Rapid Response to Crises and Emergencies: Application of Bank Safeguard and Disclosure Policies." Note.

World Bank. "Rapid Response to Crises and Emergencies: Procedural Guidelines."

A. Emergency Recovery Loan—Identification to Effectiveness

Note: The processing steps in this annex apply not only to emergency operations financed through grants and loans, but also to those financed in part or in full through trust funds. Unless borrower actions are required for completion of transactions, service standards are here provided for final clearance, including resolution of any outstanding issues with the task teams. In the vast majority of cases, these targets should be adhered to and not be subject to regional variations.

Step	Guidelines	Primary responsibility	Turnaround (working days)
Identification/ approval of proposal	Team Leader (TL) obtains the agreement of the CD on the project's outline and budget and informs OPCS of the team's intention to launch a new emergency operation. At the same time, the TL alerts regional designated emergency staff (FM, Procurement [PR], Legal, Loan Department [LOA], and Safeguards).	CD/TL	
Project Information Document (PID), Integrated Safeguards Data Sheet (ISDS)	TL prepares a draft PID and a draft ISDS, both of which are updated throughout the process.	TL	
Combined preparation/ appraisal mission	During a combined preparation-appraisal mission, the task team (TT) assists the borrower in preparing the new project.	TL/TT	
Drafting of Emergency Project Paper (EPP) and Simplified Procurement Plan (SPP)	The TT prepares the EPP[1] with the relevant annexes on procurement (including a SPP), financial management, and safeguards.	TL/TT	
Drafting of legal agreement Review by safeguards coordinator	The TL provides the designated lawyer with a copy of the EPP for drafting the legal agreement and the safeguards coordinator with a copy of the draft ISDS for confirming environmental assessment (EA) category review, comment, and clearance authority. Input from the safeguards coordinator may include, for example, key safeguard issues to consider, other safeguard policies triggered (e.g. cultural properties, natural habitats) and the form of EA document (framework, Cat A environmental impact assessment, environmental management plan (EMP), etc.).	Lawyer Safeguards Coordinator	2
Review of draft legal agreement and EPP by PR, FM, and LOA.	The TL shares a copy of the EPP and the draft legal agreement with assigned staff from PR, FM, and LOA for their inputs and preparation of necessary documentation, including the disbursement letter and procurement provisions of legal agreement.	PR FM LOA	2
Finalization of review package	The lawyer finalizes the package based on inputs from PR, FM, and LOA.	Lawyer	1
Submission of review package to CD	When TL determines that the information reflected in project documents (draft EPP, draft legal agreement, SPP, and draft disbursement letter) forms a sufficient basis to enter into negotiations, the TL submits the entire package to the CD for a formal decision meeting.	TL	
Decision meeting	The RVP or designee convenes a decision meeting to review the package. Unless the meeting concludes that the project is not ready for further processing, the decision meeting authorizes the TL to proceed with negotiations with the borrower. Minutes of the meeting record clearances provided by FM, PR, and LOA, and any conditions for agreement with borrower.	CD	Within 3 days of circulation of documents
Circulation of minutes of meeting	TL clears with the chair and circulates minutes of meeting on a no-objection basis.	TL	Objections submitted within 1 day

Step	Guidelines	Primary responsibility	Turnaround (working days)
Finalization of negotiations package	Based on the decision meeting's recommendations, the TL works closely with the lawyer, Finance Officer, and fiduciary staff to finalize the negotiations package, including a revised EPP, draft legal documents and the disbursement letter, ISDS, and PID.	TT and lawyer	Within 3 days of meeting, unless additional work with borrower is required
Submission of PID and ISDS to World Bank Infoshop	TL submits PID and finalized ISDS to Infoshop	TL	
Invitation to negotiate	TL sends the negotiations package to the borrower with an invitation to negotiate and informs the Secretary of the Board in writing of the schedule.	TL	
Negotiations	Draft legal agreements are agreed to and minutes of negotiations are signed. At negotiations, the TL also tries to (1) obtain from the borrower the authorization of signature, (2) arrange for signature of Statutory Committee Report/Recommendation, and (3) discuss with the borrower the format of the legal opinion. TL also obtains from borrower information about the Designated Account information.	TL	
Finalize Board package	TL and lawyer finalize the Board package based on minutes of negotiations.	TL/lawyer	2
Board approval	RVP (or CD, where delegated) submits Board package to the Secretary of the Board (SECBO) for Board approval on a streamlined basis. TL requests Trust Funds Division of Accounting Department (ACTTF) to generate information on status of borrower's services payments.	RVP's (or CD's) office	Documents to SECBO 10 days before Board
Notification of approval	TL prepares a notification of approval and sends it to the borrower.	TL	1 day after Board date
Signing	TL arranges for CD/borrower signature of legal documents, including the legal opinion. If there are no additional conditions of effectiveness, a notice of effectiveness is prepared and signed by the CD.	TL/CD CD's office	Same day as signature of legal documents
Notification of effectiveness	If there are additional conditions for effectiveness, the TL monitors progress toward them and submits to the designated lawyer the effectiveness package, including evidence of compliance with conditions. Once the lawyer clears the effectiveness package, the TL prepares for the CD's signature a letter confirming acceptance of the required evidence of compliance and declares the legal agreement effective. The notice is copied to the FM.	Lawyer TL/CD	Lawyer clearance of compliance evidence within 2 days of submission by TL

 For access to additional resources and information on this topic, please visit the handbook Web site at www.housingreconstruction.org.

B. Project Restructuring—Identification to Board

Note: These processing steps are based on the revised guidelines for project restructuring and additional financing as outlined in (1) "BP 13.05, Project Supervision," and (2) "BP 13.20, Additional Financing for Investment Lending," and their accompanying guidance to staff: (1) "Processing Restructuring for Investment Projects: Guidelines for Staff" and (2) "Processing Additional Financing: Guidance to Staff."

Step	Guidelines	Primary responsibility	Turnaround (working days)
Identification	TL prepares a proposal for restructuring/additional financing in a concept memorandum and sends it to the CD.[2]	TL	
Approval of proposal: the CD obtains the agreement of the regional management on the level of approval likely to be required for the project's restructuring and on the amount of additional resources necessary for the restructuring work (including for appraisal)	At this point, OPCS is informed of the team's intention to launch a new emergency operation and regional designated emergency staff (FM, PR, Legal, LOA, and Safeguards) are alerted.	CD/TL	2
Drafting of Restructured Project Paper (PP)/Additional Financing Project Paper (APP), ISDS, and Procurement Plan	The TT[3] completes detailed discussions and field work with the borrower and prepares (1) a PP or APP,[4] (2) a revised ISDS, and (3) a revised Procurement Plan.	TT	
Drafting of legal amendment Drafting of Procurement Plan Review by safeguards coordinator	The TL provides the designated lawyer with a copy of the appropriate project paper for drafting the necessary amendments and the designated procurement specialist with a revised Procurement Plan for clearance. If the ISDS is revised, a copy of it is shared with the safeguards coordinator for review and comment, confirmation of EA category, and a decision regarding delegation of authority.	Lawyer PAS Safeguards Coordinator	2[5]
Review of draft amendment by PR, FM, and LOA	The TL shares a copy of the project paper and the draft amendments with assigned staff from PR, FM, and LOA for their inputs and preparation of necessary documentation, including the disbursement letter and procurement provisions of legal amendment.	PR FM LOA	1
Finalization of review package	When amendments to the legal documents are required, the lawyer finalizes the amendments to the legal agreement.	Lawyer	1
Submission of review package to CD	When TL determines that the information reflected in project documents (draft APP, draft legal amendment, revised procurement plan, and revised Disbursement Letter) forms a sufficient basis to enter into negotiations, the TL submits the entire package to the CD for a formal decision meeting.	TL	
Decision meeting	The RVP or designee convenes a decision meeting to review the package and authorize agreement with the borrower. Minutes of the meeting record clearances provided by FM, PR, and LOA and any conditions for agreement with borrower.	CD	Within 3 days of circulation of documents
Circulation of minutes of meeting	TL clears with the chair and circulates minutes of meeting on a no-objection basis.	TL	Objections submitted within 1 day
Finalization of negotiations package	The lawyer finalizes the draft agreement, taking into account the minutes of the decision meeting.	Lawyer	1
Agreement with borrower on PP/APP and legal agreement	Agreement is reached with the borrower on the legal amendment/ PP and APP.	TL	

Step	Guidelines	Primary responsibility	Turnaround (working days)
Submission of PID and ISDS to Infoshop	As necessary, TL submits the revised PID and ISDS to Infoshop.		
Finalize Board package	TL prepares the PP package, consisting of the Memorandum and Recommendation of the President (MOP), data sheet, and PP.	TL	1
Board approval	RVP (or CD, where delegated) submits Board package to SECBO for Board approval on a streamlined basis.	RVP's (or CD's) office	For additional financing projects, documents sent to SECBO 10 days before Board
Signing	Upon approval, the CD signs amendment letter. Signed amendment letter sent to borrower for countersigning.	CD CD's office	1
PID	If necessary, TL revises (and CD clears) the PID, and TL sends revised PID to Infoshop.	TL	2

Annex Endnotes

1. A format for the EPP is available in Bank files.
2. The memo includes an outline of the restructuring, a proposed budget, and a definition for additional staffing for discussion/preparation.
3. Designated emergency staff from FM, PR, Legal, Loan, and Safeguards should be copied on all correspondence related to project documentation.
4. For a template and guidelines on documentation used for restructuring and additional financing, staff may refer to (1) "Processing Restructuring for Investment Projects: Guidelines for Staff" and (2) "Processing Additional Financing: Guidance to Staff."
5. If more than one amendment is necessary, additional time may be needed.

SAFER HOMES, STRONGER COMMUNITIES: A HANDBOOK FOR RECONSTRUCTING AFTER NATURAL DISASTERS

21
SAFEGUARD POLICIES FOR WORLD BANK RECONSTRUCTION PROJECTS

This chapter is intended to provide guidance on the application of World Bank safeguards in post-disaster projects. It contains (1) a review of environmental and social safeguards procedures for normal operations, (2) a review of environmental and social safeguards procedures for post-disaster operations, (3) observations about the implementation of safeguards, and (4) case studies related to specific operations. It also includes links to the formats needed to prepare various documents required in the post-disaster environmental and social review process.

The World Bank's environmental and social safeguard policies are a cornerstone of its support to sustainable poverty reduction. The objective of these policies is to prevent and mitigate undue harm to people and their environment in the development process. These policies provide guidelines for Bank and borrower staffs in the identification, preparation, and implementation of programs and projects.

The Bank believes that the effectiveness and development impact of projects and programs it supports has substantially increased as a result of attention to these policies. Safeguard policies also provide a platform for the participation of stakeholders in project design and have been an important instrument for building a sense of ownership among local populations.

In essence, the safeguards ensure that environmental and social issues are evaluated in decision making, help reduce and manage the risks associated with a project or program, and provide a mechanism for consultation and disclosure of information. The safeguards are listed below. More detailed summaries of selected safeguard policies are included in the annex, Safeguard Policy Summaries.

World Bank Environmental and Social Safeguards and Their Policy Objectives

OP/BP	Safeguard	Policy objectives
4.01	Environmental Assessment*	Help ensure the environmental and social soundness and sustainability of investment projects. Support integration of environmental and social aspects of projects in the decision-making process.
4.04	Natural Habitats*	Promote environmentally sustainable development by supporting the protection, conservation, maintenance, and rehabilitation of natural habitats and their functions.
4.09	Pest Management	Minimize and manage the environmental and health risks associated with pesticide use and promote and support safe, effective, and environmentally sound pest management.
4.11	Physical Cultural Resources (PCR)*	Assist in preserving PCR and in avoiding their destruction or damage. PCR includes resources of archeological, paleontological, historical, architectural, religious (including graveyards and burial sites), aesthetic, or other cultural significance.
4.12	Involuntary Resettlement*	Avoid or minimize involuntary resettlement and, where this is not feasible, assist displaced persons in improving or at least restoring their livelihoods and standards of living in real terms relative to pre-displacement levels or to levels prevailing prior to the beginning of project implementation, whichever is higher.
4.20	Indigenous Peoples*	Design and implement projects in a way that fosters full respect for indigenous peoples' dignity, human rights, and cultural uniqueness and so that they (1) receive culturally compatible social and economic benefits, and (2) do not suffer adverse effects during the development process.
4.36	Forests*	Realize the potential of forests to reduce poverty in a sustainable manner, integrate forests effectively into sustainable economic development, and protect the vital local and global environmental services and values of forests.

OP/BP	Safeguard	Policy objectives
4.37	Safety of Dams	Ensure quality and safety in the design and construction of new dams and the rehabilitation of existing dams, and in carrying out activities that may be affected by an existing dam.
7.50	Projects on International Waterways	Ensure that the international aspects of a project on an international waterway are dealt with at the earliest possible opportunity and that riparians are notified of the proposed project and its details.
7.60	Projects in Disputed Areas	Ensure that other claimants to the disputed area have no objection to the project, or that the special circumstances of the case warrant the Bank's support of the project notwithstanding any objection or lack of approval by the other claimants.

* Safeguards most likely to apply in post-disaster situations.

Environmental Safeguard Requirements for Normal Operations

The normal World Bank Policy for Environmental Assessment is guided by Operational Policy/Bank Procedure (OP/BP) 4.01 and consists of seven basic elements[1]:

1. Screening
2. Environmental assessment (EA) documentation requirements
3. Public consultation
4. Disclosure
5. Review and approval of EA documentation
6. Conditionality in loan agreements
7. Arrangements for supervision, monitoring, and reporting

The table below outlines the requirements for each of these elements.

1. World Bank, 1999, OP 4.01 "Environmental Assessment," http://go.worldbank.org/ K7F3DCUDD0 and BP 4.01 "Environmental Assessment," http://siteresources.worldbank. org/INTFORESTS/Resources/ OP401.pdf; *and* "Environmental Assessment Sourcebook and Updates," http://go.worldbank. org/LLF3CMS1I0.
2. World Bank, 1999, OP 4.01 "Environmental Assessment, Annex C, Environmental Management Plan" http:// go.worldbank.org/B06520UI80.

Elements of World Bank Environmental Assessment

EA policy element	Policy requirement	Comment
1. Screening	Projects are categories as:	Project assessed a priori, depending on estimated environmental risk
	Category A (high risk-- likely to have significant adverse environmental impacts that are sensitive, diverse, or unprecedented)	
	Category B (modest risk-- potential adverse environmental impacts on human populations or environmentally important areas--including wetlands, forests, grasslands, and other natural habitats--are less adverse than those of Category A projects)	
	Category C (likely to have minimal or no adverse environmental impacts), or	
	Financial Intermediary (FI) operation (involves investment of Bank funds through a financial intermediary, in subprojects that may result in adverse environmental impacts)	
2. Documentation	Category A, Detailed Environmental Impact Assessment (EIA)	Format presented in OP 4.01 (Annex B)
	Category B, Environmental Management Plan (EMP)	Format presented in OP 4.01 (Annex C)[2]
	Category C, No requirement	
	Category FI, Environmental Framework	SSpecific investments unknown before project implementation.
		Documentation includes requirements for subproject EA.
		Environmental Framework describes EA process.
		Loan conditions include obligation for effective supervision and monitoring of EMP implementation.
		Sector investment loans may have similar requirements.

 For access to additional resources and information on this topic, please visit the handbook Web site at www.housingreconstruction.org.

314 SAFER HOMES, STRONGER COMMUNITIES: A HANDBOOK FOR RECONSTRUCTING AFTER NATURAL DISASTERS

EA policy element	Policy requirement	Comment
3. Consultation	Category A At least two consultations Category B At least one consultation	Consultations are conducted to receive input from local affected groups on their views of important environmental issues
4. Disclosure	Category A At the World Bank Infoshop (English) In-country, accessible to local affected groups (local language) Category B In-country, accessible to local affected groups (local language) Category FI Framework disclosed at the World Bank Infoshop and appropriate in-country Web site (e.g. Ministry of Environment). Individual subproject disclosure requirements defined in Framework	
5. Review and approval	Category A Regional Safeguards Coordinator Category B Sector Manager or Regional Safeguards Coordinator Category FI Framework reviewed/approved by Regional Safeguards Coordinator; individual subproject review and approval arrangements defined in Environmental Framework	Depends on whether project is "delegated"
6. Conditionality	Borrower is obligated to implement EMP (Category A or B)	
7. Supervision, monitoring, and reporting	Category A, B, or FI Institutional arrangements defined in EA documentation (EIA, EMP, or Framework)	

Environmental Safeguard Policy Requirements for Emergency Loans

As described in 🔖 Chapter 20, World Bank Response to Crises and Emergencies, World Bank response to emergencies (including natural disasters) is guided by OP/BP 8.00, "Rapid Response to Crises and Emergencies"[3] and the "Rapid Response to Crises and Emergencies: Procedural Guidelines."

Generally speaking, Emergency Recovery Projects are not exempt from the World Bank EA Policy (see OP 4.01, paragraph 13). Under unusual circumstances, a project may be exempted, but this requires a formal process and the justification must be recorded in the loan documents. If any waivers or exemptions from OP/BP 8.00 are required, the Task Team Leader should seek approvals prior to loan negotiations. The 🔖 case studies, below, show how the policy was applied in Bank operations after the 1999 Marmara, Turkey, earthquake and the 2004 Indian Ocean tsunami in Sri Lanka.

The EA policy requires that the World Bank operation determine:

■ the extent to which the emergency was precipitated or exacerbated by inappropriate environmental practices prior to the emergency; and
■ any measures to correct these practices to be incorporated into the project or a future lending operation, for example:
  ■ mudslides destroying residences, villages, and infrastructure because of excessive rainfall, but made worse by deforestation; and
  ■ flooding caused by hurricanes, typhoons, etc., made worse by poor coastal management practices (destruction of wetlands, removal of mangrove swamps etc.).

3. World Bank, "OP 8.00, Rapid Response to Crises and Emergencies," http://go.worldbank.org/IKGMVADFB0 *and* "BP 8.00, Rapid Response to Crises and Emergencies," http://go.worldbank.org/IE6E6NYJG1.

Combined Preparation-Appraisal Mission

Prior to the departure of the combined preparation-appraisal mission, the Task Team (ideally the Task Team Safeguards Specialist) should perform the following:

- **Prepare Draft Integrated Safeguard Data Sheet (ISDS)[4]**
  New loan
  Prepare a draft ISDS for the project. The ISDS will be revised and updated during project preparation.
  Project restructuring
  Revise the ISDS from the original project design, as appropriate.

- **Meet with the Regional Safeguards Coordinator**
New loan or project restructuring
1. Discuss the project scope.
2. Review the draft ISDS.
3. Agree on a preliminary EA Category rating for the overall project (A, B, C, or FI).
4. Define EA documentation requirements.
5. Establish requirements for consultation, disclosure, review, and approval of EA documents. OP 4.01 has detailed procedures for consultation, disclosure, review, and approval of EA documentation during normal project preparation. However, when OP 8.00 applies, these procedures are subject to being streamlined, consolidated, and simplified (OP 8.00, 7(a)). Therefore, agreement should be reached with the Regional Safeguards Coordinator as to how these procedures are to be modified.
6. Determine if the project is or is not delegated.

- **Meet with the Environment and International Law Unit (LEGEN)**
  Determine country-specific policies and regulations for environmental safeguards (primarily EA) in emergency/disaster situations. If such information is not available, LEGEN should provide the Task Team with the primary government contacts who have this information.

During the combined preparation-appraisal mission, the Task Team Safeguards Specialist will take the following steps.
1. Meet with government environmental officials to determine country-specific policies and regulations for environmental safeguards (primarily EA) in emergency/disaster situations.
2. Conduct consultations with locally affected groups.
3. Revise and update ISDS as necessary.
4. Begin preparation of EA documents.

Upon mission completion, the Task Team Safeguards Specialist will take the following steps.
1. Meet with the Regional Safeguards Coordinator to finalize the ISDS and EA category and receive clearance of the ISDS.
2. Finalize EA documents (see below) as agreed upon with Regional Safeguards Coordinator.
3. Disclose EA documents as agreed upon with the Regional Safeguards Coordinator. The Environmental and Social Screening and Assessment Framework (ESSAF) (see below) must be disclosed as a condition of loan approval.

Legal agreements must include obligations of the borrower to implement the requirements specified in the EA documents.

EA Documentation Requirements

Normally, an emergency operation will require two procedural approaches: one for known subprojects to be implemented immediately, and another for projects that would be identified in the future in different time horizons (immediate, transitional, and long term).

For subprojects known at the time of loan approval. Either a detailed EIA report (OP 4.01 Annex B-subprojects considered Category A) or an EMP (OP 4.01 Annex C-subprojects considered Category B) is required for each subproject. The decision is related to degree of environmental risk associated with the individual subproject. Agreement should be reached with the Regional Safeguards Coordinator on the environmental risk of each subproject and thus which EA document

4. For samples of project ISDS documents, search in World Bank, "Documents and Reports," under "Project Documents," http://go.worldbank.org/XFNFIE0SO0.

is appropriate. Since information requirements of an EIA report are considerably greater than an EMP, the effort required to prepare an EIA is usually greater and requires more time. Therefore, unless there is an urgent need, it is strongly recommended that subprojects requiring an EIA be financed during either the medium- or long-term phases of the project.

For subprojects not known at the time of loan approval. An ESSAF is required for these loans. The environmental portion of the ESSAF describes EA safeguard review procedures to be followed as subprojects are identified and considered for financing. This framework should have the following characteristics.

- Be consistent with both the host country and the World Bank environmental safeguard requirements.
- Adopt a sequenced approach, describing different procedures for subprojects to be supported:
    - Immediately (2-4 month time frame)
    - Transitional (1-year time frame)
    - Long-term (beyond one year)
- Describe procedures and responsible organizations for each of the following actions:
    - Subproject screening
    - EA documentation
    - Public consultation
    - Disclosure
    - Review and approval
    - Conditionality
    - Supervision, monitoring, and reporting

Challenges in Developing ESSAF Documents

The following observations are based on discussions with Task Managers of post-disaster operations.

ESSAF preparation. The ESSAF document is a unique World Bank safeguard requirement.[5] Unless the host country already has had a World Bank disaster operation, host country institutions involved with disaster operations (Ministry of Housing, Ministry of Finance, etc.) are normally not familiar with World Bank safeguard requirements and would likely take a very long time to produce the ESSAF document by themselves, likely involving several iterations. Furthermore, host countries do not usually place a high priority on environmental and social safeguard issues during disaster situations, and the ESSAF document is often viewed as an obstacle to receiving the immediate assistance they need.

Forcing attention on these concerns may be extremely frustrating to all parties concerned and could affect the relationship between the host government and the World Bank. Therefore, the Bank team should either prepare the draft ESSAF for the host country review and approval or work closely with the host country in preparing the ESSAF document.

At least one public consultation with affected groups should be conducted as part of ESSAF preparation to ascertain priority issues. This will help identify the need for safeguard policies other than EA and resettlement being triggered (e.g., natural habitats, cultural properties).

ESSAF capacity assessment. As part of the ESSAF preparation, the World Bank team should conduct a capacity assessment of the institutions that will be responsible for ESSAF implementation to determine if there is sufficient staff/expertise/authority to implement ESSAF requirements.

It is strongly recommended that such an assessment be done ex ante in countries prone to natural disasters (in this way valuable time in a disaster operation will not be spent on preparing a capacity assessment evaluation).

ESSAF implementation. The ESSAF requirement is relatively new, and implementation experience in practice has so far not been exemplary. It is a critical aspect of World Bank safeguards to ensure proper supervision and follow-up of ESSAF implementation.

In disaster situations, host governments will generally agree initially to the Bank's environmental requirements in the interest of getting access to the resources that are needed to address the disaster. Without guidance on ESSAF implementation and attention from the World Bank, ESSAF requirements may be forgotten during project implementation.

5. See World Bank, "Rapid Response to Crises and Emergencies: Procedural Guidelines."

If there is a Project Implementation Unit (PIU), the project team should require an environmental and/or a social safeguards specialist be included on the PIU staff, either a staff person or an experienced consultant. The PIU should issue regular, frequent reports to affected groups and implementing institutions on any environmental or social issues that arise, measures taken to address these issues, parties responsible for addressing the issues, and a schedule for their resolution. The PIU should also issue regular and frequent information to affected groups regarding vital services, such as safety of water supply, and interim arrangements for wastewater management and solid waste disposal.

Case Studies

2004 Indian Ocean Tsunami, Emergency Recovery Program, Sri Lanka

Context. In December 2004, a massive earthquake registering 9.0 on the Richter scale struck the coast of Sumatra, Indonesia, and triggered a series of tsunami waves that directly affected coastal areas of many countries around the Indian Ocean, including Sri Lanka. The tsunami waves struck more than 1,000 km. of coastline and penetrated inland as far as 500 meters.

The government of Sri Lanka asked, *inter alia,* the World Bank for assistance in conducting a damage assessment and, simultaneously, worked with the World Bank to prepare a restructuring operation: "Tsunami Emergency Recovery Program – Phase I."

Design and preparation. Safeguard policies as required by the World Bank were adequately designed into the project framework. An Environmental and Social Screening and Assessment framework (ESSAF) was prepared. The framework was designed to help government properly address and mitigate safeguard issues. For environmental risks, this included an assessment of governments' review and approval process for EIAs and of its capacity to monitor implementation of environmental mitigating measures.

Implementation. The period after the tsunami saw a boom in reconstruction activities across the country. Government adopted a policy of "build back better." As a consequence, the opportunity to integrate cross-cutting ecological and environmental concerns was lost. After the disaster, government announced the use of a buffer zone as a disaster prevention mechanism. This was most likely done as an immediate response and was not based on sound technical judgment or on public consultation. The resulting effects on the environment were profound. With physical reconstruction prohibited in the "no build zone," vast extents of new hinterland (including some natural areas) were cleared for proposed housing schemes. No system of EA was involved with the site selection and construction process; environmental planning took a low priority. The policy was later withdrawn and the Coast Conservation Department developed a more reasonable Coastal Zone Management Plan.

Key environmental issues included those associated with extraction of natural resources as construction materials. Reconstruction created a building boom of unprecedented scale and a high demand for sand, timber, rubble, and clay, among other resources. There was no system in place to verify the origin of these materials, even though sources were identified in the EIAs. As a consequence, much of this material was extracted illegally. Although the EIAs discussed removal of debris, by the time the EIAs were mobilized, debris had already been removed and used for roads and landfill. This gave rise to adverse drainage issues in some locations.

In summary, the project did include environmental safeguards as required. However, post-clearance monitoring and secondary impacts were not properly anticipated or addressed.

1999 Marmara Earthquake Emergency Reconstruction Project, Turkey

Context. On August 17, 1999, an earthquake measuring 7.4 on the Richter scale devastated the Marmara region of Turkey. More than 15,000 lives were lost and about 200,000 people were left homeless.

The World Bank undertook an assessment to outline the likely impact of the earthquake on the economy and estimated the fiscal burden for reconstruction and recovery in the range of US$1.8–US$2.2 billion. The largest direct cost (US$0.7–US$1.2 billion) was for reconstruction and repair of the region's housing stock.

Design and preparation. The main objectives of the program were to restore living conditions in the affected region of Marmara, support economic recovery and growth, and develop an institutional framework for disaster risk management and mitigation.

Investments included housing replacement or reconstruction and restoration of sports fields, playgrounds, and other common spaces. Feedback from housing reconstruction beneficiaries received during public consultations soon after the disaster led to design changes in the housing units. Monitoring and evaluation activities of the Project Implementation Unit (PIU) and consultant reports resulted in additional public outreach efforts to the beneficiaries of the rural housing reconstruction component. A monthly newsletter was published and distributed by the PIU at the earthquake reconstruction project sites. This allowed for further adjustments and reallocations during project implementation.

Implementation. The project was developed with full compliance with the World Bank safeguard policies in effect at the time of preparation (EA and involuntary resettlement). During preparation, sites proposed by government for urban housing were planned on public land, but early in the implementation process it was clear that some expropriation would be needed. The Task Team did due diligence and proceeded in compliance with the requirements of OP 4.12, including adherence to the Resettlement Plan and close supervision of compensation. Regular site visits were made by the PIU's social scientist.

Environmental safeguards were also monitored closely by the PIU, its local branches, and the World Bank Task Team with respect to compliance with the Environmental Management Plan. The PIU had well-qualified engineers, provided environmental training to the contractors, and had a constant presence at the construction sites. Monthly environmental reports were prepared and submitted to the World Bank for review. Both the PIU and the World Bank were actively involved in ensuring high environmental standards for the reconstruction, exemplified by careful attention to the construction of a sewage treatment plant to ensure adequate treatment before municipal sewage from the housing complex near Golcuk was discharged into the Bay of Izmit. Also, as a result of careful site monitoring, additional measures for erosion control were introduced.

1999 Armenia Earthquake Recovery Project, Colombia

Context. On January 25, 1999, an earthquake measuring 6.2 on the Richter scale struck the coffee-growing region of Colombia. This was followed by an aftershock measuring 5.8 on the Richter scale. As a result of the earthquake, there were more than 1,000 deaths and more than 150,000 people left homeless. The most important physical loss was housing, but the region's infrastructure (schools, health centers, primary and secondary roads, electric power facilities, water supply and sewerage systems, and the airport) also suffered significant loss/incapacitation.

Design and Preparation. Within a week of the disaster, international donors and nongovernmental organizations assisted with immediate needs (clearing debris and temporary shelter). The World Bank was involved in the medium- and longer-term reconstruction program. Four existing World Bank loans (Municipal Health Services, Secondary Education, Agricultural Technology Development and Urban Environmental Management) were restructured for this project (totaling US$93 million).

The government of Colombia established a Reconstruction Fund for the Coffee Region (FOREC) reporting to the country's president. FOREC was to finance, execute, and coordinate the economic, social, and environmental reconstruction of the affected region. FOREC's functions were to design operational guidelines for implementation of reconstruction activities, work with local mayors to provide a framework for reconstruction activities, and oversee the reconstruction effort.

The World Bank prepared the Earthquake Recovery Loan with an additional US$225 million to continue the reconstruction effort for repair of 509 schools, rebuilding of 142 schools, and repair of 74 hospitals and health centers.

Implementation. Government recognized immediately that the earthquake had caused a number of environmental problems, and the reconstruction process offered a number of opportunities to strengthen local environmental institutions and improve environmental management. A Regional Environmental Management Plan was ordered by presidential decree (1999) that was intended to ensure the reconstruction process (including debris removal) followed environmental safeguards and ensured environmental sustainability of natural resources. Environmental standards for reconstruction work were established, land use plans were prepared which, inter alia, identified high-risk areas that were not to be developed. The process included public participation. As a result of this effort, approximately 13,000 families had to be relocated from high-risk areas. This approach enhanced local government capacity for environmental management (e.g., debris handling, soil stabilization, drainage management). Municipal administrations had a greater role in land use for public and social infrastructure works, and new housing was not placed in high-risk areas.

For access to additional resources and information on this topic, please visit the handbook Web site at www.housingreconstruction.org.

320

SAFER HOMES, STRONGER COMMUNITIES: A HANDBOOK FOR RECONSTRUCTING AFTER NATURAL DISASTERS

Natural Habitats (OP/BP 4.04)

This policy prohibits Bank support for projects that would lead to the significant loss or degradation of any Critical Natural Habitats, whose definition includes those natural habitats that are:

- legally protected;
- officially proposed for protection; or
- unprotected but of known high conservation value.

The policy is "triggered" if a subproject could result in any one or more of the following four events:

- A loss of natural habitats
- Construction of "linear features" (e.g., roads, transmission lines, pipelines) that might cut through natural habitats
- An effect on the water supply to or drainage from natural habitats
- A direct or indirect result in resettlement or migration of people in a way that would adversely affect natural habitats

If, as part of the EA process described above and/or discussions with the Regional Safeguards Coordinator, the potential for significant conversion or degradation of critical or other natural habitats is identified (in accordance with one or more of the indicated criteria), the subproject is classified as Category A; projects otherwise involving natural habitats are classified as Category A or B, depending on the degree of their ecological impacts.

During the combined preparation-appraisal mission, the Task Team Safeguards specialist should meet with government environmental officials and verify whether or not natural habitats would be affected by the project. If natural habitats are involved, the manner in which the issue would be addressed should be described in the EA documentation.

Physical Cultural Resources (OP/BP 4.11)

This policy addresses PCR, which are defined as movable or immovable objects, sites, structures, groups of structures, and natural features and landscapes that have archeological, paleontological, historical, architectural, religious, aesthetic, or other cultural significance. They may be located in urban or rural settings, and may be above or below ground or under water. Their cultural interest may be at the local, provincial, or national level, or within the international community.

If the EA process described above or discussions with the Regional Safeguards Coordinator indicate a subproject (1) will involve significant excavations, demolition, movement of earth, flooding, or other environmental changes; or (2) will be located in, or in the vicinity of, a physical cultural resources site recognized by competent authorities of the borrower, the policy would be tentatively considered "triggered."

During the combined preparation-appraisal mission, the Task Team Safeguard Specialist should meet with government competent authorities and verify whether physical cultural resources would be affected by the project. If it is verified that the project has any of the characteristics set out in (1) or (2) above, the policy is triggered and assigned to either Category A or B. The manner in which the issue would be addressed should be described in the EA documentation.

Indigenous People (OP/BP 4.20)

This policy contributes to the Bank's mission of poverty reduction and sustainable development by ensuring that the development process fully respects the dignity, human rights, economies, and cultures of indigenous peoples. For all projects that are proposed for Bank financing and affect indigenous peoples, the Bank requires the borrower to engage in a process of free, prior, and informed consultation.

A project proposed for Bank financing that affects indigenous peoples requires:

- screening by the Bank to identify whether indigenous peoples are present in, or have collective attachment to, the project area (see paragraph 8 of the policy);
- a social assessment by the borrower (see paragraph 9 and Annex A of the policy);
- a process of free, prior, and informed consultation with the affected indigenous peoples' communities at each stage of the project, and particularly during project preparation, to fully identify their views and ascertain their broad community support for the project (see paragraphs 10 and 11 of the policy);
- the preparation of an indigenous peoples planning framework; and
- disclosure of the draft indigenous peoples planning framework.

Forests (OP 4.36)

This policy applies to the following types of Bank-financed investment projects:

- Projects that have or may have impacts on the health and quality of forests

- Projects that affect the rights and welfare of people and their level of dependence on or interaction with forests

- Projects that aim to bring about changes in the management, protection, or utilization of natural forests or plantations, whether publicly-, privately-, or communally-owned.

The Bank does not finance projects that, in its opinion, would involve significant conversion or degradation of critical forest areas or related Critical Natural Habitats. If a project involves the significant conversion or degradation of natural forests or related natural habitats that the Bank determines are not critical, and the Bank determines that there are no feasible alternatives to the project and its siting, and comprehensive analysis demonstrates that overall benefits from the project substantially outweigh the environmental costs, the Bank may finance the project, provided that it incorporates appropriate mitigation measures.

The policy is "triggered" if any one of the following criteria is applicable.

- The project could result in direct or indirect loss of forests of high ecological value (e.g., through improving access for logging).

- The project would finance commercial logging operations or purchase of logging equipment.

- The host country is committed to sustainable management of forests.

- Early in project processing, the Task Team consults with the Regional Safeguards Coordinator and, as necessary, with Environmentally and Socially Sustainable Development (ESSD) and other networks to determine if any forest issues are likely to arise during the project.

For each project covered under the scope of the policy, World Bank staff ensure that an EA category is assigned in accordance with the requirements of OP/BP 4.01, Environmental Assessment. A project that is likely to have significant adverse environmental impacts with potential for conversion or degradation of natural forests or other natural habitats that that are sensitive, diverse, or unprecedented is classified as Category A; projects otherwise involving forests or other natural habitats are classified as Category B, C, or FI, depending on the type, location, sensitivity, and scale of the project and the nature and magnitude of its environmental impacts.

Resettlement (OP/BP 4.12)

This policy covers direct economic and social impacts that both result from Bank-assisted investment projects and are caused by:

- the involuntary taking of land resulting in:
  - relocation or loss of shelter;
  - loss of assets or access to assets; or
  - loss of income sources or means of livelihood, whether or not the affected persons must move to another location; or

- the involuntary restriction of access to legally designated parks and protected areas resulting in adverse impacts on the livelihoods of the displaced persons.

This policy applies to all components of the project that result in involuntary resettlement, regardless of the source of financing. It also applies to other activities resulting in involuntary resettlement that in the judgment of the Bank, are:

- directly and significantly related to the Bank-assisted project;

- necessary to achieve its objectives as set forth in the project documents; and

- carried out, or planned to be carried out, contemporaneously with the project.

To address the impacts above, the borrower ordinarily prepares a resettlement plan or a resettlement policy framework (see paragraphs 25-30 of the policy) that covers the following:

- Measures to ensure that the displaced persons are informed about their options and rights; consulted on, offered choices among, and provided with technically and economically feasible resettlement alternatives; and provided prompt and effective compensation at full replacement cost for losses of assets

- If the impacts include physical relocation, measures to ensure that the displaced persons are provided assistance (such as moving allowances) during relocation; and provided with residential housing, or housing sites, or, as required, agricultural sites for which a combination of productive potential, locational advantages, and other factors is at least equivalent to the advantages of the old site

- Where necessary to achieve the objectives of the policy, measures to ensure that displaced persons are offered support after displacement, for a transition period, based on a reasonable estimate of the time likely to be needed to restore their livelihood and standards of living; and provided with development assistance in addition to compensation measures, such as land preparation, credit facilities, training, or job opportunities

Requests for guidance on the application and scope of this policy should be addressed to the Resettlement Committee (see BP 4.12, paragraph 7).

22 FINANCIAL MANAGEMENT IN WORLD BANK RECONSTRUCTION PROJECTS

The World Bank recognizes that financial management is an integral part of the development process. In the public sector, it ensures accountability and efficiency in the management of country resources; in the private sector, it promotes investment and growth. Therefore, the first objective of the Bank's attention to financial management is **to improve borrowing countries' financial management performance**. At the same time, if the Bank is to sustain the confidence of its shareholders, other stakeholders, and the public at large, it must be able to show that its funds are used appropriately. Thus, the second objective of the Bank's financial management work is **to provide acceptable assurance on the use of Bank loan proceeds**. While these objectives are sought even in emergency operations, including post-disaster reconstruction projects, adjustments to normal procedures are sometimes required.

For the World Bank, financial management arrangements are the budgeting, accounting, internal control, funds flow, financial reporting, and auditing arrangements of the government borrower or other agency responsible for implementing Bank-supported loan operations.[1] Under OP/BP 10.02, Financial Management, the Bank requires that for each Bank-funded operation the borrower maintain acceptable financial management arrangements to provide reasonable assurance that the proceeds of the loan are used for the purposes for which the loan was granted.[2]

This chapter provides (1) an overview of the Bank's project cycle, (2) a discussion of the elements of the Bank's financial management practices as it prepares for and oversees lending, and (3) a presentation on some of the special arrangements related to Bank financial management that may be used in emergency operations. For a full discussion of the Bank's response to emergencies, see 📖 Chapter 20, World Bank Response to Crises and Emergencies.

The World Bank Project Cycle

What Is a World Bank Project?

The World Bank lends money to low- and middle-income countries to support development and change. Development projects are implemented by borrowing countries following certain rules and procedures to guarantee that the money reaches its intended target.

What Is the World Bank's Project Cycle?

A series of activities carried out by the Bank in collaboration with government ensure that Bank support is addressing the most important development issues for the country and that loans are used for purposes for which they were intended. These activities collectively are referred to as the Bank's "project cycle." The Bank's project cycle includes the following steps.

This Chapter Is Especially Useful For:

- Policy makers
- Lead disaster agency
- Financial specialists
- Project managers

---

1. The policies and procedures summarized here apply to all loans, credits, advances under the Project Preparation Facility, and grants financed from World Bank resources, including International Development Association grants and Institutional Development Fund and other Development Grant Facility grants, with the exception of Development Policy (previously known as adjustment) Loans and Guarantees. They also apply to recipient-executed grants financed from trust funds, unless the donor agreement has different terms.

2. See "Guidelines: Financial Management Aspects of Emergency Operations Processed under OP/BP 8.00" *and* "Financial Management in Operations Processed under New OP/BP 8.00: FM for TTLs," January 16, 2008 (PowerPoint presentations), internal Bank documents.

 For access to additional resources and information on this topic, please visit the handbook Web site at www.housingreconstruction.org.

323

Stage	What it entails
Country Assistance Strategy	The World Bank project cycle begins with the elaboration of a **Country Assistance Strategy**. The Bank works with a borrowing country's government and other stakeholders periodically to determine (or to update) how financial and other Bank assistance can have the largest impact. This is followed by the preparation of strategies and priorities for reducing poverty and improving living standards. Examples of nearly all the project documents mentioned in this section, including Country Assistance Strategies, are available on the Bank's Web site.[3]
Project identification	Identified projects can be for infrastructure, housing, education, health, and government financial management, among others. The World Bank and government agree on an initial project concept and its beneficiaries, and the Bank's project team outlines the basic elements in a **Project Concept Note**. Also generated during this phase are the **Project Information Document** and the **Integrated Safeguards Data Sheet**, which identifies environmental and social issues that may be raised by the project.
Project preparation	The borrower government and its implementing agency or agencies conduct feasibility studies and prepare engineering and technical designs, to name only a few of the work products required. Government contracts with consultants and other public sector companies for goods, works, and services, as necessary, not only during this phase but also later in the project's implementation phase. Beneficiaries and stakeholders are consulted to obtain their feedback and enlist their support for the project. Due to the amount of time, effort, and resources involved, the full commitment of government to the project is vital. Bank staff may determine that a proposed project could have environmental or social impacts that are included under the **World Bank's Safeguard Policies**. If so, the borrower prepares an **Environmental Assessment Report** that analyzes the planned project's likely environmental impact and describes steps to mitigate possible harm. An **Environmental Action Plan** may also be prepared. The recommendations are integrated into the project design. 📖 Chapter 21, Safeguard Policies for World Bank Reconstruction Projects, provides a detailed description of the Bank's safeguard policies.
Project appraisal	Appraisal gives stakeholders an opportunity to review the project design in detail and resolve any outstanding questions. Government and the World Bank review the work done during the identification and preparation phases and confirm the expected project outcomes, intended beneficiaries, and the system that will be used to monitor progress. Once the Bank team confirms that all aspects of the project are consistent with World Bank operations requirements and that government has the institutional arrangements ready for implementation, the project is negotiated and is ready for approval.
Project approval	Once all project details are negotiated and accepted by both sides, the Bank prepares the **Project Appraisal Document** (for investment lending) or the **Program Document** (for development policy lending), along with other financial and legal documents, for submission to the Bank's Board of Executive Directors for consideration and approval. The **Project Information Document** is updated and publicly released when the project is approved. When funding approval is obtained, conditions for effectiveness are met, and the legal documents are accepted and signed, the implementation phase begins.
Project implementation	The borrower government implements the project with funds from the World Bank. With assistance from the Bank, the implementing agency prepares the specifications for the project and carries out the procurement of goods, works, and services, as well as any environmental and social impact mitigation agreed to during preparation. Once under way, the implementing government agency reports regularly on project activities. The project's progress, outcomes, and impact on beneficiaries are monitored by government and the Bank to obtain data to evaluate the results of the operation and the project. Government and the Bank also prepare a mid-term review of project progress. Full loan disbursement and project completion can take 1–10 years.
Project completion	As the project is completed, the World Bank and the borrower government document the results achieved, problems encountered, lessons learned, and knowledge gained from carrying out the project. The World Bank team prepares an **Implementation Completion and Results Report**, using input from the implementing government agency, co-financiers, and other partners and stakeholders. The information gained is used to determine if there is additional assistance needed to sustain the benefits derived from the project. The evaluation team also assesses how well the operation complied with the Bank's operations policies, and accounts for the financial resources.
Project evaluation	The Bank's Independent Evaluation Group (IEG) assesses the performance of a selection of projects every year, measuring outcomes against the original objectives, sustainability of results, and institutional development impact. From time to time, IEG also produces **Impact Evaluation Reports** to assess the economic worth of projects and the long-term effects on people and the environment.

3. World Bank, "Documents and Reports," http://go.worldbank.org/ H1Q3T60M80.

Financial Management in Bank Operations

World Bank Operating Policies (OPs) establish the parameters for the conduct of operations and describe the circumstances under which exceptions to policy can be made. They are based on the Bank's Articles of Agreement, the general conditions, and policies. Bank Procedures (BPs) explain the procedures and documentation required to carry out the policies set out in the OPs. This section summarizes OP/BP 10.02, "Financial Management," and some related financial management policy issues.

Operational Policies

For each operation, the Bank requires the borrower to maintain financial management arrangements that are acceptable to the Bank and that provide assurance that the loan proceeds are used for the purposes for which they were lent. Where feasible, the Bank expects these arrangements to be the same ones that the institution normally uses. As mentioned above, financial management arrangements include those for budgeting, accounting, internal control, funds flow, financial reporting, and auditing. The Bank financial management operational policy requires the following.:

Assessments of financial management arrangements. The Bank assesses the adequacy of a borrower's financial management arrangements during the preparation and implementation of each operation and requires the borrower to undertake appropriate measures to strengthen any identified weaknesses in its financial management systems and processes.

Interim financial reporting. The Bank normally requires a borrower to submit interim financial reports in a form agreed with the Bank.

Audited financial statements. The Bank requires that borrowers provide audited financial statements, within six months of the end of the reporting period, that reflect the activities of the operation supported by the Bank loan. The financial statements must be prepared using accounting standards acceptable to the Bank.[5] As for the audit, the auditing standards,[6] the scope of the audit, and the auditors who conduct it must be acceptable to the Bank as well. If the borrower fails to maintain acceptable financial management arrangements, or to submit the required financial reports by their due dates, the Bank can take action against the borrower.

Bank Procedures

Throughout the preparation and implementation of a Bank-supported operation, qualified and experienced financial management staff are assigned to the Bank's project team. Where feasible, these staff ensure that the financial management requirements for individual projects are adapted to the country's circumstances, make use of the country's normal systems where capacity permits, and involve common arrangements with other donors, in order to simplify the borrower country's obligations.

Project preparation. During the preparation of each operation proposed for Bank financing, financial management staff carry out the following tasks:

- Assess the proposed financial management arrangements to identify any weaknesses and assess the risks these weaknesses pose
- Agree with the borrower on the format and content of interim and annual audited financial statements to be provided throughout the implementation of the operation
- Agree on the scope of audit work to be carried out for each operation, and the identity of the auditor or the process for selecting the auditor

Staff prepare a financial management assessment report based on the results of the assessment. A summary of their assessment is included in the **Project Appraisal Document**. The assessment also includes actions agreed with the borrower to mitigate any risks identified. If necessary, information related to financial management issues may also be recorded in the minutes of negotiations or in the legal agreements.

Project implementation. During project implementation, financial management staff review the continuing adequacy of a borrower's financial management arrangements. The extent, manner, and timing of these reviews is decided by the Bank on the basis of risk and actual implementation

4. World Bank, 2007, OP 10.02 "Financial Management," http://go.worldbank.org/YHF8Y8UF30 and BP 10.02 "Financial Management," http://go.worldbank.org/26MM8GUCU0.
5. Accounting standards acceptable to the Bank include International Public Sector Accounting Standards issued by the Public Sector Committee of the International Federation of Accountants and the International Financial Reporting Standards/International Accounting Standards issued by the International Accounting Standards Board. The Bank may accept national accounting standards that it considers to be equivalent to international standards.
6. Auditing standards acceptable to the Bank include the Auditing Standards issued by the International Organization of Supreme Audit Institutions and the International Standards on Auditing issued by the International Federation of Accountants. The Bank may accept national auditing standards that it considers to be equivalent to international standards.

performance. In reviewing the arrangements, financial management staff undertake, as necessary, visits to project locations to meet with appropriate project staff, observe the performance of the financial management system, and check the application of controls or individual transactions. Financial management staff also (1) monitor the receipt and the timeliness of, (2) acknowledge receipt of, and (3) review the interim and annual audited financial statements that the borrower is required to provide. They pay particular attention to the quality of the auditor's performance and the substance of the audit report findings.

When financial management staff note deficiencies in the arrangements, including failure to send timely audited financial statements to the Bank, poor auditor performance, or indications in the audit of weak internal controls, they discuss these matters with the borrower and make recommendations to the Bank country director. The borrower is notified of any actions taken by the country director.

Project completion and evaluation. Significant financial management performance issues during implementation are recorded in the Implementation Completion and Results Report.

Financial Management Issues That May Arise in Emergency Operations[7]

Use of country systems. The Bank believes that the use of "country systems," that is, the use of a country's national, subnational, or sectoral institutions and applicable financial management laws, regulations, rules, and procedures for the operation being supported by the Bank, can potentially improve the impact of its operations. In fact, except where country systems are assessed by the Bank as not being adequate or for situations where the context in the country may dictate the use of a special purpose implementing entity or special implementation arrangements, the Bank tries first to use existing financial management institutional arrangements for implementing Bank-supported operations. However, the use of these arrangements may be subject to capacity-strengthening measures. Note that emergency (post-disaster and post-conflict) operations are examples of where the context may dictate the use of special implementation arrangements.

Harmonization. The Bank is committed to harmonizing its financial management arrangements with other donors and aligning these around a country's own systems. Accordingly, Bank staff will seek out opportunities for "delegated cooperation" (where one donor places reliance on the work of others) and ensure that, as far as possible, particularly in cases where multiple donors are involved in co-financing the same project or program, common arrangements are agreed to among all donors and government. Where a project is funded jointly by the Bank and other donors—a common situation for emergency operations—the Bank will seek to agree, to the extent practicable, on common formats, content, and reporting periods for reports to be submitted to all donors.

Analysis of risk. The implementation arrangements satisfactory to the Bank and the extent of Bank involvement during implementation will be a function partially of the Bank's evaluation of the risk of an operation. For various reasons, emergency operations may be evaluated as having higher risk.

The Bank's financial management risk model is qualitative and based on principles embodied in internationally recognized good practices for risk management.[8] The financial management risk rating is expressed as high, substantial, modest, or low, and provides a benchmark against which various aspects of project design, supervision, and other actions that may be taken by the Bank can be established. The risk model incorporates the following concepts.

7. World Bank Financial Management Sector Board, 2005, "Financial Management Practices in World Bank-Financed Investment Operations," internal Bank report.
8. In particular, Committee of Sponsoring Organizations (COSO), *Enterprise Risk Management – Integrated Framework;* , *and* International Federation of Accountants, *ISA 315, Understanding the Entity and Its Environment and Assessing the Risks of Material Misstatement* and *ISA 330, The Auditor's Procedures in Response to Assessed Risks.* For further discussion of the COSO Framework, see Chapter 19, Mitigating the Risk of Corruption, Annex 1, How to Do It: Conducting a Corruption Risk Assessment.

 For access to additional resources and information on this topic, please visit the handbook Web site at www.housingreconstruction.org.

326

SAFER HOMES, STRONGER COMMUNITIES: A HANDBOOK FOR RECONSTRUCTING AFTER NATURAL DISASTERS

Risk	Description
Inherent Risk	Inherent risk arises from the environment in which the project is located. It is the risk that the project financial management system does not operate as intended due to such factors as country governance environment, rules, and regulations. Inherent risk comprises three elements: Country-level risk. This rating is determined at a portfolio level, for each fiscal year, and is the same for all projects for which a risk assessment is prepared in during the same fiscal year. Entity-level risk. When entities have implemented Bank-financed operations in the past, the Bank may determine this risk using internal sources, such as Implementation Completion and Results Reports and Country Portfolio Performance Reviews. If the entity is new to implementing Bank-financed operations, a risk assessment of the entity is undertaken. Project-level risk. This risk is project-specific and is assessed for each project.
Control Risk	The risk that the project's financial management system is inadequate to ensure project funds are used economically and efficiently, and for the purpose intended. Control risk is measured for all six elements of financial management: budgeting, accounting, internal control, funds flow, financial reporting, and auditing.
Detection Risk	The risk that a material misuse of loan proceeds takes place and is not detected. Detection risk is lowered by (1) capacity-strengthening measures for the weaknesses identified as posing unacceptable levels of risk, and/or (2) increasing Bank supervision.
Residual Risk	Residual risk is the combination of the project's inherent and control risks as mitigated by borrower control frameworks and Bank supervision.

Financial Management Aspects of OP/BP 8.00

The Bank's OP/BP 8.00, Rapid Response to Crises and Emergencies, is explained in detail in
📖 Chapter 20, World Bank Response to Crises and Emergencies. OP/BP 8.00 addresses the need to focus Bank assistance for emergencies on its core development and economic competencies while remaining within its mandate. This section explains some of the financial management aspects of operations implemented under OP/BP 8.00.

As discussed above, OP/BP 10.02, Financial Management requires that, for each Bank-funded operation, the borrower maintain acceptable financial management arrangements that can provide reasonable assurance that the proceeds of the loan are used for the purposes for which the loan was borrowed. Consistent with this requirement, one of the guiding principles of OP/BP 8.00 is the provision of appropriate oversight arrangements, including corporate governance and fiduciary oversight, to ensure appropriate scope, design, speed, and monitoring and supervision of rapid response operations.[9]

For financial management staff, the main difference between preparing "normal" and rapid-response operations lies in the timing of the financial management arrangements. To respond quickly to an emergency, financial management staff streamline and simplify ex ante requirements while relying more heavily on such ex post requirements as additional fiduciary controls and reviews. They need to ensure that risk-mitigating measures suitable to available capacity are in place during implementation and, as appropriate, they may rely more heavily than usual on partner institutions. Key considerations are the following.

1. Include in project design, and agree on at negotiations, only the most critical ex ante controls; noncritical mitigating measures can be implemented during the course of the project.
2. Plan carefully for intensive supervision, particularly early in implementation, when financial management arrangements are being put in place, because it is the principal mitigating measure.
3. Appoint a seasoned senior financial management staff, along with the regional point person for the implementation of OP/BP 8.00, to work on the operation and to integrate lessons from similar regional/Bank operations.

The table below shows some examples of financial management arrangements for operations under OP/BP 8.00.

9. See "Guidelines: Financial Management Aspects of Emergency Operations Processed under OP/BP 8.00" *and* "Financial Management in Operations Processed under New OP/BP 8.00: FM for TTLs," 2008, World Bank PowerPoint presentations.

Examples of Financial Management Arrangements for Operations Processed under OP/BP 8.00[10]

Area	Ex ante arrangements	Ex post arrangements
Budget	■ Support 100% financing of activities to avoid delays in counterpart financing. ■ Provide adequate funds for essential initial operations even if sound estimates are not completed. ■ Reevaluate existing operations to find "excess" funds that can be quickly mobilized for the emergency operation. ■ Encourage Bank and other donors to align reporting requirements with government's cycle.	Detailed budget can be prepared later.
Accounting and reporting	■ Use existing reporting frameworks from government or other projects. ■ Use manual systems or computer spreadsheets until on-line systems can be implemented. ■ Use a commercially available off-the-shelf accounting package that is quick to install and easy to use, especially if technical support is available in-country. ■ Outsource accounting functions to private sector or international firms, as needed. ■ Use United Nations agencies/programs and/or local and international nongovernmental organizations with sufficient financial management capacity. ■ Simplify reports, limiting them to a list of expenditures.	Disseminate project reports to the lowest level beneficiary possible to help build in social accountability.
Staffing	■ Outsource key operations to provide the needed staff in the short run. The Terms of Reference (TORs) could include training and capacity development of country staff and systems so that, over time, the country is gradually able to assume full responsibility for the financial management aspects of the activities. ■ Use staff from other parts of the implementing entities of the same project or from other projects.	Train staff, even those with a limited accounting background, on simple cash accounting to provide the minimum records to get things moving quickly.
Internal controls	To compensate for weak controls in low-capacity environments, consider: ■ internal, concurrent audits conducted by government or outsourced to private firms; ■ additional controls exercised by independent persons from different parts of government, implementing entity, or community, to help ensure that duties are separated; that transactions are budgeted, authorized, executed, and recorded properly; and that services are delivered as specified; and ■ using financial management agents to review implementing entity transactions and/or to process transactions in the short run to help ensure due diligence; TORs could include training and capacity development of country staff and systems so that, over time, the country is gradually able to assume full responsibility for the financial management aspects of the activities.	Increase reliance on interim audits and/or more frequent (3-month or shorter period) external audits, including requesting an opinion on the internal controls and on-agreed procedures. Conduct performance audits to track the execution of project activities and deliverables.
Funds flow and disbursement arrangements	■ If country financing parameters allow, finance 100% of project expenditures and limit the number of expenditure categories to one, or at most two. ■ As much as possible, use retroactive financing and reimbursement of expenditures. ■ Use output-based disbursements.[11] ■ Ensure that the designated account[12] is funded quickly and adequately. ■ Use simplified report-based disbursements. ■ Pool financing with other donors/government.	If necessary, the Bank may disburse primarily through direct payments.
External audit	The frequency, scope, and quality of audits are extremely important factors in helping ensure that funds are used for the intended purposes. ■ In consultation with other sector and procurement colleagues, expand audit scope as needed to cover technical, institutional, and financial reviews. ■ When national audit institutions have weak capacity, complement their teams with private sector auditors to help improve the quality of the audit and also build capacity gradually. (See ⬛ Chapter 19, Mitigating the Risk of Corruption, Annex 2, How to Do It: Conducting a Construction Audit, for an audit methodology that can be used for a concurrent or ex post audit of a reconstruction project.) ■ In the short run, use international auditors in some projects to substitute for low country capacity. ■ For project preparation advances (PPAs), consider the use of annual audits. ■ Subject to procurement approval, amend contracts of audit engagements for existing projects (either in the same sector or in others) to cover the work of the emergency operation.	Audits should be carried out more frequently than annually, and financial management staff should follow up closely with the project implementing entity in a shorter time frame (i.e., from 6 months to 45 days).

10. World Bank, n.d., "Guidelines: Financial Management Aspects of Emergency Operations Processed under OP/BP 8.00," internal Bank report.
11. Global Partnership for Output-Based Aid, "Checklist for Designing Output-Based Aid Schemes," http://www.gpoba.org/designing/index.asp.
12. The "designated account" is the account of the borrower that is held in a financial institution acceptable to the Bank and operated on terms and conditions acceptable to the Bank, into which the Bank disburses proceeds from the loan account.

23

PROCUREMENT IN WORLD BANK RECONSTRUCTION PROJECTS

Carrying out procurement efficiently under World Bank-financed projects is critical to good project implementation, to the attainment of the objectives of the projects, and to their sustainability. Equally, the Bank, as part of its developmental role, is interested in strengthening the capacity of its borrowers to administer public procurement in an effective and transparent way as part of sound governance and good project management.[1]

To this end, the Bank has established procurement rules to be followed by borrowers for the purchase of goods, works, and services required for the projects financed by the Bank, and procedures for Bank review of the procurement decisions made by borrowers. World Bank project teams are required as an integral part of project preparation and appraisal, to make an assessment of the capacity of the project implementing agency or project implementation unit to administer procurement. For a description of the World Bank project cycle, see 📖 Chapter 22, Financial Management in World Bank Reconstruction Projects.

Procurement can become particularly challenging in emergency (post-disaster and post-conflict) operations, even for a government with established procurement capacity. Therefore, assessment of procurement capacity takes on special importance in emergency operations. For a full discussion of the Bank's response to emergencies, see 📖 Chapter 20, World Bank Response to Crises and Emergencies.

This chapter provides (1) a review of the Bank's procurement policies and procedures, (2) a summary of the Bank's procurement assessment process, and (3) a brief discussion of procurement issues that the Bank may need to address in emergency operations.

Public Procurement in Bank Operations

It is the Bank's fiduciary responsibility to ensure that the proceeds of its loans are used only for specified purposes, and that attention is paid to economy and efficiency, without regard to political and other non-economic influences or considerations. Therefore, the Bank has established procurement rules to be followed by borrowers for the procurement of goods, works, and services required for the projects financed by the Bank, and procedures for Bank review of the procurement decisions made by borrowers.

World Bank Operating Policies (OPs) establish the parameters for the conduct of operations and describe the circumstances under which exceptions to policy can be made. They are based on the Bank's Articles of Agreement, the general conditions, and policies. Bank Procedures (BPs) explain the procedures and documentation required to carry out the policies set out in the OPs. This section summarizes OP/BP 11.00, "Procurement."[2]

Operational Policies

Procurement rules and instructions. The rules that apply to the procurement of all goods, works, and services financed with Bank loan proceeds are detailed in the Procurement Guidelines[3] and the Procurement Policies and Procedures,[4] and those that apply to the selection and employment of consultant services are detailed in the Consultant Guidelines[5] and the Consulting Services Manual.[6] The guidelines are incorporated by reference in the Loan Agreement, and are binding on the borrower.

This Chapter Is Especially Useful For:

- Policy makers
- Lead disaster agency
- Procurement specialists
- Project managers

1. World Bank, 2002, "Revised Instruction for Carrying out Assessment of Agency's Capacity Assessment to Implement Procurement; Setting of Prior-Review Thresholds and Procurement Supervision Plan," http://siteresources.worldbank. org/PROCUREMENT/Resources/ Assessment-all.pdf. "Loan" in this Operating Policy/Bank Procedure means International Development Account (IDA) credits and IDA grants and Project Preparation Facility (PPF) advances to which the Bank's Procurement Guidelines are applicable according to the provisions of the relevant agreement with the Bank for the credit, grant, or PPF advance, but excludes development policy lending, unless the Bank agrees with the borrowers on specified purposes for which the loan proceeds may be used. "Procurement" refers to the purchase of goods, works, or services (e.g., the hiring of consultants); "borrower" includes the recipient of a grant or PPF advance, or the project implementing agency, when it is different from the borrower.
2. World Bank, 2001, OP 11.00 "Procurement," http://go.worldbank. org/Y66EAJUGL1 and BP 11.00 "Procurement," 2001, http:// go.worldbank.org/Z33TBIUH90.
3. World Bank, 2006, "Guidelines: Procurement under IBRD Loans and IDA Credits," http://go.worldbank. org/RPHUY0RFI0.
4. World Bank, "Procurement Policies and Procedures," http:// go.worldbank.org/JXJZSH4F50.
5. World Bank, 2006, "Guidelines: Selection and Employment of Consultants by World Bank Borrowers," http://go.worldbank. org/U9IPSLUDC0.
6. World Bank, 2006, *Consulting Services Manual: A Comprehensive Guide to the Selection of Consultants* (Washington, DC: World Bank), http://siteresources.worldbank. org/INTPROCUREMENT/ Resources/2006ConsultantManual. pdf.

 For access to additional resources and information on this topic, please visit the handbook Web site at www.housingreconstruction.org.

329

Principles of procurement. Four basic principles guide the Bank's procurement requirements:

- Ensuring economy and efficiency in the procurement of goods, works, and services, as mandated by the Articles
- Giving eligible bidders from developed and developing countries a fair opportunity to compete in providing goods, works, and services financed by the Bank
- Encouraging the development of domestic industries—contracting, manufacturing, and consulting industries—in borrowing countries
- Providing for transparency in the procurement process

Competition, economy, and efficiency. Competition is the basis for economic and efficient procurement. The Bank prefers procurement methods that maximize competition. Procurement of goods and works normally requires the use of international competitive bidding, and, for the selection of consultants, it normally requires the use of quality and cost-based selection(QCBS). Some exceptions are permitted.

Eligibility to compete. Any firm from any member country is eligible to compete for Bank-financed contracts except in any of the following circumstances.

- The borrower country prohibits commercial relations with the firm's country.
- The firm has a conflict of interest.
- The firm is owned by government, unless it is legally and financially autonomous, operates under commercial law, and is not a dependent agency of the borrower.
- The firm is under sanction by the Bank for having engaged in corrupt or fraudulent practices.

Domestic preference. To encourage the development of domestic industries, the Bank permits:

- preference to bids offering goods manufactured within its country;
- preference to bids for works contracts from eligible domestic contractors in countries below a specified per capita income threshold; and
- credit to proposals for consulting services that include nationals as key staff.

Transparency. Transparency is an essential part of the Bank's efforts to ensure effective use of loan funds and to combat fraud and corruption. To promote transparency, the Bank:

- requires public notification of procurement opportunities;
- favors the use of open competitive procedures that include public bid opening;
- provides a specific mechanism by which a losing bidder may request, and receive, an explanation as to why its bid was not selected; and
- discloses the results of bidding processes, including the names of firms or individuals awarded contracts and the value of the contracts.

Role of the borrower and the Bank. The borrower is responsible for all aspects of project implementation, including procurement. For each project, the Bank assesses the capacity of the implementing agencies to carry out the required procurement and determines the level of associated risk. The borrower prepares a procurement plan that covers the activities necessary to ensure that project procurement will be carried out efficiently and professionally. The Bank assists the borrower in planning for procurement, including preparation of the procurement plan, and it supervises and monitors procurement decisions throughout project implementation.

If a borrower fails to carry out procurement in accordance with the procedures agreed to in the Loan Agreement, the Bank can cancel the amount of the loan allocated to the goods, works, or services that have been misprocured. The Bank may also apply other legal remedies.

Country procurement assessments. The Bank and the borrower's government together periodically assess the effectiveness of the borrower's procurement system and identify reforms to address deficiencies in the system. The findings of this assessment are incorporated into the Country Assistance Strategy. This chapter includes a description of the country procurement assessment process, below,

Fraud and corruption. The Bank requires that borrowers and bidders observe the highest standards of ethics during the procurement and execution of Bank-financed contracts. Firms found to have participated in fraudulent or corrupt practices or activities are declared ineligible to be awarded future Bank-financed contracts, either indefinitely or for a stated period of time. If a representative of the borrower is found to be engaging in such corrupt or fraudulent practices, the Bank cancels the amount of the loan allocated to the contract in question, unless the borrower takes action to remedy the situation that is satisfactory to the Bank.

Bank Procedures

Project preparation, appraisal, and implementation. For each project proposed for Bank financing, a procurement specialist (PS) is included in the task team from its inception.

Procurement capacity assessment and planning. The PS assesses the capacity of the agencies that will implement the operation to carry out project procurement, and the risks associated with procurement under the operation. The PS uses the most current applicable Country Procurement Assessment Report (CPAR) for this assessment. If the assessment reveals deficiencies, the Bank works with the borrower to formulate an action plan to strengthen capacity (including training or technical assistance, as appropriate) and mitigate the identified risks.

As soon as the nature and main components of the proposed project are identified, the PS assists the borrower in preparing the project procurement plan for an initial period of at least 18 months, taking into account any technical, financial, or management constraints the borrower may be facing. The procurement plan, which is updated annually or as needed during project implementation, covers:

- the list of contract packages;
- the project procurement program, including timing of the contracts;
- the methods for procuring the necessary goods, works, and services;
- the required Bank standard bidding documents; and
- the institutional arrangements to carry out the procurement.

For projects for which the contracting schedule and specific contracts cannot be precisely defined, the procurement plan consists of a description of all administrative aspects of procurement and consultant selection, including:

- criteria for efficient contract packaging and appropriate procurement methods;
- timing of all procurement activities and the system to monitor procurement progress; and
- actions to keep the business community informed of opportunities and outcomes of project procurement.

Project appraisal and negotiations. During appraisal, the Bank develops a procurement supervision plan and agrees with the borrower on standard bidding documents to be used for the project. The Bank and the borrower also agree on any activities to strengthen the procurement capacity of the borrower during implementation. These agreements are incorporated in the Loan Agreement.

Project implementation. During project implementation, the Bank evaluates whether the borrower's procurement actions comply with the provisions of the Loan Agreement, and monitors adherence to the procurement plan and progress with the strengthening of the implementing agency. If major deficiencies occur, corrective actions are proposed.

Role of Bank staff in procurement activities. In working with borrowers on procurement matters, Bank staff maintain strict neutrality and impartiality. Staff do not:

- recommend to borrowers that they use particular consulting firms, suppliers, or contractors;
- undertake activities that are the responsibility of borrowers; or
- participate in evaluating bids or proposals.

Allegations of fraud and corruption and misprocurement. Complaints alleging fraudulent or corrupt practices by a bidder, supplier, contractor, or consultant in the procurement process of a Bank-financed contract are referred by Bank staff to the Department of Institutional Integrity. When Bank staff determine that the borrower has followed procurement procedures that are not in accordance with those set out in the Loan Agreement, the borrower is notified in writing. The notice brings the violation to the borrower's attention and advises that, if the situation is not rectified, the Bank may declare misprocurement. One of the key Bank specialists involved in these situations is the Regional Procurement Adviser. Bank procedures that apply in these cases are detailed in BP 11.00.

The World Bank Procurement Assessment Process

One of the main responsibilities of the Bank with respect to procurement is to help borrower countries improve their procurement systems. Sound public procurement policies and practices are essential to good governance. Procurement assessment takes place at two levels: for the country as a whole and for the individual government agency that is being proposed as the executing agency for a World Bank project. This section briefly describes the methods for assessing the procurement capacity of both entities.

World Bank Assessment of Country Procurement Capacity[7]

Using the country procurement assessment, the Bank assists its member countries in analyzing the quality of their public procurement policies, organization, and procedures. The result of this assessment is the CPAR. Many countries' CPARs are available on the World Bank Web site.[8]

Purpose of the country assessment. The main purpose of the country procurement assessment is to establish the need for and guide the development of an action plan to improve a country's system for procuring goods, works, and consulting services. To accomplish this, the primary objectives of a country procurement assessment are to:

- analyze the country's public sector procurement system, and how well it works in practice;
- identify institutional, organizational, and other risks associated with the procurement process, including procurement practices unacceptable for use in Bank-financed projects;
- develop a prioritized action plan to bring about institutional improvements; and
- assess the competitiveness and performance of local private industry participation in public procurement and the commercial practices that relate to public procurement.

Scope of the country assessment. The Bank considers a country's procurement system to be composed of the following elements, each of which is analyzed in the country procurement assessment:

- Legal framework
- Procurement system organizational framework
- Procurement capacity building system/institutions
- Procurement procedures/tools
- Decision-making and control system
- Anticorruption initiatives and programs
- Private sector participation in the system
- Contract administration and management
- System for addressing complaints

Beside examining these elements, the country procurement assessment also examines how procurement is supposed to be carried out for goods, works, and consultant services, including, where applicable, large and/or complex turnkey, supply/install, management of public utilities, concession, information technology, and other contracts. More importantly, the country procurement assessment examines the application of the rules and enforcement in practice. Poor dissemination of rules, inadequate training of personnel, lack of enforcement, failure to maintain good records, endemic corruption, and a variety of other factors create risks that can undermine an otherwise seemingly adequate system.

Outcome of the country assessment. The result of the country assessment is the CPAR. Each CPAR will be different and reflect the scope and content as agreed to in the approved memorandum that is developed before the initiation of the assessment. The CPAR will generally include a discussion and analysis of findings for both the public sector and the private sector components of the assessment, a recommended action plan and sequence of actions, recommendations on technical assistance that will be needed to implement the action plan and sources of financing for it, and a monitoring plan for the project.

7. World Bank, "Assessment of Country's Public Procurement System," http://go.worldbank.org/RZ7CHIRF60.
8. World Bank, "Documents and Reports," http://go.worldbank.org/L5OGDXGTR0.

 For access to additional resources and information on this topic, please visit the handbook Web site at www.housingreconstruction.org.

SAFER HOMES, STRONGER COMMUNITIES: A HANDBOOK FOR RECONSTRUCTING AFTER NATURAL DISASTERS

World Bank Assessment of Agency Procurement Capacity[9]

Bank project teams are required as an integral part of project preparation and appraisal to make an assessment of the capacity of the implementing agency or project implementation unit designated to administer project procurement. This agency capacity assessment is useful both for Bank-financed procurement and other types of operational arrangements, such as multi-donor financing arrangements.

Purpose of the agency assessment. The objectives of the agency capacity assessment are similar to those of the country procurement assessment, specifically to:

- evaluate the capability of the implementing agency and the adequacy of procurement and related systems in place to administer procurement in general and Bank-financed procurement in particular;
- assess the risks (institutional, political, organizational, procedural, etc.) that may negatively affect the ability of the agency to carry out the procurement process;
- develop an action plan to be implemented as part of the project, as necessary, to address the deficiencies detected by the capacity analysis and to minimize the risks identified by the risk analysis; and
- propose a suitable Bank procurement supervision plan for the project considering the relative strengths, weaknesses, and risks revealed by the assessment.

The agency capacity assessment is carried out by a PS assigned to the project during the project preparation stage of the project cycle. The aim is to have the assessment and the agreed-upon action plan finalized by the time of project appraisal.

Scope of the agency assessment. The capacity review includes an assessment of the capacity of the agency to carry out all phases of procurement. Typically, it includes a review of the following:

- Legal aspects and procurement practices
- Procurement cycle management, whose key elements are:
  - Procurement planning
  - Preparation of bidding documents
  - Management of bidding process, from advertisement to bid opening
  - Bid evaluation
  - Contract award
  - Preparation and signing of contract
  - Contract management during implementation, including dispute resolution methods
  - General handling of procurement cycle (duration, actors, reviews, etc.)
- Organizational structure of the procurement unit, including:
  - Organization of procurement unit and allocation of functions
  - Internal procedural manuals and instructions and historical compliance
- Support and control systems
- Record-keeping
- Staffing
- General procurement environment
- Private sector viewpoint, including
  - General efficiency and predictability of the system
  - Transparency of the procurement process
  - Quality of contract management
  - General reputation of the agency as free of corruption

Outcome of the agency assessment. The outcome of the agency capacity assessment is an action plan to build the agency's capacity. The assessment includes a detailed description of the actions to be taken and the associated timetable. Actions may address any or all of the topics listed above, and recommendations may include additional capacity, staffing, training, support by consultants, and improvements required in facilities, organization, record-keeping, reporting, and planning and monitoring. Actions that are essential for project implementation are prioritized and are initiated before procurement starts. Others are implemented during the life of the project. Terms of reference for any consulting assignments and cost estimates are also included in the plan. The detailed plan forms part of the project implementation documentation and is agreed to with the borrower as part of project negotiations.

9. World Bank, 2002, "Revised Instruction for Carrying out Assessment of Agency's Capacity Assessment to Implement Procurement; Setting of Prior-Review Thresholds and Procurement Supervision Plan," http://siteresources.worldbank.org/PROCUREMENT/Resources/Assessment-all.pdf.

Characteristics of a Good Public Procurement System

According to the CPAR framework, a public procurement system is well functioning if it achieves the objectives of transparency, competition, economy and efficiency, fairness, and accountability. The following key elements help determine whether a particular system meets these objectives:

- A clear, comprehensive, and transparent legal framework with easily identifiable rules that govern all aspects of the procurement process, including:
  - Advertising of bidding opportunities
  - Maintenance of records related to the procurement process
  - Predisclosure of all criteria for contract award
  - Contract award based on objective criteria to the highest-ranked evaluated bidder
  - Public bid opening
  - Access to a bidder complaints review mechanism
  - Disclosure of the results of the procurement process
- Clarity on functional responsibilities and accountabilities for the procurement function, including:
  - Those who implement procurement, including preparation of bid documents and the decision on contract award
  - Those who are accountable for proper application of the procurement rules
  - Means of enforcing these responsibilities and accountabilities, including the application of appropriate sanctions
- An institutional framework that differentiates between those who carry out the procurement function and those who have oversight responsibilities
- Robust mechanisms for enforcement
- Well-trained procurement staff

This not an exhaustive list of issues and the CPAR is customized to the issues present in each country, including the manner in which all the above actually work in practice.

Source: World Bank, "CPAR Instruction," http://go.worldbank.org/J2H75S2RB0.

Procurement Issues in Emergency Operations and Potential Solutions

The Bank's OP/BP 8.00, Rapid Response to Crises and Emergencies, is explained in detail in Chapter 20, World Bank Response to Crises and Emergencies. OP/BP 8.00 addresses the need to focus Bank assistance for emergencies on its core development and economic competencies while ensuring the Bank remains within its mandate. This section explains some of the specific procurement issues of operations implemented under OP/BP 8.00.

Emergency operations may give rise to unique procurement issues. This is due to two situations in particular: (1) risks inherent in the post-disaster procurement environment, such as the scale of procurement and the time pressure under which it may be taking place; and (2) the complexity of the institutional arrangements, particularly if there are numerous funding sources and special arrangements, such as multi-donor funds.

These issues have been addressed in the broader context of reconstruction project financial management in Chapter 22, Financial Management in World Bank Reconstruction Projects. Included below is a list of specific measures that the Bank and the borrower government may want to consider to lower the risk and increase the efficiency of procurement in post-disaster reconstruction projects.

- The Bank should assign experienced emergency PSs to provide real-time advice to the borrower on post-disaster procurement. This should include developing a Simplified (6-month) Procurement Plan (SPP), which is permitted by OP/BP 8.0. The SPP will simplify procurement while ensuring borrower compliance with the Bank procurement procedures explained above.
- The following special procurement actions are permitted under OP/BP 8.0 and should be evaluated:
  - Using rapid procurement methods (direct contracting or simple shopping) for the procurement of services of qualified United Nations agencies/programs and/or suppliers (for goods) and civil works contractors already mobilized and working in emergency areas (for works)
  - Using single sourcing or Consultant's Qualification Selection for contracting firms that are already working in the area and that have a proven track record for the provision of technical assistance
  - Extending contracts issued under existing projects for similar activities by increasing their corresponding contract amounts
  - Where alternative arrangements are not available, using Force Account for delivery of services directly related to the emergency
  - Using national competitive bidding, accelerated bidding, and streamlined procedures, and applying Bank provisions on elimination, as necessary, of bid securities
- Recent CPARs or agency procurement assessments should be used to design the reconstruction financial management and procurement system, or a simplified assessment of procurement capacity should be immediately conducted.
- If a special arrangement for financial management or procurement is being considered, strive to incorporate as many donors and other agencies as possible in both the analysis of options and the use of the arrangement, to reduce transaction costs for the government.
- Carry out a procurement assessment focused specifically on the multi-donor arrangement for procurement jointly with other donors and agencies.
- If major nongovernmental agencies involved in reconstruction are not part of the multi-donor arrangement, use the framework proposed in Chapter 19, Mitigating the Risk of Corruption, Annex 1, How to Do It: Conducing a Corruption Risk Assessment, to assess nongovernmental organization procurement capacity, since their capacity will affect the overall efficiency of the reconstruction program.
- Technical assistance can be provided to the public procurement system at various levels to increase the efficiency and speed of procurement, without undermining transparency and control.
- Technical assistance can also be provided to agencies that may not be directly involved in procurement but have a coordination or indirect role—such as the treasury department, ministry of finance, line departments, and others.
- Properly managed and staffed concurrent audits can maintain control while allowing the volume of procurement to increase.
- Procurement and project management agents are other options to provide additional procurement and implementation capacity. The Bank maintains a list of prequalified procurement and project management agents with the ability to rapidly deploy to emergency areas. Borrowers can access companies from this list under appropriate authorized single source arrangements.

PART 4
TECHNICAL REFERENCES

Guiding Principles

1   A good reconstruction policy helps reactivate communities and empowers people to rebuild their housing, their lives, and their livelihoods.

2   Reconstruction begins the day of the disaster.

3   Community members should be partners in policy making and leaders of local implementation.

4   Reconstruction policy and plans should be financially realistic but ambitious with respect to disaster risk reduction.

5   Institutions matter and coordination among them improves outcomes.

6   Reconstruction is an opportunity to plan for the future and to conserve the past.

7   Relocation disrupts lives and should be kept to a minimum.

8   Civil society and the private sector are important parts of the solution.

9   Assessment and monitoring can improve reconstruction outcomes.

10 To contribute to long-term development, reconstruction must be sustainable.

The last word: Every reconstruction project is unique.

DISASTER TYPES AND IMPACTS

This chapter presents a range of data and a brief discussion on the nature, distribution, and impact of disaster events. It also covers the emerging understanding of the nexus between disaster risk and poverty. This information is provided to help policy makers and World Bank task managers who may need data, concepts, or policy arguments to justify attention to disaster risk reduction (DRR) in reconstruction or to define DRR policy objectives in the context of public investment planning in general.

Data collection on disasters has improved markedly in recent years and several authoritative sources are listed in the resources section below. The International Strategy for Disaster Reduction (ISDR) was launched in 2000 as a framework to coordinate actions to address disaster risks at the local, national, regional, and international levels. The Hyogo Framework for Action 2005–2015 (HFA), endorsed by 168 United Nations (UN) member states at the World Conference on Disaster Reduction in Kobe, Hyogo, Japan, in 2005, urges all countries to make major efforts to reduce their disaster risk by 2015.

In 2009, the United Nations International Strategy for Disaster Reduction Secretariat (UNISDR) published the *2009 Global Assessment Report on Disaster Risk Reduction*, the first biennial global assessment of disaster risk reduction, prepared in context of the implementation of the ISDR. The report, entitled *Risk and Poverty in a Changing Climate: Invest Today for a Safer Tomorrow*, urges a radical shift in development practices, and a major new emphasis on resilience and disaster planning. The report's authors express the concern that response mechanisms after the event are never enough.[1]

This chapter summarizes a number of points from the *Global Assessment Report*, as well as data from the ISDR's Disaster Statistics, 1991–2005 and from the Centre for Research on the Epidemiology of Disasters (CRED) 2008 Annual Disaster Statistical Review.[2] (See sidebar for a description of CRED.)

Natural Disaster Definitions, Frequencies, and Impacts

CRED defines a disaster as "a situation or event [which] overwhelms local capacity, necessitating a request to a national or international level for external assistance; an unforeseen and often sudden event that causes great damage, destruction and human suffering."

Disasters are the convergence of hazards with vulnerabilities. As such, an increase in physical, social, economic, or environmental vulnerability can mean an increase in the frequency of disasters. The complete EM-DAT divides disasters into 2 categories (natural and technological), and further divides the natural disaster category into 5 subcategories, which in turn cover 12 disaster types and more than 30 subtypes. The principal categories and subcategories are shown below.

The Centre for Research on the Epidemiology of Disasters (CRED)

CRED is a nonprofit institution with international status under Belgian law, located in Brussels at the School of Public Health of the Université Catholique de Louvain (UCL). CRED has been active for more than 30 years in the fields of international disaster and conflict health studies, with research and training activities linking relief, rehabilitation, and development, and in 1980 became a World Health Organization (WHO) Collaborating Centre as part of WHO's Global Program for Emergency Preparedness and Response.

Since 1988, with support from the United States Agency for International Development's (USAID) Office of U.S. Foreign Disaster Assistance (OFDA), CRED has maintained the Emergency Events Database (EM-DAT), a worldwide database on disasters. The database contains data on the occurrence and effects of almost 16,000 natural and technological disasters in the world from 1900 to the present. Its main objective is to assist humanitarian action at both national and international levels and aims at rationalizing decision making for disaster preparedness and at providing a more objective base for vulnerability assessment and priority setting. The database is compiled from various sources, including UN agencies, nongovernmental organizations (NGOs), insurance companies, research institutions, and press agencies. CRED consolidates and updates data on a daily basis, checks it at 3-month intervals, and conducts annual revisions at the end of each calendar year.

EM-DAT data are used by a range of international agencies, including ISDR and the World Bank, for reporting and analyzing disaster statistics.

1. UNISDR, 2009, *Risk and Poverty in a Changing Climate: Invest Today for a Safer Tomorrow. 2009 Global Assessment Report on Disaster Risk Reduction,* (Geneva: United Nations), http://www.preventionweb.net/gar09.
2. Jose Rodriguez et al., 2009, *Annual Disaster Statistical Review 2008: The numbers and trends,* (Brussels: CRED), http://www.cred.be/publication/annual-disaster-statistical-review-numbers-and-trends-2008.

 For access to additional resources and information on this topic, please visit the handbook Web site at www.housingreconstruction.org.

Disaster subcategory definitions

Geophysical: Events originating from solid earth

Meteorological: Events caused by short-lived/small to meso-scale atmospheric processes (in the spectrum from minutes to days)

Hydrological: Events caused by deviations in the normal water cycle and/or overflow of bodies of water caused by wind set-up

Climatological: Events caused by long-lived/meso- to macro-scale processes (in the spectrum from intraseasonal to multi-decadal climate variability)

Biological: Disaster caused by the exposure of living organisms to germs and toxic substances

Natural Disaster Categories, Types, and Subtypes

Biological	Geophysical	Hydrometeorological	
		Hydrological	**Meteorological**
Epidemic	**Earthquake**	**Flood**	**Storm**
■ Viral infectious disease	**Volcano**	■ General flood	■ Tropical cyclone
■ Bacterial infectious disease	**Mass movement (dry)**	■ Storm surge/coastal flood	■ Extra-tropical cyclone
■ Parasitic infectious disease	■ Rockfall	**Mass movement (wet)**	■ Local storm
■ Fungal infectious disease	■ Landslide	■ Rockfall	**Climatological**
■ Prion infectious disease	■ Avalanche	■ Landslide	**Extreme temperature**
Insect infestation	■ Subsidence	■ Avalanche	■ Heat wave
Animal stampede		■ Subsidence	■ Cold wave
			■ Extreme winter condition
			Drought/wildfire
			■ Forest fire
			■ Land fire

Source: UCL, "EM-DAT: The OFDA/CRED International Disaster Database," UCL, http://www.emdat.be.

Disaster victims in CRED data include both those killed and those otherwise affected. Using a different set of data to separate only those killed provides another striking indicator of the impact of disasters in recent years.

Disaster Fatalities by Type of Disaster and Level of Development, 1991–2005

Country type	Hydrometeorological				Geophysical		Biological	Total
	Flood	Windstorm	Drought*	Slide	Earthquake & tsunami	Volcano	Epidemic	
Organisation for Economic Co-operation and Development member country	2,150	5,430	47,516	426	5,910	44	442	**61,918**
Central and Eastern Europe and Commonwealth of Independent States	2,635	512	3,109	1,176	2,412	-	568	**10,412**
Developing countries	97,061	65,258	12,599	9,369	397,303	900	47,616	**630,106**
Least developed countries	20,127	149,517	3,320	1,739	9,247	201	70,588	**254,739**
Countries not classified	99	767	57	23	2,277	-	104	**3,327**
Total	**122,072**	**221,484**	**66,601**	**12,733**	**417,149**	**1,145**	**119,318**	**960,502**

Source: ISDR Disaster Statistics, http://www.unisdr.org.   *Drought-related disaster category includes extreme temperatures.

Disasters are frequently classified according to their frequency and their impact, as measured by number of victims and economic damage. The following tables show disaster data for 2008 and averages for the 2000–2007 time period.

Natural Disasters: Frequency by Region

No. of natural disasters	Africa	Americas	Asia	Europe	Oceania	Global
Climatological						
2008	10	4	9	9	0	32
2000-07 (Average)	9	14	13	19	2	57
Geophysical						
2008	3	8	18	2	1	32
2000-07 (Average)	3	7	22	3	2	37
Hydrological						
2008	48	39	73	9	9	178
Avg. 2000-07	42	39	82	28	5	196
Meteorological						
2008	10	44	43	13	2	112
Avg. 2000-07	9	34	42	15	7	107
Total						
2008	71	95	143	33	12	354
Avg. 2000-07	63	94	160	65	16	397

No. of victims (in millions)	Africa	Americas	Asia	Europe	Oceania	Global
Climatological						
2008	14.5	0.1	91.1	0.0	0.0	105.6
Avg. 2000-07	9.6	1.1	68.4	0.3	0.0	79.5
Geophysical						
2008	0.0	0.1	47.6	0.0	0.0	47.8
Avg. 2000-07	0.1	0.4	3.6	0.0	0.0	4.2
Hydrological						
2008	1.0	15.9	27.7	0.2	0.1	44.9
Avg. 2000-07	2.5	1.3	101.7	0.4	0.0	105.9
Meteorological						
2008	0.8	3.7	11.4	0.0	0.0	15.9
Avg. 2000-07	0.4	2.8	38.0	0.4	0.0	41.7
Total						
2008	16.2	19.9	177.8	0.3	0.1	214.3
Avg. 2000-07	12.6	5.6	211.8	1.1	0.1	231.2

Damages (billions of 2008 US$)	Africa	Americas	Asia	Europe	Oceania	Global
Climatological						
2008	0.4	2.0	21.9	0.0	0.0	24.4
Avg. 2000-07	0.0	2.4	1.1	3.5	0.4	7.4
Geophysical						
2008	0.0	0.0	85.8	0.0	0.0	85.8
Avg. 2000-07	0.8	1.0	9.5	0.3	0.0	11.6
Hydrological						
2008	0.3	12.1	3.7	1.3	2.1	19.5
Avg. 2000-07	0.4	1.9	9.7	7.7	0.3	19.9
Meteorological						
2008	0.1	50.0	6.8	3.4	0.5	60.7
Avg. 2000-07	0.1	38.6	10.7	3.0	0.3	52.6
Total						
2008	0.9	64.0	118.2	4.7	2.5	190.3
Avg. 2000-07	1.3	43.8	31.0	14.5	1.0	91.6

Source: UCL, "EM-DAT: The OFDA/CRED International Disaster Database," http://www.emdat.be. [Original tables contain rounding errors.]

The following table shows that disasters affect people—as well as regions—unequally.

Average Number of People Affected by Continent and Disaster Origin, 1991–2005 (per million inhabitants)

	Hydrometerological	Geological	Biological
Africa	22,803	81	951
Americas	5,186	374	149
Asia	56,486	794	63
Europe	2,404	46	17
Oceania	39,817	585	16

Source: "ISDR Disaster Statistics," http://www.unisdr.org.

Understanding Intensive versus Extensive Disaster Risks

The 2009 *Global Assessment Report on Disaster Risk Reduction* points out the distinction between intensive and extensive disaster risks. Intensive risks are those that produce high mortality disaster events. The report notes that between January 1975 and October 2008, 0.26 percent of the 8,866 disaster events recorded accounted for 78.2 percent of the mortality. These included the 1983 drought in Ethiopia; the 1976 earthquake in Tanshan, China; and, more recently, the Indian Ocean tsunami in 2004 and Cyclone Nargis in Myanmar in 2008.

At the same time, losses from low-intensity, but more extensive disaster events continue to affect housing, local infrastructure, and large numbers of people. The report states that "99.3% of local loss reports in 12 Asian and Latin American countries that were sampled accounted for only 16% of the mortality but 51% of housing damage. These losses caused by 'extensive risk' are pervasive in both space and time…"

ISDR states that the drivers of both types of risk are similar: locally specific increases in exposure, vulnerability, and hazard due to broader urbanization, economic and territorial development, and ecosystem decline, exacerbated by poor urban governance and the vulnerability of rural livelihoods.

Poverty and Exposure to Risk

The correlation between poverty and risk is becoming clearer as disaster data collection and analysis improves. Empirical evidence from all regions of the world shows that disasters produce measureable declines in income, consumption, and human development indicators, and that these effects are disproportionally concentrated in poor households and communities. The effects of disasters are especially pronounced in some of the indicators of human development most important to poverty reduction: productivity, health, and education.

Poor households have a limited capacity to buffer themselves against disaster losses, whether the risks are intensive or extensive. They may also have limited social protection, depending largely on whatever public measures are available during disaster recovery.

For access to additional resources and information on this topic, please visit the handbook Web site at www.housingreconstruction.org.

This discussion points out the importance of investment in measures to prevent and reduce disaster risk. By the time a disaster strikes, it may seem too late to interrupt the negative feedback loop between poverty and disaster risk. But this is not the case; there are numerous opportunities for users of this handbook to contribute to the effort of reducing poverty by addressing disaster risk factors in reconstruction. These include:

- Ensuring that financial assistance for housing and community reconstruction reaches the poor
- Insisting that investments are made in disaster risk reduction in the reconstruction of housing, infrastructure, and other community assets
- Involving local professionals (builders, architects, engineers) in training and post-disaster planning oriented toward risk reduction
- Making permanent improvements in instruments such as planning guidelines, building codes, and housing designs that will continue to be used after reconstruction
- During reconstruction, encouraging government, academic institutions, the private sector, and civil society to think proactively about measures they can take to reduce future community exposure to hazards
- Working with government to establish social protection mechanisms that help different social groups prepare for and recover from disasters

The following figure attempts to capture some of the interactions of poverty and disaster risk.

The Disaster Risk-Poverty Nexus

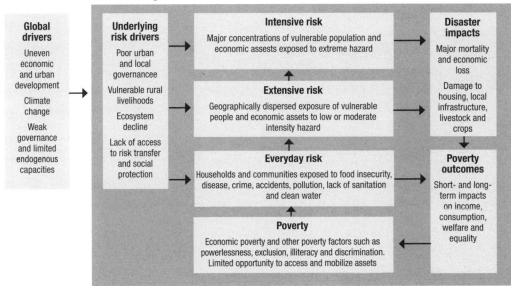

Source: ISDR, 2009, *Risk and Poverty in a Changing Climate: Invest Today for a Safer Tomorrow*, Global Assessment Report on Disaster Risk Reduction, (Geneva: United Nations), http://www.preventionweb.net/gar09.

A Note on the Interpretation of Disaster Data

Over the last 30 years, the development of telecommunications and the media and increased international cooperation have played a critical role in the number of disasters that are reported internationally. In addition, increases in humanitarian funds have encouraged reporting of more disasters, especially smaller events.

CRED has concluded that the increase in the number of disasters until about 1995 is explained partly by better reporting of disasters in general, partly due to active data collection efforts by CRED, and partly due to real increases in both the frequency and the impact of certain types of disasters. They estimate that the data in the most recent decade present the least bias and reflect a real change in numbers. This is especially true for floods and cyclones.

CRED has warned users of its data that although climate change could affect the severity, frequency, and spatial distribution of hydrometeorological events, users need to be cautious when interpreting disaster data and take into account the inherent complexity of climate and weather related processes—and remain objective scientific observers. The figure below shows trends in frequency and impact of disasters over the 1989–2008 time frame.

Trends in Occurrence of Disasters and Number of Victims, 1989–2008

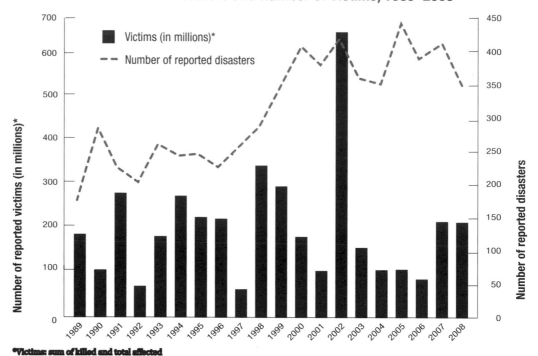

*Victims: sum of killed and total affected

Source: EM-DAT, UCL: Brussels, http://www.emdat.be.

Resources

Center for Hazards and Risk Research (CHRR). "Hotspots." http://www.ldeo.columbia.edu/chrr/research/hotspots/.

Centre for Research on the Epidemiology of Diseases (CRED). Université Catholique de Louvain, Ecole de Santé Publique. Brussels. http://www.cred.be/.

ISDR. 2009. *Risk and Poverty in a Changing Climate: Invest Today for a Safer Tomorrow.* Global Assessment Report on Disaster Risk Reduction. Geneva: United Nations. http://www.preventionweb.net/gar09.

Rodriguez, Jose et al. 2009. *Annual Disaster Statistical Review 2008: The numbers and trends.* Brussels: CRED. http://www.cred.be/publication/annual-disaster-statistical-review-numbers-and-trends-2008.

DISASTER RISK MANAGEMENT IN RECONSTRUCTION

Disaster risk management (DRM) is a systematic process of using administrative directives, organizations, and operational skills and capacities to implement strategies, policies, and improved coping capacities to lessen the adverse impacts of hazards and the possibility of disaster. Disaster risk reduction (DRR), a related but narrower concept, is the practice of reducing disaster risks through systematic analysis and management of the causal factors of disasters, including reduced exposure to hazards, lessened vulnerability of people and property, wise management of land and the environment, and improved preparedness.[1] This chapter uses the broader concept of DRM.

Disaster risk is the potential losses, in lives, health status, livelihoods, assets, and services that could occur in a particular community or a society over some specified future time period due to disasters. Disaster risk is created by a complex interaction of factors, both natural and human-generated, that expose people and the environment to hazards. The following types of interventions are used to manage disaster risk: (1) policy and planning measures, (2) physical preventive measures, (3) physical coping and adaptive measures, and (4) capacity building at the community level.

Policy makers and reconstruction project task managers will probably never conduct a risk analysis, but they may have to evaluate a mitigation plan for a neighborhood or infrastructure system or make a relocation decision. The commitment to reducing disaster risk must drive such decisions.

Specific DRM actions that can be taken are discussed throughout this handbook as they apply to the chapter topics. This chapter gives users of the handbook a working understanding of what disaster risk analysis entails and of how both short- and long-term mitigation measures are used to reduce disaster risk in reconstruction. It focuses on the basic principles, policies, and instruments of DRM, and on their application in a reconstruction program. Because of its post-disaster focus, this chapter principally addresses interventions 1 and 3, above.

Disaster Timeline

DRM as a discipline can be conceived of as a program of interventions whose focus and relative importance changes from the pre-disaster period to the disaster period to the post-disaster period. The figure below attempts to show the relative importance of each of these interventions at different points in time relative to a disaster event. This handbook focuses on the post-disaster reconstruction period, as does the discussion in this chapter.

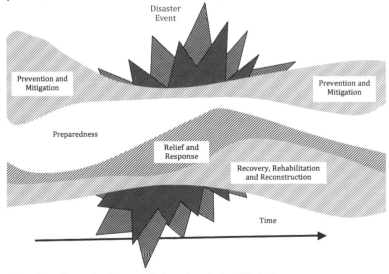

(Adapted from International Recovery Platform, *Learning from Disaster Recovery – Guidance for Decision Makers*, 2007, p. 14, Fig H.)

1. United Nations International Strategy on Disaster Reduction (UNISDR), 2009, "UNISDR Terminology on Disaster Risk Reduction," http://www.unisdr.org/eng/terminology/terminology-2009-eng.html. Some institutions use these two terms synonymously.

Applying Disaster Risk Management Principles in Reconstruction

The key DRM decisions related to housing and community reconstruction, and the handbook chapters where these issues are discussed, are the following:

- Whether and where to relocate households (📖 Chapter 5, To Relocate or Not to Relocate)
- The housing technology, construction procedures, and norms to be used in construction, retrofitting, and reconstruction (📖 Chapter 6, Reconstruction Approaches, and 📖 Chapter 10, Housing Design and Construction Technology)
- How to restore infrastructure services, including site selection and mitigation measures in both new construction and retrofitting (📖 Chapter 7, Land Use and Physical Planning, and 📖 Chapter 8, Infrastructure and Services Delivery).

When making these decisions, it is important to look for opportunities to promote both short- and long-term DRM measures.

In the immediate term, there is an opportunity to analyze risk and use the outputs of the analysis to identify cost-effective risk mitigation measures to implement in the reconstruction program. In an area of habitual flooding, for example, housing may be reconstructed in-situ on stilts.

At the same time, the disaster may create an opportunity—while the public's consciousness of disaster risk is heightened—to identify and begin to implement longer-term DRM measures. Longer-term mitigation includes strengthening DRM institutions and other measures that have a more systematic and far-reaching impact but require time to plan and implement. For example, after a flood, a commitment might be made to begin planning an early warning system to ensure evacuations under certain flooding conditions that includes rainwater monitoring and radio announcements, even though it could take time to fully implement.

Conducting a Risk Analysis

An all-hazards risk assessment (or risk analysis) is a determination of the nature and extent of risk developed by analyzing all potential hazards and evaluating existing conditions of vulnerability that could pose a potential threat or harm to people, property, livelihoods, and the environment on which they depend. The risk analysis shows vulnerabilities in a particular location and quantifies the potential impact of a disaster on a community. These factors are crucial when selecting among various mitigation options or deciding whether to relocate a community. Project managers should investigate whether a risk analysis has already been conducted for the location. It may be available from regional or international bodies. Four steps of a risk analysis and the issues they address are described below.

> *DRR is particularly important in developing countries: 90 percent of disaster-related injuries and deaths are sustained in countries with per capita income levels below $760 per year. In addition, losses from natural disasters are 20 times greater (as a percentage of gross domestic product) in developing countries than in industrialized countries.*
>
> *UNISDR (United Nations International Strategy for Disaster Reduction), 2004, Living with Risk: A Global Review of Disaster Reduction Initiatives. Vol 1. (Geneva: UNISDR).*

Step 1: Identify hazards and analyze their probability.
How frequently do different types of disasters occur here? What is the probability that they will recur?

Hazard identification. To help predict the magnitude and duration of a potential hazard, a record of similar previous hazards is developed and characteristics of those hazards are collected and compared. The data collected should show magnitude, duration, impact, date, and extent. (The table in the annex provides additional information about potential sources of data.) Changes in temperature and rainfall projected from climate change should be factored into a risk analysis using projections or global models. The 📖 case study below explains the functions of the Central American Probabilistic Risk Assessment (CAPRA), an example of a regional organization that can provide risk assessment data.

Hazard probability. Using hazard data, the return period of a disaster in a specific area can be estimated. Recent trends, such as those produced by climate change, may not be included in historic data, but should be taken into consideration. The output of the probabilistic hazard analysis is a map of the hazard for various return periods.[2] Specific outputs include (1) wind speeds, (2) inundation depths and extents, and (3) ground motion.

2. A return period (or recurrence interval) is an estimated interval of time between hazard events of a certain intensity or size. It is a statistical measurement averaged over an extended period of time. The trauma of the disaster tends to cause people to underestimate recurrence intervals (i.e., assume the disaster will recur sooner than historical information would suggest).

 For access to additional resources and information on this topic, please visit the handbook Web site at www.housingreconstruction.org.

346   SAFER HOMES, STRONGER COMMUNITIES: A HANDBOOK FOR RECONSTRUCTING AFTER NATURAL DISASTERS

Step 2: Create an inventory of exposure and vulnerability.

What assets of this community might be affected if the disaster recurs? How would they be affected? How likely are various outcomes?

Develop an asset inventory. Identify the buildings or infrastructure at risk, including information on the structure's use, materials, age, and dimensions. The information should be collected at a geographic level relevant to the analysis (e.g., city block, neighborhood, or region). Sources of data may include government census reports, community-level surveys, and high-resolution satellite images. (Satellite data used in any process should be validated using a second method, such as a site survey. See 📖 Chapter 17, Information and Communications Technology in Reconstruction, for a discussion of these issues.)

Develop valuation data. Estimate replacement costs of the assets identified. If valuation data are not available, estimates based on gross domestic product and comparative country-level data can be used as proxies.

Catalogue vulnerability characteristics. Some structures withstand specific types of disasters better than others. The factors that contribute to a building's vulnerability include roof type, roof-wall connection, construction type, window protection, height, foundation type, and elevation. The prevalence of these factors must be catalogued in order to develop estimates of loss.

Identify or develop damage and loss functions. Physical vulnerability is described as the degree to which an asset may sustain damage when exposed to a hazard. A vulnerability analysis quantifies the susceptibility of an asset type to damage for each magnitude of hazard. Develop damage and loss functions for buildings, content, and infrastructure for different return periods and hazards, based on the information above, local damage data, existing vulnerability curves developed for similar structures, and expert or heuristic judgments. Historic information and community experience from past events help predict the effect of a disaster on a community, including identifying undamaged areas, hazard durations, and cascading hazards. Potential for damage is measured using the mean damage ratio (MDR), the ratio of damage incurred to the asset's replacement cost. Two outputs from this analysis include the following.

- Vulnerability or damage function: The curve that relates the MDR to the magnitude of a hazard.
- Loss function: The curve that relates the repair cost to the magnitude of the hazard.

Step 3: Estimate the probability of losses.

What could losses cost us?

A computer model is usually used to overlay the hazard and vulnerability data (using a geographic information system [GIS]) and to map loss estimates for each hazard probability developed above. Data for this step are often collected and posted by the United Nations Office for the Coordination of Humanitarian Affairs (http://www.unocha.org/). After this step, an at-risk community should be able to better understand what the impact could be should disaster strike. Two outputs from this analysis include the following.

- Average annual loss (AAL): The sum of all monetary losses over all return periods multiplied by the probability of a disaster occurring. Expressed mathematically, AAL = ($ loss) x $\sum$ (probability of occurrence).
- Loss exceedance curve (LEC): A curve that shows the correlation between the average recurrence interval and losses. It is used to predict losses for different recurrence intervals.

Step 4: Develop a risk atlas.

Where are losses likely to happen?

A risk atlas illustrates hazard areas and corresponding community damages and losses for a series of probable events over different return periods. A separate map is generated for each return period event. The atlas is used to identify which mitigation measures need to be considered. Specific examples of mitigation measures are provided in the next section.

Types of DRM Measures

- *Policy and planning: e.g., institutional, policy, and capacity-building measures designed to increase the abilities of public and private institutions to manage disaster risks.*
- *Physical preventative: e.g., building sea-walls as part of flood defense mechanisms.*
- *Physical coping and adaptive: e.g., flood shelters for use during a disaster event.*
- *Capacity building at the community level: e.g., developing a community-based hazard mitigation plan.*

Source: Department for International Development, 2005, "Natural Disaster and Disaster Risk Reduction Measures, A Desk Review of Costs and Benefits," http://www.dfid.gov.uk/ Documents/publications/disaster-risk reduction-study.pdf.

Identifying and Selecting Mitigation Measures

Hazard mitigation is any action taken to reduce or eliminate the risks from natural hazards. Once the risk analysis has been carried out, the information can be used to define and implement hazard mitigation activities and projects. To do this, the mitigation options must be identified, and the costs and benefits of each option evaluated. Based on that analysis, implementation decisions can be made.

Various mitigation measures may be considered when planning housing and infrastructure reconstruction, but the most feasible will be short-term measures that minimize the destructive and disruptive effects of disasters on the built environment. Longer-term measures should also be initiated. These are discussed in the next section.

The principal mitigation measures are:

- locational mitigation, in which damage or loss is reduced by avoiding the physical impacts of an event;
- structural mitigation, in which damage is resisted through bracing of buildings or construction of a levee;
- operational mitigation, in which damage or loss is minimized by interventions such as emergency planning, tsunami warning, or other temporary measures; and
- risk sharing, in which the cost of the damage is shared.[3]

The 📖 case study on Pupuan, Indonesia, below, shows how the full range of mitigation options should be considered, even those that are political difficult.

Short-Term Mitigation for Housing

Based on the risk assessment described above, alternative mitigation measures for housing can be considered. These measures are not mutually exclusive; more than one may apply. Information in other handbook chapters can be used to support the evaluation of the options, as noted below. Site selection for housing is likely to take place in an extremely decentralized manner (at the household and village levels); therefore, communication with the public should be considered an important mitigation tool.

Choose hazard-resistant housing designs and construction technologies. For housing, design standards exist internationally and are readily available for various types of construction and disasters. Building codes are the most common regulatory instrument for ensuring safe construction methods, although they may not be promulgated or enforced. An authoritative source of model codes for residential and commercial buildings is the International Code Council.[4] Also see 📖 Chapter 10, Housing Design and Construction Technology, for a discussion of housing construction issues.

Relocate housing. DRM considerations should be applied in site selection for both temporary and permanently relocated housing. While reconstruction should not occur in areas frequently affected by hazards, this is admittedly difficult where nonvulnerable alternatives are scarce or land use regulations do not prevent it. 📖 Chapter 5, To Relocate or Not to Relocate, discusses the range of issues that arise in evaluating the relocation option. Reconstruction guidelines should include the topic of site selection, as should the reconstruction communication program, so that both agencies and individuals are educated about the importance of these decisions.[5] 📖 Chapter 3, Communication in Post-Disaster Reconstruction, explains the principles of communication with the affected community and the general public.

Rehabilitate and retrofit housing. Rehabilitation deals with structural and nonstructural modification of buildings and infrastructure facilities. Since new zoning laws and updated design and construction codes usually can't be applied retroactively, it is important that, to reduce the impact of disasters, the safety and structural integrity of existing buildings and infrastructure facilities is improved during the rehabilitation process.

Train builders in DRM. The training program should provide an understanding of how the hazards may affect the household and community and of recommended mitigation strategies for the specific affected region. 📖 Chapter 16, Training Requirement in Reconstruction, describes some of the specific content in the training programs for the builders.

3. Charles Scawthorne, 2009, "Disaster Reduction and Recovery: A Primer for Development Managers" (Washington, DC: World Bank).
4. International Code Council, http://www.iccsafe.org/.
5. U.S. Federal Emergency Management Agency (FEMA), 2001, *Telling the Tale of Disaster Resistance: A Guide to Capturing and Communicating the Story* (Denver: FEMA Region VIII), http://www.fema.gov/library/viewRecord.do?id=1762.

Mitigate the existing site. The location or structure of a building can greatly increase its vulnerability. Mitigation measures should address the specific causes of a building's or infrastructure's vulnerability. For example, it is illogical to invest in expensive reinforcement of a structure resting on unstable soil. Removal, relocation, or elevation of in-place structures in highly hazardous areas, especially those built before building codes were established, is frequently the only option. A community must prioritize options based on the importance of a structure and its relative vulnerability. For instance, a venerated historic religious building with a high potential loss may take priority over other buildings and infrastructure.

Short Term Mitigation for Infrastructure

Based on the risk assessment described above, alternative mitigation measures for infrastructure can be considered. These measures are not mutually exclusive; more than one may apply. The information in ✍ Chapter 8, Infrastructure and Services Delivery, complements this section, providing guidance on a DRM-oriented infrastructure project development process.

Select or change the site. DRM considerations should be applied in site selection for new infrastructure. Reconstruction should not occur in areas frequently affected by hazards, although it may be impossible to avoid if housing settlement has already taken place and services are needed and where nonvulnerable alternatives are scarce. Where site selection cannot be used to avoid risk, other mitigation measures are applied.

Mitigate the existing site. It is often difficult to relocate infrastructure to a site that does not experience hazards. For example, a road may have to cross a river or stream and therefore enter a floodplain. In this example, mitigation might consist of designing a bridge with a proper elevation and span based on an analysis of the floodplain. Using information from the risk analysis, the design of the bridge is fine-tuned to the hazards and vulnerabilities at the site (e.g., soft soils, liquefaction potential, etc.). A community must prioritize options based on the importance of the facility and its relative vulnerability. For instance, a water system with a high potential loss may take priority over other infrastructure. The ✍ case study on Bamako, Mali, below, explains how solid waste management and storm water management were used to reduce flooding in an urban area.

Redesign or reengineer the infrastructure. Design and engineering improvements are used to retrofit in-place infrastructure. Because construction techniques and technologies are constantly improving, one should research the most recent recommended practices when considering engineering improvements for infrastructure.

Use protection and control measures (applies to both housing and infrastructure). Protective and control measures focus on protecting structures by erecting protective barriers (e.g., dams and reservoirs, levees, discharge canals, floodwalls and sea-walls, retaining walls, safe rooms or shelters, and protective vegetation belts) and deflecting the destructive forces from vulnerable communities, structures, and people. Some of these measures may be appropriate to implement during reconstruction; others may be longer-term investments that require time to plan, finance, and implement. The requirements for these measures should be incorporated into the land use planning framework, based on a rigorous assessment of risks. See ✍ Chapter 7, Land Use and Physical Planning, for a discussion of the role of planning in risk mitigation, and the ✍ case study on the use of a coastal protection zone as a mitigation strategy in Sri Lanka, below.

Comparing Mitigation Options

To select the preferred option for mitigating risk in a particular situation, it is necessary to compare options in an objective manner according to consistent criteria.[6] Several methodologies can be used to evaluate and select mitigation options and rank the potential mitigation projects; two are discussed below. These evaluation tools are used after the potential hazards and vulnerabilities in a community have been identified using risk analysis. The selection of options, including the relative weighting of criteria, is ideally carried out with the participation of the affected community.

STAPLEE. One methodology that considers a comprehensive set of criteria is referred to as "STAPLEE." This methodology examines the **S**ocial, **T**echnical, **A**dministrative, **P**olitical, **L**egal, **E**conomic, and **E**nvironmental opportunities and constraints of implementing a particular mitigation measure. To use this methodology and other similar methodologies, the mitigation project is evaluated and scored for each criterion. It may also be necessary to weight the criteria to reflect

6. The comparison of mitigation options depends on knowing the improvement in vulnerability that will result from various mitigation options, relative to a baseline, information that may be very difficult to ascertain scientifically. Therefore, subjective judgment will often need to be exercised, which may be the judgment of the affected community itself, solicited using a participatory approach to evaluating alternative mitigation measures.

their relative importance. This scoring could be in the form of a number or a "yes/no" decision. STAPLEE helps determine whether the project is feasible and can be used to compare several mitigation options to each other.

Cost-Benefit Analysis. Another way to evaluate a mitigation project is to use a cost-benefit analysis (also known as a benefit-cost analysis) to determine cost effectiveness. The cost-benefit analysis is used to assess for which alternatives, if any, the benefits outweigh the costs. The steps in a cost-benefit analysis are, for each project:

1. Conduct a hazard risk assessment and compute the AAL before mitigation.
2. Conduct a hazard risk assessment and compute the AAL after mitigation.
3. Determine the present value of the benefit using the difference between the AALs, the project lifespan, and a discount factor for the time value of the benefits.
4. Estimate the cost to implement the mitigation measure and discount those costs as well.
5. Divide the present value of the benefit of the mitigation project by the present value of the cost to mitigate.

The project with the highest cost-benefit ratio produced by this analysis is the preferred mitigation option.

Long-Term Measures

Institutional strengthening. Government agencies at the national and local government levels in disaster-affected countries may already have in place DRM policies and regulations. Implementation of the policies may fall within the jurisdiction of the ministry of public works, the ministry of land, and/or the ministry of urban development and planning departments at different levels. Enforcement may fall within the ministry of public works, civil defense, or police departments. Most disaster-affected urban areas have some type of DRM policies and regulations in place, generally under the jurisdiction of the local planning department or planning commission. The problem is that these measures are often not fully enforced or implemented. Rural areas may not have these policies or regulations in place and may not have defined DRM responsibilities within local agencies.

Although institutional weaknesses differ from country to country, there are some shared concerns that will affect the promotion of DRM principles in reconstruction. The table below provides examples of institutional DRM issues and potential solutions. These issues should be viewed as entry points where work with DRM agencies can begin.

Institutional Weaknesses and Potential Solutions

Institutional weakness	Potential solutions
Building codes have not been established or are not being enforced.	Use the expertise gathered for disaster recovery and the global media focus to promote establishing/updating building codes. Work with the ministry of public works or municipal public works departments and involve enforcement agencies in the discussions.
	Work directly with builders to improve construction practices. Oversight of the construction is key.
Land use/zoning regulations have not been established or are not being enforced.	Use the expertise gathered for disaster recovery and the global media focus to promote establishing/updating land use regulations. Work with the ministry of planning and local planning departments.
There are no clear lines of disaster risk management responsibility among government agencies.	Build on ad hoc institutional arrangements developed for response and recovery from the current disaster to institutionalize responsibilities for prevention and response to future disasters.
Disaster response and recovery plans are limited or nonexistent.	During reconstruction, develop a response and recovery plan using lessons learned from the disaster to determine needs and division of responsibilities.
Incentives for disaster-resistant building practices are weak.	Use computer models or case studies to demonstrate mitigation benefits.
	Perform cost-benefit analysis.
	Promote incentive-based, disaster-resistant programs (insurance programs, government catastrophic pools).

Regulatory measures. It is generally not realistic to implement major regulatory reforms in the immediate aftermath of a disaster; however, a revision to a key ordinance or issuance of guidelines is often feasible. At the same time, the disaster may raise awareness among decision makers such that they become motivated to begin the process for implementing more substantive reforms after the immediate recovery and reconstruction issues have been dealt with.

In most cases, regulatory measures should be considered before other measures because they provide the framework for mitigation decision making, organizing, and financing. Regulatory measures are the legal and other regulatory instruments that governments use to prevent, reduce, or prepare for the losses associated with hazard events. Examples include:

- legislation that organizes and distributes responsibilities to protect a community from hazards;
- insurance regulations that reduce or transfer the financial and social impact of hazards;
- new and/or updated design and construction codes, and land use and zoning regulations (land use planning is detailed in 📖 Chapter 6, Reconstruction Approaches); and
- regulations that provide incentives for implementing mitigation measures.

In post-disaster situations where regulatory measures do not exist, reconstruction and rehabilitation (at a minimum) should reflect the experience and standard practices and guidelines used internationally for similar disasters. For housing, such standards are readily available and can be adapted to the local conditions and environment in an emergency. See 📖 Chapter 10, Housing Design and Construction Technology, for more detail on housing standards.

Community-based hazard mitigation planning. Creating disaster-resistant communities requires community involvement. The figure at right shows the steps in a participatory hazard mitigation planning process. It is similar to the steps described for reconstruction; however, the planning process allows for more participation and longer-term thinking about priorities and options.

Stakeholder workshops conducted during reconstruction can be opportunities for local officials and the community to begin developing the outlines of the longer-term hazard mitigation strategy and planning process. The communications program related to the disaster is a valuable tool for two-way communication between the public and government about DRM. For more information on the use of community involvement in planning and reconstruction, see 📖 Chapter 12, Community Organization and Participation.

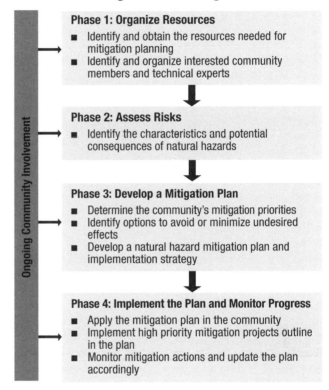

The Hazard Mitigation Planning Process

Phase 1: Organize Resources
- Identify and obtain the resources needed for mitigation planning
- Identify and organize interested community members and technical experts

Phase 2: Assess Risks
- Identify the characteristics and potential consequences of natural hazards

Phase 3: Develop a Mitigation Plan
- Determine the community's mitigation priorities
- Identify options to avoid or minimize undesired effects
- Develop a natural hazard mitigation plan and implementation strategy

Phase 4: Implement the Plan and Monitor Progress
- Apply the mitigation plan in the community
- Implement high priority mitigation projects outline in the plan
- Monitor mitigation actions and update the plan accordingly

Ongoing Community Involvement

Case Studies
1999 Landslide, Pupuan, Indonesia
Not Considering All Potential Risk Mitigation Applications

In January 1999, a landslide in the village of Pupuan, Bali, killed 38 people. Local residents said the cause of the disaster was a combination of high rainfall, significant slope modifications for rice agriculture, housing construction in high-risk areas, lack of infrastructure, and removal of forest cover. The DRR strategies that had been implemented included structural approaches (e.g., levee construction, hillside terracing, hazard-resistant housing) and nonstructural approaches (e.g., strengthening communications networks, human settlement rezoning, strengthened cooperation between nongovernmental organizations [NGOs] and government agencies). However, resource management actions were not pursued (e.g., abandoning hillside rice agriculture and reforesting slopes). Land use and population pressures, in addition to a 1,000-year-old tradition of terraced rice growing, led to strong local resistance to changing the resource use practices that might have avoided the landslide.

Source: Brent Doberstein, 2006, "Human Dimensions of Natural Hazards: Adaptive Management of Debris Flows in Pupuan, Bali and Jimani, Dominican Republic," University of Waterloo, Canada, http://www2.bren.ucsb.edu/~idgec/papers/Brent_Doberstein.doc.

Disaster-Related Data Sharing and Coordination
Central American Probabilistic Risk Assessment Platform
Central America is vulnerable to a wide variety of natural hazards that present a challenge to the region's sustainable social and economic development. In response, the region has taken a proactive stance on risk prevention and mitigation. The CAPRA platform represents an opportunity to strengthen and consolidate methodologies for hazard risk evaluations supporting this stance and existing initiatives. Led by the Center for Coordination for the Prevention of Natural Disasters in Central America (Centro de Coordinación para la *Prevención de los Desastres Naturales en América Central* [CEPREDENAC]), in collaboration with Central American governments, the International Strategy for Disaster Reduction, the Inter-American Development Bank, and the World Bank, CAPRA provides tools to communicate and support decisions related to disaster risk at local, national, and regional levels in Central America. It uses a GIS platform and probabilistic risk assessment to support decisions in such sectors as emergency management, land use planning, public investment, and financial markets. Current CAPRA applications use data for (1) the creation and visualization of hazard and risk maps, (2) cost-benefit analysis tools for risk mitigation investments, and (3) the development of financial risk transfer strategies. Future applications by CAPRA partners may include real-time damage estimates, land use planning scenarios, and climate change studies.

Sources: CAPRA, http://www.ecapra.org/en; *and* CEPREDENAC, http://www.sica.int/cepredenac/.

2004 Indian Ocean Tsunami, Sri Lanka
Delays in Defining Coastal Risk Strategy Affect Housing Reconstruction and Land Ownership
In the housing damage assessment conducted in Sri Lanka in February 2005, after the Indian Ocean tsunami, it was estimated that nearly 98,500 housing units had been damaged. The government of Sri Lanka (GOSL) announced the use of a coastal buffer zone as a disaster prevention mechanism. Based on the buffer zone policy, government initially estimated 55,525 housing units could be reconstructed *in-situ* through the homeowner-driven cash grant program financed by the World Bank and other donors, but that all other households would need to be relocated elsewhere. The buffer zone decision was based more on the need for government to provide an immediate response than on well-researched technical considerations and public consultation, and applying the decision to a densely populated coastal belt had profound implications on the environment, on livelihoods, and on the economy.

Almost immediately, the prohibition of reconstruction in the buffer zone set off a wave of land clearing for housing schemes in the hinterland (some in environmentally sensitive areas). No environmental assessment methodology or environmental management practices were enforced for site selection and construction. As a result, crucial environmental planning practices were ignored. Subsequently, due to many problems with implementation, the GOSL withdrew the buffer zone policy and reverted to the coastal protection zone (CPZ) setbacks stipulated in the Coastal Zone Management Plan already established by the Coast Conservation Department using scientific investigation.

Reverting to the CPZ was positive. It reduced the coastal population that needed to be relocated; the number of owner-driven in-situ grants was revised upward to 78,500 housing units. However, combined with poor communications with the public regarding the change, it also had negative consequences, delaying reconstruction by six months for many families who thought they would have to relocate. It also had a differential economic impact on families in the CPZ: they were offered a donor-built house irrespective of prior land ownership status when it was thought they would have to relocate, whereas families outside the CPZ were eligible for the cash grant, and only if they could document land ownership. Additionally, some poor families inside the buffer zone reportedly sold their land cheaply thinking they could not reconstruct in-situ. If this were widespread, it might have caused a redistribution of wealth in the coastal areas, although there is no documented evidence that this occurred.

Sources: World Bank, 2005, "Sri Lanka Post-Tsunami Recovery Program Preliminary Damage and Needs Assessment," http://go.worldbank.org/BSJBQ6RHI0; *and* World Bank, 2009, "Tsunami Emergency Recovery Program, Implementation Completion and Results Report," Report No. ICR00001105.

1999 Floods, Bamako, Mali
Disaster Risk Management as Sustainable Local Development

Flash flooding throughout Bamako, Mali, in August 1999, caused death, destruction, and significant economic losses for several thousand families. The United States Agency for International Development Office of U.S. Foreign Disaster Assistance (OFDA), in collaboration with Action Contre la Faim, an international NGO that works to provide safe water, analyzed the causes of the flooding and launched a 4-year, US$525,000 mitigation project in the city's most flood-affected district. One of the primary causes of flooding in Bamako, as in many cities, was the disposal of solid waste in waterways, which reduced the storm water capacity of waterways. The project, which aimed to reduce flooding risks by improving storm water management and solid waste management, was part of a larger effort to help local governments improve services, including flood mitigation, which was one of the most critical. Watershed management techniques included improving storm water retention, removing debris from the drainage system, and expanding solid waste management using local collection teams. The project generated livelihood opportunities for unemployed youth, and quickly became self-sustaining, with fees more than offsetting costs. As a consequence, Bamako has not since had a similar flood disaster. The project had other unanticipated impacts, including the reduction in the incidence of water- and mosquito-borne illnesses by 33 percent to 40 percent in the project area.

In a similar project in Kinshasa, Democratic Republic of Congo, in 1998, OFDA calculated that the program, rather than having a cost, produced a projected net savings of US$426 per household, the equivalent of more than 50 percent of annual household income. The Ministry of Health of the Democratic Republic of Congo showed that the project, which included a public health education component, reduced the incidence of cholera in the community by more than 90 percent.

This model of reducing risk by improving local public services, which shows how risk reduction can contribute to broader development goals, can easily be replicated in other cities with similar challenges.

Source: Charles A. Setchell, 2008, "Multi-Sector Disaster Risk Reduction as a Sustainable Development Template: The Bamako Flood Hazard Mitigation Project," Monday Developments (April 2008), http://www.usaid.gov/our_work/humanitarian_assistance/disaster_assistance/sectors/files/Multi_Sector_Disaster_Risk_Reduction.pdf.

Resources

FEMA. 2004. *Communication Strategy Toolkit.* Washington, DC: FEMA. http://www.fema.gov/library/viewRecord.do?id=1774.

FEMA. 2004. *Primer for Design Professionals: Communicating with Owners and Managers of New Buildings on Earthquake Risk* (FEMA 389). Washington, DC: FEMA. http://www.fema.gov/library/viewRecord.do?id=1431.

FEMA. 2001. *Telling the Tale of Disaster Resistance: A Guide to Capturing and Communicating the Story.* Denver: FEMA Region VIII.

FEMA. 2004. *Using HAZUS-MH for Risk Assessment: How-To Guide (FEMA 433).* Washington, DC: FEMA. http://www.fema.gov/library/viewRecord.do?id=1985.

United States Agency for International Development Office of U.S. Foreign Disaster Assistance. "Preparedness and Mitigation Programs." http://www.usaid.gov/our_work/humanitarian_assistance/disaster_assistance/publications/prep_mit/index.html.

United Nations International Strategy for Disaster Reduction (UNISDR). "Library on Disaster Risk Reduction." http://www.unisdr.org/eng/library/lib-index.htm.

World Bank, Global Facility for Disaster Reduction and Recovery, http://gfdrr.org.

World Bank, "Disaster Risk Management," http://go.worldbank.org/BCQUXRXOW0.

For access to additional resources and information on this topic, please visit the handbook Web site at www.housingreconstruction.org.

Hazard Data

Hazard	Type of data/use	Potential data sources
Cyclone	**Land cover data**/Wind barriers (trees, buildings); damage (flying objects, fallen trees)	A, E, NR, FS, RSA, PRSF, PL
	Elevation data/Wind acceleration; coastal surge intrusion	A, E, PW, WR, RSA, PSIP
	Bathymetry (shoreline water depth)/Storm-surge hazard modeling	A, E, MA, NR, PW, WR
	Wind speed maps	PL, PW
	Coastline and still-water elevation maps/Storm-surge hazard modeling	A, E, MA, NR, PW, WR, PL
Drought	**Precipitation and rain gauge data**/Rainfall records and trends	A, ME, WR
	Global humidity index	UNEP/GRID
		University of East Anglia/ Climatic Research Unit
Earthquake	**Soil maps**/Ground motion patterns	A, E, I, L, NR, SS
	Soil and ground conditions maps/Liquefaction susceptibility	DM, E, SS
	Landslide potential data/Post-earthquake landslide potential	DM, E, SS
	Fault line maps	A, DM, E, I, L, NR, SS
Fire	**Fuel maps, land cover maps**/Fire fuel sources	A, E, F, NR, RSA, PRSF
	Critical weather data (low humidity, wind)	A, ME, WR
	Land elevation/Predict fire speed	A, E, PW, WR, RSA, PSIP
Flood	**Digital Elevation Model (DEM) or Digital Terrain Model (DTM) for bare earth**/Predict water flow	A, E, PW, WR, RSA, PSIP
	Contour data/Complements DEM/DTM data	PW, SW
	Historic precipitation data	A, ME, WR, PL
	Soil data/Areas of water infiltration	A, E, I, L, NR, SS
	Locations of river and hydraulic structures (bridges, dams, levees)	A, E, I, L, NR, PW
Landslide	**Slope data (DEM, DTM)**/Areas of susceptibility	A, E, PW, WR, RSA, PSIP
	Soils maps/Areas of high susceptibility	A, E, I, L, NR, SS
	Land cover	A, E, F, NR, RSA, PRSF, PL
Tsunami	**Bathymetry (shoreline water depths)**/Tsunami hazard modeling	A, E, F, MA, NR, PW, WR
	Coastline still-water elevations/Tsunami hazard modeling	A, E, F, MA, NR, PW, WR
	Elevation data/Tsunami intrusion	A, E, PW, WR, RSA, PSIP

KEY: Public ministry, department, or agency: A=Agriculture and Fisheries, DM=Disaster Management, E=Environment, I=Irrigation, L=Land Management, F=Fisheries, MA=Maritime Affairs, ME=Meteorological, NR=Natural Resources, PL=Local Planning, PW=Public Works, WR=Water Resources, FS=Forestry, RSA=Remote Sensing Agencies (such as IKONOS or NASA's ASTER), SW=Storm Water Management, SS=Soil Survey. Private sources: PRSF=Private Remote Sensing Firm, PSIP=Private Satellite Imagery Provider.

Vulnerability Data

Asset	Type of Data/Use	Potential Data Sources
Population	**Census data**/Population locations, vulnerable populations (e.g. young, elderly, impoverished, etc.), and demographics	CSO, MP, MS
Buildings	**Critical infrastructure – medical care**/Locations and capacities of hospitals and clinics	MH, MP
	Critical infrastructure – police and civil defense/ Locations and capacities of responders	CD, MP
	Critical infrastructure – fire/ Locations and capacities of responders	CD, MP
	Building locations/Structural damage and loss locations	CSO, MP, MS, PRSF, PSIP, RSA
	Building characteristics/Structural damage and loss quantification, building types, construction types, vulnerable characteristics (e.g. roof type, first floor elevation, foundation type, etc.)	LB, MP, PRSF, PSIP, PW, RSA
	Vulnerability functions/Structural damage and loss quantification	ACOE, FIA, U
Transportation Lifelines	**Road data**/Damage locations, road closures	MP, MT, PRSF, PSIP, RSA
	Bridge data/Damage locations, bridge closures	MP, MT, PRSF, PSIP, RSA
	Railroad data/Damage locations, rail closures	MP, MT, PRSF, PSIP, PRC, RSA
	Port data/Damage locations, port closures, economic loss	MA, MP, MT, PRSF, PSIP, PPC, RSA
Utility Lifelines	**Electrical Data**/Damage locations, power outages	MP, MPw, PW
	Potable Water Data/Damage locations, water availability	MP, MW, PW
	Communication Data/Damage locations, communication outages	MC, MP, PW

KEY: ACOE=U.S. Army Corps of Engineers, CD=Civil Defense, CSO=Central Statistical Organization, FIA=Flood Insurance Administration, LB=Local Builders, MA=Maritime Affairs, MC=Ministry of Communications, MH=Ministry of Health, MP=Ministry of Planning, MPw=Ministry of Power, MS=Ministry of Statistics, MT=Ministry of Transportation, MW=Ministry of Water, PPC=Private Port Companies, PRC=Private Rail Companies, PW=Public Works, RSA=Remote Sensing Agencies, U=Universities.

MATRIX OF DISASTER PROJECT FEATURES

Areas of comparative analysis	Gujarat (India) earthquake (2001)	Sri Lanka earthquake / tsunami (2004)	Indonesia (Aceh and Nias) earthquake / tsunami (2004)	Katrina (USA) hurricane (2005)	Pakistan earthquake (2005)
1. Scale of disaster	25,000 people dead and 200,000 injured, 600,000 displaced or homeless and 348,000 houses destroyed and 844,000 damaged as per initial survey.[1]	35,322 people dead and 21,441 injured, 500,000 displaced and 114,069 houses damaged or destroyed.	167,900 people dead or missing, 513,500 displaced and 113,500 houses damaged or destroyed as per initial survey.	1,836 dead and 705 missing, 0.6 Million displaced and 70,000 houses damaged or destroyed as per initial survey.	73,338 people dead and 128,304 injured 3.5 million people homeless, 462,363 houses destroyed and 109,956 damaged.
2. Reconstruction strategy	80% owner driven reconstruction program and 20% public-private partnerships (NGO's) driven program.	As of November 2006, policy change leading to 73% owner driven reconstruction program and 27% donors or NGO-driven program.	100% donor and NGO driven program.	100% Government sponsored contractor driven program.	100% owner driven reconstruction program.
3. Government financial assistance	Not a uniform package, leading to equity issues. Assistance disbursed in 3 tranches. Compensation ranged from INR 5,000 to 90,000 (US$126 to 2277).	Uniform assistance package. Assistance of LKR 100,000 (US$880) disbursed in 2 tranches for partially damaged houses and LKR 250,000 (US$2,200) disbursed in 4 tranches for destroyed houses.	Uniform assistance package. Assistance of IDR 20 Million (US$2,000) for repairable (damaged) house and IDR 42 Million (US$4,200) for full construction of house (destroyed)	Not a uniform package. Assistance based on actual value of house and insurance cover. Assistance of up to US$150,000 available for homeowner.	Uniform assistance package. Assistance of PKR. 75,000 (US$1,250) for partially damaged house disbursed in 2 tranches and assistance of PKR 175,000 (US$2,917) for fully destroyed house disbursed in 4 tranches.
4. Government technical assistance and training	Government providing technical assistance through formal training program under which 29,000 masons and 6,200 engineers trained. Additional training through donor technical assistance packages taking place. GSDMA initiated mason training in collaboration with Gujarat Council of Vocational Training. 450 masons certified under the program as of October 2006.	Government providing technical assistance and advice but no formal training program exists.	Done by government through employing services of supervision consultants who are trained on new building codes and disseminate this information amongst beneficiaries. Provided beneficiaries with construction checklist and trained community members on how to identify faults. A technical field officer visited sites for technical assistance. NGOs conducted their own trainings and workshops of employed contractors for compliance with building codes.	US government through Pathway Construction Initiative provided $5.0 million to Mississippi and Louisiana to assist workers to enter the construction industry, while assisting critical rebuilding efforts in those states. In each state, Reconstruction Centers of Excellence were established to provide workforce services for the construction industry. FEMA inspected an estimated 1.9 Million homes. HUD worked with Home Depot to conduct workshops for affected homeowners on hurricane preparedness and repair of houses and distributed "Tech Sets" on storm resistant roofing and wind resistant openings. HUD Field Offices were established to coordinate all HUD technical assistance requests from local elected officials.	The Government is providing technical assistance through launch of over 600 Army-led Assistance and Inspection (AI) teams as well as through establishment of 12 Housing Reconstruction Centers (HRCs) and engagement of services of over 26 NGO's. As of October 2006, over 834,324 people had received trained in seismic resistance building techniques as well as general awareness training.

Areas of comparative analysis	Gujarat (India) earthquake (2001)	Sri Lanka earthquake / tsunami (2004)	Indonesia (Aceh and Nias) earthquake / tsunami (2004)	Katrina (USA) hurricane (2005)	Pakistan earthquake (2005)
5. Government material facilitation	Government provided material facilitation through 1,082 Materials Banks opened through which subsidized steel and cement as well as excise duty and sales tax exemption for building materials in certain areas such as Kutch.	No formal material facilitation mechanisms rather interventions as and when required by District Secretaries.	BRR hired a technical advisor to assist in addressing the added stress of reconstruction on the supply chain and negate its impacts. This project was launched in 2006 to deal with the housing logistics challenge. WFP shipping service for carriage of construction materials launched at the request of BRR and UNORC in 2005.	No formal material facilitation mechanisms in place.	Government analyzed projected construction material requirements and formulated strategy for establishing a building materials supply chain. Consultations with the construction industry and transporters to establish benchmark prices. One hundred thirty two construction material hubs were established in the affected areas where materials were available at published rates.
6. Disbursement progress	As of first quarter of 2006 disbursements are calculated at INR 37.54 Billion (US$950 Million)	As of October 2006, housing disbursements stood at US$98.15 Million.	As of October 2006, housing disbursements stood at US$557 Million.	US$113 Million in housing assistance disbursed by US Department of Housing and Urban Development (HUD) in collaboration with Federal Emergency Management Agency (FEMA) as of October 2006.	ERRA had disbursed over PKR 48 Billion (US$800 Million) as of October 2006.
7. Reconstruction progress	911,000 damaged houses repaired and over 201,000 houses reconstructed as of first quarter of 2006. In 5 years almost 58% of the destroyed houses have been reconstructed.	61,019 houses reconstructed after lapse of 3 years out of 114,069 houses and 47,995 are in progress out of total of 114,069 houses. So 53% of houses have been completed while 42% of houses are under construction and 5% houses are yet to be reconstructed.	90,861 houses reconstructed after lapse of 3 years out of 113,500 houses. Overall reconstruction progress is 80%.	As of October 2006, 2,000 damaged housing units repaired and leased while 20,000 new housing units leased to affected families. Reconstruction and rehabilitation progress is approximately 28%.	As of October 2006, 208,292 houses had been reconstructed including 99,247 destroyed houses and 109,045 damaged houses. Houses are constructed as per seismic standards. 349,000 houses are in reconstruction in. After first full year of reconstruction over 39% of the housing stock damaged or destroyed in earthquake has been reconstructed.
8. Ensuring compliance and building quality / standards	Multi-hazard resistant construction ensured through payment of installments after engineer's certification. Third party quality audit by National Council for Cement and Building Materials (NCCBM).	Construction as per minimum accepted standards ensured through direct donor / NGO assistance. Third party technical quality audits conducted in most divisions.	Third Party monitoring and evaluation through UN-HABITAT to look at performance of housing program with respect to official building codes issued by the Indonesian government.	Construction as per building codes ensured through respective local governments of affected areas.	Housing grant is released after inspections and certification by AI Teams that the house is built as per seismically resistant standards. Compliance monitoring teams and third party technical audits were used.

Areas of comparative analysis	Gujarat (India) earthquake (2001)	Sri Lanka earthquake / tsunami (2004)	Indonesia (Aceh and Nias) earthquake / tsunami (2004)	Katrina (USA) hurricane (2005)	Pakistan earthquake (2005)
9. Communications strategy	Many general campaigns on safety and hazard risk awareness launched used electronic and print media and events such as festivals. However no housing specific campaign launched.	Program has suffered due to absence of a clear communications strategy and media relations strategy.	No strategic level communications rather field level interventions by BRR such as community outreach programs.	NGO's such as The Center for Faith Based and Community Initiatives (CFBCI) developed and implemented a communication strategy including the production and distribution of *Hurricane Toolkit: Recovery After the Storm*, an informational guide to federal and local resources available to hurricane victims, and the organizations serving them. To date, over 50,000 hard copies have been distributed, and the publication has been posted on HUD's Web site. HUD sponsored workshops and summits to advise local governments, non-profits, and community groups of programs and assistance available from HUD.	A comprehensive communications strategy and a public information campaign were launched. Print, electronic media, TV, and radio were used and over 600,000 posters and pamphlets distributed. Press briefings and regular media visits to the affected areas were arranged. NGO's partnered with ERRA formulated and launched information programs during which adapted materials were disseminated and information kiosks were established at the Housing Reconstruction Centers (HRCs).
10. Grievance redressal mechanism	Grievance redressal through normal legal procedure of courts and ombudsman.	Formal grievance redressal mechanism absent, only normal legal channels available.	Informal grievance redressal mechanism.	135 FEMA Disaster Recovery Centers (DRCs) and local one-stop centers were established to facilitate assistance to the public during the recovery efforts. Staff gave on-site referrals of individuals and families to specific assistance sources and acted as a liaison with state and local partners to ensure effective service delivery and minimize grievances.	ERRA established ten Data Resource Centers (DRCs) in affected areas to handle grievances and to facilitate payments. The DRCs acted as information centers for other problems. State and Province level Reconstruction Agencies were focal points for grievances related to incorrect bank account information. The respective Battalion Commanders of the Army in AJK and NWFP were the focal points for grievances related to surveys and inspections as well as requests for "Category Change".
11. Monitoring and evaluation (M&E) and data management	No formal M&E structure in GSDMA exists, however monitoring procedures established through Technical Assistance and housing beneficiary database established.	The Government's Development Assistance Database (DAD) monitored recovery and reconstruction operations. Shortcomings of the DAD were that it depended on regular inputs by donors and there are no established mechanisms to verify district level information.	M&E system established through UN-HABITAT assistance. Information management a challenge. Gender-disaggregated data not available.	Formal M&E at FEMA as well as local governments and housing authorities. All individual agencies have own databases which in many cases have restricted access.	M&E wing established at ERRA. Housing beneficiary database established at NADRA with disbursement database accessible to all on ERRA Web site. Reporting, M&E system for housing developed with UN-HABITAT assistance as well as Training Monitoring Information System (TMIS). Gender disaggregated data available.

Areas of comparative analysis	Gujarat (India) earthquake (2001)	Sri Lanka earthquake / tsunami (2004)	Indonesia (Aceh and Nias) earthquake / tsunami (2004)	Katrina (USA) hurricane (2005)	Pakistan earthquake (2005)
12. Risk transfer mechanisms	Insurance to 14 types of hazards for 10 years at premium of INR 367 (US$9.2) deducted by the State from the last housing assistance installment.	Limited individual housing insurance policies.	Limited individual housing insurance policies.	FEMA's National Flood Insurance Program (NFIP) provided funds to policyholders affected by Hurricane Katrina to help them rebuild or relocate. Virtually all claims closed by October 2006. NFIP claims represent more than $16.1 billion in payments to more than 205,000 policyholders, more than all other claims combined since NFIP inception in 1968.	No concept of housing insurance in the affected areas.
13. Ensuring transparency	Direct payments into bank accounts for which 660,000 accounts opened. Financial audits conducted.	Direct payments into bank accounts through two State Banks, Peoples Bank and Bank of Ceylon. Third Party beneficiary eligibility and financial audits conducted.	Establishment of an anti corruption unit within BRR and launch of programs such as PQAM, Procurement, Quality Assurance and Monitoring as well as introduction of a staff integrity pact at BRR voluntarily monitored by Transparency International (Indonesia).	The Office of Inspector General (OIG) developed and participated in a fraud prevention program to educate state agencies, and federal, State, and local law enforcement to identify fraud in HUD grant programs and other support programs. OIG established division to combat waste, fraud, and abuse in the Gulf Coast States. The Hurricane Recovery Audit Oversight Division audited disaster funding, worked with the HUD Office of Investigations, and other federal and state law enforcement agencies.	Direct payments into bank accounts for which 660,0 accounts opened. interna audits as well as external audits through Auditor General of Pakistan's Offic conducted. All disburseme data available on ERRA Web site.
14. Program implementation challenges	Temporary shelters became permanent, disbursement delays, owner and tenant issues, and equity issues.	Equity issues, relocation issues, weak communications strategy, land availability, environmental issues, and providing land titles.	Land tenureandownership, damaged land and relocation, construction material costs, declining donor commitments, and land and spatial planning	High construction material costs, high labor costs, lengthy application procedure, and lengthy processing times.	Further increasing compliance, focusing on no work started cases, owner tenant issue, hazardous lanc issue, and scarcity of labor.

Source of table: Pakistan, Earthquake Recovery and Reconstruction Agency (ERRA), October 2007 (modified).

Endnote: 1. Initial survey figures changed as grievance cases were resolved through subsequent resurveys.

Sources for table:
ADB Capacity Building for Earthquake Rehabilitation and Reconstruction Technical Assistance Completion Report, 2006
Earthquake Reconstruction and Rehabilitation Authority (ERRA) GoP, Progress Report, 2007
Earthquake Reconstruction and Rehabilitation Authority (ERRA) GoP Rural Housing Reconstruction Policy
Earthquake Reconstruction and Rehabilitation Authority (ERRA) GoP Web site: http://www.erra.gov.pk
Federal Emergency Management Agency (FEMA), US Government Web site: http://www.fema.gov
Gujarat State Disaster Management Authority (GSDMA) Government of India Web site: http://www.gsdma.org
Gujarat State Disaster Management Authority (GSDMA, Government of India)-World Bank Quarterly Report, 2006
Institute for Crisis , Disaster and Risk Management (Washington DC) February 2003 Report
Ministry of Finance and Planning GoSL and Reconstruction and Development Agency (RADA) GoSL Report, 2006
News Report on Web site: http://www.gujaratplus.com
Reconstruction and Rehabilitation Agency of Aceh and Nias (BRR) GoI Progress Report, 2006
Reconstruction and Rehabilitation Agency of Aceh and Nias (BRR) GoI Housing Policy Decree
Reconstruction and Rehabilitation Agency of Aceh and Nias (BRR) GoI Web site: http://www.brr.go.id
SUCI Gujarat State Committee Report, 2006
The Brookings Institution (USA) Report: Katrina One Year On, 2006
The Institute for Southern Studies Special Report on Katrina Crisis, 2007
The Urban Institute (USA) Report: After Katrina, 2006
United Nations Development Program (UNDP) Report: Responding to Rapid Change – Technical Support for BRR, 2006
United Nations Disaster Assessment and Coordination (UNDAC) Team Bhuj (Gujarat) Final Report, 2001
United Nations Settlement Programme (UN-HABITAT, Indonesia) and BRR Housing Milestone Data, 2006
Update on US Government Web site on Katrina – What the Government is Doing: http://www.dhs.gov
US Department of Housing and Urban Development Web site: http://www.hud.gov
Web site: http://www.e-aceh-nias.org
Web site: http://www.indianngos.com
Web site: http://www.reliefweb.int

GLOSSARY

Acceptable risk: The level of potential losses that a society or community considers acceptable given existing social, economic, political, cultural, technical, and environmental conditions.

Accountability: The state of being accountable; liability to be called on to render an account.

Adaptation: The adjustment in natural or human systems in response to actual or expected climatic or other stimuli or their effects, which moderates harm or exploits beneficial opportunities.

Adobe: Compressed earth, normally in the form of bricks or blocks.

Agency-driven reconstruction in relocated site (ADRRS): An agency-led approach in which an agency contracts the construction of houses on a new site, generally with little or no involvement with the community or homeowners.

Agency-driven reconstruction *in situ* (ADRIS): An agency-led reconstruction approach in which damaged houses are rebuilt, generally by a construction company, in pre-disaster locations.

All-hazards risk analysis: A determination of the nature and extent of risk developed by analyzing all potential hazards and evaluating existing conditions of vulnerability that could pose a potential threat or harm to people, property, livelihoods, and the environment on which they depend.

Apartment owner-occupant: The transitional reconstruction option in which the occupant owns his or her apartment, formally or informally.

Apartment tenant: The transitional reconstruction option in which the occupant rents the apartment, formally or informally.

Assessment: The survey of a real or potential disaster to estimate the actual or expected damages and to make recommendations for prevention, preparedness, response, and reconstruction.

Assistance scheme: A method for providing assistance to households after a disaster, allowing them to rebuild and to reestablish their way of life, which may include cash transfers, vouchers, and/or in-kind contributions.

Audit: An official examination and verification of accounts and records to analyze the legality and regularity of project expenditures and income, in accordance with laws, regulations, and contracts, such as loan contracts and accounting rules. May also analyze efficiency and effectiveness of use of funds.

Bandwidth: Capacity of ICT and telecom systems to transmit digital or analog data in a given time period. The slowest connection point can degrade bandwidth to that point referred to as a bandwidth bottleneck.

Baseline data: The initial information collected during an assessment, including facts, numbers, and descriptions that permit comparison with the situation that existed before and measurement of the impact of the project implemented.

Basic needs: The items that people need to survive. This can include safe access to essential goods and services such as food, water, shelter, clothing, health care, sanitation, and education.

Biological disaster: Disaster event caused by the exposure of living organisms to germs and toxic substances.

Bribery: Offering an inducement for a person to act dishonestly in relation to a business opportunity.

Build Back Better: Approach to reconstruction that aims to reduce vulnerability and improve living conditions, while also promoting a more effective reconstruction process.

Building code: A set of ordinances or regulations and associated standards intended to control aspects of the design, construction, materials, alteration, and occupancy of structures necessary to ensure human safety and welfare, including resistance to collapse, damage, and fire.

Building inspection: Inspections necessary to establish whether a damaged structure poses an immediate threat to life, public health, or safety, usually accompanied by a process of tagging.

Bunga houses: Structures built with compressed stabilized earth blocks.

Capacity development or capacity building: The process by which the capacities of people, organizations, and society are strengthened to achieve social and economic goals, through improvement of knowledge, skills, systems, and institutions.

Capacity: The combination of all physical, institutional, social, and/or economic strengths, attributes, and resources available within a community, society, or organization that can be used to achieve agreed-upon goals. Also includes collective attributes such as leadership and management.

Cash approach (CA): Unconditional financial assistance for housing reconstruction without technical support.

Cash transfers: Direct payments or vouchers to provide resources to affected populations to carry out housing reconstruction, in exchange for work on infrastructure projects, or for other purposes.

Catastrophe: A situation in which all or most people living in a community are affected along with the basic supply centers, making self-help impossible.

Civil society organization (CSO): National and local nongovernmental and not-for-profit organizations that express the interests and values of their members and/or others based on ethical, cultural, political, scientific, religious, or philanthropic considerations.

Climate change: Meteorological changes attributed directly or indirectly to human activity that alter the composition of the global atmosphere or to natural climate variability.

Climatological disaster: Disaster event caused by long-lived/meso- to macro-scale processes (in the spectrum from intraseasonal to multi-decadal climate variability).

Collusion: Cooperation between two or more parties to defraud or deceive a third party, usually with an anti-competitive purpose.

Community: A group of households that identify themselves in some way as having a common interest, bond, values, resources, or needs as well as physical space. A social group of any size whose members reside in a specific locality, share government, and often have a common cultural and historical heritage.

Community participation: A process whereby stakeholders can influence development by contributing to project design, influencing public choices, and holding public institutions accountable for the goods and services they provide; the engagement of affected populations in the project cycle (assessment, design, implementation, monitoring, and evaluation).

Community-based organizations (CBOs): Organizations whose principal concerns are the welfare and development of a particular community. CBOs may not represent all the households in a particular area.

Community-driven reconstruction (CDR): Approach to reconstruction that entails varying degrees of organized community involvement in the project cycle, generally complemented by the assistance of an agency that provides construction materials, financial assistance, and/or training.

Complaint mechanisms: Mechanisms that allow corruption to be reported by social actors, including public employees, ideally in a confidential manner.

Complex disasters: Multidimensional events of long duration often spawned by human-generated events, such as war and civil strife.

Conditional cash transfer: Cash given on the condition that the recipient does something (for example, rebuild a house, attend school, plant seeds, provide labor, or establish or reestablish a livelihood).

Conservation: Actions taken to secure the survival or preservation of buildings, cultural artifacts, natural resources, energy, or any other thing of acknowledged value to society.

Construction guidelines or standards: A document prepared by a recognized standard-setting organization that prescribes methods and materials for the safe use and consistent performance of specific technologies; sometimes developed by consensus of users.

Construction technology: The choice of building materials and the technique and means used to erect or repair a house.

Contingency planning: A management process that identifies and analyses potential events or situations that might threaten the society or the environment and establishes arrangements in advance to enable timely, effective, and appropriate responses to and recovery from any such events and situations.

Coping capacity: The manner in which people and organizations use existing resources to achieve various beneficial ends during unusual, abnormal, and adverse conditions of a disaster phenomenon or process.

Corruption: The misuse of an entrusted position for private gain by employing bribery, extortion, fraud, deception, collusion, and money-laundering, including gains accruing to a person's family members, political party, or institutions in which the person has an interest.

Critical services: Services required to be maintained in the event of a disaster include power, water, sewer and wastewater, communications, education, emergency medical care, and fire protection/emergency services.

Cultural asset: Building, structure, landscape, object, or artifact that help establish a society's social roots and history.

Cultural heritage: Movable or immovable objects, sites, structures, groups of structures, and natural features and landscapes that have archaeological, paleontological, historical, architectural, religious, aesthetic, or other cultural significance. May include historic buildings, historic areas and towns, archaeological sites, and the contents.

Cultural significance: The perceived value of an asset as a result of its continuity of presence and worth to society.

Damage assessment: The process utilized to determine the magnitude of damage caused by a disaster or emergency event.

Demolition: Destruction of damaged structures to (1) eliminate an immediate threat to lives, public health, safety, and improved public or private property or (2) ensure the economic recovery of the affected community to the benefit of the overall community.

Detailed assessment: An in-depth assessment of disaster impact, often of a single location or a single sector, such as housing or environment. (See "rapid assessment.")

Disaster: A situation or event which overwhelms local capacity, necessitating a request to a national or international level for external assistance; an unforeseen and often sudden event that causes great damage, destruction and human suffering.

Disaster debris: Waste items such as trees, woody debris, sand, mud, silt, gravel, building components and contents, wreckage, vehicles, and personal property that remain after a disaster.

Disaster response: Process to address the immediate conditions that threaten the lives, economy, and welfare of a community.

Disaster risk: The magnitude of potential disaster losses (in lives, health status, livelihoods, assets and services) in a particular community or group over some time period arising from its exposure to possible hazard events and its vulnerabilities to these hazards.

Disaster risk management: The systematic process of using administrative directives, organizations, and operational skills and capacities to implement strategies, policies, and coping capacities of society and communities to lessen the adverse impacts of hazards and the possibility of disaster.

Disaster risk reduction: The practice of reducing disaster risks through systematic analysis and management of the causal factors of disasters, including reduced exposure to hazards, lessened vulnerability of people and property, wise management of land and the environment, and improved preparedness.

Early recovery: A process which seeks to catalyze sustainable development opportunities by generating self-sustaining processes for post-crisis recovery. It encompasses livelihoods, shelter, governance, environment, and social dimensions, including the reintegration of displaced populations, and addresses underlying risks that contributed to the crisis.

Early-warning system: The set of capacities needed to provide timely and meaningful information to enable individuals, communities, and organizations threatened by hazards to prepare and to act appropriately and in sufficient time to reduce loss of life, injury, livelihoods, damage to property and damage to the environment.

Earthquake: A sudden motion or trembling caused by a release of strain accumulated within or along the edge of earth's tectonic plates.

Economic security: Conditions that allow a household or community to meet its essential economic needs in a sustainable way without resorting to strategies which are damaging to livelihoods, security, and dignity.

Emergency management: The organization and management of resources and responsibilities for addressing all aspects of emergencies, in particular, preparedness, response, and initial recovery.

Emergency services: The set of specialized agencies that have responsibility to serve and protect people and property in emergency situations.

Empowerment: Authority given to an institution or organization (or individual) to determine policy and make decisions. Inclusion of people who are ordinarily outside of the decision making process.

Enabling environment: The rules and regulations, both national and local, which provide a supportive environment for a specific activity, such as community participation or DRM, to take place.

Environmental degradation: The reduction of the capacity of the environment to meet social and ecological objectives and needs.

Environmental impact assessment: The process by which the environmental consequences of a proposed project or program are evaluated, undertaken as an integral part of planning and decision-making processes with a view to limiting or reducing the adverse impacts of the project or program.

Equity: The quality of being impartial and "fair" in the distribution of development benefits and costs and the provision of access to opportunities for all.

Erosion: The washing away of soil and rocks along streams and hillsides on public and private property Erosion may cause a threat to health, safety, and the environment.

Exposure: The experience of coming into contact with an environmental condition or social influence that has a harmful or beneficial effect.

Extortion: Threatening another with adverse consequences unless demands, usually for payment, are met.

Flood: A general and temporary condition of partial or complete inundation of normally dry land areas from (1) the overflow of inland or tidal waters, (2) the unusual and rapid accumulation or runoff of surface waters from any source, or (3) mudflows or the sudden collapse of shoreline land.

Floodplain: Any land area, including a watercourse, susceptible to partial or complete inundation by water from any source. Floodplain maps show inundation limits for floods of selected recurrence intervals and are used for zoning, insurance, and other regulatory purposes regarding heath and safety.

Framework data: The seven themes of geospatial data that are used by most geographic information system (GIS) applications (geodetic control, orthoimagery, elevation and bathymetry, transportation, hydrography, cadastral, and governmental units). These data include an encoding of the geographic extent of the features and a minimal number of attributes needed to identify and describe the features.

Fraud: Deceiving another person in order to gain some financial or other advantage.

Geographic Information System (GIS): A computer system for the input, editing, storage, retrieval, analysis, synthesis, and output of location-based (also called geographic or geo-referenced) information. GIS may refer to hardware and software, or include data.

Geological hazard: Geological process or phenomenon that may cause loss of life, injury, and other health impacts, property damage, loss of livelihoods and services, social and economic disruption, or environmental degradation.

Geophysical disasters: Seismic events (such as earthquakes, tsunamis, volcanic eruptions, landslides) related to the motion of the earth's tectonic plates.

Geo-referenced (or geo-spacial) information: Data, photos, or videos referenced geographically (for or by a GIS) relating to earth's physical features and attributes such as latitude, longitude, or locality/jurisdiction. Can be used to assess damage, map hazards, identify natural and materials resources and critical infrastructure at risk, plan restoration, monitor progress, and evaluate results on maps using a GIS.

GLIDE (GLobal IDEntifier): A system of unique disaster identifying numbers which when referenced in data sets will save time; create a common reference point for relating Bank projects to diverse and scattered sources of data; and eliminate confusion.

Hazard: A natural process or phenomenon, or a substance or human activity, that can cause loss of life, injury, and other health impacts, property damage, loss of livelihoods and services, social and economic disruption, or environmental degradation. The probability of occurrence, within a specific period of time in a given area, of a potentially damaging natural phenomenon.

Hazard mapping: The process of establishing geographically where and to what extent particular hazards are likely to pose a threat to people, property, or the environment.

Hazardous materials (HAZMAT): Any substance or material that, when involved in an accident and released in sufficient quantities, poses a risk to people's health, safety, and/or property. Includes explosives, radioactive materials, flammable liquids or solids, combustible liquids or solids, poisons, oxidizers, toxins, and corrosive materials.

Heritage: The combined creation and products of nature and of man that make up the living environment in time and space, including monuments, archeological sites, and movable heritage collections, historic urban areas, vernacular heritage, cultural landscapes (tangible heritage, which include natural and cultural sites), and living dimensions of heritage and all aspects of the physical and spiritual relationship between human societies and their environment (intangible heritage).

Historic preservation: A professional endeavor that seeks to preserve, conserve and protect buildings, objects, landscapes or other artifacts of historic significance.

Host families: A transitional settlement option sheltering the displaced population within the households of local families or on land or in properties owned by them.

House design: The form, dimensions, natural lighting, ventilation, and spatial organization of dwellings.

House owner-occupant: Tenure option where the occupant owns the house and land or is in part-ownership, such as when repaying a mortgage or loan. Ownership may be formal or informal.

House tenant: Tenure option where the house and land are rented by the occupant, formally or informally.

Household: Members of the same family unit sharing common income and expenditure sources. This definition may vary from context to context.

Housing: The immediate physical environment, both within and outside of buildings, in which families and households live and which serves as shelter.

Housing standard: Level of quality of a dwelling generally linked with the social level of the residents (including size, location, architecture, cost, workmanship quality).

Housing-sector assessment: An assessment to collect information such as demographic data, housing types, housing tenure situations, settlement patterns before and after the disaster, government interventions in the housing sector, infrastructure access, construction capacity, and market capacity to provide materials and labor for reconstruction.

Hydraulics: A branch of science or engineering that addresses fluids (specially water) in motion, water's action in rivers and channels, and works for raising water.

Hydrological disaster: Disaster event caused by deviations in the normal water cycle and/or overflow of bodies of water caused by wind set-up.

Hydrology: The scientific study of the waters of the earth, especially with relation to the effects of precipitation and evaporation on water in streams, lakes, and on or below the land surface.

Hydrometeorological disasters: The result from weather-related events, such as tropical water-related occurrences (hurricanes, typhoons, cyclones), windstorms, winter storms, tornadoes, and floods.

Hydrometeorological hazard: Process or phenomenon of atmospheric, hydrological or oceanographic nature that may cause loss of life, injury, and other health impacts, property damage, loss of livelihoods, and services, social and economic disruption, or environmental degradation.

Hyogo Framework for Action: The agreed framework of actions to reduce disaster risks from 2005–2015 established by more than 190 countries following the World Conference on Disaster Risk Reduction held in Kobe, Hyogo Japan, January 2005.

Indicator: Quantitative or qualitative factor or variable that provides a simple and reliable means to measure achievement or to reflect the changes connected to an operation.

Inflation: An increase in the supply of currency or credit relative to the availability of goods and services, resulting in higher prices and a decrease in the purchasing power of money.

Information and communications technologies (ICTs): The collective technology used to create, store, exchange, analyze, and process information in all its forms integrated with the procedures and resources to collect, process, and communicate data.

Infrastructure: Systems and networks by which public services are delivered, including: water supply and sanitation; energy and other utility networks; and transportation networks for all modes of travel, including roads and other access lines.

Integrity pact: An agreement between government and bidders for public contracts that neither side will pay, offer, demand, or accept bribes nor collude with competitors in obtaining or carrying out the contract.

Internally displaced persons: Persons or groups of persons who have been forced or obliged to flee or to leave their homes or places of habitual residence—in particular as a result of or in order to avoid the effects of armed conflict, situations of generalized violence, violations of human rights, or natural or human-made disasters—and who have not crossed an internationally recognized state border.

Interoperability: The capability of different ICT applications to exchange data via common exchange, file formats, and protocols. In the broadest sense, interoperability takes into account the social, political, language, and organizational factors that impact system performance.

Land tenant: Tenure option in which the house is owned but the land is rented, formally or informally.

Land use planning: The process undertaken by public authorities to identify, evaluate, and decide on different options for the use of land areas, including consideration of (1) long-term economic, social, and environmental objectives; (2) the implications for different communities and interest groups; and (3) the subsequent formulation and promulgation of plans that describe the permitted or acceptable uses. (See "physical planning.")

Landslide: Downward movement of a slope and materials under the force of gravity.

Lifelines: Public facilities and systems that provide basic life support services such as water, energy, sanitation, communications, and transportation.

Liquefaction: The phenomenon that occurs when ground shaking causes loose soils to lose strength and act like viscous fluid, which, in turn, causes two types of ground failure: lateral spread and loss of bearing strength.

Livelihoods: The ways in which people earn access to the resources they need, individually and communally, such as food, water, clothing, and shelter.

Logical framework (logframe): A conceptual tool used to define project, program, or policy objectives, and expected causal links in the results chain, including inputs, processes, outputs, outcomes, and impact. It identifies potential risks as well as performance indicators at each stage in the chain.

Loss assessment: Analyzes the changes in economic flows that occur after a disaster and over time, valued at current prices.

Management information systems: ICT-base systems used to analyze related past, present, and predictive information in conjunction with operational methods and processes to help post-disaster initiatives run efficiently.

Market analysis: Research undertaken to understand how a market functions, how a crisis has affected it, and the need for and most appropriate form of support. Research can include information on supply and demand of goods and services, price changes, and income/salary data.

Metadata: Information about data, such as content, source, vintage, accuracy, condition, projection, responsible party, contact phone number, method of collection, and other characteristics or descriptions.

Meteorological disaster: Disaster event caused by short-lived/small to meso-scale atmospheric processes (in the spectrum from minutes to days).

Microfinance: A broad range of small-scale financial services (such as deposits, loans, payment services, money transfers, and insurance) to poor and low-income households and their microenterprises.

Mitigation: The lessening or limitation of the adverse impacts of hazards and related disasters.

Money-laundering: Moving cash or assets obtained by criminal activity from one location to another in order to conceal the source.

Monitoring: The ongoing task of collecting and reviewing program-related information that pertain to the program's goals, objectives, and activities.

Morphology: The size, form and structure of an object (such as a house).

National platform for disaster risk reduction: A generic term for national mechanisms for coordination and policy guidance on disaster risk reduction that are multisectoral and inter-disciplinary in nature, with public, private and civil society participation involving all concerned entities within a country.

Natural hazard: Natural process or phenomenon that may cause loss of life, injury and other health impacts, property damage, loss of livelihoods and services, social and economic disruption, or environmental degradation.

Needs assessment: A process for estimating (usually based on a damage assessment) the financial, technical, and human resources needed to implement the agreed-upon programs of recovery, reconstruction, and risk management. It evaluates and "nets out" resources available to respond to the disaster.

Nongovernmental organization (NGO): A nonprofit, voluntary, service-oriented, and/or development-oriented organization, operated either for the benefit of its members or of other members, such as an agency. Also, civil society organization (CSO).

Nonstructural measures: Any measure not involving physical construction that uses knowledge, practice or agreement, to reduce risks and impacts, in particular through policies and laws, public awareness raising, training and education. (See "structural measures.")

Occupancy with no legal status: Occupancy option in which the occupant occupies property without the explicit permission of the owner. Also called a "squatter."

Open source: Nonproprietary software code and applications developed by a community of interested developers and made freely available (without a license) for use and further development. For example, Linux and many Google applications.

Open standards: Standards for ITC made available to the general public and are developed (or approved) and maintained via a collaborative and consensus-driven process. Open standards facilitate interoperability and data exchange among different products or services and are intended for widespread adoption.

Operating energy: The energy consumed by a building for heating, cooling, lightening and ventilating.

Owner-driven reconstruction (ODR): A reconstruction approach in which the homeowner undertakes rebuilding with or without external financial, material and technical assistance.

Participatory assessment: An approach to assessment that combines participatory tools with conventional statistical approaches intended to measure the impact of humanitarian assistance and development projects on people's lives.

Physical planning: A design exercise based on a land use plan used to propose the optimal infrastructure for public services, transport, economic activities, recreation, and environmental protection for a settlement or area. A physical plan can have both rural and urban components, although the latter usually predominates. (See "land use planning.")

Post-disaster needs assessment (PDNA): Usually a rapid, multi-sectoral assessment that measures the impact of disasters on the society, economy, and environment of the disaster-affected area.

Preparedness: The knowledge and capacities developed by governments, professional response and recovery organizations, communities, and individuals to effectively anticipate, respond to, and recover from the impacts of likely, imminent or current hazard events or conditions.

Prevention: Activities to provide outright avoidance of the adverse impact of hazards and means to minimize related environmental, technological and biological disasters; in the context of public awareness and education related to disaster risk reduction, changing attitudes and behavior to promote a "culture of prevention."

Probability: A statistical measure of the likelihood that a hazard event will occur.

Project cycle (also "project life cycle"): The sequence of activities that make up a project and how they relate to one other, generally: identification, preparation, appraisal, presentation and financing, implementation, monitoring, and evaluation.

Qualitative data: Information based on observation and discussion that can include perceptions and attitudes.

Quantitative data: Numerical information, such as numbers of intended recipients, payments disbursed, cash transferred, or days worked broken down by gender, age, and other variables.

Rapid assessment: An assessment that provides immediate information on needs, possible intervention types, and resource requirements. May be conducted as a multi-sectoral assessment or in a single sector or location. (See "detailed assessment.")

Reconstruction: The restoration and improvement, where possible, of facilities, livelihoods, and living conditions of disaster-affected communities, including efforts to reduce disaster risk factors. Focused primarily on the construction or replacement of damaged physical structures, and the restoration of local services and infrastructure.

Recovery: Decisions and actions taken after a disaster to restore or improve the pre-disaster living conditions of the affected communities while encouraging and facilitating necessary adjustments to reduce disaster risk. Focused not only on physical reconstruction, but also on the revitalization of the economy, and the restoration of social and cultural life.

Recurrence interval: The time between hazard events of similar size in a given location based on the probability that the given event will be equaled or exceeded in any given year.

Regulatory measures: Legal and other regulatory instruments established by government to prevent, reduce, or prepare for losses, such as those associated with hazard events, such as land use regulations in high-risk zones.

Relief: The provision of assistance or intervention immediately after a disaster to meet the life preservation and basic subsistence needs of those people affected.

Relocation: A process whereby a community's housing, assets, and public infrastructure are rebuilt in another location.

Remittances: Payments sent from migrant workers to family members in the country of origin.

Remote sensing: A field of study where the goal is to infer the properties of the earth's surface or the atmosphere itself without being in direct contact with them. Post-disaster remote sensing includes imagery of the disaster area captured from aircraft and satellites to study changes to the landscape or structures.

Repair: Restoration to working order following decay, damage, or partial destruction.

Repair cost: The cost associated with the replacement or restoration of damaged components. It does not include upgrades of other components triggered by codes and standards, design associated with upgrades, demolition of the entire facility, site work, or applicable project management costs.

Replacement cost: The cost for all work necessary to provide a new facility of the same size or design capacity and function as the damaged facility in accordance with all current applicable codes and standards.

Resettlement (involuntary resettlement): Direct economic and social losses resulting from displacement caused by land taking or restriction of access to land, together with the consequent compensatory and remedial measures. Generally related to infrastructure projects or changes in land use for public purposes. Relocation is one mitigation measure considered in carrying out resettlement.

Residual risk: The risk that remains in unmanaged form, even when effective disaster risk reduction measures are in place, and for which emergency response and recovery capacities must be maintained.

Resilience: The ability of a system, community, or society potentially exposed to hazards to resist, absorb, adapt to, and recover from the stresses of a hazard event, including the preservation and restoration of its essential basic structures and functions.

Response: The provision of emergency services and public assistance during or immediately after a disaster to save lives, reduce health impacts, ensure public safety, and meet the basic subsistence needs of the affected people.

Results framework: A tool for identifying and measuring objectives at the sector, country or regional level, usually laid out in diagrammatic form. It uses an objective tree to link high-level objectives through a hierarchy to program-level outcomes (and ultimately individual activities) and sets out a means by which achievement at all levels can be measured.

Retrofitting: Reinforcement or upgrading of existing structures to become more resistant and resilient to the forces of hazards.

Return period: The estimated likelihood of a disaster reoccurring in an area; a series of probable events.

Rights-based assessment: Evaluates whether people's basic rights are being met. The basis is generally considered to be the United Nations Universal Declaration of Human Rights.

Risk: The probability that a particular level of loss will be sustained by a given series of elements as a result of a given level of hazard. Elements under threat can include populations, communities, the built environment, the natural environment, economic activities, and services.

Risk analysis: A determination of the nature and extent of risk by analyzing potential hazards and evaluating existing conditions of vulnerability that could pose a potential threat or harm to people, property, livelihoods, and the environment on which they depend.

Risk assessment: A methodology to determine the nature and extent of risk by analyzing potential hazards and evaluating existing conditions of vulnerability that together could potentially harm exposed people, property, services, livelihoods, and the environment on which they depend.

Risk atlas: A series of maps showing community damages and losses as well as hazard areas for a series of probable events; a separate map is generated for each return period event. (See "return periods.")

Risk management: The systematic approach and practice of managing uncertainty and potential losses through a process of risk assessment and analysis and the development and implementation of strategies and specific actions to control, reduce, and transfer risks.

Risk transfer: The process of formally or informally shifting the financial consequences of particular risks from one party to another whereby one party (a household, community, enterprise, or state authority) will obtain post-disaster resources from another party in exchange for ongoing or compensatory social or financial benefits.

Satellite imagery: Images captured from above the earth using remote sensing technology.

Secondary hazard: A threat whose potential would be realized as the result of a triggering event that itself constitutes an emergency (for example, dam failure can be a secondary hazard associated with earthquakes).

Shelter: A habitable covered living space, providing a secure, healthy living environment with privacy and dignity for the groups, families, and individuals residing within it.

Social protection: Public measures to provide income security to the population. Use of social risk management to reduce the economic vulnerability of households and to help smooth consumption patterns.

Social safety net: Generally refers to non-contributory transfers (in cash or in kind), targeted at both populations at risk of economic destitution and the permanently poor, designed to keep their income above a specified minimum level.

Squatter: A person occupying a housing unit or land without legal title to it.

Stakeholders: All those agencies and individuals who have a direct or indirect interest in a humanitarian intervention or development project, or who can affect or are affected by the implementation and outcome of it.

Storm surge: Rise in the water surface above normal water level on the open coast due to the action of wind stress and atmospheric pressure on the water surface.

Structural measures: Any physical construction to reduce or avoid possible impacts of hazards, or application of engineering techniques to achieve hazard-resistance and resilience in structures or systems. (See "nonstructural measures.")

Sustainable development: Development that meets the needs of the present without compromising the ability of future generations to meet their own needs.

Targeting: The identification and recruiting of potential assistance recipients by local communities, government, or external agencies.

Transitional reconstruction: The processes by which populations affected but not displaced by conflict or natural disasters achieve durable solutions to their settlement and shelter needs.

Transitional settlement: the processes by which populations affected and displaced by conflict or natural disasters achieve settlement and shelter throughout the period of their displacement, prior to beginning transitional reconstruction.

Transitional shelter: Shelter that provides a habitable covered living space and a secure, healthy living environment with privacy and dignity for those within it during the period between a conflict or natural disaster and the achievement of a durable shelter solution.

Unconditional cash transfers: Cash transfers from governments or NGOs given without conditions attached to individuals or households, with the objective of alleviating poverty, providing social protection, or reducing economic vulnerability. (See "conditional transfers.").

User-driven reconstruction: Similar to owner-driven reconstruction, the approach in which the occupant of the property may not be the owner in a formal sense but may still possess sufficient property rights or sense of ownership to be willing to take on the reconstruction responsibility.

Vernacular architecture: The dwellings and other buildings that reflect people's environmental contexts and available resources, customarily owner- or community-built, utilizing traditional technologies. Vernacular architecture reflects the specific needs, values, economies, and ways of life of the culture that produces them. They may be adapted or developed over time as needs and circumstances change.

Vulnerability: The relative lack of capacity of a community or ability of an asset to resist damage and loss from a hazard. The conditions determined by physical, social, economic, political, and environmental factors or processes that increase the susceptibility of a community to the impact of hazards.

Vulnerable groups: Groups or members of groups particularly exposed to the impact of hazards, such displaced people, women, the elderly, the disabled, orphans, and any group subject to discrimination.

Warning systems: Mechanisms used to persuade and enable people and organizations to take actions to increase safety and reduce the impacts of a hazard.

Watershed: An area of land from which all of the water under it or on it drains to the same place, which may be a river, lake, reservoir, estuary, wetland, sea or ocean.

INDEX

ECO-AUDIT
Environmental Benefits Statement

The World Bank is committed to preserving endangered forests and natural resources. The Office of the Publisher has chosen to print *Safer Homes, Stronger Communities* on recycled paper with 30 percent postconsumer fiber in accordance with the recommended standards for paper usage set by the Green Press Initiative, a nonprofit program supporting publishers in using fiber that is not sourced from endangered forests. For more information, visit www. greenpressinitiative.org.

Saved:

- 41 trees
- 13 million Btu of total energy
- 3,874 lb. of net greenhouse gases
- 18,659 gal. of waste water
- 1,133 lb. of solid waste